# 4th International Conference on Advanced Engineering and Technology (4th ICAET 2017)



IOP Conference Series: Materials Science and Engineering
Volume 317

Shanghai, China
15 – 17 December 2017

ISBN: 978-1-5108-5983-8
ISSN: 1757-8981

**Printed from e-media with permission by:**

Curran Associates, Inc.
57 Morehouse Lane
Red Hook, NY 12571

CURRAN ASSOCIATES INC. proceedings.com

**Some format issues inherent in the e-media version may also appear in this print version.**



Printed by Curran Associates, Inc. (2018)



Institute of Physics
Dirac House, Temple Back
Bristol BS1 6BE UK

Phone: 44 1 17 929 7481
Fax: 44 1 17 920 0979

techtracking@iop.org

**Additional copies of this publication are available from:**

Curran Associates, Inc.
57 Morehouse Lane
Red Hook, NY 12571 USA
Phone: 845-758-0400
Fax: 845-758-2633
Email: curran@proceedings.com
Web: www.proceedings.com

# Table of contents

## Volume 317

**4th International Conference on Advanced Engineering and Technology (4th ICAET)**

**15–17 December 2017, Shanghai, China**

**Accepted papers received: 13 February 2018**
**Published online: 13 March 2018**

### Preface



### Papers

IOP Conf. Series: Materials Science and Engineering **317** (2018) 011001 doi:10.1088/1757-899X/317/1/011001

# Preface

The 4th ICAET 2017 is the 2017 4th International Conference on Advanced Engineering and Technology (4th ICAET 2017) took place in Incheon, South Korea, on December 15-17, 2017.

The conference program covered invited, oral, and poster presentations from scientists working in similar areas to establish platforms for collaborative research projects in this field. This conference will bring together leaders from industry and academia to exchange and share their experiences, present research results, explore collaborations and to spark new ideas, with the aim of developing new projects and exploiting new technology in this field.

The committee of ICAET expresses their sincere thanks to all authors for their high-quality research papers and careful presentations. All reviewers are also thanked for their careful comments and advices. Thanks are finally given to Trans Tech Publication as well for producing this volume.

The Organizing Committee of ICAET 2017

Committee Chair

Prof. Dongyan SHI

College of Mechanical and Electrical Engineering

Harbin Engineering University

IOP Conf. Series: Materials Science and Engineering **317** (2018) 011001 doi:10.1088/1757-899X/317/1/011001

# Conference Photograph

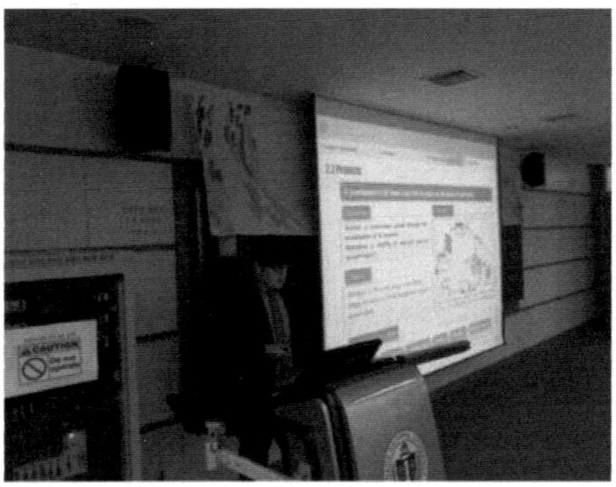

IOP Conf. Series: Materials Science and Engineering **317** (2018) 011001 doi:10.1088/1757-899X/317/1/011001

# Sponsors

| Incheon Disaster Prevention Research Center (IDPRC) | I D P R C 인천대학교 방재연구센터 Incheon Disaster Prevention Center |
|---|---|

Recently, various efforts to prevent and prepare are vitally needed for prevention of Disasters and calamities. So we understand the necessity for technology of disaster and we built up the Incheon Disaster Prevention Research Center (IDPRC) in Incheon National University (INU) in 1997.

Accordingly, Incheon Disaster Prevention Research Center (IDPRC) in Incheon National University has progressed of research on the prevention of disasters and calamities through the various seminars, conference and lectures. These research could be conducted cooperation with Incheon National University (INU) in various fields structure, soil, hydraulic and environment.

Incheon Disaster Prevention Research Center (IDPRC) will try to be a leader in the disaster of industry through the various research activities and global conference.

| Incheon National University | INU 인천대학교 INCHEON NATIONAL UNIVERSITY |
|---|---|

Incheon National University established a global campus by integrating two city council funded colleges into one in March 2010 and became a national university operated by the legal entity. With the start of attracting the branch of Lawrence Berkeley Research Center that is a world famous US national policy research center for education and research, Incheon National University will attract St. Petersburg University in Russia, Polymers University in Britain, and Kent University in Belgium and other foreign universities and research centers. These efforts will make INU spring to a world-wide competitive university.

As a local development leading university, Incheon National University established INU VISON 2020 and improved its competition actively through investment in selection and concentration. In compliance with the geographic feature and areal specialty, it will concentrate on the international trade, goods distribution, applied technology convergence, BNT-based life science, urban science and the sinology- based local humanity fields.

IOP Conf. Series: Materials Science and Engineering **317** (2018) 011001 doi:10.1088/1757-899X/317/1/011001

# Lists of Committees

| Committee name | Organizer name | Organization |
|---|---|---|
| Honorary Chair | Dongyan SHI | Harbin Engineering University |
| Conference General Chairs | J.W. Jung | Chungbuk University,<br>South Korea |
| Honorary Chair | D.K. Kim | Dong-A University |
| Technical Committees | I. Mansouri | Birjand University of Technology |
| | J.W. Hu | Incheon National University |
| | J.K. Ahn | Incheon National University |

IOP Conf. Series: Materials Science and Engineering **317** (2018) 011002 doi:10.1088/1757-899X/317/1/011002

# Peer review statement

All papers published in this volume of *IOP Conference Series: Materials Science and Engineering* have been peer reviewed through processes administered by the proceedings Editors. Reviews were conducted by expert referees to the professional and scientific standards expected of a proceedings journal published by IOP Publishing.

# Effect of Graphene Oxide on the Damping Capability of Recycled Mortar

**Jing-Jie Wei[1], Wu-Jian Long[1*], Chang-Le Fang[1], Hao-Dao Li[1], Yue-Gui Guo[1]**

[1] Guangdong Provincial Key Laboratory of Durability for Marine Civil Engineering, Shenzhen Durability Center for Civil Engineering, College of Civil Engineering, Shenzhen University, Shenzhen 518060, Guangdong, P.R. China

**Abstract.** The use of recycled aggregate as replacement of natural aggregate has increased in recent years in order to reduce the high consumption of natural resources in construction industry. This paper presents an experimental investigation on the effect of graphene oxide (GO) on the damping capability of recycled mortar. The effect of GO on damping capability was examined by using dynamic mechanical analyzer (DMA), It is showed that the recycled mortar with GO has a better damping capability than the recycled mortar without GO. Microstructural analysis of the recycled mortar with GO showed to have much denser and better crystallization of hydration products.

## 1. Introduction

The rapid economic growth in China has encouraged a number of construction activities. Construction and demolition (C&D) wastes produced during new construction, renovation, and demolition of buildings and structures have become a serious problem in many countries [1]. A large amount of the recycled aggregate are derived from this waste. Owing to the increasing cost of landfill, the scarcity of natural resources coupled with the greater demand for aggregates in construction, the use of recycled aggregate to partially or totally replace the natural aggregate has, therefore, become a common practice.

A number of publications on the use of recycled fine aggregate in mortar or concrete concluded that strength and durability decreased [2, 3]. Moreover, the porosity of mortar containing recycled fine aggregate is normally larger than the mortar made with natural sand [4, 5]. This is attributed to the large surface area and poor surface quality of the recycled fine aggregate, which creates a weak interface transition zone (ITZ) and thus decreases bonding between the matrix and the fine aggregate [6].

Based on the abovementioned studies, there are not only very few studies on the properties of recycled mortar with GO, but also on the variation of the dynamic mechanical properties of recycled mortar with GO at different temperatures. As important performance indicators of materials, dynamic mechanical properties are critical to the structural dynamic response, materials damage and damping energy consumption. Currently, most of the techniques aim to increase the damping properties of structures by installing with various damping devices such as machines or hydraulic pressure systems in order to reduce the vibration of structures. Therefore, the attempt of this paper is to investigate the effect of GO on the properties of recycled mortar, including damping capability and microstructure.

## 2. Materials and experimental procedure

### *2.1. Cement and chemical admixture*

Ordinary Portland Cement (OPC) type 42.5R was used as the binder material, conforming to the requirements of Chinese Standard GB175-2007 [7]. The chemical compositions and physical properties of the cement used are shown in table 1. In order to ensure the uniform dispersion of GO, this study uses the polycarboxylate-based superplasticizer type of Sika TMS-YJ-1, confirming to the requirements of JG/T223-2007[8].

**Table 1.** Chemical compositions and physical properties of cement.

| Ingredient | CaO | $SiO_2$ | $Al_2O_3$ | $Fe_2O_3$ | MgO | $SO_3$ | $K_2O$ | $Na_2O$ | LOI |
|---|---|---|---|---|---|---|---|---|---|
| Content (mass %) | 64.42 | 20.52 | 5.62 | 3.78 | 2.11 | 2.10 | 0.28 | 0.20 | 0.87 |

### *2.2. Recycled fine aggregate*

The recycled fine aggregate used in the experimental study reported here was obtained from a recycling plant, where mixed demolition wastes were processed by mechanical crushing, sieving and sorting operations. All of the recycled fine aggregate was sieved first to obtain various grading sizes and the different fractions were kept in sealed containers to prevent humidity exchange with the surrounding environment. The recycled fine aggregate used is conformed to the requirements specified in GB/T25176-2010[9], with fineness modulus at 2.39, 1.2% mud, and 0.8% clay lump. The grading curve of the recycled fine aggregate is shown in figure 1.

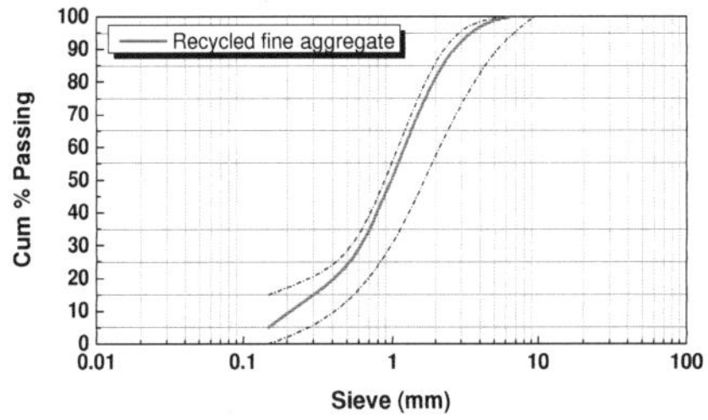


**Figure 1.** Particle-size distribution of recycled fine aggregate.

### *2.3. Mixtures proportioning and preparation*

The focus of this work is not on the development of recycled mortar, but on the use of advanced techniques to examine the effect of GO on the damping capability of recycled mortar. The mix proportions of various recycled mortar are given in table 2. Based on previous investigations and the high water absorption rate of recycled sand, the water/cement ratio (*w/c*) was selected as 0.66 to ensure the adequate workability of recycled mortar. Based on the work [10] the amounts of GO to be used in the recycled mortar are 0.00%, 0.05%, 0.10% and 0.20% by mass of cement respectively. while the PCE/GO ratio used was 1.3. The mixture was poured into the customized molds of 30*30*30 mm for the DMA test.

**Table 2.** Mix proportions of recycled Mortar

| Sample | Recycled sand (g) | Cement (g) | Water (g) | W/C | GO (g) | GO/Cement | PCE/GO |
|---|---|---|---|---|---|---|---|
| $R_0$ | 1350 | 450 | 300 | 0.66 | 0.00 | 0.00% | 1.3 |
| $R_1$ | 1350 | 450 | 300 | 0.66 | 0.225 | 0.05% | 1.3 |
| $R_2$ | 1350 | 450 | 300 | 0.66 | 0.45 | 0.10% | 1.3 |
| $R_3$ | 1350 | 450 | 300 | 0.66 | 0.90 | 0.20% | 1.3 |

Note: R is the recycled mortar, P is the cement paste.

### *2.4. Test methods*

The dynamic mechanical properties of materials are often measured by using the dynamic mechanical analyzer. The loss factor can be obtained from DMA testing. In this study, the dynamic mechanical property test was carried out by using a dynamic mechanical analyzer at a frequency of 1Hz, a heating rate of 5°C/min, temperature ranging from −50 to 60°C, a maximum dynamic force of 80 N, and a static force of -100 N. The tests were carried out on the specimens at the age of 14 days and 28 days. Based on the repeatability of DMA tests proved by Yuan et al. [11], DMA tests were carried out on the same sample.

The element analysis of cement paste samples was carried out using X-ray diffraction (XRD). The XRD analysis was performed by using X-ray diffractometer (XRD, type PANalytical) equipped with monochromatic Cu-Ka radiation at 40 kV and 40 mA. The internal morphology of the recycled mortar was assessed by ESEM at 7 days. Moreover, the fracture surface morphology after the 28-day DMA test was assessed by environmental scanning electron microscopy (ESEM, type Quanta 200 FEG).

## 3. Results and discussions

### *3.1. Damping capability*

The results of the tan δ analysis for the $R_0$, $R_1$, $R_2$ and $R_3$ recycled mortar specimens are shown in figures 2(a) and (b). It can be observed that the damping capability properties of $R_0$ are different from those of other specimens because of the different GO/Cement ratios. When the w/c and sand-cement ratio (*s/c*) were the same, the loss factor of GO recycled mortar ($R_1$, $R_2$ or $R_3$) was higher than that of recycled mortar without GO ($R_0$) at the temperature between -50 °C and 60 °C; the loss factor of recycled mortar increased with the amount of added GO. Thus, GO addition has a positive effect on the loss factor of recycled mortar. The ability of GO to enhance the damping capability is due to its nano-material nature, which improves the interface between individual components in the mixture, and non-uniform stress distribution under external force [12].

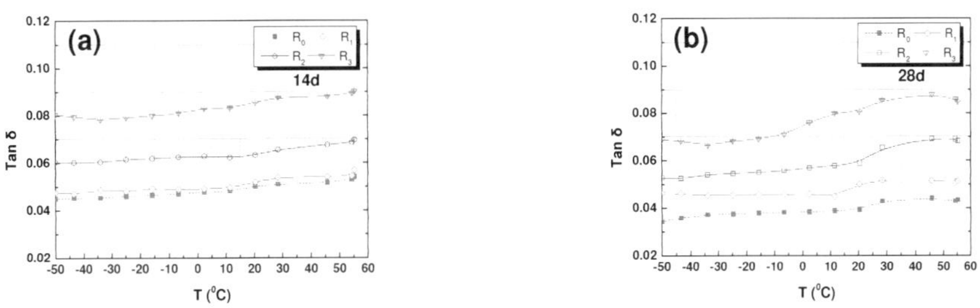


**Figure 2.** DMA analysis results of tan δ versus temperature: (a)14-day; (b) 28-day.

It is very important for the selection of damping material that is able to reach its maximum damping ability in the service environment [13]. In other words, the loss factor of recycled mortar at various temperatures need to be considered. It was found that the curves of the loss factor of recycled mortar showed different values at various temperatures and the tan δ increased with the increase of temperature from -50 °C to 60 °C. Comparing figures 2(a) and (b), it can be observed that a higher compressive strength lead to a lower loss factor for both at14 and 28 days. The loss factor of the recycled mortar at 28 days was found to be lower than that at 14 days. It is noticed that the decrease in loss factor is corresponding to the increase in flexural and compressive strengths. The reason for this is likely due to the porosity of the recycled mortar at 28 days that is lower than its porosity at 14 days. This result is again in good agreement with what is reported by Ping [14].

### *3.2. Microstructure characterization*

*3.2.1. X-ray diffraction analysis.* Figure 3 shows the XRD patterns of the cement composites in the recycled mortar with different GO additions after 28-day of curing. The phases detected are the usual cement hydrates as CSH, CH and Calcium carbonate, Calcium carboaluminate hydrate [15]. Based on XRD analysis, phases of CH, CSH+CC and CCA+CH have the highest characteristic peaks, which are located around 18, 29 and 48, respectively.

Since the weight fractions of GO used in the recycled mortar were all rather small, there are no peaks corresponding to GO in the XRD patterns of GO-recycled mortar samples. Therefore, no major difference could be observed in the diffraction patterns of the different samples, indicating they all have similar mineralogical compositions. However, the increase of peaks was seen in the by-product of the cement hydration created through the reaction of the –COOH functional groups with $C_2S$, $C_3S$ phase. The quantity of CH at early ages can be evaluated as the degree of hydration and the amount of C-S-H generation; while the amorphous phase can be considered as the contributor to strength [16]. These mineralogical results indicate that GO can promote hydration reactions. As a result, the increase of calcium carboaluminate hydrated phase may contribute to enhanced energy absorption and storage modulus of the recycled mortar, which is in good correlation with the results of damping capability analysis.

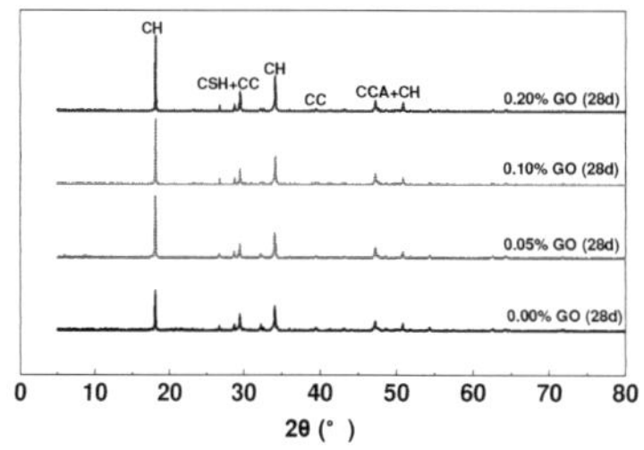


**Figure 3.** XRD results at 28 days.

*3.2.2. SEM analysis.* Figure 4 A-B shows the crystal morphology of recycled mortar without GO cured for 7-day. It can be seen that many pores and cracks exist in CSH and there are many lower CSH density areas. In addition, it can also be found that cracks usually pass through dense hydration products in a straight-through manner. Figure 4 C-D shows the crystal morphology of the recycled mortar with GO cured for 7-day. It was found that the production of thin non-uniform platelets and entangled network of rod like crystals were observed at various locations in the GO sample after 7 days. The high density CSH can be seen with the addition of GO. Besides, the area of high density increased with the increase of added GO. In particular, as compared with other fillers, GO exhibits unique two- dimensional structure which can effectively deflect, or force cracks to tilt and twist around GO. The process may help to impede fine cracks. The results of the microanalysis agree well with those examined in dynamical mechanical analysis tests. Because capillary pores, CSH density, cement hydrates, sand are the main factors affecting loss factor of recycled mortar [17].

**Figure 4.** SEM images in recycled mortar (A, B) and in GO- recycled mortar (C,D) after 7days curing.

## 4. Conclusion

In this study, the workability, dynamical mechanical properties and microstructure of the recycled mortar incorporating GO at different amounts of 0.05%, 0.1%, 0.2% by mass have been investigated. Based on the presented experimental results, the following conclusions can be drawn.

1. The loss factor (damping capability) of the GO-recycled mortar ($R_1$, $R_2$ and $R_3$) was higher than that of the recycled mortar without GO ($R_0$) at the temperature between -50 °C and 60 °C. The loss factor of the GO-recycled mortar increased with the increased of GO addition. The ability of GO to enhance the damping capability is due to its nano-material nature which improves the interface in the mixture, and the non-uniform stress distribution under external force.
2. The addition of GO in cementitious materials can promote the hydration of cement.

## References

[1] Nagapan S, Rahman I A, Asmi A. 2012 *Int. J. Adv. Appl. Sci.* **1(1)** 2252-8814.
[2] Dhir R K, Limbachiya M C, Leelawat T, et al. 1999 *Proc. Inst. Civil Eng. Struct. Build.* **134(3)** 257-74.
[3] Ryu J S 2002 *Mag. Concr. Res.* **54(1)** 7-12.
[4] Le T, Saout G L, Garcia-Diaz E, et al. 2017 *Constr. Build. Mater.* **141** 479-90.
[5] Kou S C, Poon C S, Etxeberria M 2011 *Cem. Conc. Comp.* **33(2)** 286-91.
[6] Bosque I F S D, Zhu W, Howind T, et al. 2017 *Cem. Concr. Comp.* **81** 25-34.
[7] GB/T 175-2007, Common Portland cement. Chinese national standard; 2007.
[8] JG/T 223-2007, 2007 *Polycarboxylates high performance water-reducing admixture,* Chinese construction industry standard.
[9] GB/T 25176-2010, 2010 *Recycled fine aggregate for concrete and mortar,* Chinese national standard.
[10] Lei B, Wang X, Rao C, et al. 2015 *Recycled masonry mortar used for civil engineering applications, comprises recycled aggregate, cement, fly ash, aluminum slag, graphene oxide, composite additive and water, in preset weight ratio*, China, CN104402351-A; 2015.
[11] Yuan Q, Liu W, Pan Y R, et al. 2015 *J. Mater. Civ. Eng.* **28(3)** 04015154.
[12] Zhang H, Zhao Y, Meng T, et al. 2015 *Constr. Build. Mater.* **95** 721-35.
[13] Shang Y, Zhang D, Yang C, et al. 2015 *Constr. Build. Mater.* **96** 20-8.
[14] Ping S 2010 *Poly. Comp.* **8(5)** 308-13.
[15] Fang L, Yuan Q, Deng D, et al. 2107 *J. Mater. Civ. Eng.* **29(8)** 04017080.
[16] Yang C, Huang R 1996 *Cem. Concr. Res.* **26(1)** 83–91.
**[17]** Fenu L, Forni D, Cadoni E 2016 *Comp. Part B Eng.* **92(8)** 142-50.

**Acknowledgement**
The authors gratefully acknowledge the financial support provided by the National Natural Science Foundation of China (No.51578341, No. 51278306).

IOP Conf. Series: Materials Science and Engineering **317** (2018) 012002 doi:10.1088/1757-899X/317/1/012002

# Investigation of the Effect of Small Hardening Spots Created on the Sample Surface by Laser Complex with Solid-State Laser

**O Nozdrina[1], I Zykov[1], A Melnikov[1], V Tsipilev[1], S Turanov[1]**

[1]Tomsk Polytechnic University, Tomsk, Russia

**Abstract.** This paper describes the results of an investigation of the effect of small hardening spots (about 1 mm) created on the surface of a sample by laser complex with solid-state laser. The melted area of the steel sample is not exceed 5%. Steel microhardness change in the region subjected to laser treatment is studied. Also there is a graph of the deformation of samples dependence on the tension. As a result, the yield plateau and plastic properties changes were detected. The flow line was tracked in the series of speckle photographs. As a result we can see how mm surface inhomogeneity can influence on the deformation and strength properties of steel.

## 1. Introduction

It is known that the plastic deformation of metallic samples is accompanied by evolving patterns of macroscopic deformation localization in accordance with the stadiality of the loading curve. It is nonuniform at each stage. At the pre-destruction stage localized deformation foci again begin to move with a constant speed for each of them. However, at this stage, the speed of movement depends on the coordinate of the origin of each focus. And the closer this place to the zone of future destruction, the more slowly such a focus moves. The velocities of the foci are automatically synchronized from the beginning of this stage so that their simultaneous "arrival" to the center - "collapse" of the autowave is ensured. The place of destruction and the lifetime of the specimen before destruction prove to be deterministic phenomena occurring at earlier stages of plastic deformation. Slip band (or Chernov-Luders band) is the most notable examples. It arise at the stage of easy slip and the “neck” formed at the stage of pre-destruction of the sample [1, 2].

In this work we conducted experimental attempt to influence the formation of slip band by changing the surface layer of the sample. Laser hardening of large surfaces is not appropriative. In this case sample deformation can lead to the separation of the hardened layer from the base metal [3]. Also, hardened layer can become a stress concentrator with crack forming, which leads to an early destruction of the sample [4, 5]. It was decided to create inhomogeneities system of equidistant hardened surface foci by spot laser hardening on the surface of steel (X45CrSi9-3). Thus, created deformation traps and boundaries of the foci constrain the propagation of cracks in the sample.

The aim of the research is to study the influence of the surface on the deformation processes in the sample volume and to study the collective effects of two-dimensional structure equidistant foci.

## 2. Materials and methods

Steel samples (grade X45CrSi9-3) with the gage size 60x8x3 mm were prepared for the experiments. The matrix of hardening foci with a pitch of 0.12 mm on both sides of the workpieces was created with laser system BlackLight. The pulse duration is 12 ms, the power density is 23.4 kW / cm$^2$ and the spot diameter is 0.35 mm.

3 types of samples were prepared: whole gage area treated, half-treated and check test specimens. The entire surface treated specimen is shown in figure 1.

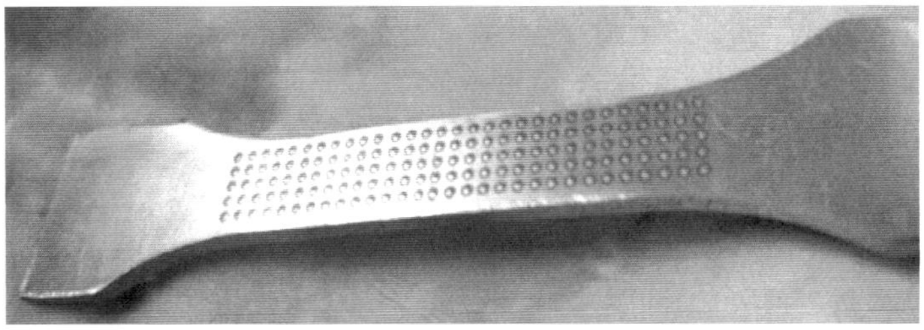

**Figure 1.** Samples after laser treatment.

An universal testing machine LFM-125 kN was used for carrying out the tensile tests. The microhardness was measured with the microhardness tester PMT-3.

The method of double-exposure speckle photography was used to study the inhomogeneity of plastic flow. As a result, a series of double-exposure specklegram of gage area of the specimen were obtained [6].

## 3. Results and discussion

Figure 2 shows photographs of the laser beam influenced area on the sample surface. In figure 2a it is possible to clearly distinguish two characteristic zones for laser radiation - laser impact zone (LIZ) and heat affected zone (HAZ). The absence of scratch marks in LIZ left by the grinding tool prior to laser treatment shows that the material has melted. There was a quenching from the liquid phase. Areas without etching indicate a very fine-grained structure. Figure 2b shows the cross section of the LIZ. The depth of the non-etching region is about 300 μm.

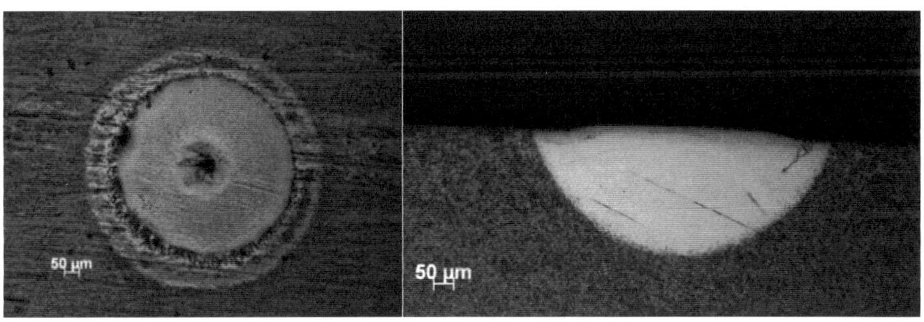


a) b)

**Figure 2.** Hardening focus: a) top view; b) sectional view.

Figure 3 shows the results of microhardness measuring in LIZ and near it. The measurement was made from the center of the hardening surface of spot to the edge of the sample. It can be seen from the graph that in the treated area the microhardness increase from 280 MPa to 1050 MPa. The higher microhardness at the periphery of the quenching spot appears due to high cooling rates compared to the center.

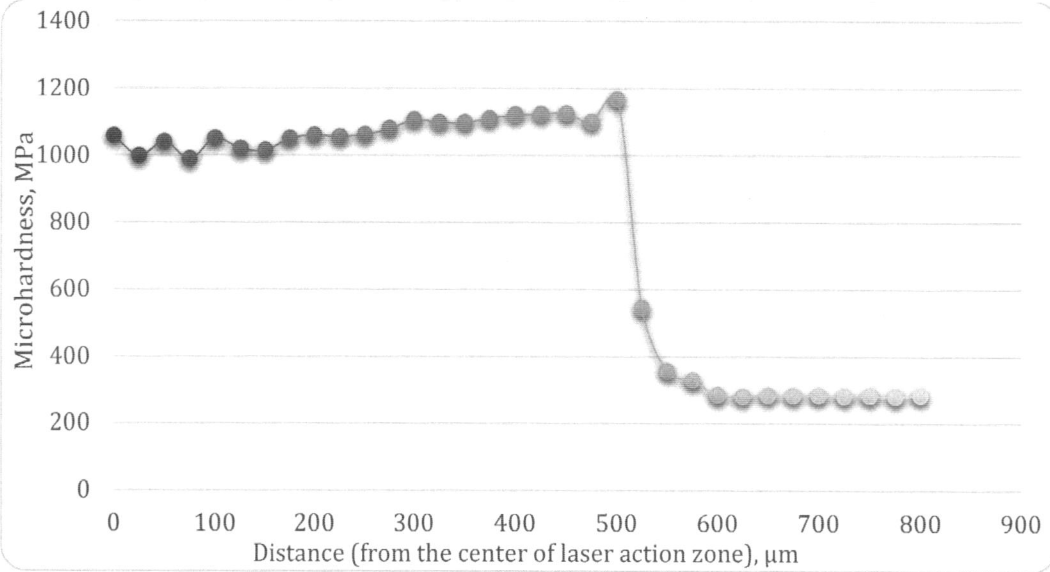


**Figure 3.** Change in steel microhardness in the area subjected to laser treatment.

Under tension plastic deformation begins with the forming and moving of the slip bands inside the specimen.

Figure 4 shows the characteristic tensile curves for the initial and treated samples. In the tensile diagrams of the initial and half-treated samples yield plateau is visible, while in a fully treated sample it is almost absent. Also on the diagram the dependence of the sample plasticity on the number of consolidation foci is traced, increasing of its number sharply reduces the plasticity. On the diagram of the fully treated sample there is no "necking" stage, which is clearly visible on the diagram of the initial sample.

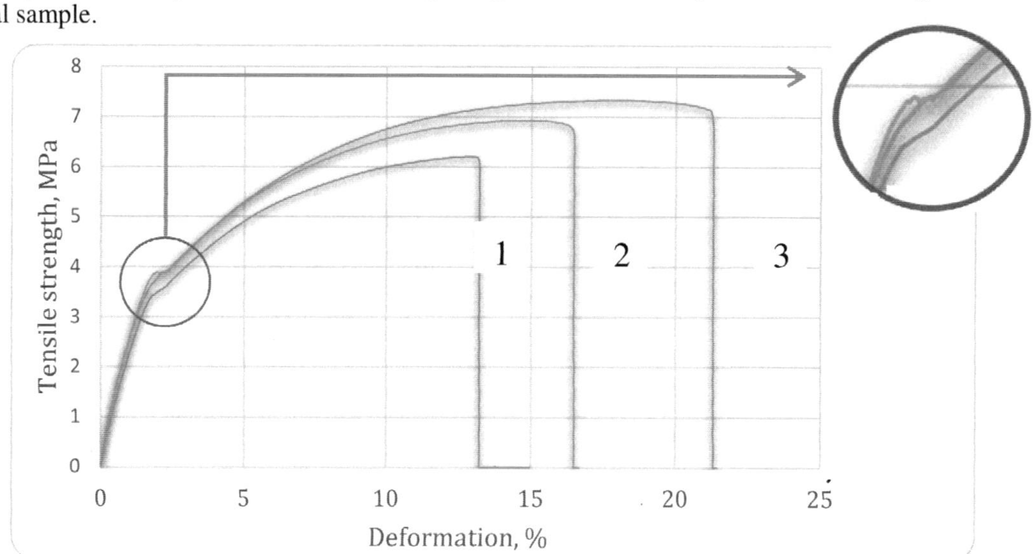


**Figure 4.** Tensile diagrams of samples with different areas of treatment: 1 - fully treated; 2 - half-treated; 3 - initial, without treatment.

Tensile diagrams of the initial and half-treated samples are almost identical up to that time (deformation is about 6%), when the motion of the slip band does not reach the processed part of half-treated sample. Then the stresses in the samples reach such values when cracks form in the treatment spots. Stresses begin to decrease and the sample breaks down in the focus area.

In a fully treated sample an easy glide stage is observed slightly. The slip band originates on the unprocessed part of the sample near mobile grip and disappears to the processed part. The increasing stresses in the foci lead to cracking, as a result the module decreases.

The bands appear in all samples at the hundredth second after the beginning of stretching. In the initial sample three slip bands (figure 5a) are formed at once and freely moving along the entire length of the sample. After that specklegram of initial sample doesn't change. The sample was ruptured through the forming of "neck" in the middle of gage.

In a half treated specimen the band starts from grip with the unprocessed sample part and moves toward the center of the sample (figure 5b). 12 seconds after the beginning of the movement one more slip band has formed on the border with treated area and start moving towards the first. After the collapse of these bands in the tension diagrams of samples 2 and 3 (up to now almost identical) distinction begins to occur. In a half treated specimen on the edges of the gage in untreated areas two more bands appear. It moves to the grips, after which a rupture occurs.

At the same time, in a fully treated sample only one localization band forms in a narrow transition zone between the fixed grip and the treated area.

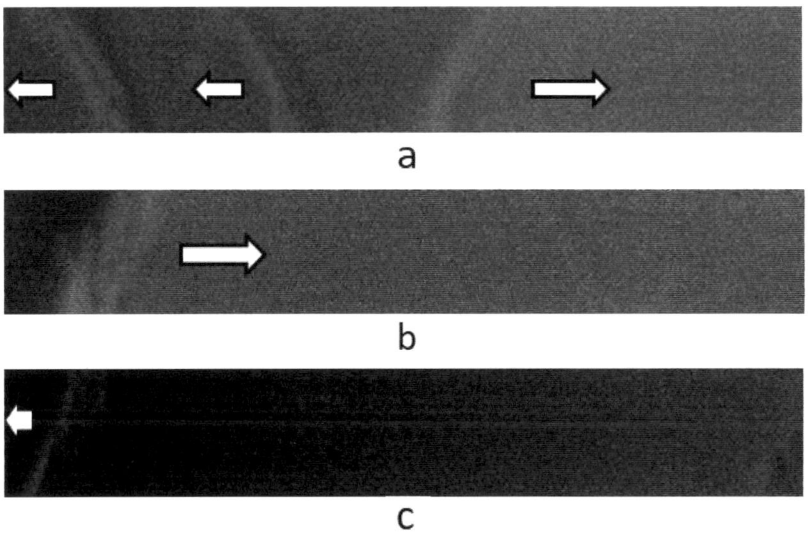


**Figure 5.** The origin and motion directions of slip bands in the samples:
a - initial; b - half-treated; c - fully treated.

With further increase in the load in the treated samples the foci were not deformed due to their high hardness and low plasticity. Cracks appeared in foci and moved forward, but stopped at the boundary of treated and untreated areas. This subsequently led to the rupture. Untreated sample was destroyed with the formation of a pronounced "neck".

## 4. Conclusion

This is the surface type of steel treatment, total volume of hardening foci is much smaller than the sample volume. The foci sectional area is not more than 5% of the sectional area of the sample. When the samples are stretched, the slip band cannot pass through the region of the hardening points, which leads to its suppression (see the tensile diagram of samples 2 and 3). Due to the suppression of localization the loading diagram has changed quite significantly. In the fully treated sample easy glide stage was almost completely possible to get rid. This indirectly indicates that the staginess of the loading curve can be a consequence of the localization of the deformation. At the same time, the hardening foci are stress concentrators and areas of formation of brittle cracks that reduce the strength and plasticity of the samples. This indicates necessity of optimization the structure and, as a consequence, the processing mode of the material.

IOP Conf. Series: Materials Science and Engineering **317** (2018) 012002 doi:10.1088/1757-899X/317/1/012002

## References

[1] Zuev L 2011 *Phys. Mesomech.* **3** 275-82.
[2] Zuev L 2015 *Adv. Met. Phys.* **1** 35-60.
[3] Stanford N, Dunne D and Ferry M 2003 *Acta. Mat.* **51** 665–76.
[4] Panin V, Egorushkin V and Panin A 2012 *Phys. Uspekhi* **55** 1260.
[5] Gilman J 1994 *Mech. Mater.* **17** 83–96.
[6] Zuev L, Danilov V and Barannikova S 2008 *Nauka, Novosibirsk* **328**.

## Acknowledgements

The research is funded by the Ministry of Education and Science of the Russian Federation in the frame of State Program in the field of scientific activities No. 13.3647.2017/ПЧ (Project part). The experimental calculations are carried out at Tomsk Polytechnic University within the framework of Tomsk Polytechnic University Competitiveness Enhancement Program grant.

# The Effect of Direct Rolling of Without-heating process on the Microstructure and Properties of a 400 MPa Rebar

**Liu Xin[1], Feng Guanghong[1], Qi Yuanhong[2]**

[1]Central Iron and Steel Research Institute, Beijing 100081, China
[2]State Key Laboratory of Advanced Steel Processes and Products, Central Iron and Steel Research Institute, Beijing 100081, China)

**Abstract.** In this study, the mechanical properties and phase composition of HRB400 rebars produced via a novel direct rolling of without-heating (DROW) process and the traditional hot rolling (THR) process were compared experimentally, and the effect of the temperature gradient on the phase distribution in the rebar was investigated by combining phase transformation dynamics with the finite element method. Compared with the THR process, the grains of the produced rebar were more refined when using the DROW process, and the average yield strength and tensile strength increased by 17 and 5 MPa, respectively. Moreover, by adopting the DROW process, the volume fraction of pearlite could be increased, thereby inhibiting the growth of ferrite. The DROW process was demonstrated to not only reduce the energy consumption, but also enhance the mechanical properties of the rebar.

## 1. Introduction

With the development of modern materials science and technology and the change of market demand, the requirements for the microstructure and the mechanical properties of rebars have continuously improved. Paul et al. [1] studied and compared the fatigue performance of microalloyed and thermo-mechanically treated (TMT) rebars, and discovered that the microalloyed rebar possessed a uniform ferrite-pearlite microstructure, which contributed to their improved anti-fatigue performance. Mukherjee et al. [2] predicted the rim hardness limit of tempered martensite in a TMT rebar. Zaky et al. [3] discovered that the temper-softened bainite phase was generated in a BST500S rebar microalloyed with V/Nb when the bar was first hot-rolled, then water-quenched to 600 °C and then cooled in air, thereby ensuring the desired mechanical properties. Chen et al. [4] studied the anti-corrosion properties of 400 MPa graded rebar, and Wei et al. [5,6] investigated the effect of a novel chemical reagent on the anti-corrosion properties of rebar after a quenching treatment. Furthermore, You et al. [7] explored a kind of novel glass fiber rebar composite with a tensile strength of up to 900 MPa.

In addition, many research groups adopted numerical computation methods to study the performance of rebar. For instance, Paul et al. [8] utilized the finite element method (FEM) to investigate the generation and expansion of fatigue cracks, whereas German et al. [9] employed the FEM to study the effect of the rebar's corrosion on the concrete cover. Rocha et al. [10,11] studied the fatigue behavior of rebar utilizing the Navarro-De Los Rios model and predicted the crack initiation stages including the growth stage of short cracks and the expansion stage of long cracks.

At present, studies on the enhancement of the anti-corrosion properties, fatigue life and mechanical properties of rebar mainly focus on the adjustment of the microalloying components, the rolling control and the cooling control, etc. In contrast, only a few studies focus on improving the existing process for the production of hot-rolled rebars. The direct rolling of without-heating (DROW) is a novel process for fabricating bars and rods. It is based on an optimization of the continuous casting process parameters to significantly increase the temperature of the casting billet after it is transported out of the continuous caster, and then directly transporting the casting billet to the rollers, thereby eliminating the heating process and saving energy. Compared with the traditional hot rolling (THR) process, this novel process endowed the casting billet with a lower surface temperature and a higher core temperature. In order to understand the effect of the temperature distribution on the microstructure and mechanical properties of the final products, we studied the variation of the temperature in the rolled-piece and the variation of the microstructure and mechanical properties of the products during the DROW process employing the FEM and field measurements.

## 2. Experimental and Simulation Procedure

### *2.1. Rebar Fabrication*

For this study, rebars were fabricated using the DROW process and the THR process, respectively, and figure 1 shows a schematic illustration of production line setup for the rolling of the rebar. The dimensions of the casting billet were 150mm × 150mm × 10000mm, and the casting billet was rolled to obtain a rebar with a diameter of 32mm after 12 passes. For the THR process, the casting billet was first transported into the heating furnace for the heating process, and was then transported to the rolling mill for rolling. For the DROW process, the overall temperature of the casting billet was significantly higher when it was transported out of the continuous caster due to an optimization of the continuous casting parameters. Thus, the casting billet could be directly transported to the rolling mill for the rolling process and did not require a heating step.

Portable IR thermometers were used to measure the temperatures before the rolling process, and the fixed thermometers on the back of the last stand of the finishing mill were used to measure the finish rolling temperature (as showed in figure 1).

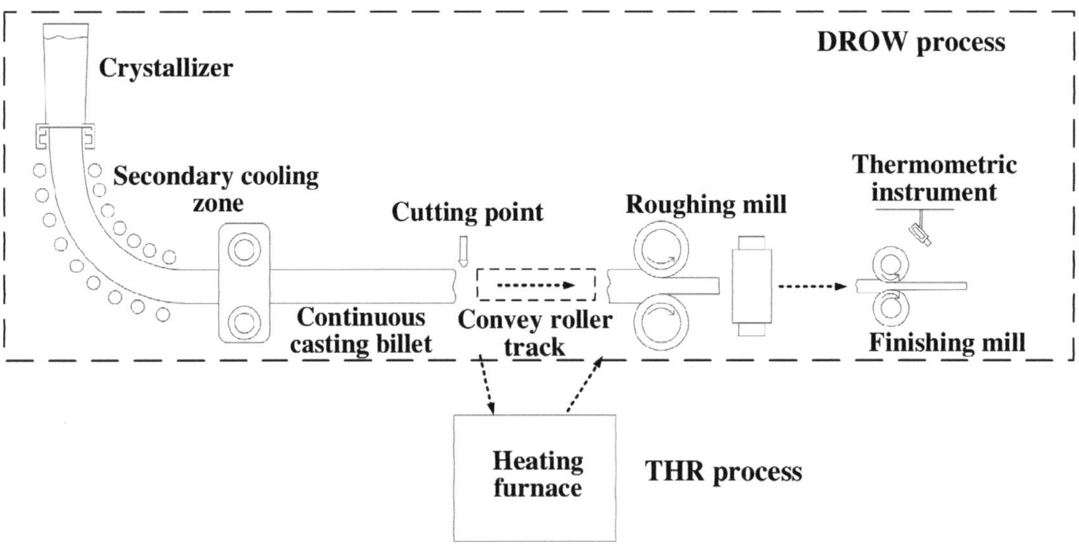


**Figure 1.** Schematic illustration of the production line setup for the rolling of the rebars.

### *2.2. Material and Experimental Procedure*

The experimental specimens were derived from commercial HRB400 rebar, and the chemical composition is provided in table 1.

**Table 1.** Chemical composition of the rebar (in wt%).

| C | Si | Mn | P | S | V | Others |
|---|---|---|---|---|---|---|
| 0.23 | 0.42 | 1.32 | 0.021 | 0.020 | 0.025 | Trace residual Cr, Ni, Cu, etc. |

After the rebar samples were produced using the DROW process and the THR process, respectively, the corresponding microstructure morphology of the samples was observed using a metalloscope, and the mechanical properties were investigated using a tensile testing machine.

The thermal simulation experiments for the DROW process and the THR process were carried out using a Gleeble-3500 thermal simulator. First, the specimen was heated to 1200 °C at a heating rate of 20 °C/s, and then held at this temperature for 5 min. Afterwards, the specimen was cooled down to the deformation temperature of 880 °C (970°C) at a cooling rate of 10 °C/s, and then left at this temperature for 1 min. The specimens were compressed by 70% at a deformation rate of 10 $s^{-1}$. Finally, the specimens were cooled down to room temperature at a cooling rate of 0.1, 0.5, 1, 2, 5, 10, 20 and 30 °C/s, respectively. The dynamic CCT curves were obtained utilizing the thermal dilation method and the metallographic method.

### *2.3. Simulation of the Thermal Deformation Process*

The material phase transformation was investigated by FEM. Considering the geometric symmetry of the casting billet. A quarterly geometric model was adopted for the analysis.

*2.3.1. Heat transfer equation.* The equation describing the temperature of the rolled-pieces during hot rolling process can be written as follows:

$$\rho c\acute{T} = \nabla \cdot (k\nabla T) + \sigma_{ij}\acute{\varepsilon}_{ij}^{p} + \sum L_{AB}\acute{\xi}_{AB} + \acute{Q} \tag{1}$$

where, $\rho$, $c$, $k$ and $Q$ denote the density, specific heat, thermal conductivity and the heat provided by the external heat source. $L_{AB}$ is the latent heat of phase transformation from phase A to phase B, $\acute{\xi}_{AB}$ is the derivative of the volume fraction transforming from phase A to phase B, ζAB, with respect to time, also referred to as the transformation rate, and $\sigma_{ij}$ and $\varepsilon_{ij}^{p}$ are the stress and the plastic strain tensor, respectively. Finally, $\sigma_{ij}\acute{\varepsilon}_{ij}^{p}$ and $\sum L_{ab}\acute{\xi}_{ab}$ are the heat resulting from plastic deformation and the heat variation resulting from phase transformation.

*2.3.2. Boundary conditions.* The contact heat transfer between the rolled-piece and the rolls was calculated using the following equation:

$$q = h_c(T_b - T_r) \tag{2}$$

where $q$ is the heat flux density, $h_c$ is the contact heat transfer coefficient, and $T_b$ and $T_r$ are the surface temperature of the rolled-piece and the rolls, respectively.

The heat transfer between the rolled-piece and air can be divided into convective heat transfer and radiation heat transfer:

$$\begin{cases} q = h_a(T_b - T_a) \\ q = \varepsilon\sigma(T_b^4 - T_a^4) \end{cases} \tag{3}$$

where $h_a$ is the convective heat transfer coefficient, and $\varepsilon$, $\sigma$ and $T_a$ denote the emissivity, the Stefan-Boltzmann constant and the environment temperature, respectively.

*2.3.3. Phase transformation model.* When computing the phase transformation via continuous cooling transformation curves, there is a linear relationship between the transformed phase volume and the temperature difference:

$$V = V_0\frac{T_s - T_i}{T_s - T_f} \tag{4}$$

where $T_s$, $T_i$ and $T_f$ are the starting temperature, the actual cooling temperature and the final temperature during the phase transformation process, respectively. $V_0$ is the maximum transformed phase volume fraction for a given cooling rate, and $V$ is the instantaneously transformed volume fraction of a specific phase structure.

*2.3.4. Simulation parameters.* The relevant computational parameters are listed in table 2.

**Table 2.** Parameters required for the simulation of the rolling process.

| Physical property | Unit | Value |
|---|---|---|
| Ambient temperature | °C | 25 |
| Emissivity | — | 0.85 |
| Contact heat transfer coefficient | $W/(m^2 \cdot K)$ | 5500 |
| Thermal conductivity | $W/(m \cdot K)$ | 32 |
| Convective heat transfer coefficient | $W/(m^2 \cdot K)$ | 30 |

## 3. Results and Discussion

### *3.1. Analysis of the Temperature of Rolled-Piece*

Before rolling process, since the temperature difference between the surface and the core of rolled-piece is different in the two rolling processes. However, the difference gradually decreases during the rolling process as revealed by the FEM results. Table 3 shows, at the head region of rolled-piece, the simulated temperature before rolling and after finishing rolling for the two different rolling processes.

**Table 3.** Comparison of the temperature in the DROW and the THR process.

| Rolling process | Head region of rolled-piece | Surface temperature (°C) | Core temperature (°C) |
|---|---|---|---|
| DROW | Before the start of rolling | 947 | 1027 |
| | After the final rolling | 892 | 925 |
| THR | Before the start of rolling | 1043 | 1090 |
| | After the final rolling | 975 | 1014 |

In the THR process, the surface and core temperatures of head region of rolled-piece before rolling are 1043 and 1090 °C, respectively, corresponding to a difference of 47 °C. When rolled-piece had moved to the exit of the final rolling mill, the difference between the surface and core temperature of head region of rolled-piece had dropped to 39°C.

In the DROW process, the surface and core temperatures of head region of rolled-piece are 947 and 1027 °C, respectively, corresponding to a temperature difference of 80 °C; by the time the rolled-piece arrived at the exit of the final rolling mill, the temperature difference had dropped to 33 °C. Thus, during the rolling process, the temperature gradient between the surface and the core of rolled-piece more rapidly decreases in the DROW process.

The experimental results and the calculated temperatures of the rebar head surface before and after the rolling process are compared and listed in Table 4. In comparison, the computational results obtained for the temperature are in better agreement with the experimental values.

**Table 4.** Comparison of the experimentally and the calculated temperatures.

| Rolling process | Head surface position of rolled-piece | Experimental temperature (°C) | Calculated temperature (°C) |
|---|---|---|---|
| DROW | Before the start of rolling | 951 | 947 |
| | After the final rolling | 880 | 892 |
| THR | Before the start of rolling | 1048 | 1043 |
| | After the final rolling | 971 | 975 |

### *3.2. Analysis of the Phase Transformation*

Figure 2 compares the experimentally obtained dynamic CCT curves for the two different rolling processes, and it can be seen that the phase transformation curves are lower for higher holding temperatures. In figure 2, Ac1 is 718 °C, and Ac3 is 820 °C.

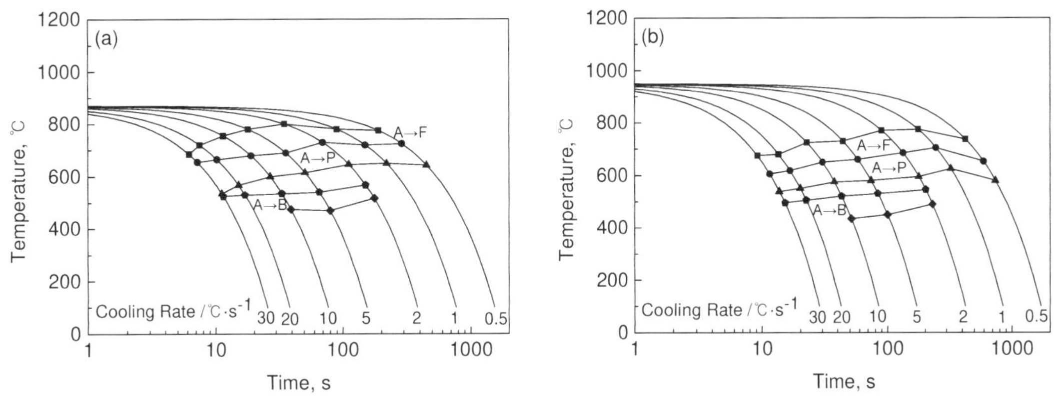


**Figure 2.** Dynamic CCT curves for the two different rolling processes:
(a) DROW process; (b) THR process

Figure 3 compares the morphology of the grains of the rebar for the two different rolling processes.

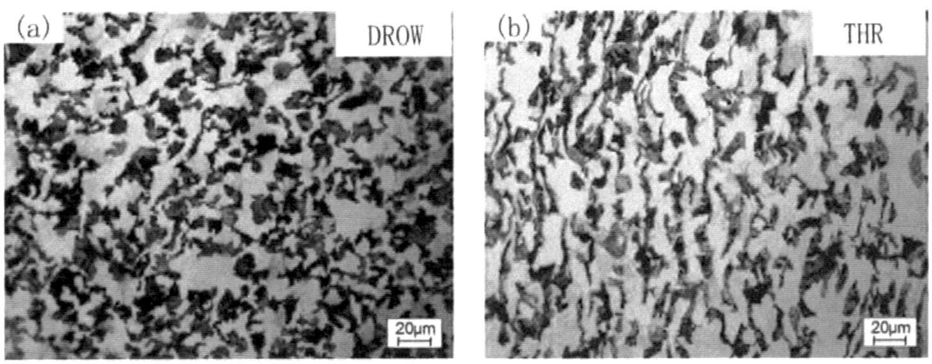


**Figure 3.** Grain morphology obtained for the rebar after the two different rolling processes:
(a) DROW process; (b) THR process

Because the temperature of rolled-piece is lower in the DROW process than in the THR process, the grains of the rebar are more refined.

Based on the CCT curves shown in figure 2 and the equations (1)-(4), the volume fractions of the different phases in the radial direction of the rebar at room temperature were simulated using the FEM, and the results were compared with experimental data. Figure 4 shows the phase distribution and the volume fraction of the cross-section of the rebar.

The results show that the volume fraction calculated for ferrite, pearlite and bainite show little variation along the radial direction. Compared with the THR process, in the DROW process, the ferrite content decreased by 6.42%, the pearlite content increased by 6.44% while the bainite content decreased by 0.02%.

The experimental and calculated volume fraction of each phase are compared in Table 5 for rebar fabricated via the DROW process, and the calculated values are in good agreement with the experimental data.

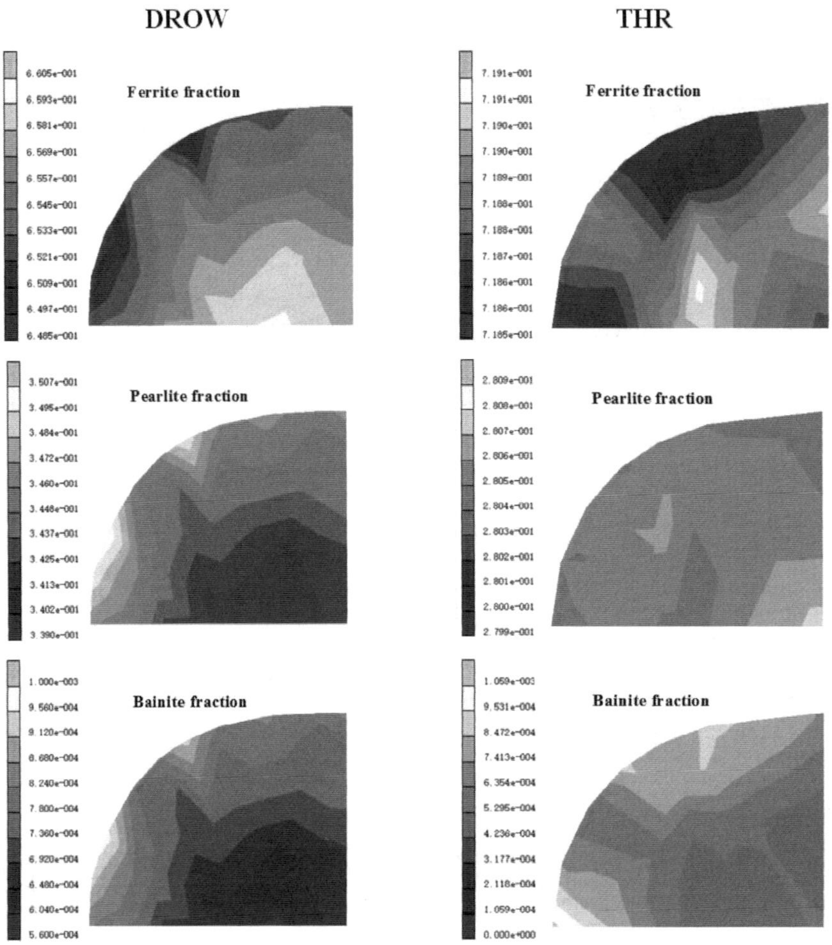


**Figure 4.** Volume fraction of each phase found in the rebar.

**Table 5.** Comparison of the experimental and calculated volume percentages for each phase.

| | Experimental value (vol%) | Calculated value (vol%) |
|---|---|---|
| Ferrite | 63.23 | 65.45 |
| Pearlite | 36.59 | 34.47 |
| Bainite | 0.18 | 0.08 |

### *3.3. Experimental Comparison of the Mechanical Properties of the Rebars*

The experimental results obtained for the mechanical properties of the rebars fabricated via the DROW and the THR process, respectively, are compared in Table 6.

**Table 6.** Mechanical properties of the rebars fabricated via the two different rolling processes.

| | Average yield strength (MPa) | Average tensile strength (MPa) |
|---|---|---|
| DROW | 449 | 612 |
| THR | 432 | 607 |

The average yield strength and tensile strength of rebar produced via the DROW process showed the enhancement, and they were found to be improved by 17 and 5 MPa, respectively, compared with the average yield strength and tensile strength of the rebar produced via the THR process.

IOP Conf. Series: Materials Science and Engineering **317** (2018) 012003 doi:10.1088/1757-899X/317/1/012003

## 4. Conclusions

(1) When changing to the DROW process, the pearlite content increased by 6.44%, and the ferrite content decreased by 6.42%. In addition, compared with the THR process, the DROW process did not have a negative effect on the uniformity of the cross-sectional phase distribution.

(2) The temperature of rolled-piece is lower in the DROW process than in the THR process, the grains of rebar are more refined, and the average yield strength and tensile strength increased by 17 and 5 MPa, respectively.

(3) The rebars produced by the DROW process can improve mechanical properties without requiring a heating furnace. Thereby reducing production costs and increasing productivity.

## References

[1] Paul S K, Rana P K, Das D, ChaDROWa S, Kundu S 2014 *Constr. Build. Mater.* **54** 170-9.
[2] Mukherjee M, Dutta C, Haldar A 2012 *Mater. Sci. Eng.: A.* **543** 35-43.
[3] Zaky A I, El-Morsy A, El-Bitar T 2009 *J. Mater. Proc. Tech.* **209(3)** 1565-9.
[4] Chen Y, Yang Z M, Wang H M 2012 *J. Iron Steel Res. Int.* **19(12)** 48-52.
[5] Wei J, Dong J H, Ke W 2011 Constr. Build. Mater. **25(3)** 1243-7.
[6] Wei J, Dong J H, Ke W 2010 Constr. Build. Mater. **24(3)** 275-82.
[7] You Y J, Jay Kim J H, Kim S J, ParkY H 2015 Constr. Build. Mater. **75(2015)** 54-62.
[8] Paul S K, Majumdar S, Kundu S 2014 *Mater. Des.* **58** 402-11.
[9] German M, Pamin J 2015 *Arch. Civ. Mech. Eng.* **15(4)** 1160-72.
[10] Rocha M, Brühwiler E, Nussbaumer A 2015 Int. J. Fat. **75** 198-204.
[11] Rocha M, Brühwiler E, Nussbaumer A 2015 Int. J. Fat. **75**: 205-12.

IOP Conf. Series: Materials Science and Engineering **317** (2018) 012004 doi:10.1088/1757-899X/317/1/012004

# Characteristics of Crushing Energy and Fractal of Magnetite Ore under Uniaxial Compression

**F Gao[1, 2]* D Q Gan[1, 2] and Y B Zhang[1, 2]**

[1]College of Mining Engineering, North China University of Science and Technology, Tangshan 063210, China;
[2]Mining Development and Safety Technology Key Lab of Hebei Province, Tangshan 063210, China

**Abstract.** The crushing mechanism of magnetite ore is a critical theoretical problem on the controlling of energy dissipation and machine crushing quality in ore material processing. Uniaxial crushing tests were carried out to research the deformation mechanism and the laws of the energy evolution, based on which the crushing mechanism of magnetite ore was explored. The compaction stage and plasticity and damage stage are two main compression deformation stages, the main transitional forms from inner damage to fracture are plastic deformation and stick-slip. In the process of crushing, plasticity and damage stage is the key link on energy absorption for that the specimen tends to saturate energy state approaching to the peak stress. The characteristics of specimen deformation and energy dissipation can synthetically reply the state of existed defects inner raw magnetite ore and the damage process during loading period. The fast releasing of elastic energy and the work done by the press machine commonly make raw magnetite ore thoroughly broken after peak stress. Magnetite ore fragments have statistical self-similarity and size threshold of fractal characteristics under uniaxial squeezing crushing. The larger ratio of releasable elastic energy and dissipation energy and the faster energy change rate is the better fractal properties and crushing quality magnetite ore has under uniaxial crushing.

## 1. Introduction

Magnetite ore is an important raw material in iron and steel industry, because of the low grade of iron ore in many countries in the world, the ore can be used for smelting after being crushed and grinded and chose to become concentrate. According to the law of thermodynamics, material failure is driven by energy fundamentally [1]. Under the influence of the complex action from geologic activity and ore blasting, the internal structure of the magnetite ore is discontinuous and the mechanical property has obvious non-linear characteristic, the traditional rock mechanics method is difficult to explain the essence of the crushing process. Energy dissipation promotes damage evolution and strength reduction, and energy release leads to failure of specimen. Therefore, it is closer to the actual situation to study the breaking law of magnetite in the aspect of energy. Based on the energy principle, many scholars use statics test method to explore the energy law and failure criterion of rock destabilization failure [3-11].

Crushing is a critical process for magnetite ore processing, the rock fragmentation has the self-similarity of geometrical shape, and its distribution is a fractal structure [12-16]. The energy and material are exchanged with the ones in outside world during the crushing process of magnetite ore, and the distribution of fragmentation has an important influence on the subsequent grinding and monomer dissociation, and clarifying the energy and fractal characteristics in the crushing process of magnetite ores can help to reveal the breaking law and solve the problem of high energy consumption in magnetite grinding, which has important theoretical value and engineering significance. However, the literature on the mechanism of magnetic iron ore fragmentation is still rare. One of the main functions of the mechanical crushing of magnetite is uniaxial crushing. Therefore, on the basis of the physical characteristics of magnetite ore, standard cuboid specimens were processed to carry out uniaxial crushing tests to analyse the essential laws of energy absorption and transformation during the crushing process of magnetite ore, and study the mechanism of energy evolution during crushing, fractal characteristics and their related relationships.

## 2. Uniaxial crushing test scheme

The random sampling was carried out at a blasting site in Shui Chang Iron Mine which is affiliated to Shou Gang Group Mining Company in China. The average density of magnetite ore is $3.26\times10^3$kg/m$^3$ in natural state, the grade is 28%. It can be known that the main components of magnetite ore are O, Si and Fe, while the magnetite iron ore contains small proportions of Mg and Al and a lot of existed fissures and pores through scanning electron microscope tests. The magnetite ore with the same block size of the same mine pile were selected for being cut, polished and processed to be specimens with 70mm width, 70mm length and 35mm thickness. The size accuracy of the specimen meets the requirements of the Rock Mechanics Test Code (SL264-2001) of Hydraulic and Hydroelectric Engineering in China, and the uniaxial crushing tests of the standard cuboid specimens were carried out by using ATW-3000 electro-hydraulic servo test machine.

The uniaxial compressive test of magnetite is operated in strictly accordance with the rules of rock mechanics test. The preload 3KN is applied firstly, then the flexible displacement control is used, and the loading rate is 0.0025mm/s. The servo control machine records the load, displacement, loading time and so on, and simultaneously draws the load-displacement curve. When the specimen breaks down, the loading stopped, the states of the specimen before and after loading are shown in figure 1.

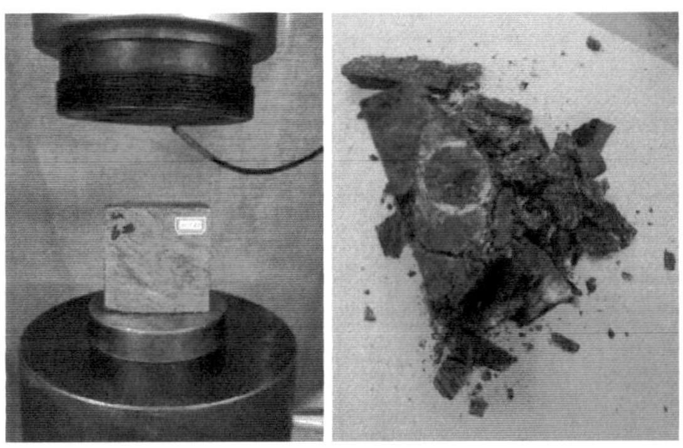

**Figure 1.** The states of specimen before and after loading.

## 3. Deformation characteristics during uniaxial crushing process

The mechanical properties have a certain difference from the original ore mass because of the effect of blasting shock wave and high-speed impact. According to the recorded results of the load displacement, the stress-strain curves of the crushing of magnetite were plotted, as shown in figure 2. According to the modified Hock Brunn Criteria of jointed rock mass studied by E HOEK and D WOOD [17], and the GE X R's division on uniaxial stress-strain curve [18], the stress-strain curve of the uniaxial crushing of the magnetite ore is divided into the compaction stage (OA), elastic stage (AB), plasticity and damage stage (BC), unstable crushing stage (CD) based on the variation of stress-strain curve slope. O is the starting point of the curve, the slope of the OA segment is always increasing, the slope of AB remains unchanged, and the stress and strain are linearly changing. The curve slope becomes smaller after B, and the whole curve of BC segment is concave. C is the curve vertex, CD segment drops rapidly, D is the end point of loading. The proportions of loading time in each stage are OA: 45.96%~53.83%, AB: 5.88%~10 69%, BC: 62%~10.69%, CD: 7.03% ~ 8.87%. The average value of the strain range in compaction phage is 53.74% in total strain, and the ratio of the maximum stress to the peak stress is 46.55%. The elastic stage range is smaller than that of ordinary hard brittle rock, but the ratio of maximum stress in elastic stage to peak stress is much greater than that of ordinary hard brittle rock [19]. The time ratios in each stage are corresponding to strain ratios.

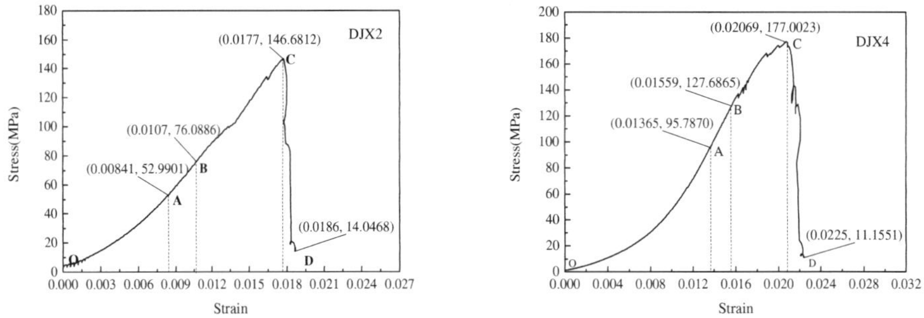


**Figure 2.** Strain-stress curves of magnetite ore in uniaxial crushing.

The compaction stage and the plasticity and damage stage account for the main part of the crushing process of the uniaxial crushing of the magnetite ore. The scope of compaction stage shows that the cracks in the magnetite ores are abundant. The curves recovery plasticity rises after significant stress damage, and the sticky-slip phenomenon occurs in the plastic and damage state [20]. It is indicated that the new crack initiation is significant and the pore expands after the elastic stage, plastic deformation and sticky slip are the main transitional forms of the crushing process of magnetite ores. With the increase of the stress, the slip degree of ore particle and rock particle and the expansion degree of porosity and fissure increase, the stress concentration of pores and fissures reaches the limit at the peak stress, and a large number of pores and fissures are rapidly connected to form shear and tensile failure inner magnetite ore, and then the stress drops quickly.

## 4. Energy evolution analysis of crushing process

### *4.1. Energy mechanism of uniaxial compression*

The energy absorption, transformation and dissipation occur during the uniaxial crushing process of magnetite specimen. During the loading process, the test machine has done the mechanical work to the magnetite, which causes the specimen deformed, damage, destroy and break. Assuming that the test machine loading system is a closed-energy system, the mechanical work done by the press machine can be regarded as the total energy absorbed by the specimen. According to the first law of thermodynamics, in the case of no heat exchange with the outside world, the total energy absorbed by the specimen $U$ equals the sum of dissipation energy $U_d$ and the releasable elastic strain energy $U_e$, namely:

$$U = U_d + U_e \tag{1}$$

$$U = \int_0^{\varepsilon_1} \sigma_1 d\varepsilon_1 + \int_0^{\varepsilon_2} \sigma_2 d\varepsilon_2 + \int_0^{\varepsilon_3} \sigma_3 d\varepsilon_3 \tag{2}$$

$$U^e = \frac{1}{2E_0}\left[\sigma_1^2 + \sigma_2^2 + \sigma_3^2 - 2\mu(\sigma_1\sigma_2 + \sigma_2\sigma_3 + \sigma_1\sigma_3)\right] \tag{3}$$

In uniaxial compression tests, confining pressure $\sigma_2$=0 and $\sigma_3$=0, the total absorbed energy and the releasable elastic strain energy of the specimen under uniaxial compression are respectively calculated as follows [1]:

$$U = \int_0^{\varepsilon_1} \sigma_1 d\varepsilon_1 = \sum_{i=0}^{n} \frac{1}{2}\left(\varepsilon_{i+1} - \varepsilon_i\right)\left(\sigma_{i+1} + \sigma_i\right) \tag{4}$$

$$U^e = \frac{\sigma_i^2}{2E_O} \tag{5}$$

Where $\varepsilon_i$, $\sigma_i$ are the stress value and strain value of specimen at a time in uniaxial compression process, according to rock strength and overall failure criterion based on the principle of energy dissipation and release, $E_0$ is initial elastic modulus of cell without damage for simplified computing, the elastic modulus at the end of the elasticity stage is the initial elastic modulus [2, 4].

### *4.2. Energy evolution analysis of crushing process*

The energy curves during uniaxial compression of magnetite ore are drawn, as shown in figure 3. The peak stresses of magnetite ore were counted, as well as absolute absorption energy, releasable elastic energy and dissipation energy of critical points (A, B and C) between deformation stages, and the absolute absorption energy, the residual releasable elastic energy and the maximum dissipated energy at the end of loading (D). Absolute absorption energy is the value of each point in absorption energy curve. The ratios of the change of the absorption energy ($\Delta U$), the change ratios of the releasable energy ($\Delta U_e$) and dissipation energy ($\Delta U_d$) to the total absorption energy $U_0$ were calculated in the four deformation stages as shown in table 1. The average change rates of absorption energy, the releasable elastic energy and the dissipation energy are calculated in each stage, as shown in table 2, the positive change rates indicates the increase of energy, and the negative change rates indicates energy reduction. The ratios of releasable elastic energy ($U_e$), dissipation Energy ($U_d$) accounted for the absorption energy ($U$) were calculated at the transition points and the loading end moment, as shown in table 3.

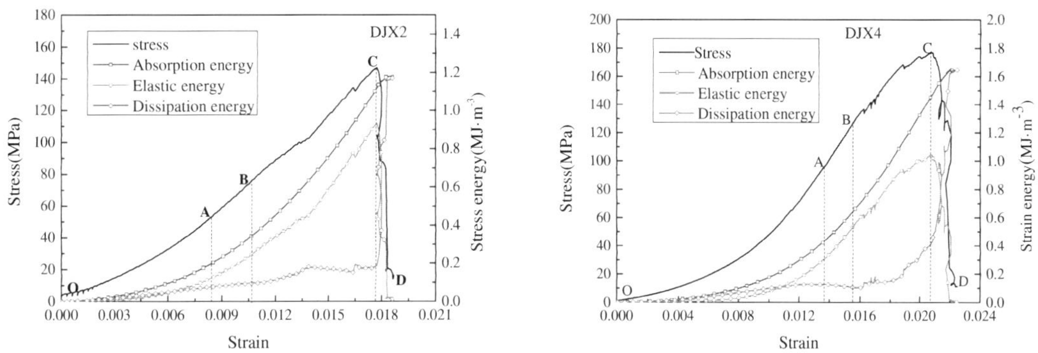


**Figure 3.** Energy evolution curves of raw magnetite ore in uniaxial squeezing crushing.

**Table 1.** The change ratios of strain energy in total absorption energy in stages.

| Number | $P_{max}$ ( MPa ) | Compression stage | | | Elastic stage | | | Plasticity and damage stage | | | Crushing stage | | |
|---|---|---|---|---|---|---|---|---|---|---|---|---|---|
| | | $\Delta U/U_0$ | $\Delta U_e/U_0$ | $\Delta U_d/U_0$ | $\Delta U/U_0$ | $\Delta U_e/U_0$ | $\Delta U_d/U_0$ | $\Delta U/U_0$ | $\Delta U_e/U_0$ | $\Delta U_d/U_0$ | $\Delta U/U_0$ | $\Delta U_e/U_0$ | $\Delta U_d/U_0$ |
| DJX6 | 113.18 | 0.232 | 0.143 | 0.084 | 0.066 | 0.069 | 0.001 | 0.549 | 0.415 | 0.138 | 0.153 | -0.608 | 0.760 |
| DJX2 | 147.69 | 0.169 | 0.105 | 0.065 | 0.121 | 0.099 | 0.016 | 0.653 | 0.580 | 0.075 | 0.058 | -0.776 | 0.833 |
| DJX4 | 177.24 | 0.262 | 0.184 | 0.077 | 0.129 | 0.147 | -0.013 | 0.488 | 0.298 | 0.184 | 0.121 | -0.627 | 0.747 |
| Average | 146.04 | 0.22 | 0.14 | 0.08 | 0.11 | 0.11 | 0.00 | 0.56 | 0.43 | 0.13 | 0.11 | -0.67 | 0.78 |

**Table 2.** The change rate of strain energy in stages (KPa·m$^{-3}$s$^{-1}$)

| Number | $P_{max}$ (MPa) | Compression stage | | | Elastic stage | | | Plasticity and damage stage | | | Crushing stage | | |
|---|---|---|---|---|---|---|---|---|---|---|---|---|---|
| | | $\Delta U/\Delta t$ | $\Delta U_e/\Delta t$ | $\Delta U_d/\Delta t$ | $\Delta U/\Delta t$ | $\Delta U_e/\Delta t$ | $\Delta U_d/\Delta t$ | $\Delta U/\Delta t$ | $\Delta U_e/\Delta t$ | $\Delta U_d/\Delta t$ | $\Delta U/\Delta t$ | $\Delta U_e/\Delta t$ | $\Delta U_d/\Delta t$ |
| DJX6 | 113.18 | 0.76 | 0.47 | 0.28 | 1.98 | 2.09 | 0.02 | 3.10 | 2.34 | 0.78 | 3.06 | -12.2 | 15.19 |
| DJX2 | 147.69 | 0.81 | 0.50 | 0.31 | 2.49 | 2.05 | 0.33 | 3.92 | 3.48 | 0.45 | 1.82 | -24.35 | 26.13 |
| DJX4 | 177.24 | 1.13 | 0.80 | 0.33 | 3.97 | 4.52 | 0.41 | 5.66 | 3.46 | 2.14 | 3.92 | -20.39 | 24.28 |
| Average | 146.04 | 0.90 | 0.59 | 0.31 | 2.81 | 2.89 | 0.25 | 4.23 | 3.09 | 1.12 | 2.93 | -18.98 | 20.66 |

**Table 3.** The ratios of strain energy in absorption energy at the transition points and the end of loading

| Number | $P_{max}$ (MPa) | A | | B | | C | | D | |
|---|---|---|---|---|---|---|---|---|---|
| | | $U_e/U$ | $U_d/U$ | $U_e/U$ | $U_d/U$ | $U_e/U$ | $U_d/U$ | $U_e/U$ | $U_d/U$ |
| DJX6 | 113.18 | 0.618 | 0.363 | 0.715 | 0.285 | 0.742 | 0.264 | 0.020 | 0.983 |
| DJX2 | 147.69 | 0.623 | 0.385 | 0.706 | 0.279 | 0.833 | 0.166 | 0.008 | 0.989 |
| DJX4 | 177.24 | 0.704 | 0.295 | 0.848 | 0.163 | 0.716 | 0.282 | 0.003 | 0.995 |
| Average | 146.04 | 0.65 | 0.35 | 0.76 | 0.24 | 0.76 | 0.24 | 0.01 | 0.99 |

*4.2.1.* Analysis *on the change of absorption energy.* The results of the calculation in table 1 shows that the average ratios of absorption energy in each stage accounted for the total absorption energy indicate that the absorption energy fluctuation changes during the compression process, and the plasticity and damage stage is the main link of energy absorption. It can be seen from table 2 that the change rate of the absorption energy increases in turn in the compaction stage, the elastic stage, the plasticity and the damage stage, and the change rate of absorption energy in the crushing stage is reduced, indicating that the capacity of the specimen absorbing mechanical energy tends to saturation near the peak strength, and the breakage happens when the absorption energy reaches the limit in the specimen.

*4.2.2.* Analysis *on the change of the releasable elastic energy.*The ratios of the amount of releasable elastic energy to the total absorption energy in the process of specimen compression and deformation were shown in table 1. Because of the mineral $Fe_3O_4$ and $Al_2O_3$ in the magnetite specimen in the test, the plastic deformation and damage occur in the plastic and damage stage, and the elastic deformation of the specimen was still in a certain degree at the same time. The absorption energy can be converted into the releasable elasticity energy, which is mainly caused by internal stress and elastic modulus of the specimen. The elastic strain energy is higher in the plastic and deformation stage than that in the elastic stage. The ratio of elastic strain energy released in the crushing stage to total absorption energy is 60.79% ~ 77.61% (average value 67%). The remaining absorption energy dissipated in the crack compaction, initiation, expansion, acoustic emission, infrared radiation before the peak stress, and so on. The concentrated releasing of elastic-strain energy is the main reason for the fragmentation of magnetite specimen. According to the calculation results of table 2, the absolute values of the average change rate of the releasable elastic energy in the crushing stage are respectively 32 times, 6.56 times and 6.14 times than the ones in former stages before crushing.

Table 3 shows that the absorption energy is stored in the form of the releasable elastic energy, and the less absorption can be consumed before the peak stress in the aspects of fissure and pore compaction, mineral particle movement, volume element deformation, internal local damage, specimen temperature change, thermal radiation, acoustic emission, etc. When the stress reaches the peak value, the energy storage ability of the specimen reaches limit, and the most of releasable elastic energy in the specimen is released and consumed by the crack extension and the block splitting. However, the specimen has been separated into blocks at the end of loading and the mechanical energy produced by the press machine can be directly consumed in the further broken of blocks. It should be noted that the average ratio of releasable elastic energy stored in the specimen accounts for 72.3% to the total absorption energy before the peak stress, which is higher than the average ratio (67%) of the releasable elastic energy consumed in specimen crushing to the total absorbed energy, indicating that elastic strain energy stored in the specimen before the peak stress can't be used in the specimen crushing thoroughly. A small portion of the stored elastic strain energy can be dissipated by the shape recovery of many crystals during unloading, which is the result of rich containing $Fe_3O_4$, $Al_2O_3$ and other minerals in specimens.

*4.2.3. Analysis on the change of dissipation energy.* Energy dissipation always happens from deformation beginning to the end of loading, the ratio of dissipation energy to the total absorption energy can reflect the rock deformation and damage degree [1, 14]. It can be seen from table 1 that the damage degree of the magnetite ore is very small in the elastic deformation stage, and the dislocation and slippage of mineral particles occur in the plastic and damage stage. After the peak stress, the dissipation energy ratio in the crushing stage is 78% to the total absorption energy, which is higher than released energy, whose ratio is 67% to the total absorption energy. The cause of this phenomenon is that, the deformation still occurs after the peak stress, the mechanical work from the press machine directly acts on the further crushing of the specimen, the dissipated energy of the crushing stage should be the sum of the releasable elastic energy and the partial mechanical energy.

The change rate of dissipation energy can reflect the activity from deformation to damage inner specimens. According to the calculation results of table 2, the average change rates of dissipation energy indicates that the fracture compression and less crack initiation happen in the compaction stage and the elastic stage, granular slip dislocation and large fissure initiation and expansion are active in the plastic and damage stage, the crack through the junction is very rapid after the peak stress.

## 5. Fractal characteristics of uniaxial extrusion crushing of magnetite ore

### *5.1. Fractal theory of rock fragmentation*

Self-similar characteristics of rock failure and fragmentation are usually described with fractal dimension theory [12, 21]. Fractal dimension is the main geometrical parameter describing the self-similarity of rock fragmentation statistics. At present, the method of double logarithm fitting of granularity-accumulative mass percent is widely used in the research of fragmentation fractal [18, 21, 22].

$$Log\frac{M(r)}{M} = bLogr \tag{6}$$

$$D = 3 - b \tag{7}$$

Where, $M(r)/M$ is a percentage of the quality of rock masses with a block degree less than $r$ in total crushing mass, $r$ is feature size of fragment, $D$ is fractal dimension of fragment size, $b$ is the slope of the regression line in the double logarithm coordinate system.

### *5.2. Fractal characteristics of magnetite ore block size*

The fragments of magnetite ore were sieved; the cumulative mass distribution curves were drawn. When the feature sizes of magnetite fragments are smaller than 20mm, the cumulative mass distribution curves show good linearity, however the quality of fragments whose feature sizes is larger than 20mm occupies main part in total fragment quality.

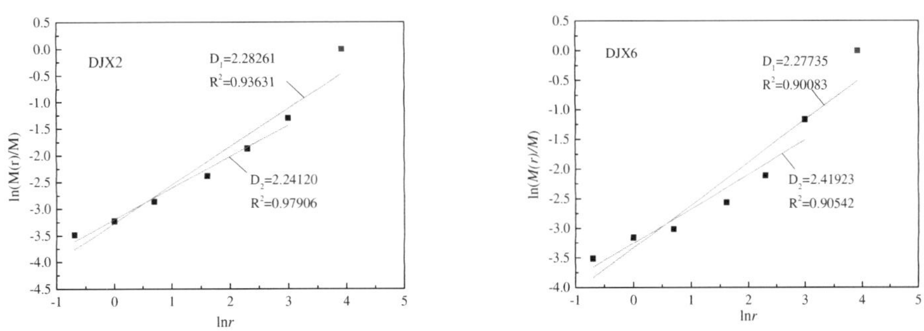


**Figure 4.** Double logarithm curves between quality percentage and granularity.

According to the sieving results of fragment, the double logarithm scatter plots between the cumulative mass percent (M(r)/m) and the granularity (r) are drawn, as shown in figure 4. The two-logarithm scatter graphs for all fragments and fragments smaller than 20mm were fitted on fractal dimension respectively to gain the fractal dimension of all fragments, ie $D_1$, and the fractal dimension of less than 20mm fragments, ie $D_2$, as shown in table 4. The results of fractal dimension calculation show that the uniaxial crushing of magnetite ore is statistically self-similar, and the statistical self-similarity in the small scale range is higher than that in all the fragmentation, which is consistent with fractal characteristics of rock fragmentation [12-15]. According to figure 4, the fitting results of fragmentation distribution and fractal change significantly at 20mm, which reflects that magnetite ore has the dual-scale self-similarity of under uniaxial compression. The ore fragments of magnetite have self-similar characteristics in the small size range and the whole size range, but the significance is different. According to the research results by HUANG D, et al [14], 20mm can be used as the size threshold of fractal dimension of magnetite ore under uniaxial crushing, and the fractal characteristic of magnetite iron ore was weakened when fragment sizes are larger than 20mm.

**Table 4.** Fractal dimensions of raw magnetite ore under uniaxial squeezing crushing.

| Specimen number | Fractal dimension calculation results of all fragments | | | Fractal dimension calculation results of fragments smaller than 20mm | | |
|---|---|---|---|---|---|---|
| | Fitting formulas | $D_1$ | $R^2$ | Fitting formulas | $D_2$ | $R^2$ |
| DJX2 | y=0.71739x-3.26922 | 2.28261 | 0.93631 | y=0.58802x-3.1974 | 2.41198 | 0.97906 |
| DJX3 | y=0.69117x-3.16396 | 2.30883 | 0.93233 | y=0.56259x-3.09257 | 2.43741 | 0.97712 |
| DJX4 | y=0.72053x-3.34569 | 2.27947 | 0.90691 | y=0.5732x-3.2639 | 2.4268 | 0.93498 |
| DJX5 | y=0.69444x-3.19903 | 2.30556 | 0.92849 | y=0.55958x-3.12416 | 2.44042 | 0.97996 |
| DJX6 | y=0.72265x-3.33449 | 2.27735 | 0.90083 | y=0.58077x-3.25573 | 2.41923 | 0.90542 |

### *5.3. Correlation analysis of fractal features and energy*

The fragmentation of materials is the result of energy evolution, and fractal dimension is the quantitative describe of fragmentation characteristics. Therefore, the fractal characteristics of magnetite ore fragments after uniaxial crushing are related to the energy characteristics. The magnetite ore are broken down integrally under compression in dressing, combining the energy characteristic data of table 1 and table 2, the relation curves between fractal dimension of all fragments ($D_1$) and strain energy ratios ($\Delta U_e/U_0$ and $\Delta U_d/U$)accounting for total absorption energy in the crushing stage were plotted, the relation curves between the fractal dimension of all fragments and the strain energy change rate ($\Delta U_e/\Delta t$ and $\Delta U_d/\Delta t$) in the crushing stage were plotted as well, as shown in figure 5. It can be seen that in the crushing stage, the ratio of the elastic energy reduction to the absorption energy and the absolute value of the reducing rate increase with the increase of the integral fractal dimension, and the dissipation energy ratio to the absorption energy and the change rate are basically increase with the increase of the integral fractal dimension.

**Figure 5.** Relational curves between whole fractal dimension and strain energy.

The release value and release rate of elastic energy affect the fragmentation degree, as well as dissipation energy in the crushing process of magnetite ores. The higher ratios of energy release and dissipation to the total energy and the faster the change rates of energy are, the more obvious the fractal characteristics of all magnetite ore is, and the higher crushing degree can be gained.

## 6. Conclusions

Following conclusions can be obtained according to the present study:

(1) During the magnetite ore crushing process under uniaxial compression, the compaction stage and plasticity and damage stage are the main deformation stages, plastic deformation and visco-slip are the main transitional forms from elastic deformation stage to plasticity and damage stage, the proportion of strain range in different stages corresponds well with time ratio.

(2) The plasticity and damage stage is the main stage on energy absorption and transformation inner specimen. The absorption energy value varies with the loading process. The average ratio of releasable elastic energy stored in the specimen before the peak stress is higher than that consumed in specimen crushing of the total absorbed energy. After the peak stress, the press machine continues to work to make some blocks much more broken, the average ratio of the energy consumed in specimen crushing accounting for the total absorption energy is 78%. The damage evolution of magnetite can be realized by energy dissipation, and the rapidly release of elastic energy in the crushing stage is the fundamental reason of the disintegration of magnetite.

(3) The uniaxial crushing of the magnetite ore is characterized by dual-scale fractal characteristics, and the fractal characteristics of fragments smaller than 20mm are more obvious than that of all fragments. The uniaxial compression crushing of magnetite ore has a size threshold of fractal dimension. The crushing energy of magnetite ore is related with the fractal dimension of all blocks. The more the releasable elastic energy ratio and release rate, dissipation energy ratio and dissipation rate are, the higher the fractal dimension is , the better fragmentation effect of magnetite ore can be get.

## References

[1] Zhang L M, Gao S and Wang Z Q 2013 *J. Rock. Soil. Mech.* **11** 3071-76.
[2] Xie H P, Ju Y and Li L Y 2005 *Chin. J. Rock. Mech. Eng.* **17** 3003-10.
[3] Zhao Z H and Xie H P 2008 *J. J. Sichuan. Univ: Eng. Sci. Ed.* **02** 26-31.
[4] Yang S Q, Xu W Y and Su C D 2006 *J. Acta. Mech. Solida. Sin.* **02** 213-16.
[5] Meng Q B, Hang L J, Pu H, Weng S Y 2015 *J. China. Coal. Soc.* **40** 2386-97.
[6] Xu X L, Chen L, Gao F and Zhang Z Z 2015 *J. Acta. Mech. Solida. Sin.* **36** 154-63.
[7] Li Y L, Liang W, He G C and Zhang Z J 2015 *J. Univ. South. China: Sci and Tech.* **29** 31-36.
[8] Zhang Z Z, Gao F 2015 *Chin. J. Rock. Mech. Eng.* **1** 1-11.
[9] Liu X R, Guo J Q, Wang J B, Li P and Zhang Q Q 2013 *J. Rock. Soil. Mech.* **34** 305-15.
[10] Guo J Q, Liu X R, Wang J H and Yu Y 2013 **44** 5046-49.
[11] Zhou H, Li Z, Yang Y C, Zhang C Q and Sheng Q 2013 *Chin. J. Rock. Mech. Eng.* **32** 2170-84.
[12] Liu S, Xu J Y, Bai E L and Gao Z G 2013. *J. Vib. Shock.* **32** 163-66.
[13] He M C, Yang G X, Miao J L, Jia X N, Jiang T T 2009 *Chin. J. Rock. Mech. Eng.* **28** 1521-29.
[14] Huang D, Tan Q, Huang R Q 2012 *Chin J Rock Mech Eng.* **31** 1379-89.
[15] Wang L and Gao Q 2007 *Chin. J. Rock Mech Eng.* **26** 1202-11.
[16] Deng T, Yang L D and Han W F 2007 *J. Tongji. Univ: Nat Sci Ed,* **35** 10-14.
[17] Hoek E and Wood D 1992 *Proc. Int. Conf. of Eurok. (Chest: England)* p 202-204.
[18] Ge X R, Zhou B H and Liu M G 1992 *J. China. Min. Mag.* **1** 57-60.
[19] Zhao W 2010 R*ock Mechanic* Chang Sha: Central South University Press, pp. 12-4.
[20] Xi D Y and Xu S L 2012 *Fundamentals of rock physics* He Fei: University of Science and Technology of China Press, pp. 144-63.
[21] Gao F, Xie H P, Wu J B 1999 *Chin. J. Rock. Mech. Eng.* **18** 497-502.
[22] Xie H P 1997 *Introduction to fractal rock mechanics* Beijing: Science Press, pp. 242-3.

## Acknowledgements

The authors wish to acknowledge the financial support from two National Natural Science Foundations of China (Grant No. 51374087 and Grant No. 51774187) and the Natural Science Foundation of Hebei Province (Grant No. 2016209220).

IOP Conf. Series: Materials Science and Engineering **317** (2018) 012005 doi:10.1088/1757-899X/317/1/012005

# The Influence of Oscillatory Fractions on Mass Transfer of Non-Newtonian Fluid in Wavy-Walled Tubes for Pulsatile Flow

**Donghui Zhu[1*], Yongning Bian[1]**

[1]State Key laboratory of structural analysis for industrial equipment, Dalian university of technology, 116024, China

**Abstract.** The shape of pipeline structure, fluid medium and flow state have important influence on the heat transfer and mass effect of fluid. In this paper, we investigated the mass transfer behavior of Non-Newtonian fluid CMC solution with 700ppm concentration in five different-sized axisymmetric wave-walled tubes for pulsatile flow. It is revealed that the effect of mass transfer is enhanced with the increase of oscillatory fractions $P$ based on the PIV measurements. Besides, mass transfer rate was measured by the electrochemical method in the larger oscillatory points rate range. It is observed that mass transfer rate increases with the increase in $P$ and reached the maximum mass transfer rate at the most optimal oscillatory fractions $P_{opt}$. After reaching the optimal oscillatory fractions $P_{opt}$, the mass transfer rate decreases with increasing $P$.

## 1. Instruction

The flow of fluid and heat and mass transfer in various flow paths become an important part in the study of fluid mechanics. It is very important to clarify the mechanism of flow characteristics and transfer process for the development and designment of various heat transfer equipment. Changing flow field can enhance the heat transfer effect. Currently, there is less research on the oscillatory fractions of the pulsating flow in the wave-walled tube.

Nishimura et al. studied the characteristics of fluid flow and mass transfer in sinusoidal wavy-walled tube. Their results show that the rapid fluid mixing produced by eddy motion can achieve better quality transfer enhancement in the basin with inertial influence. The results of the study of the flow field of the unopposed flow in the wall tube showed that the most effective mass transfer intensification occurred before the net flow was about to enter the transitional flow. The optimal number of frequencies with flow rate is non-linear. They point out that because there is no self-excited oscillatory fractions in the three dimensional tube, resonance enhancement cannot be used to explain the strengthening mechanism. However, the results of the study that accompanied the pulsating flow of the reverse flow were not given. In this paper, the effect of oscillatory fractions on mass transfer is studied. It provides some reference for the development of high efficiency and low consumption heat transfer device.

## 2. Experimental Setup and Methods

The experimental equipment of this study mainly consists of three parts: PIV equipment shown in figure 1, mass transfer rate measurement system illustrated in figure 2 and pulsatile flow system plotted in figure 3 and 4. Here, PIV technology is applied to the experimental system of pulsatile flow field, which enables us to illustrate the complex flow characteristics. The working fluid with tracer particles is extracted from the storage tank by centrifugal pump, and its net flow is controlled by the rotor flow meter. Before entering the wave wall tube, the pulsation pump superimposes the sine vibration of the working fluid. Pulse laser to produce light irradiation on the wave wall tube central axis, vertical plane of the high-speed digital camera to catch the image data of the tested area, at the same time through collecting transmission to the computer's memory. The synchronous controller is used to control the optical fiber sensor, digital camera, and image acquisition board, so that they can work in a strictly synchronous signal and ensure the coordination of all parts.

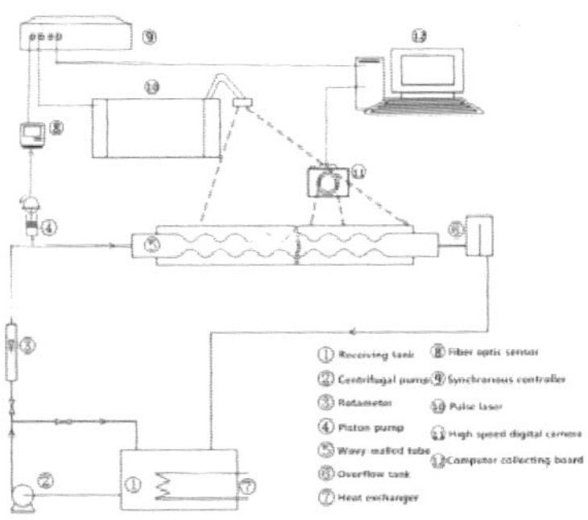

**Figure 1.** Experimental setup.

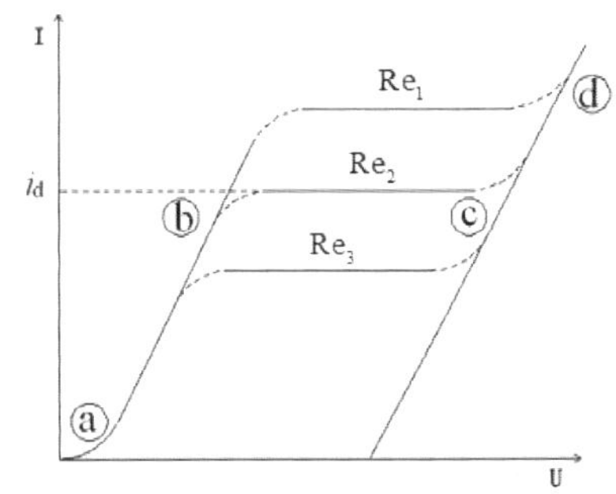


**Figure 2.** Limiting current id vs. Voltage u.

Electrochemical technology is one kind of limiting current technology based on the electrochemical reaction, namely in solid electrodes on the surface to maintain a voltage causes the electrochemical reaction rate as large as possible, reaching the electrochemical polarization. Polarization current relatively become stable at this time, and its size is determined by the mass transfer rate as is shown in the figure 2.

The measurement system of mass transfer rate is shown in figure 3. Nickel cathode module was set in the ninth wave band, the anode was set at 6 and 12 wavelength, respectively. At the same time, in order to ensure that the ratio of anode and cathode surface area is large enough, choosing entrance section and export section of nickel pipe as auxiliary anode. In this experiment, the electrolyte with special preparation was used as the working fluid. The number of Schmidt of this solution was 1570, which meets the diffusion control conditions.

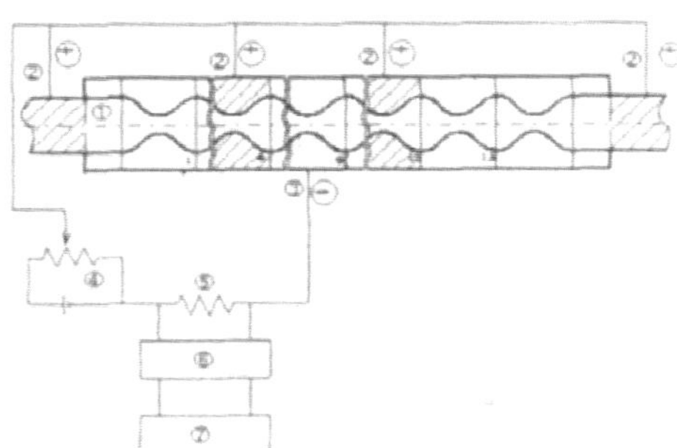

**Figure 3.** Mass transfer rate measurement system.

## 3. Definition and Choice of Parameters

### *3.1. Definition of Parameters*

Quantity of pulsation flow is steady flow combine to oscillatory flow, and it can be defined as following:

$$Q_i = Q_S + Q_O \sin(2\pi t / T) \quad (1)$$

where $Q_i$is the oscillatory flow, $Q_s$ is steady flow, and $Q_o$ is peak of oscillatory flow. Here, $Q_o$ is defined as:

$$Q_O = 2\pi fs(\pi D_P^2 / 4) \quad (2)$$

Here, $f$ is the frequency of oscillation, $s$ is the useful length of the piston, and $D_p$is the diameter of the piston.

The Reynolds number for steady flow:

$$Re_S = u_s D_{max} / v \quad (3)$$

The oscillatory fraction of the flow rate:

$$P = Q_O / Q_S \quad (4)$$

As a non-dimensional expression of oscillatory frequency, the Strouhal number is defined:

$$St = D_{max}(2\pi f / u_s) \quad (5)$$

To express the waviness degree of the wavy-walled tubes, the wave factor Fw is defined as:

$$F_W = \frac{D_{max}}{D_{min}} \times \frac{2a}{\lambda} \quad (6)$$

### *3.2. Selection of Parameters*

Figure 4 shows the pattern of wavy-walled tube used in this study, and it is clear to see that the tube is axial symmetry. The sizes of them are listed in table 1.

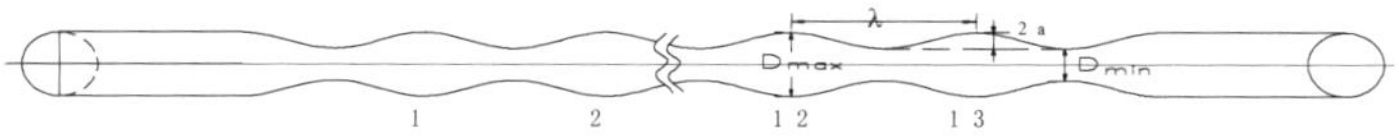

**Figure 4.** Dimensions of wavy-walled tube.

**Table 1.** Size of five wavy-walled tubes.

| $F_W$ | 0.83 | 0.69 | 0.58 | 0.36 | 0.15 |
|---|---|---|---|---|---|
| $D_{max}$(mm) | 10 | 10 | 10 | 10 | 10 |
| $D_{min}$ (mm) | 3 | 3 | 3 | 5 | 7 |
| $2a$(mm) | 3.5 | 3.5 | 3.5 | 2.5 | 1.5 |
| $\lambda$(mm) | 14 | 17 | 20 | 14 | 14 |

### *3.3. Selection of Vibration Fraction*

Since PIV can only study flow characteristics in qualitative research, the study scope of vibration fraction $P$ is limited, and $P=1/\sqrt{2}$, $P=1$, and $P=\sqrt{2}$ are selected under PIV experiment. The electrochemical mass transfer experiment was selected as $P=1/\sqrt{2}$, 1, $\sqrt{2}$, 2, 2.5, 3, 3.5, 4, 4.5, 5, 5.5 and 6.

## 4. Results and Discussion

### *4.1. PIV Experimental* Results

Pulsating flow field separately on the five sets of wave wall tube PIV image velocimetry experimental study, the PIV technology was applied to the pulsating flow field of tube flow field, the wave wall

IOP Conf. Series: Materials Science and Engineering **317** (2018) 012005 doi:10.1088/1757-899X/317/1/012005

tube in the period of a vibrating flow visualization pictures, software Tecplot processing image data to obtain the flow chart, further analysis concludes two kinds of stable flow structure and an unstable flow structure as is shown in figure 5.

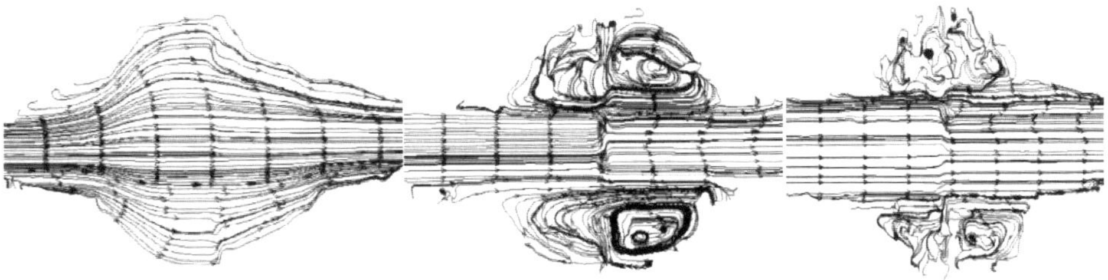

**Figure 5(a).** "I" **Figure 5(b).** "II" **Figure 5(c).** "III".

Integrated the three kinds of flow pattern, flow by clear layers to form vortex, finally the flow and vortex damage, through the comparisons of mixed flow can be found that three kinds of flow pattern corresponding to the mass transfer effect must be "I < II <III" relationship.

As is shown in table 2, the experiment obtained the flow structure corresponding to each of the eight equal parts of each vibration cycle of each of the three vibration points, and the number of the second flow form was denoted by $n$.

**Table 2.** The influence of $F_w$ and p on flow structure.

| *Fw=0.83/0.69/0.58 n=32* | | | | | | | | |
|---|---|---|---|---|---|---|---|---|
| *St* | (*t/T*) | | | | | | | |
| | 0 | 0.125 | 0.25 | 0.375 | 0.5 | 0.625 | 0.75 | 0.875 |
| 0.59 | II | II | II | II | II | II | II | II |
| 1.0 | II | II | II | II | II | II | II | II |
| 1.82 | II | II | II | II | II | II | II | II |
| 3.33 | II | II | II | II | II | II | II | II |
| *Fw=0.36 n=23* | | | | | | | | |
| 0.59 | I | I | I | II | II | II | II | II |
| 1.0 | I | II | II | II | II | II | I | II |
| 1.82 | I | I | II | II | II | II | II | II |
| 3.33 | II | I | I | II | II | II | II | II |
| *Fw=0.15 n=13* | | | | | | | | |
| 0.59 | I | I | I | I | II | II | II | II |
| 1.0 | I | I | I | II | II | II | II | II |
| 1.82 | I | I | I | I | I | II | II | I |
| 3.33 | I | I | I | I | I | II | II | I |

To sum up, the second type of flow number of each oscillatory fraction is shown in table 3.

**Table 3.** The number of the second flow type of each oscillatory fraction.

| $F_w$ | $P=1/\sqrt{2}$ | $P=1$ | $P=\sqrt{2}$ |
|---|---|---|---|
| 0.15 | 5 | 9 | 13 |
| 0.36 | 15 | 21 | 23 |
| 0.58 | 27 | 28 | 32 |
| 0.69 | 28 | 29 | 32 |
| 0.83 | 30 | 31 | 32 |

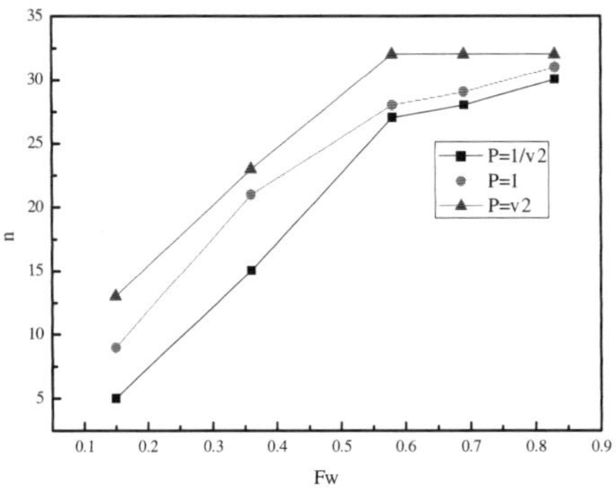


**Figure 6.** The number of the second flow type of each vibration fraction.

It can be seen from figure 6 that the number of the second type of flow in each vibration is increased with the increase of $F_w$, indicating that the mixing degree of flow increases with the increase of $F_w$; In addition, in the same pipeline, n increases with the increase of the oscillatory fraction, indicating that the mixing of the fluid increases with the increasing of the oscillatory fraction. It can be concluded that the mass transfer effect of fluid is related to the oscillatory fraction, and the mass transfer effect is enhanced with the increase of *P* in the test range of the PIV.

*4.2. Electrochemical Mass Transfer Experiment*

To measure the average current of each oscillatory fraction, the strengthening coefficient $E=I_0/I_s$ is compared with the limiting current $I_0$ under the steady flow field. The mass transfer coefficient of different vibration fraction P is shown in the table below.

**Table 4.** The mass transfer coefficient of different oscillatory fraction p.

| P | $1/\sqrt{2}$ | 1 | $\sqrt{2}$ | 2 | 2.5 | 3 | 3.5 | 4 | 4.5 | 5 | 5.5 | 6 |
|---|---|---|---|---|---|---|---|---|---|---|---|---|
| U | 3.65 | 4.5 | 5.42 | 6.5 | 7.19 | 7 | 7 | 7.3 | 6.7 | 5.9 | 6.16 | 6 |
| E | 1.04 | 1.28 | 1.54 | 1.85 | 2.05 | 2 | 2 | 2.08 | 1.91 | 1.68 | 1.76 | 1.04 |

Figure 7 shows that mass transfer effect is related to the rate of oscillatory fraction in the pulsating flow field. In the range of oscillatory fraction of electrochemical mass transfer experiment, the mass transfer coefficient increases with the increase of the vibration fraction, reaching the maximum when *P* is equal to 3. Then it decreases with the increase of the vibration fraction.

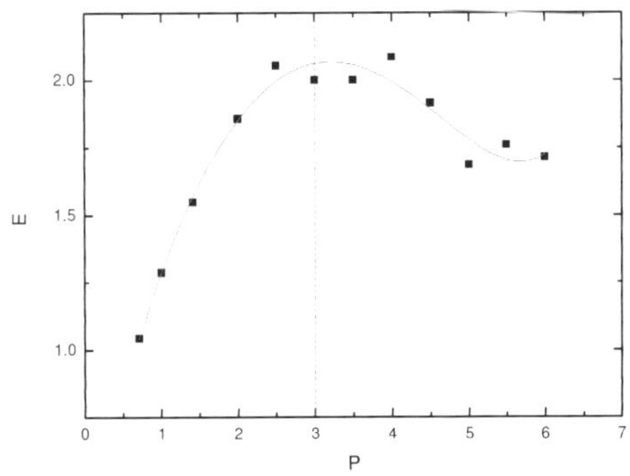


**Figure 7.** E VS. P.

## 5. Conclusions

1) The mass transfer effect of the fluid is related to the vibration fraction in the pulsating flow field, and the mass transfer effect is enhanced with the increase of P in the scope of the PIV test.

2) In the pulsating flow field, mass transfer effect is related to the rate of oscillatory fraction. In the range of vibration fraction of electrochemical mass transfer experiment. The mass transfer coefficient increases with the increase of the vibration fraction, reaching the maximum when P is equal to 3. Then it decreases with the increase of the vibration fraction. Therefore, it makes sense to increase the value of the vibration fraction in the range of $P \leq 3$.

## References

[1] Nishimura T, Murakami S, Awamura Y 1993 *Chem. Eng. Sci.* **48** 1793-800.
[2] Sobey I J, Drazin P G 1986 *J. Fluid Mech.* **171** 263-87.
[3] Nishimura T 1995 *Heat and Mass Transf.* **30** 269-78.
[4] Zhang L 2011 Exp. Fluid Mech. **25(6)** 64.
[5] Lee C L 2014 *Effects of the operating condition and the geometric parameter on the mass transfer rate in the wavy-walled tubes for pulsatile flow* Master's Thesis.
[6] Nishimura T, Kajimoto Y, Kawamura Y 1986 *J. Chem. Eng. Japan,* **19** 142-4.
[7] Nishimura T, Murakami S, Arakawa S, et al. 1990 *Int. J. Heat Mass Transfer,* **33(5)** 835-45.

## Acknowledgment

This study is sponsored by the Natural Science Foundation of China (No.11172059).

# The Study on Grinding Ratio in Form Grinding with White Fused Alumina (WA) Grinding Wheels

**Wang Junming[1], Wang Jiong[2], Lou Deyuan[1]**

[1]School of Mechanical Engineering, Hubei University of Technology, Wuhan 430068, China
[2] School of Mechatronics Engineering, Harbin Institute of Technology, Harbin, China

**Abstract.** The study is carried out based on an experiment of form grinding spur rack with white fused alumina (WA) grinding wheels. In the experiment, SOV-3020A type tri-axial image mapper is utilized to measure the profile of the tooth space in the rack, and the curve equations between the sectional area of the tooth space and the tooth sequence under different grinding depths are established by nonlinear curve regress using software of origin8.0. Then, it deduces the prediction equations for current grinding ratio and cumulative grinding ratio under different grinding depths. The result shows that the grinding ratio is exponential decline relationship with the increase of the number of the tooth to be ground under the same grinding depth, and the decline speed is fast in the initial stage. With the increase of grinding depth, the grinding ratio increases gradually. The cumulative grinding ratio is about twice as high as the current grinding ratio. Thus, large grinding depth is generally used in rough grinding to improve grinding efficiency.

## 1. Introduction

Grinding ratio is defined as the volume of material removed from the work per unit volume of wheel wear. It is related with Young's modulus, the elastic failure energy and the visco-elastic properties of bond.

It is obvious that the large grinding ratio can obtain high grinding efficiency, so many researchers have done many works about how to enhance the grinding ratio.

In 2011,Shi et al.suggested that the improvement of the heat resistance of the modified resin binders could enhance their adhesive force to diamond grits, thereby the grinding ratio of the grinding wheels was enhanced.And Duan et al. (2015) suggested resin abrasive with diamond coated corundum had a relatively high grinding ratio, reaching to 4.074, higher than Tig by 17.7%.But Liu et al. (2016) found that the metal bonded diamond wheel could achieve the same level dimensional accuracy consistency and higher grinding ratio in 2016, compared with resin bonded diamond wheel.

Actually, the change of the grinding ratio in a grinding cycle was confirmed experimentally, and the grinding ratios seem to vary widely (Izumi and Ochi, 2006).

In a word, there are many achievements for grinding ratio, but a few research findings involved form grinding in previous studies. Additionally, most people only researched the grinding ratio based on grinding performance loss because of grinding wheel wear, without considering the changes of grinding ratio during grinding process. Actually, the changes play a very important role in form grinding. To some extent, grinding ratio changes reflect the changes of grinding wheel wear, which affects the form grinding precision largely.

## 2. Form Grinding Experiment Design

The main equipment of the form grinding experiment is NC rack grinding machine produced by HuaZhong University of Science and Technology. The parameters of which are as follow: The spindle motor: Y132M-4 type three-phase asynchronous motor produced by Simo; Power: 7.5KW; Rated speed: 1440r/min. The parameters of the form grinding wheel are as follow : WA/F120K type white fused alumina (WA) grinding wheel manufactured by Qingdao Sisha Taiyi Abrasion Corporation limited; Diameter : 350mm; The maximum lineal speed: 25m/s.

This subject is mainly a single-factor experiment, which aims to measure the sectional area of tooth space under different grinding depth, and to explore the grinding ratio and the cross-sectional shape error of grinding racks. Table 1 shows the design of the grinding experiment.

**Table 1.** The grinding experiment design.

| | |
|---|---|
| NC rack grinding machine | Precision : 6 grade; Maximum grinding length : 1000mm; Maximum lineal speed: 25m/s |
| Form grinding wheel | Type : WA/F120K; Diameter : 350mm;Dressed according to the module about the test rack (*m*=2mm) |
| Workpiece | Steel 45 (HRC52), 350mm×50mm*50mm |
| Rack parameters | Module :m=2mm; pressure angle :20 degree; tooth width: B=50mm |
| Grinding depth | $a_p = 60\mu m$, $a_p = 50\mu m$, $a_p = 40\mu m$ |
| Grinding fluid | Dry grinding |
| Wheel dressing parameters | Diamond truing pen, lineal speed: 25m/s. feed speed :0.25m/min; per dressing amount: $5 \sim 20\mu m$ |
| Measure equipment | SOV-3020A type image mapper |

In the experiment design, tri-axial image mapper SOV-3020A is used to measure any geometry dimension on the plane. It can output real-time image based on the figure of real work-piece shape to engineering drawings in AutoCAD. The grating ruler resolution can reach 0.0005mm.

A schematic diagram of the experimental setup is shown in figure 1, and the experiment is conducted as follows:

1. Test rack blanks are placed on the workbench, its sectional area is projected by mapper through LED light source.
2. The profile of projection is depicted by MV8 measurement software and then is processed into CAD drawings.
3. The sectional area of tooth space is calculated by the enquiry area function of CAD.

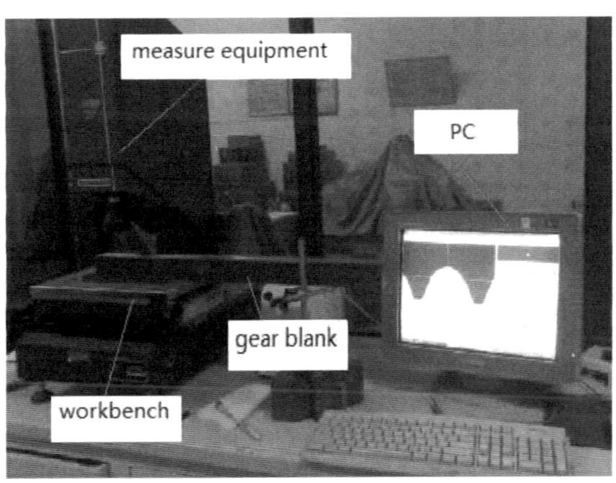


**Figure 1.** Measuring diagram.

## 3. The Experimental Data Analysis

Figure 2 is a picture of three test specimen ground. Each of them is unslotted for ease of measurement and analysis, and the cross-sectional shape is tested one time at five-tooth intervals. The grinding experimental data is presented in table 2.

**Figure 2.** A picture of the grinding test specimen.

**Table 2**. Grinding experiment data.

| tooth sequence (i) | sectional area measurement data ($mm^2$) | | |
|---|---|---|---|
| | $a_p = 60\mu m$ | $a_p = 50\mu m$ | $a_p = 40\mu m$ |
| 1 | 0.0796 | 0.0672 | 0.053 |
| 6 | 0.0781 | 0.0663 | 0.0521 |
| 11 | 0.078 | 0.066 | 0.0508 |
| 16 | 0.0777 | Singular point | 0.0506 |
| 21 | Singular point | 0.065 | 0.0493 |
| 26 | 0.077 | 0.0646 | 0.0483 |
| 31 | 0.0756 | 0.0636 | 0.0479 |
| 36 | 0.0746 | Singular point | 0.0457 |
| 41 | 0.0739 | | Singular point |
| 46 | 0.0734 | | 0.0436 |
| 51 | Singular point | | 0.0435 |

As is shown in table 2, it has deleted some singular points. Then the curve equations between the sectional area of the tooth space and the tooth sequence under different grinding depths are established by nonlinear curve regress using software of origin8.0. Three curve equations are as the following:

$$A_{a_p=60\mu m} = 0.0851 - 0.0058e^{i/64.5} \tag{1}$$

$$A_{a_p=50\mu m} = 0.07509 - 0.00799e^{i/88.49486} \tag{2}$$

$$A_{a_p=40\mu m} = 0.0929 - 0.03973e^{i/224.43095} \tag{3}$$

From above equations, it is concluded that these two factors distribute exponentially, and the maximum relative error among three fitting curves is less than 1%.

After the equations above are partly adjusted, the unified equation of removal sectional area is concluded as follow:

$$A_{Wi} = (0.442 + 5\times10^{-8} a_p^{-3} - 14.3a_p + 139.1a_p^{2}) - (0.462 - 16.472a_p + 147.75a_p^{2})e^{i/(1887-63967a_p+559706a_p^{2})} \tag{4}$$

In the equation (1)-(4), $a_p$ is the grinding depth, i is the tooth sequence and i=1, 2, 3.......

## 4. Research on the Grinding Ratio in Form Grinding

In form grinding rack, the changes of the cross-sectional shape reflect the changes of the wear loss of the grinding wheel at the same grinding depth to some extent. After the experimental data processed by CAD, it is found that the first tooth space image is overlapped by the thirty-sixth tooth space image, as is shown in figure 3. It is clear that the area of the latter one is smaller than that of the former one because wheel has been wear after grinding many times. The shadow is the sectional shape error of the thirty-sixth tooth space to be ground in figure 3.

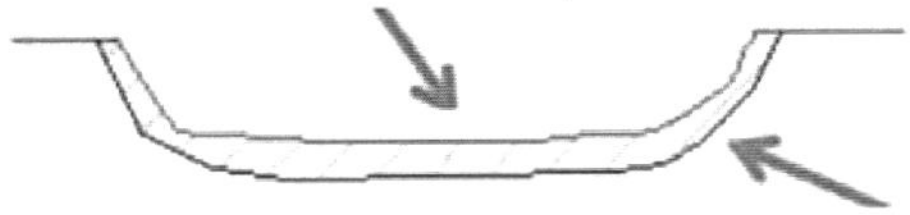


**Figure 3.** The sectional shape error of the tooth space.

Based on the grinding process of the rack, a series of calculation formulas would be obtained as follow:

Current volume of grinding wheel wear:

$$V_{Si} = (A_{W1} - A_{Wi})\pi D \quad (i = 1,2,3,\cdots,n) \tag{5}$$

Cumulative volume of grinding wheel wear:

$$V_S = \sum_{i=1}^{i=n} (A_{W1} - A_{Wi})\pi D \quad (i = 1,2,3,\cdots,n) \tag{6}$$

Current volume of the removal material of the work-piece:

$$V_{Wi} = BA_{Wi} \quad (i = 1,2,3,\cdots,n) \tag{7}$$

Cumulative volume of the removal material of the work-piece:

$$V_W = \sum_{i=1}^{i=n} BA_{Wi} \quad (i = 1,2,3,\cdots,n) \tag{8}$$

Current grinding ratio:

$$G_i = \frac{V_{Wi}}{V_{Si}} \quad (i = 1,2,3,\cdots,n) \tag{9}$$

Cumulative grinding ratio:

$$G = \frac{V_W}{V_S} = \frac{\sum_{i=1}^{i=n} BA_{Wi}}{\sum_{i=1}^{i=n} (A_{W1} - A_{Wi})\pi D} \quad (i = 1,2,3,\cdots,n) \tag{10}$$

The definition for the symbols in above equations as follow:

$A_{wi}$ : current sectional area of the i[th] tooth space to be ground

$A_{w1}$ : the sectional area of the first tooth space to be ground

D: the diameter of grinding wheel
B: the width of the rack blank

According to the equation (4), (9) and (10), the predicting formulas of the current grinding ratio and the cumulative grinding ratio under different grinding depth would be obtained. Figure 4 and figure 5 are the predicting charts for the current grinding ratio and the cumulative grinding ratio respectively.

According to figure 4 and figure 5, it is clear that the grinding ratio is exponential decline relationship with the increase of tooth number, and the decline speed is fast in the initial stage. With the increase of the grinding depth, the grinding ratio increases gradually. The cumulative ratio is about twice as high as the current grinding ratio. Therefore, it concludes that large grinding depth is generally used in rough grinding to improve grinding efficiency.

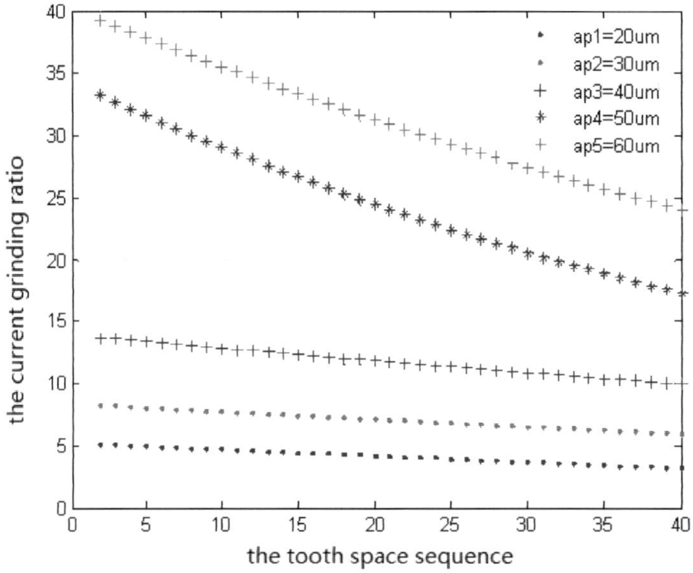


**Figure 4.** The simulation diagram of the current grinding ratio.

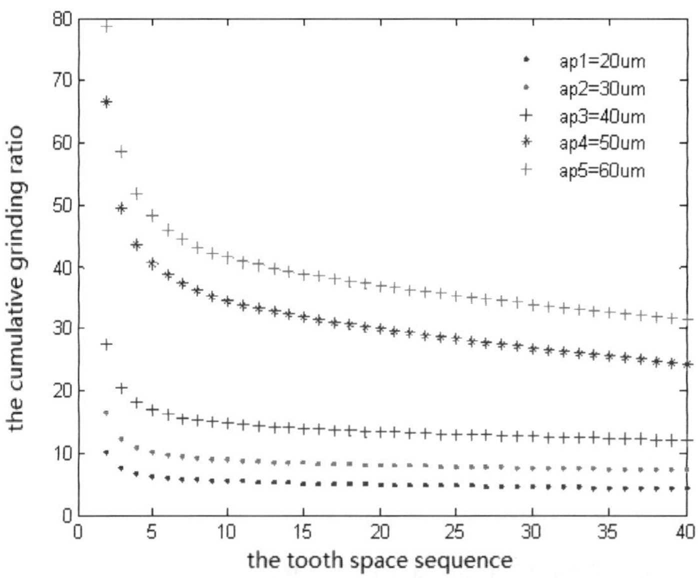


**Figure 5.** The simulation diagram of the cumulative grinding ratio.

## 5. Conclusion

1. The study is carried out based on an experiment of form grinding spur rack with white fused alumina (WA) grinding wheels. And the paper deduces the predicting equations of the current grinding ratio and the cumulative grinding ratio under different grinding depths.

2. The results show that the grinding ratio is exponential decline relationship with the increase of tooth sequence, and the decline speed is fast in the initial stage. With the increase of grinding depth, the grinding ratio increases gradually. The cumulative ratio is about twice as high as the current grinding ratio. Therefore it concludes that large grinding depth is generally used in rough grinding to improve grinding efficiency.

## Reference

[1] Shi D L, Li K H, Zhao Y J, Liu Q W, Ding Y L, Ding C S 2011 *Diamond Abras. Eng.* **31(185)** 56-9.
[2] Duan W Y, Yin Y H, Xue Q H, Li Y L 2015 *Bull. Chinese Ceram. Soc.* **34(3)** 808-12.
[3] Liu W Y, Xie H, Zhang Y X, Gong J F 2016 *Cem. Carbl.* **33(1)** 38-41.
[4] Izumi M, Ochi O 2006 *Japan Soc. Abras. Tech.* **50(8)** 471-6.

# The Beam Characteristics of High Power Diode Laser Stack

**Yuanyuan Gu[1], Yueming Fu[1], HuiLu[1], Yan Cui[1]**

[1]Tianhua College of Shanghai Normal University, Section of Practice and Training, Shanghai, 201815, China

**Abstract.** Direct diode lasers have some of the most attractive features of any laser. They are very efficient, compact, wavelength versatile, low cost, and highly reliable. However, the full utilization of direct diode lasers has yet to be realized. However, the poor quality of diode laser beam itself, directly affect its application ranges, in order to better use of diode laser stack, need a proper correction of optical system, which requires accurate understanding of the diode laser beam characteristics. Diode laser could make it possible to establish the practical application because of rectangular beam patterns which are suitable to make fine bead with less power. Therefore diode laser cladding will open a new field of repairing for the damaged machinery parts which must contribute to recycling of the used machines and saving of cost.

## 1. Introduction

Various concepts for providing solutions for scaling up power in the application focus for high power diode lasers have been investigated and realized in industrial solutions in the past years. This task is specified in relatively simple terms in bringing the optimum beam quality and power transfer efficiency from the high-power diode laser to the application focus. High-power single-mode diode laser arrays provide a scaling solution based on the inherent high beam quality of the individual single mode emitter source and the assembly with high quality micro-optics with two micro-optical elements per array in fast and slow axis direction. Thus, very high power laser systems with stringent demands on the beam quality rely on one additional active laser cavity and, in some cases, additional power amplifying stages to improve the beam quality of the diode lasers, with fiber lasers and disk lasers being the two most prominent examples [1-4]. Laser systems with high powers at moderate beam quality can also be obtained by directly combining diode lasers, spatially and spectrally into a fiber. Such a direct diode laser system is very compact, very efficient and it can be modulated from dc to the GHz range by solid state electronics. Thus, direct diode lasers can be easily used in processes with an active feedback loop to regulate the laser power. Free space beam shaping with precision micro-optics is a key element to access the size, the quality and stability needed in commercial applications. Aspheric cylindrical lenses in high index precision glass can be tailor-made in micro to macro dimension for optimum form accuracy of the aspheric surface and optimized to the respective task of the applications. Micro-optical cylindrical slow-axis lens arrays in standard quartz glass material are available in high precision in pitch and form accuracy by reflow lithography [5].

## 2. Astigmatism of Diode Laser Stack

Astigmatism is a kind of aberration. When using the optical system imaging of near field on the surface of the diode laser cleavage, Astigmatism would be found, as a result of the existence of astigmatism, there would appear on the focal line two image point. Diode lasers are used on both sides of the active layer in transverse refractive index difference of optical waveguide which is formed by the effect of active area to restrict the photons. There are two types of optical limiting including gain

IOP Conf. Series: Materials Science and Engineering **317** (2018) 012007 doi:10.1088/1757-899X/317/1/012007

and refractive index waveguide. Early diode laser is gain waveguide laser bar. For the lateral refractive index of waveguide, they are widely used in the direction perpendicular to the junction plane of the Gaussian beam waist in the cleavage plane, and in front of the waist in the plane wave, as shown in figure 1 (a). When the lateral waveguide structure was gain waveguide, then the light field distribution as shown in figure 1 (b), there was the waist inside the cavity apart from the cavity surface for D (called a amount of astigmatism), which is also the minimum width of the near field, the real waist in the center of the cavity observed by external observer. Therefore, looked from the propagation direction, the direction of the two synthesis of wave front is a cylinder, the output light is astigmatism. Its influence is spherical lens imaging of the cleavage cavity surface, the virtual image plane and the cavity of the waist like the surface of the waist (that is, the transverse optical field like surface) is not correct, which should be in the same place. The consequence is far field distribution in "rabbit ear", in the early days of oxidation strip laser has appeared the situation in the far field. At the same time, the existence of the aberration would cause the aberration of the lateral mode increased and widen the spectrum width. This has done a lot of difficulties to the application, unless the aberration of the elimination measures, otherwise difficult to general optical system focus is very small spot, focal spot polishing, uneven distribution of field is difficult to make the laser and single mode fiber coupling efficiency. Even lateral refractive index waveguide limitation, it is hard to make the size of the above characterization of astigmatism value is zero, be in commonly 2 $\mu m$ above.

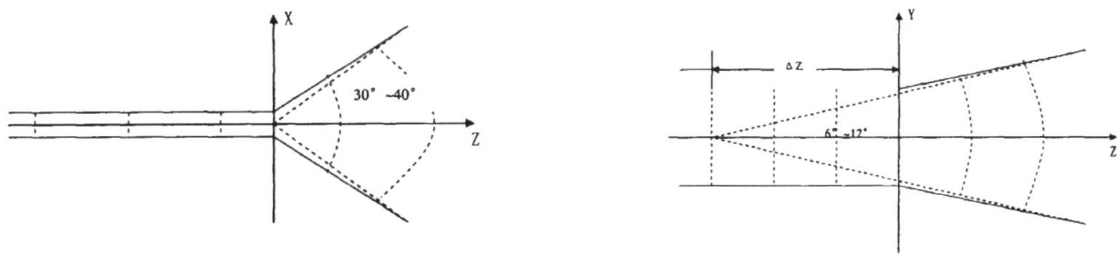


(a)Perpendicular to the junction plane (the direction of fast axis divergence angle)

(b) Divergence angle of Parallel to the junction plane (slow axis) direction

**Figure 1.** The divergence angle junction plane in fast and slow axis direction.

In the application, generally, there are two ways to eliminate assignation, adopts a diode laser without astigmatism, or by optical system to correct it.

### 3. Divergence Angle of Diode Laser Stack in Far-Field

If the diode laser emission is the ideal of Gauss beam, the intensity distribution should be as follows formula

$$I(r) = I_{\max} \exp[-2(\frac{r}{\omega})^2] \tag{1}$$

where I(r) is the intensity which radius is $\omega$ apart from the gauss beam waist size with r, $I_{max}$ is the maximum intensity at the waist. Obviously, when $r=\omega$, the intensity is the $I_{max}/e_2$, as shown in figure 2. The divergence of the Gauss peak beam is

$$\theta = \frac{4\lambda}{\pi\omega} = \frac{1.27\lambda}{\omega} \tag{2}$$

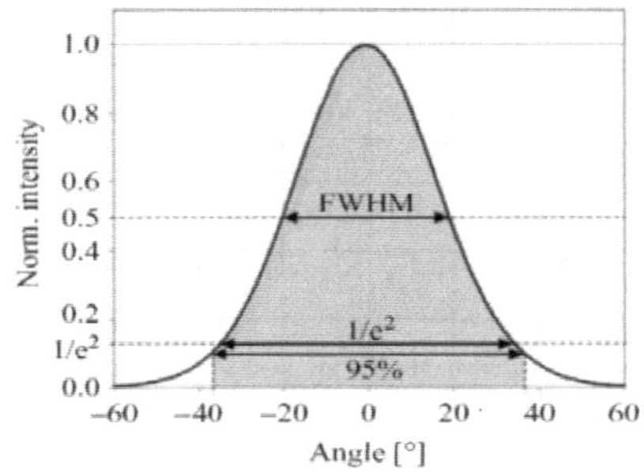


**Figure 2.** The ideal Gauss distribution.

Diode laser far field is not strictly the Gauss distribution, the beam in the horizontal and lateral asymmetry has large divergence angle. Due to the thinner active layer of the diode laser, the transverse with large divergence angle can be expressed as

$$\theta_{\perp} = \frac{4.05\left(\bar{n}_2^2 - \bar{n}_1^2\right)d/\lambda}{1+\left[4.05\left(\bar{n}_2^2 - \bar{n}_1^2\right)/1.2\right](d/\lambda)^2} \tag{3}$$

In the formula, $\bar{n}_2$ and d are active layer of laser refractive index and the thickness, respectively. $\bar{n}_1$ is the layer limited refraction index, is the lasing wavelength.

Apparently, when d is very small, this can be neglected for second in the denominator.

$$\theta_{\perp} \approx \frac{4.05\left(\bar{n}_2^2 - \bar{n}_1^2\right)d}{\lambda} \tag{4}$$

From the formula, $\theta_{\perp}$ increased with the increase of d and Ga1-xAlxAs/GaAs, the diode laser and d curve in the first half was consistent. This can be interpreted as decrease with increasing d, the optical field to both sides of the active layer, equivalent to the thickening of the active layer, and reduce the $\theta_{\perp}$. When the active layer thickness is comparable with the wavelength, but still work in fundamental transverse mode, can ignore 1 in the denominator approximately.

$$\theta_{\perp} \approx \frac{1.2\lambda}{d} \tag{5}$$

The consistency in style show in the active thickness within a certain range, the transverse optical field has a good characteristic of Gauss beam. Within this range, and reduce with the increase of d, which can be explained by the theory of diffraction. As diode laser with active layer width W is larger in the lateral, the divergence angle $\theta_{\parallel}$ is small, which can be expressed as the follow formula

$$\theta_{\parallel} \approx \frac{\lambda}{W} \tag{6}$$

Hence, although the slow axis of the active region is large in size, the divergence angle is relatively small, about 8-10 degree of the active region, and the smaller size of the fast axis direction, the divergence angle is large about 25-40 degree. The far field spot is an elliptical. The long and short axes are corresponding to the fast axis and slow axis. In many applications, optical systems need to shape the far field beam of the non-circular symmetrical circular.

IOP Conf. Series: Materials Science and Engineering **317** (2018) 012007 doi:10.1088/1757-899X/317/1/012007

## 4. Beam Parameter Product (BPP) Evaluation Method

According to the ISO-11146 standard, laser beam quality is described by the beam parameter product (BPP) through the following equation [6].

$$BPP = w_0 \cdot \theta / 2 \tag{7}$$

where, $w_0$ is the beam waist radius of lasers, $\theta$ is far-field divergence angle of laser.

The beam parameter product (BPP) is used to measure the beam quality of diode optical parameter. The minimum value of product is called the diffraction limit, to a certain wavelength of lambda $\lambda$, the diffraction limit as shown in follow equation.

$$BPP_d = \lambda / \pi \tag{8}$$

Beam parameter product is defined as follow

$$BPP = w_0 \times \theta_0 / 2 \tag{9}$$

where $w_0$ is the beam waist radius, $\theta_0$ is the far field divergence angle. Beam parameter product with another expression for the beam quality parameter $M^2$

$$M^2 = \frac{BPP}{\lambda / \pi} = \frac{\omega_0 \cdot \theta_0}{\lambda / \pi} \tag{10}$$

## 5. Conclusion and Outlook

The characteristic of diode laser stack beam is introduced through astigmatism, based on two order moments of the intensity of laser diode stack divergence angle, beam radius and beam quality evaluation methods, manly focuses on the relationship between the concept of factor and beam parameter product between BPP and $M^2$, At present, we plan to make polarization coupling experiment for the laser diode laser stacks which contain more diode laser bars and try other kinds of beam coupling methods. We continue to bend ourselves to improve the coupling efficiency. In addition, we are also studying on the diode laser beam shaping, expecting the consistent beam quality of fast axis and slow axis by rearrangement of beam and focusing the beam into single fiber.

## References

[1] Hemenway M et al. 2017 *SPIE* **10086** 10086051-7.
[2] Ebert C, et al. 2017 *Proc. SPIE* **10086** 10086071-8.
[3] Huber, M H et al. 2017 *Proc. SPIE* **10086** 100850B1- 8.
[4] Perales M et al. 2017 *Proc. SPIE* **10086** 100860G1- 8.
[5] Pietrzak A et al. 2017 *Proc. SPIE* **10086** 100860J1- 10.
[6] ISO/DIS-Standard 11 146, International organization for Standardization.

# High Power High Efficiency Diode Laser Stack for Processing

**Yuanyuan Gu[1], HuiLu[1], Yueming Fu[1], Yan Cui[1]**

[1]Tianhua College of Shanghai Normal University, Section of Practice and Training, Shanghai, 201815, China

**Abstract.** High-power diode lasers based on GaAs semiconductor bars are well established as reliable and highly efficient laser sources. As diode laser is simple in structure, small size, longer life expectancy with the advantages of low prices, it is widely used in the industry processing, such as heat treating, welding, hardening, cladding and so on. Respectively, diode laser could make it possible to establish the practical application because of rectangular beam patterns which are suitable to make fine bead with less power. At this power level, it can have many important applications, such as surgery, welding of polymers, soldering, coatings and surface treatment of metals. But there are some applications, which require much higher power and brightness, e.g. hardening, key hole welding, cutting and metal welding. In addition, High power diode lasers in the military field also have important applications. So all developed countries have attached great importance to high-power diode laser system and its applications. This is mainly due their low performance. In this paper we will introduce the structure and the principle of the high power diode stack.

## 1. Introduction

Many volume industrial applications call for high power efficiency laser sources. The new product line by IPG offers fiber lasers with wall plug efficiency in excess of 50%, with output power ranging from 1kW to 10kW. This ECO-laser product line is based on the improved or redesigned set of optical and electronic components. In the case of single emitter pumps the ex-fiber power efficiency has been increased up to 68% from the industry standard of 50-60%. Similar progress has been done in the development of fibers, fiber blocks and other optical components. Several applications in the field of material processing – for instance sheet metal cutting and deep penetration welding –require laser systems with both high output power and high brightness [1, 2]. Cutting machines for high quality cutting are usually equipped with 2 kW to 8 kW of optical output power [3, 4] and a beam parameter product in the range of 4 mm mrad. Due to the limited brightness of state of the art direct diode lasers, today's cutting machines are equipped with disk, fiber or CO2 lasers [3, 4], all developed countries have attached great importance to high-power diode laser system and its applications. Germany is developing initiative "BRIOLAS" (brilliant diode lasers) program, the German Federal Ministry of Education and Research (BMBF) is supporting the BRIOLAS-initiative with about 30 M€. The United States has launched "SHEDS" (High Efficiency Diode Sources) from 2005. Now Germany laser line corporation has products, that direct diode laser system has the largest output power of 10000w, and fiber coupled diode laser system have the largest output power of 8000W into the fiber, with the1.5mm diameter [5].

## 2. The Principle of the Laser Diode Stack

Stimulated radiation in a certain frequency, reflected back and forth between the two cleavages, the formation of the two columns in the opposite direction of propagation wave superposition, finally forming a standing wave in the resonant cavity. A resonant cavity length, refractive semiconductor

IOP Conf. Series: Materials Science and Engineering **317** (2018) 012008 doi:10.1088/1757-899X/317/1/012008

crystal rate of N, then lambda /n laser propagation in a semiconductor crystal when the wavelength of the resonator, the result is the presence of standing wave is allowed only semi integral multiple of wavelength is exactly equal to the harmonic cavity length, the condition is

$$m(\frac{\lambda}{2n}) = l \tag{1}$$

where m is integer.

Oscillation which does not meet this condition will gradually loss, where a series of specific wavelength meet stimulated radiation form a stable oscillation in the cavity [6].

If V is the positive voltage on the laser diode, I is the current of laser diode, $P_{ex}$ is the output power of the laser diode, r is the series resistance of the laser diode, power efficiency $\eta_p$ is

$$\eta_p = \frac{P_{ex}}{IV + I^2 r} \tag{2}$$

The internal quantum efficiency of the diode laser is

$$\eta_i = (1 + \frac{2v_r \tau_r}{d} + \frac{\tau_r}{\tau_{nr}})^{-1} \tag{3}$$

where $\tau_r$ the lifetime of carrier radioactive recombination is, $\tau_{nr}$ is the lifetime of non-radioactive carrier recombination, $v_r$ is the interface recombination velocity caused by the lattice mismatch, $d$ is the active area thickness.

Because of the loss itself in cavity, the resulting photons are absorbed, and then define the external quantum efficiency $\eta_{ex}$

$$\eta_{ex} = \frac{P_{ex} / hv}{I / e} \tag{4}$$

The structure of laser diode stack

Welded several laser diode bars to one in the front of micro channel heat sink, the micro channel heat sinks are arranged vertically, thus forming a diode laser stack. At present, the main problem of high power diode laser stacks is heat dissipation, where the micro channel heat sink cooling technology is one of the most advanced stack packing technology.

The micro channel heat sink was realized early in silicon, initially mainly used in microelectronic integrated circuit radiator. For high power diode laser stack, oxygen free copper is usually used as the material of the micro channel heat sink with water as the coolant. The utility model had the advantages of expanding the area of the solid - liquid contact surface, improving the heat conduction efficiency. Micro channel orientation and the laser stack strip are transversely vertical; the structural parameters between two layers of micro channel designed same, so that the laser stack strip was as uniformly as possible refrigeration. The micro channel heat sink of the thickness was approximately 1.2-2.5mm. When water flowed in micro channel, heat was taken away which generated by diode laser in order to protect the mirror catastrophic thermal injury by diode laser bar. As the micro channel heat sink thermal impedance is very low, which could achieve $4.982\times10^{-3}$K·cm$^2$/W, the heat transfer efficiency is dozens of times of the ordinary cooler. Therefore, endurance the thermal stability of diode laser bar, output power of high power diode laser stack wouldn't loss energy because of integrated diode laser bar, but directly proportional to the number of semiconductor laser bar, the micro channel structure diagrams are shown in figure 1 and figure 2.

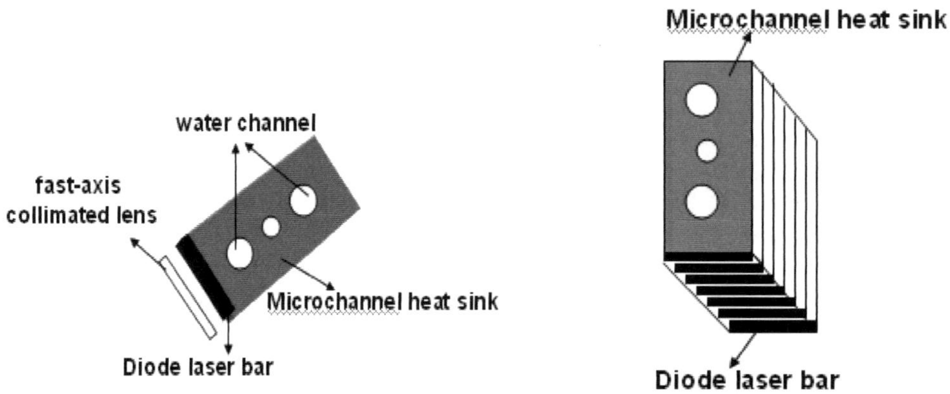


**Figure 1.** Scheme of a diode laser bar (left) and a diode laser stack (right). The diode laser bar is mounted p-side down to the edge of an actively cooled Cu microchannel heatsink. The n-contact is realized by a Cu foil, a fast-axis collimated lens is placed in front of the diode laser bar.

The basic elements of high power diode lasers are laser bar, mounted onto water-cooled heat sink (shown in figure 1). The microchannel cooling technology was originally based on silicon anisotropic etching; today these heat sinks are typically manufactured from copper. They contain a network of small channels. Water is driven through these microchannels, removes the excess heat, thus, prevents the bar mirror facets from thermal destruction. The efficient cooling is the precondition for high average output power and high long-term stability. The output power of single laser Bar of 1cm width is in the general 40-100W range. Now many of the German company's products can achieve the lifetime of 20,000 hours or more. Each microchannel heat sink is then vertically stacked into a rigid structure to produce the final two-dimensional high power diode laser stack. From one bar to thirty bars in a variety of configurations are readably available. The high-power output of laser stack does not necessarily reduce, but it is directly proportional to the number of laser bar [7].

## 3. Accurate Positioning Designed among Stacking Layers

Laser stack overlapped by several to dozens of micro channel cooling diode laser bars, layers in the vertical, horizontal direction need accurate positioning. If there was location error between layers, fast and slow axis beam collimation would cause directivity deviation, such as shown in figure 2 and figure 3, where, (a) expressed without positioning error of beam alignment between vertical stack, graph (b) indicates the presence of positioning error of beam alignment between the vertical direction of the lamination.

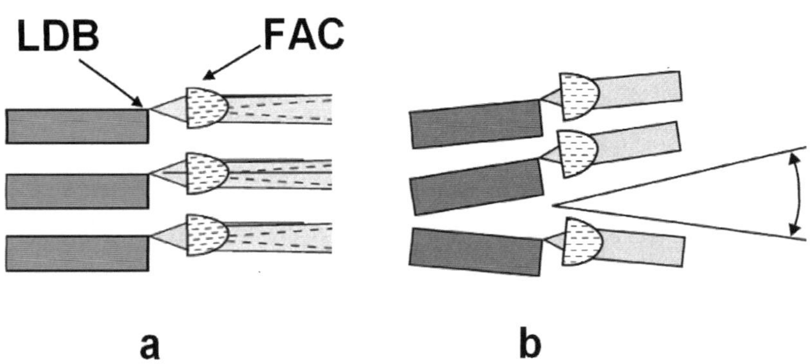


**Figure 2.** Positioning error among layers in the vertical stack.

IOP Conf. Series: Materials Science and Engineering **317** (2018) 012008 doi:10.1088/1757-899X/317/1/012008

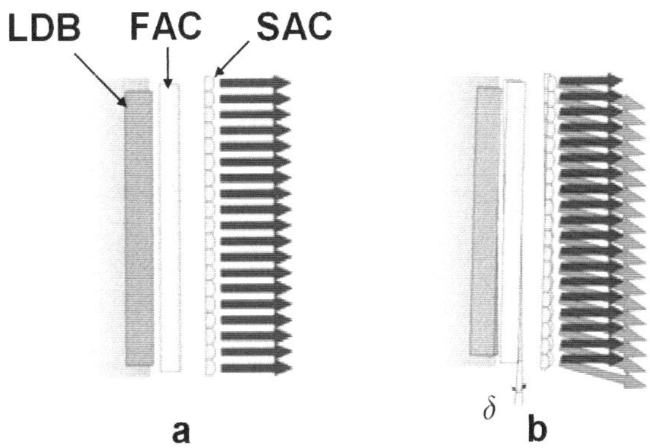


**Figure 3.** Positioning error among layers in horizontal stack.

The laser emission direction is inconsistent among the stack, and this deviation would be further increased by the laser propagation, resulting in decreased overall laser stack beam quality, which was difficult for the beam shaping focus. In order to improve the positioning precision of stack in the vertical direction, two location of a top wire was applied based on positioning screw, at the bottom of the top stack, respectively.

To ensure accurate positioning laminated horizontal direction, two positioning column were applid in the lateral plate stacked array so that precise positioning was done in the horizontal direction of the laminate

## 4. Conclusion and Outlook

The paper proceeds from the principle of high power diode laser stack, the analysis of the main performance parameters of diode laser stack, around the high power diode laser stack package, heat dissipation and positioning design difficulty gradually introduced, expounding the importance of micro channel heat dissipation

At present, we plan to make polarization coupling experiment for the laser diode laser stacks which contain more diode laser bars and try other kinds of beam coupling methods. We continue to bend ourselves to improve the coupling efficiency. In addition, we are also studying on the diode laser beam shaping, expecting the consistent beam quality of fast axis and slow axis by rearrangement of beam and focusing the beam into single fiber.

## References

[1] Hengesbach S, Poprawe R, Hoffmann D 2015 *SPIE Proc.* **93480B** doi: 10.1117/12.2085068.
[2] Brauch U, Loosen P and Opower H 2000 *High-Power Diode Lasers for Direct Applications*, Springer, Berlin & Heidelberg, 303–368.
[3] Bystronic Laserquellen, http://www.bystronic.ch/de/produckte/laserchneidsysteme/bylaser. php (1 August2016).
[4] TRUMPF Werkzeugmaschinen GmbH + Co. KG, [TruLaser: Cost-effective cutting through thick and thin] (2016).
[5] Berk Y. et al., 2011 *Proc. SPIE* **7918** 79180W-1—W12.
[6] Friedheim D, Franz X. Daiminger, Petra H, Veit B 2000 *Proc. SPIE* **3889** 45-53.
[7] www.laserline.de.

IOP Conf. Series: Materials Science and Engineering **317** (2018) 012009 doi:10.1088/1757-899X/317/1/012009

# Analysis and Design of the Innovation and Entrepreneurship Training Management System based on School Enterprise Cooperation (Taking the School of Computer and Information Engineering of Beijing University of Agriculture as an example)

**Zhang Qianyi[1], Li Xiaoshun[1], Hu Ping[1], Ning Lu[1]**

[1]Beijing University of Agriculture, Beijing 102206, China

**Abstract.** With the promotion of undergraduate training mode of "3+1" in Beijing University of Agriculture, the mode and direction of applied and compound talents training should be further visualized, at the same time, in order to make up for the shortage of Double Teachers in the school and the lack of teaching cases that cover the advanced technology in the industry, the school actively encourages the cooperation between the two teaching units and enterprises, and closely connects the enterprise resources with the school teaching system, using the "1" in "3+1" to carry out innovative training work for students. This method is beneficial for college students to integrate theory into practice and realize the purpose of applying knowledge in Higher Education. However, in the actual student training management, this kind of cooperation involves three party units and personnel, so it is difficult to form a unified management, on the other hand, it may also result from poor communication, which leads to unsatisfactory training results. At the same time, there is no good training supervision mechanism, causes the student training work specious. To solve the above problem, this paper designs a training management system of student innovation and Entrepreneurship Based on school enterprise cooperation, the system can effectively manage the relevant work of students' training, and effectively solve the above problems. The subject is based on the training of innovation and entrepreneurship in the school of computer and information engineering of Beijing University of Agriculture. The system software architecture is designed using B/S architecture technology, the system is divided into three layers, the application of logic layer includes student training management related business, and realized the user's basic operation management for student training, users can not only realize the basic information management of enterprises, colleges and students through the system, at the same time, it also realizes the information operation of student training management [1]. The data layer of the system creates database applications through Mysql technology, and provides data storage for the whole system.

## 1. Research significance

Beijing University of Agriculture is a cultivation of Applied Talents in University of Beijing, the school attaches great importance to the cultivation of students' practical skills and innovation ability, the "3+1" training system, it is very important to the students during the school training last year internship, which is directly related to the realization of the school for the industry of Beijing city culture "use, applied talents stay". So the school to promote cooperation, to build a "undergraduate innovation training base", through the introduction of "double" teachers, teaching resources and realize the integration of enterprise, deepen the content of personnel training, broaden employment channels for students to. Since 2015, School of computer and information engineering has been carrying out the school enterprise training and training base with famous enterprises such as Neusoft group, soft group and An Bo group, and achieved gratifying results. But in practice a lot of difficulties, such as the number of students, teaching content is complex, before the school take the traditional manual management mode, this means a lot of drawbacks, first of all, the three party enterprise school, students cannot do effective communication, information cannot be timely communication, but also because of a variety of human factors on practice progress, leading to efficiency low practice, low interest student practice a series of drawbacks. For the modern student internship management system, these drawbacks must be overcome. Therefore, it is urgent to establish a "college student's innovative training management system", and realize the concise, efficient and scientific management of the training work [2]. According to the problems mentioned above, combined with the specific situation of the school of computer and information engineering of Beijing University of Agriculture, the design of a complete student internship management system is the best solution to the above problems.

This paper designs a set of cooperative base students training management system, the system responsible for related information entry, preservation and management practice and innovation of various aspects of the final synthesis of simple and intuitive the chart information. At the same time, when designing the system, we also consider that the system must have great flexibility and flexibility. Because each intern for six months or a year, basically have to change a student information, along with the development of computer and Information Engineering College internship policy adjustment, the number of interns are more and more, will increase the workload of the system, increase the operating pressure. Only the establishment of students' innovation and Entrepreneurship Management System of scientific and effective, can effectively improve the efficiency of practice management, reduce management costs, and fundamentally change the status quo, the school students of the three together, let the teachers perform better supervision, increase the teachers and students to communicate, at the same time, internships can also better to understand the situation of students, improve the training effect to a new height [3].

## 2. Demand analysis

### *2.1. Business requirements analysis*

The main purpose of the training system of student enterprise cooperation base is to provide a better training platform for students. In this system, we can query the emphasis of teaching and learning at the knowledge level at any time. It can improve the learners' practical ability and improve the quality of College education. For cooperative enterprises, it can also promote communication and gain benefits. The system achieves win-win results through the cooperation of schools and enterprises.

In accordance with the normal teaching arrangements to guide student's innovation and entrepreneurship training work, when teaching service cooperative enterprises have additional requirements, can apply to the school teaching guidance committee, make appropriate adjustment.

The whole sequence of 2.1.2 process is completed in the College Teaching Guidance Committee under the guidance of the training before the start of training the students by establishing the direction of College Teaching Guidance Committee, students enter the school enterprise joint training base, you can choose the training direction, then in accordance with the instructor and training required to complete the training task, training manuals, final submission training and training report appraisal table of materials can be calculated in practice performance to the college [4].

The main form of 2.1.3 students' training is the form of collective organization, and the overall principle is to closely link the actual training with the training of skills.

To school designated enterprise training students to sign the training agreement and acceptance of two-way student training enterprise agreement should negotiate the responsibilities of both parties, rights and obligations, working conditions, working time, labor conditions, and health insurance and other issues, and meet the requirements of labor law rules and regulations.

Assisted undergraduate and graduate students in completing the task of innovation and entrepreneurship. School of innovation and entrepreneurship guidance team in this platform, according to the demand for the construction of innovation and entrepreneurship training system innovation resources, MOOC released a variety of innovative entrepreneurial activities and lectures online information, so as to realize the innovation of teaching.

Achieve online management of innovation and entrepreneurship teams. The school also need to innovation and entrepreneurship project audit, each team needs to fill in the "innovation platform for innovation and entrepreneurship in the project application form", the school online review of the project, the final release of the audit opinion, and according to the results of the audit office and other office resource allocation. In the process of project development, the system will record the progress of each project; after the project is completed, the innovation and entrepreneurship team will be required to complete the project acceptance according to the requirements of the Institute, and the results will be publicized on the platform.

College will be released according to the actual need, at all levels of innovation and entrepreneurship competition and discipline competition in information platform, encourage students to actively participating students according to the enrollment situation, to contest the organization and training work [5].

### *2.2. functional requirements analysis*

*2.2.1. Administrator module.* The most basic role of the administrator module is the administrator. The specific work includes the administrator user information management, department management, class information management and enterprise information management not only that, the system administrator also relates to the implementation of the plan of training management, training management and system management announcement. The aspects of its management are as follows:

*2.2.1.1. User information management.* The user information management mainly includes the teacher who undertakes the training instruction task

Refers to the teachers who undertake this training task, add, delete, change, check and so on. Student personal information, student training data management is also part of the user information management system. This function is mainly to facilitate the training team leader, understand the specific situation of students to participate in training, you can see each student's training data. When both the school and enterprise teaching leaders need to understand the professional situation of students, they can search the corresponding training direction through the system.

*2.2.1.2. Department management.* The system provides the management function of two levels of information about the hospital and department. When the information is changed, it needs to be updated in a timely manner. This process includes the work of increasing, deleting, modifying and checking the information of hospitals and departments. The reason is that the basic information of the college and department and the requirements for the training position are always changing.

*2.2.1.3. Student training management.* Student training management system is an important part of the system. Administrators can manage students through the system training process, and some basic operations include: add, delete, print, search and so on. Students can choose from the system for their own training direction, the administrator will be based on the college teaching steering committee approval results for operation.

*2.2.1.4. Information release management.* Information management is for the convenience of all personnel involved in the training work in time to see the latest training information, policies related to the latest release of new entrepreneurial activities and competition information, entrepreneurial projects, venture capital support and school.

*2.2.2. College teaching steering committee management module*

The management of college teaching guidance committee mainly includes the inquiry and statistics of training situation, the direction of training, the arrangement of training work schedule and the overall arrangement of related resources of college innovation and entrepreneurship.

On the one hand, we should arrange the work on campus, and on the other hand, we should contact the enterprises outside the campus.

*2.2.2.1. The student Instruction Committee account login module.* The College Teaching Steering Committee has a higher level of account and requires higher authority. It is authorized by the administrator. For the sake of security, the college should assign a special person to

*2.2.2.2. Arrangement module of training work.* The purpose of this module is to effectively evaluate the quality of training and to distribute the teachers' strength both inside and outside the school. The first has been identified in the premise of the students under the direction of the training (has been teaching guidance committee approval), a reasonable allocation of school teachers, so each student has fixed the guidance of teachers, to guide the students' training work. After the internship, the instructor will give the results according to the training report submitted by the students, and give the evaluation and scoring according to the students' performance in the process of training.

*2.2.2.3. Innovation and entrepreneurship guidance module.* Institute for review in the declaration of this, master students of grade two innovative entrepreneurial team project innovation, according to the content and project budget, the hardware and software environment to ultimately determine the funding support and carry out the work needed to support.

*2.2.3. Faculty functions within the school*

The school guidance teacher needs to review the student's training plan, training suggestions and requirements, including safety education, etc.. The school guidance teacher login to the system, modify personal information, and communicate with students, after the training, also need to work with students to complete training guidance records and training reports. Finally, the training opinions and evaluation results are given.

*2.2.3.1. Login module.* The school teacher completes the registration process of the system account. The user name of the account number is the teacher worker number. After entering the login screen, the tutor can change personal information.

*2.2.3.2. Problem communication module.* Problem communication is a platform for students and teachers to communicate with each other. Through this platform will put forward their own problems, the teacher regularly browses this, read the students' message, if the problem can be solved, by the teacher contact students to solve the problem. If the problem is beyond the scope of the teacher's competence, he may respond to the steering committee of the hospital and give him a final answer. Through this module to record the problems encountered in the training process and the methods to solve these problems, which will play an important role in improving the quality of training.

*2.2.3.3. Assessment results.* After completing the training, students need to submit relevant materials, including training content, purpose and experience, etc.. On the basis of the performance of students during the enterprise internship, the teachers from outside and inside the school are graded and given comments on the training reports submitted by the students. After assessing the results, the instructor will provide students with training results to the hospital teaching steering committee, in the hospital for publicity, without objection, according to the results of student training achievements.

*2.2.4. Training enterprise module*

*2.2.4.1. Information release module.* The enterprise releases the training information, the enterprise culture, the teacher's introduction at any time, and has the direct correlation information with the training practice

*2.2.4.2. Training resource module.* According to the training schedule, upload training course resources at any time, so that students can use fragmentation time to carry out self-study.

*2.2.4.3. Training process management module.* Arrange practical training, carry out online communication, answer questions, publish daily attendance, training schedule, publicity and so on.

*2.2.5. Student module*

Students are the most important role in the school enterprise cooperation training management system. Students need to manage and maintain personal data, communicate with teachers, check scores and training opinions, and write personal experiences and so on. Therefore, student modules should include the following functional modules:

*2.2.5.1. Login module.* Since the number of students is relatively large, it is necessary for administrators to register each student's account number and distribute the registered account to every student participating in the training. The student list is imported by the administrator and no longer need to be registered.

*2.2.5.2. Personal data modification module.* Personal information modify module provides the personal information editing function for the user, the student user in the system not only contains the information related to school students, in addition, some information such as number, professional class, teacher information unified entry after no longer change, if there is an error, contact the teacher to revise the background business.

*2.2.5.3. Student training module.* Student training module is a platform for students to choose the direction of training, students through the module to choose the direction of professional practice, or the formation of innovative entrepreneurial team.

*2.2.5.4. Problem communication module.* Problem communication is a platform for students and teachers to communicate with each other. Students can express their experience and related business questions in the "my question" section.

*2.2.5.5. View training evaluation and achievement module.* Each student can log in to the system for a week after the end of the training session to review his evaluation, opinions and achievements during his training period, but he cannot modify the information. If there is any objection, it is necessary to make a review to the school guidance teacher and the enterprise instructor in the period of publicity.

*2.2.5.6. Competition module.* Students can according to published information on the contest, directly through the system to enroll, through the audit, namely learning materials and training notice through the platform to receive a college about the contest, and enjoy the competition related services.

*2.2.5.7. Innovation and entrepreneurship module.* Students can submit their own ideas, innovation and entrepreneurship projects through the platform, and give the project plan. After the audit, the company will provide financial, environmental and software and hardware support according to the specific circumstances of each project. At the same time, a team supported by the Institute will be required to submit project progress on time by platform.

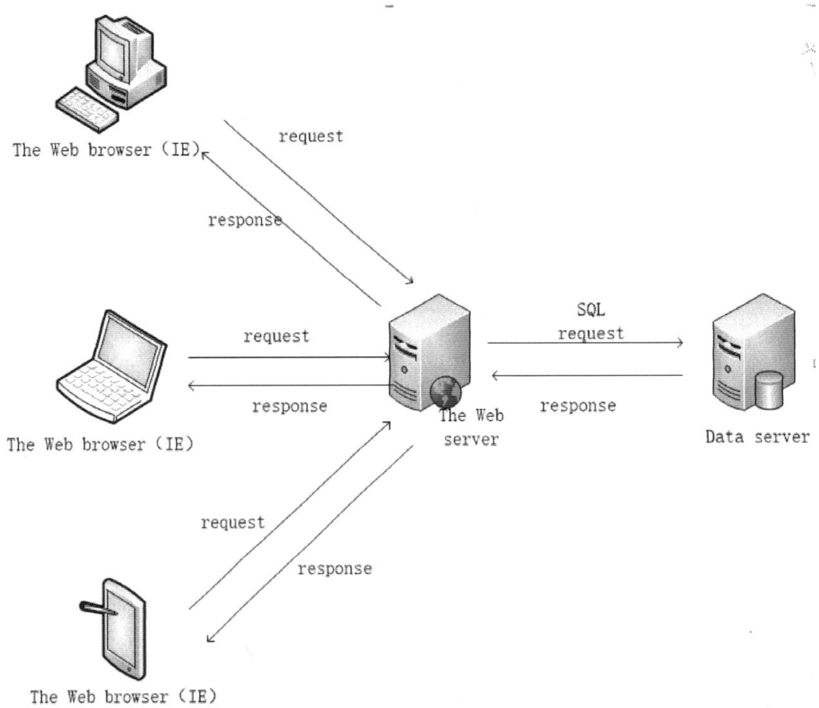


**Figure 1.** System architecture diagram.

## 3. System design

### *3.1 Overall architecture design*

System software architecture shown in figure 1, each year's graduates are large, according to the voluntary choice of these students, more than half of the students through the school enterprise cooperation channels to participate in training. Because the user has the dispersion, the visit quantity to be big, moreover requests the system operation simple and so on the characteristic, we synthesize the above analysis, decided to adopt the B/S structure to carry on the design [6].

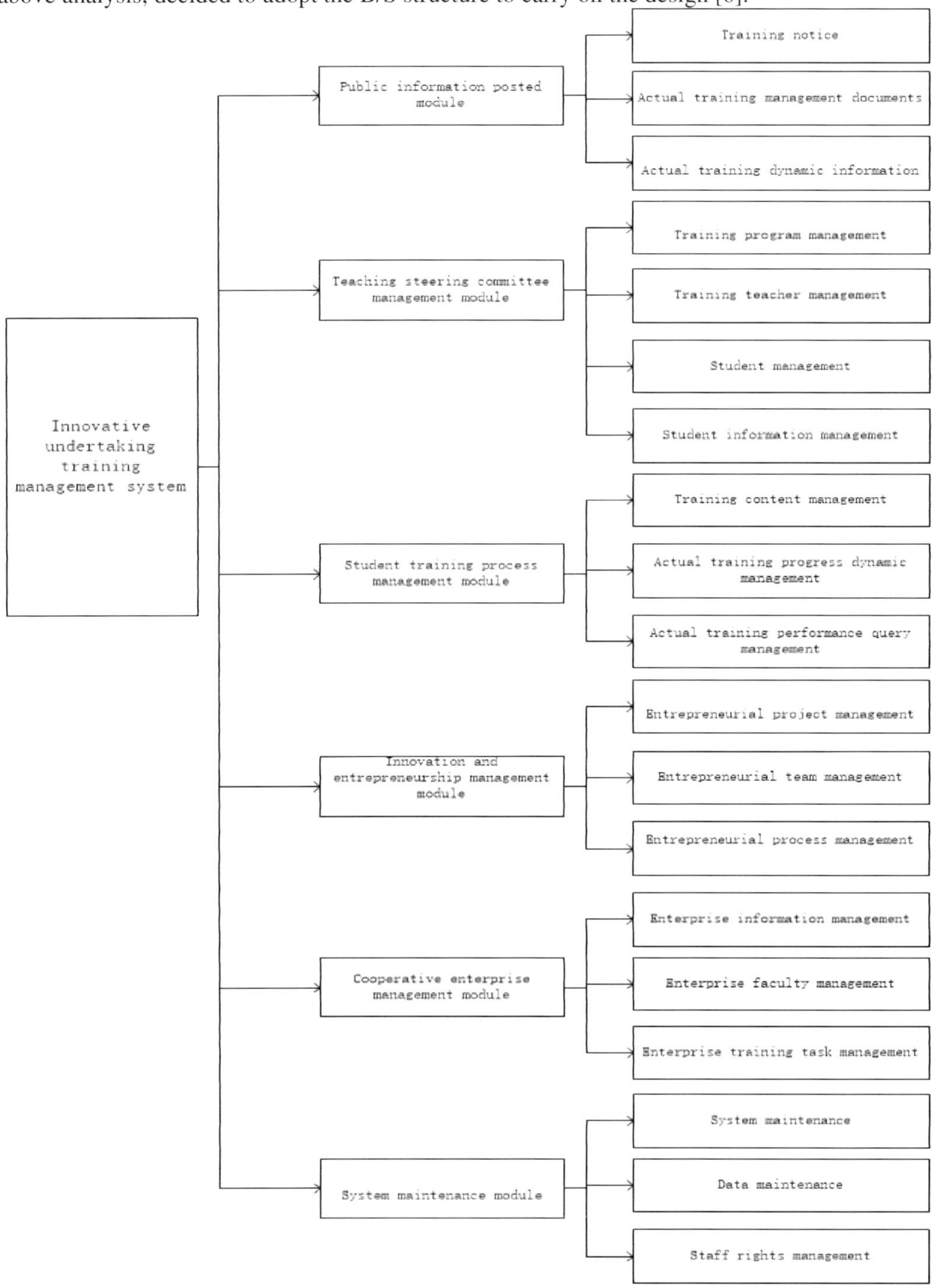


**Figure 2.** System function diagram.

### *3.2 Function module design*

All of the above analysis, the main direction of this paper is to design the system of students enrolled in the training and Training Institute released direction information and both sides together on the training process of management, the school guidance teachers training evaluation and scoring, the tutor gives enterprise training evaluation and suggestions, training students' training results online inquiry and other business processes as the main line, and to the students, school teachers, school administrators and Teaching Guidance Committee for the four main users, they constitute the main users of the system type, we targeted for the following detailed design.

#### *3.2.1. Instruction Committee*

*3.2.1.1. Management module design.* Beijing University of Agriculture Institute of computer and information engineering student training arrangements is the Institute of Teaching Guidance Committee and the relevant person in charge of enterprises to fully communicate, and ultimately determine the content of training. School leadership and Teaching Guidance Committee in charge of the teaching needs to be combined with the actual situation, the enterprise can provide practical information for a reasonable analysis, but also a comprehensive training for students living conditions, work pressure, work environment and other factors into consideration, we can determine the specific programs and processes. All aspects of the program are completed, the two sides signed the relevant contract. Contract category is generally more, mainly including personal safety, accommodation and other living conditions, be sure to provide protection. Completed the work, to administrators and school guidance teacher appointment work, they are responsible for the information management system and training the students to guide the work of the Department, to complete the business enterprise leadership can guide teachers, school teachers query, query student information and training students training quality evaluation work from the system.

*3.2.1.2. Function design of department management module.* The training direction management needs the system information manager to create the corresponding module for the different team, then enters and manages the student's information. Department management, mainly administrators will be related to the college information input system, and classified management. In the management of teachers and students, the information administrator guides the teachers and their students into the information system and relates each other. All of these information management work is unified by the system information administrator. In the aspect of innovation and entrepreneurship management, the information administrator will publish and distribute the relevant resource information according to the requirements of the instruction committee.

#### *3.2.2 Enterprise information management design*

*3.2.2.1. Workflow of information management system.* Before the training begins, the enterprise will send decision information to the University, which is classified by the system administrator and stored in the system database.

*3.2.2.2. Function design of enterprise information management module.* Company introduction, the company introduced by system administrators, administrators will showcase the company's introduction and classification of information to students, can help students better understand the specific situation of the company, increase between enterprise and University students. The enterprise will pass the instructor's information to the administrator. The information administrator will be responsible for sorting the information and importing it into the system and it can also be modified.

*3.2.2.3. Practical teaching management module.* Training enterprises will be based on the progress of the course, appropriate upload corresponding sound, video learning materials, to facilitate students pre class preparation and review after class. At the same time, students will be able to control their learning progress through periodic publication, achievement and attendance.

#### *3.2.3 Student training management module*

*3.2.3.1. Analysis of the main management during the training period.* In the school enterprise cooperation relationship, students are the main role, in this system is no exception, they constitute the whole system is the most important core component, their demand is the core direction of the system design direction. Therefore, students should be the first to design the system. This section mainly includes writing training reports, training online registration, inquiries, training results and basic information management operations.

*3.2.3.2. Function design of student management module.*

A student online registration

According to the training direction and the training enterprise, the student union will sign up on the system according to their own interests and professional skills.

B team formation and project reporting

Students can form various types of innovation and entrepreneurial teams through the platform, and declare the project in the form of a team, and apply for funds.

C training report management

Students need to log in their written reports so that they can see them. In addition, the teacher to students scoring and evaluation work is based on this internship report carried out.

D training quality evaluation

This part of the work is done by the teacher and is usually carried out after the student submits the report.

E problem communication

Mainly refers to students in the face of difficulties, in the system with the guidance of teachers to communicate and communicate.

### *3.2.4 Information announcement management module*

This module is mainly made for students, and it is convenient for students to get the latest information about training, innovation and entrepreneurship at any time. There are many information bulletins, large and small, for both individuals and teachers.

*3.2.4.1. Announcement management.* In the announcement management, users can delete, query, edit and create the announcements

*3.2.4.2. Issue a notice.* If you want to publish the announcement, must be ahead of the announcement is created, but it does not mean that the announcement must be issued immediately in place, this time is flexible, the user can set the release date in the system, but also related to property announcement. The system releases bulletins for all users of the system, including enterprise personnel, university personnel, and training students.

Design of 3.2.5 system management module

System management work analysis: This includes the basic maintenance of system services, the distribution and management of different user rights, as well as the regular backup and maintenance of system data. These are specifically responsible for the system administrator.

A user rights management

More than one in the whole system in the role, according to the above analysis, the system contains four types of users, design according to different character set different permissions, so that different users can enjoy the convenience of information systems and related operations.

B data backup

Inevitably, there are inevitable emergencies, and data backup is an urgent measure to deal with these situations.

C system maintenance

The regulation of system performance is called system maintenance, and the performance of the server can be adjusted according to the running status of the system. In addition, you can always pay attention to the memory usage of the processor and the server, and do the corresponding operation.

## *3.3 Database design*

### *3.3.1 Creates the base table*

*3.3.1.1. Admins table.* As shown in table 1 for the main table administrator, administrator users to record relevant information, contains the following attributes: Office (type), name (name), ID (sex), gender, Department (dept-id), account number (account) and telephone (TEL) etc..

**Table 1.** Administrator table.

| NO. | Field name | Field length | Data structure | Remarks | Primary |
|---|---|---|---|---|---|
| 1 | id | | int | | yes |
| 2 | name | 30 | nvarchar | name | |
| 3 | account | 6 | nvarchar | Account number | |
| 4 | tel | 15 | nvarchar | Telephone | |
| 5 | sex | | bit | Gender | |
| 6 | tape | 20 | nvarchar | type | |
| 7 | Dept-id | 5 | nvarchar | Department id | |

*3.3.1.2. Enterprise table.* As shown in table 2, the departmental information sheet records the relevant information of the enterprise department.

**Table 2.** Department table.

| NO | Field name | Field length | Data structure | Remarks | Primary |
|---|---|---|---|---|---|
| 1 | id | | int | | yes |
| 2 | name | 30 | nvarchar | Department name | |
| 3 | num | 5 | nvarchar | Department number | |
| 4 | statum | | bit | state | |

*3.3.1.3. Student form.* As shown in table 3 for students to record student information table, mainly, the name (name), where the professional (college-id), gender (sex), QQ (QQ), telephone (TEL), mail (email), the class (class-id), where the school department (college-id), major-id (the professional), ID (account), account etc..

**Table 3.** Student table.

| NO | Field name | Field length | Data structure | Remarks | Primary |
|---|---|---|---|---|---|
| 1 | id | | int | | yes |
| 2 | name | 30 | nvarchar | name | |
| 3 | account | 10 | nvarchar | Account number | |
| 4 | college-id | 5 | nvarchar | system -id | |
| 5 | sex | | bit | Gender | |
| 6 | class-id | 5 | nvarchar | class-id | |
| 7 | qq | 30 | nvarchar | QQ | |
| 8 | major-id | 5 | nvarchar | major-id | |
| 9 | tel | 15 | nvarchar | Telephone | |
| 10 | Train-company-id | 6 | nvarchar | Train-company | |
| 11 | email | 50 | nvarchar | email | |

*3.3.1.4. All school tables (colleges).* As shown in table 4 for all the school table, mainly used to record the relevant information of each department, a responsible person (leader), ID (depart-ID), office (office-phone), a (depart-mobil), ID, head of mobile phone (leader_mobi) etc..

IOP Conf. Series: Materials Science and Engineering **317** (2018) 012009 doi:10.1088/1757-899X/317/1/012009

**Table 4.** All school tables/

| NO | Field name | Field length | Data structure | Remarks | Primary |
|---|---|---|---|---|---|
| 1 | id | | int | | yes |
| 2 | name | 50 | nvarchar | Professional name | |
| 3 | Dpart-id | 5 | nvarchar | Department of | |

*3.3.1.5. The school has a professional table (majors).* As shown in table 5, the school table is used primarily to record school specific information.

**Table 5.** The school has a professional table.

| NO | Field name | Field length | Data structure | Remarks | Primary |
|---|---|---|---|---|---|
| 1 | id | | int | | yes |
| 2 | Leader-mobi | 20 | nvarchar | Leader-mobi | |
| 3 | Dpart-id | 5 | nvarchar | Dpart | |
| 4 | leader | 10 | nvarchar | leader | |
| 5 | Dpart-name | 50 | nvarchar | Dpart-name | |

*3.3.1.6. As shown in table5.* Innovation and entrepreneurship team list;

**Table 6.** Innovation and entrepreneurship team.

| NO | Field name | Field length | Data structure | Remarks | Primary |
|---|---|---|---|---|---|
| 1 | ID | | int | | yes |
| 2 | team_name | 30 | nvarchar | team_name | |
| 3 | team_lead | 20 | nvarchar | team_lead | |
| 4 | team_memb | 20 | nvarchar | team_memb | |

*3.3.2 Establish E-R diagram*

In the E-R diagram, there are several relational graphs that appear frequently, namely, many to many (N:M), one to one (1:1) and one to many (1:N) three relations. Because of the limitations of space, we cannot show them here, so we will integrate the five tables mentioned above. As shown in figure 3.

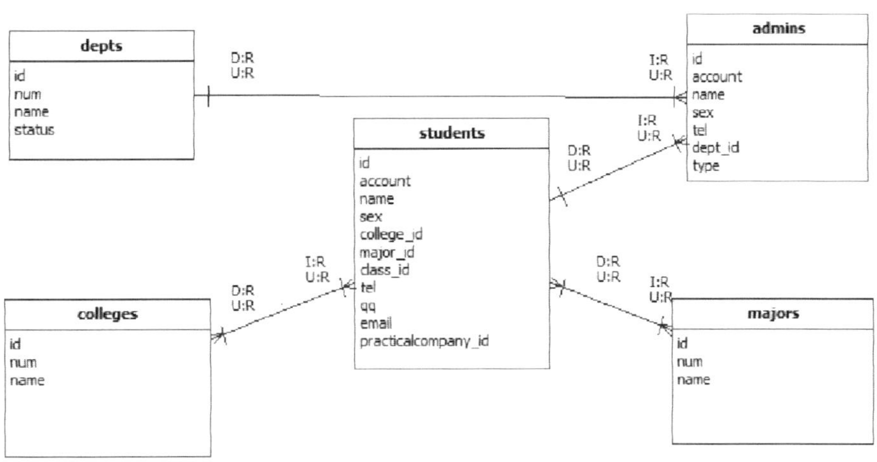


**Figure 3.** E-R diagram.

## 4. Summary and Outlook

As an agricultural college in Beijing, Beijing University of Agriculture has attached great importance to the cultivation of students' practical ability. In order to expand teaching model, it actively promotes the school enterprise cooperation mode of running schools, through this way of cooperation to improve students' practical ability, but also to give students the opportunity to understand the society ahead of time. Students learn theoretical basic knowledge in school, and practical ability in practical training base [7]. Through this period of study and practice, students can feel the work method and work atmosphere before graduation, and it is a very beneficial experience for college students.

The main purpose of this study is to provide a very effective communication platform for enterprises and schools. With the Ruby development language and RubyMine development tool, the basic information of school teachers, teaching guidance committee function module, enterprise settings, user login and students to participate in the training module (including internship report, fill in registration etc.) can be put into practice. The innovation and entrepreneurship training management system based on school enterprise cooperation has a very practical and high scalability characteristic, such characteristics make it more close to the real teaching management work.

## Reference

[1] Jie X L, Yan H M, Zhu W F, Yin G C, 2017 *J. Shangqiu Teachers College*, **33(01)** 34-41.
[2] Hong J 2013 *Zhejiang Soc. Sci.* **05** 140-143+160.
[3] Li W N 2017 *Liaoning Econ. Manag. Inst. Liaoning Vocat. College Econ.* **04** 113-115.
[4] What Z L 2015 *Design and implementation of information management system of training room,* Guangxi University.
[5] Yang Y 2015 *Research and analysis of network teaching practice management system of Computer Department.* Yunnan University.
[6] Liang Y C, Wang K C, Zhao F Z, Yang J 2014 *Exp. Tech. Manag.* **31(06)** 210-212.
[7] Zhang Y D 2013 *Design and implementation of experiment and practice teaching management system in Vocational Colleges.* Nanjing University of Science and Technology.

IOP Conf. Series: Materials Science and Engineering **317** (2018) 012010 doi:10.1088/1757-899X/317/1/012010

# The Strength Analysis of Differential Planetary Gears of Gearbox for Concrete Mixer Truck

**M H Bae[1], T Y Bae[2] and D J Kim[3]**

[1]Dept. of Aviation Maintenance, Changwon Moonsung University, 51410, Changwon city, Korea
[2]Dept. of Mechatronics, Changwon Moonsung University, 51410, Changwon city, Korea
[3]Research Director, Dai Ho Hydraulics Co Ltd, 51568, Changwon city, Korea

**Abstract.** The power train of mixer gearbox for concrete mixer truck includes differential planetary gears to get large reduction ratio for operating mixer a drum and simple structure. The planetary gears are very important part of a mixer gearbox where strength problems namely gear bending stress, gear compressive stress and scoring failure are the main concern. In the present study, calculating specifications of the differential planetary gears and analyzing the gear bending and compressive stresses as well as scoring factor of the differential planetary gears gearbox for an optimal design of the mixer gearbox in respect to cost and reliability are investigated. The analyses of actual gear bending and compressive stresses of the differential planetary gears using Lewes & Hertz equation and verifications of the calculated specifications of the differential planetary gears evaluate the results with the data of allowable bending and compressive stress from the Stress-No. of cycles curves of gears. In addition, we also analyze actual gear scoring factor as well as evaluate the possibility of scoring failure of the differential planetary gear.

## 1. Introduction

A concrete mixer is a device that homogeneously combines cement, aggregate such as sand or gravel, and water to form concrete. A typical concrete mixer uses a revolving drum to mix the components. The drum is traditionally made of steel but on some trucks as a weight reduction measure, fiberglass has been used. Special concrete transport trucks are made to transport and mix concrete up to the construction site.

(a) Photograph of mixer truck (b) Gearbox

**Figure 1.** Drum capacity, 8m³ grade concrete mixer truck and the mixer gearbox with hydraulic motor.

The concrete mixer truck with drum capacity, 8m³ grade concrete mixer truck and the mixer gearbox with hydraulic motor has been shown in figure 1. Table 1 indicates the specifications of the mixer gearbox that calculated by developed program. Mixer gearbox is driven by a hydraulic motor, as an important device to rotate the mixer drum and to convert the required torque and rotational speed.

**Table 1.** Specifications of the mixer gearbox.

| Hydraulic. motor Max. input torque/speed | Gear ratio | Max. output torque |
|---|---|---|
| 397N·m/1,320rpm | 132:1 | 52,400 N·m |

In spite of the increasing initial torque resulting to the inertia moment increases of output section, the compound differential planetary gear system applies the rotating motion that makes the mixer drum run smoothly which consists of the sun gear, the differential planetary gear and two ring gears. Gear teeth are damaged due to the lack of fatigue strength, compound planetary gears for mixer and severe operating conditions of a concrete mixer truck that can become a problem. The schematic diagram of an analytical model of mixer gear box has been shown in figure 2.

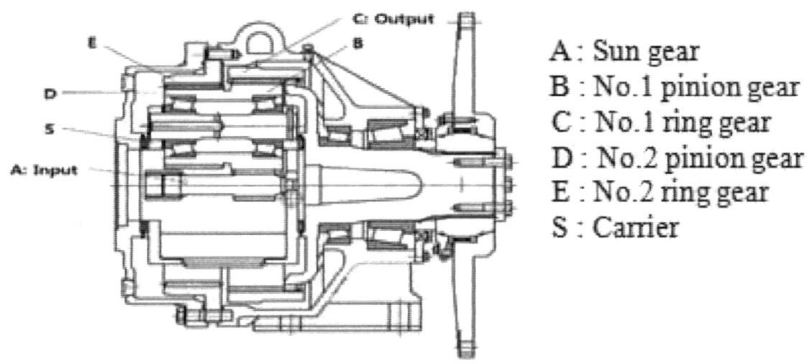


**Figure 2.** Schematic diagram of analytical model.

Several investigations have been reported, as cited by D.E. Imwalle et al [2]. D.L. Seager et al [3] in their paper established load distribution calculation of the planetary gears. F. Cunliffe et al [4] analyzed the dynamic tooth loads in epicyclic gears for planetary gears. Castellani G. et al [5] also cited the gear strength analysis method. Coy, J.J. et al [6] further emphasized the dynamic capacity and surface pressure durability life of spur and helical gears. Oda et al [7] similarly stressed the effect of bending endurance strength for addendum modification of spur gears and it was likewise investigated. There is also an inclusion of a typical bending strength calculation of planetary gears AGMA 218.01[8] and Gear Handbook by D.W. Dudley [9] that shows the bending strength calculation method of planetary gears.

In this study, developing the gear specifications calculation program and producing detailed specifications of the differential planetary gear system for mixer gearbox are based on Gear Handbook by D.W. Dudley [9]. Developing the stress analysis program of differential planetary gear system by Lewes [1-5] & Hertz [6, 9] equation and analyzing the safety factor of gear bending and compressive stresses considering required life time of mixer gearbox and the S/N curve are presented in the Gear Handbook by D.W. Dudley.

The predictive validity has also been verified with respect to the developed programs. Figure 3 shows the equation system solving with gear specifications calculation and strength analysis of the differential planetary gear system for mixer gearbox. It has also been verified the predictive validity with respect to the development and estimation of the scoring failure for differential planetary gears by scoring factor analysis.

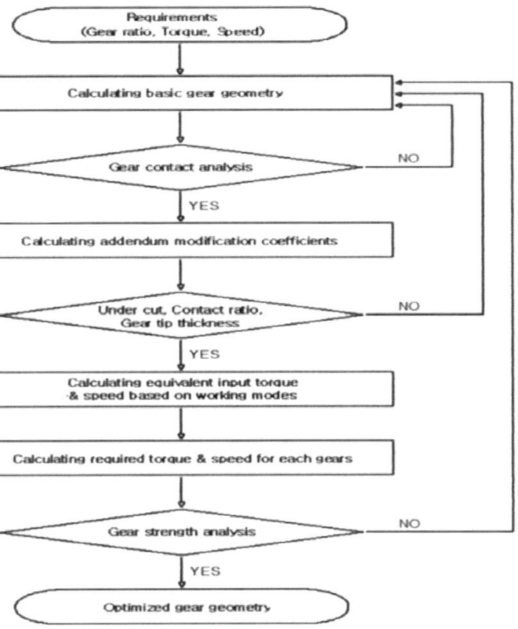


**Figure 3.** Equation system solving with gear specifications calculation and stress analysis.

## 2. Material and data analysis

### *2.1. Calculation of gear specifications*

Table 2 shows the specifications of the planetary gears of mixer gearbox after calculation. Figure 4 presents the results of the gear specification calculation program.

**Table 2.** Specifications of the planetary gear system.

| Item | Sun gear(A) | No.1pinion gear(B) | No. 1 ring gear(C) | No. 2 pinion gear(D) | No. 2 ring gear(E) |
|---|---|---|---|---|---|
| Module | 4 | 4 | 4 | 4 | 4 |
| Pressure angle(°) | 27 | 27 | 27 | 27 | 27 |
| Helix angle(°) | 0 | 0 | 0 | 0 | 0 |
| No. gear teeth | 10 | 35 | 80 | 31 | 76 |
| Tooth mod. factor | 0 | 0 | 0 | +0.5220 | +0.5220 |
| Pitch dia. | 40 | 140 | 320 | 124 | 304 |
| Outside dia. | 48 | 148 | 312 | 136.176 | 300.176 |
| Over pin measurement | $52.501^{-0.087}_{-0.183}$ ($\Phi$8) | $151.338^{-0.104}_{-0.196}$ ($\Phi$7.5) | $309.851^{+0.416}_{+0.257}$ ($\Phi$7) | $140.320^{-0.094}_{-0.177}$ ($\Phi$8) | $298.081^{+0.395}_{+0.244}$ ($\Phi$7) |
| Face width | 56 | 70 | 72.5 | 71 | 22 |
| Backlash | 0.117 ~ 0.220 | | 0.184 ~ 0.314 | 0.184 ~ 0.314 | |
| Center distance | 90 | | 90 | 90 | |

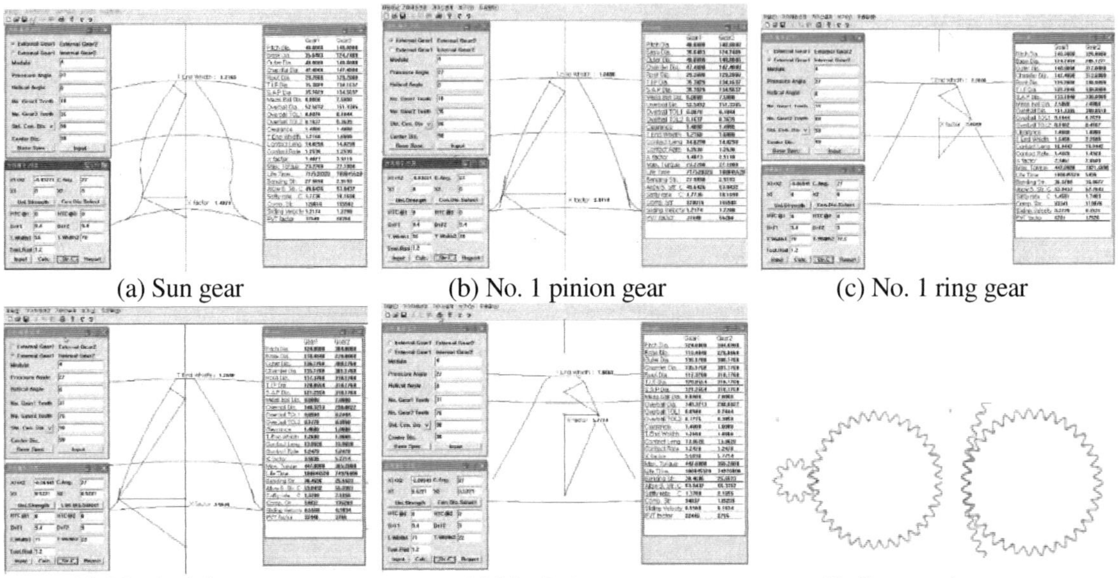

(a) Sun gear (b) No. 1 pinion gear (c) No. 1 ring gear

(d) No. 2 pinion gear (e) No. 2 ring gear (f) Gear tooth contact

**Figure 4.** The results of the gear specifications calculation program.

### *2.2. Input equivalent torque/ rotation speed analysis*

The required service period of life for a concrete mixer truck is 15 years with the vehicle operation rate of 70%, operating time is set 12 hours for a day, based on the total 28,400 hours, as shown in table 3.

**Table 3.** Operating mode and the required life period.

| Working mode | Frequency of use (%) | Working time (h) | Input | | Duty cycle | Cycle ratio |
|---|---|---|---|---|---|---|
| | | | Torque (N·m) | Speed (rpm) | | |
| Input concrete | 4 | 1,136 | 189.9 | 1320 | 89971200 | 0.125391849 |
| Driving | 41 | 11,644 | 189.9 | 264 | 184440960 | 0.257053291 |
| Normal working | 12 | 3,408 | 241.7 | 660 | 134956800 | 0.188087774 |
| Maximum working | 1 | 284 | 284.8 | 132 | 2249280 | 0.003134796 |
| Driving | 38 | 10,792 | 52.4 | 264 | 170945280 | 0.238244514 |
| Washing | 4 | 1,136 | 52.4 | 1980 | 134956800 | 0.188087774 |
| Total | 100 | 28,400 | - | - | 717,520,320 | 1 |

Equivalent mean torque for the average equivalent load of mixer reducer, $T_{mi}$ is as follows:

$$T_{mi} = \left[ \frac{\sum N_i t_i T_i^n}{\sum N_i t_i} \right]^{\frac{1}{n}} \tag{1}$$

where $T_i$ is working torque, N is rotating speed, t is working time, n is power index (n=20.8)
Equivalent mean rotating speed for the average rotating speed of mixer reducer, $N_{mi}$ is as follows:

$$N_{mi} = \left[ \frac{\sum N_i t_i}{\sum t_i} \right] \tag{2}$$

where, $N_{mi}$ is equivalent rotating speed for the average equivalent rotating speed, $N_i$ is rotating speed, $t_i$ is working time.

From the equation (1) and (2), the equivalent mean torque/rotating speed was calculated as 227.6N.m/421.08rpm.

*2.3. Calculation of torque and number of rotation*

From schematic diagram in figure 2, the gear ratio of mixer reducer calculated by relative speed diagram method [10] is as follows:

$$\Upsilon = \left\{ \frac{1 + Z_E Z_B / Z_A Z_D}{1 - Z_E Z_B / Z_D Z_C} \right\} \tag{3}$$

The number of rotation for each planetary gear calculated by relative speed diagram method is as follows:

$$N_B = N_D = \left\{ \frac{Z_A Z_C (N_A - N_C)}{Z_B (Z_A + Z_C)} \right\} \tag{4}$$

$$N_C = N_A / \Upsilon \tag{5}$$

$$N_S = \left\{ \frac{Z_A Z_D N_A}{Z_E Z_B + Z_A Z_D} \right\} \tag{6}$$

From the above equations, the torque and rotation speed is shown in table 4.

**Table 4.** Torque and number of rotation (N·m/rpm).

| | |
|---|---|
| $T_A/N_A$ (Torque/Number of rotation of sun gear) | 227.6 / 421.08 |
| $T_B/N_B$ (Torque/Number of rotation of No.1 pinion gear) | 4380.6(265.6) / 106.13 |
| $T_C/N_C$ (Torque/Number of rotation of No.1 ring gear) | 10,012.5 / 3.19 |
| $T_S/N_S$ (Torque/Number of rotation of carrier) | 30,264.9 / 43.95 |
| $T_D/N_D$ (Torque/Number of rotation of No.2 pinion gear) | 4,380.6 / 106.13 |
| $T_E/N_E$ (Torque/Number of rotation of No.2 ring gear) | 3,579.7 / 43.95 |

where, $Z_A$ is the number of teeth of sun gear, $Z_B$ is the number of teeth of No.1 pinion gear, $Z_C$ is the number of teeth of No.1 ring gear, $Z_D$ is the number of teeth of No. 2 pinion gear, $Z_E$ is the number of teeth of No. 2 ring gear.

*2.4. Calculation of gear bending stress*

The actual gear bending stress equation by Lewes formula is as follows:

$$S = \frac{29{,}400\pi T}{N_a F X Z} \tag{7}$$

where, S is the actual gear bending stress (N/mm$^2$), T is torque on gears (N·m), $N_a$ is the contact length of action (mm), F is the face width of gear (mm), X is the Lewes bending factor (mm), Z is the number of teeth in gear.

Allowable gear bending stress equation by Gear Handbook of D.W Dudley and AGMA Standard 218.01 [8] including gear bending S/N curve is as follows:

$$Sab = \frac{C_1}{N_F^{\frac{1}{20.8}}} \tag{8}$$

where, Sab is the allowable gear bending stress (N/㎟), $N_F$ is the No. of cycles, $C_1$ is the coefficient.

### *2.5. Calculation of gear compressive stress*

The actual gear compressive stress $P$ (N/mm$^2$) applied at the tip of the planetary gears based on contact formula of Hertz is as follows:

In the case of external gear contact, the actual gear compressive stresses of sun gear and pinion gear are,

$$P_S = 19.43\sqrt{\frac{2\pi T_S \times CDSIN\alpha}{A_S(CDsin\Phi - A_S) \times F_c \times N_a \times Z_S}} \tag{9}$$

$$P_P = 19.43\sqrt{\frac{2\pi T_S \times CDSIN\alpha}{A_P(CDsin\Phi - A_P) \times F_c \times N_a \times Z_S}} \tag{10}$$

In the case of internal gear contact, the actual gear compressive stresses of pinion gear and ring gear are,

$$P_P = 19.43\sqrt{\frac{2\pi T_P \times CDSIN\alpha}{A_P(CDsin\Phi + A_P) \times F_c \times N_a \times Z_P}} \tag{11}$$

$$P_R = 19.43\sqrt{\frac{2\pi T_P \times CDSIN\alpha}{A_R(A_R - CDsin\Phi) \times F_c \times N_a \times Z_P}} \tag{12}$$

where, α is the normal pressure angle, Φ is the transverse pressure angle, T is the torque on driving gear (N·m), $F_c$ is the active face width in contact (mm), Z is the No. of gear teeth, CD is the operating center distance, Na is the contact length of action (mm), A=, OR is the outside radius of gear, BR is the base radius of gear.

Allowable gear compressive stress equation by Gear Handbook of D.W Dudley and AGMA Standard 218.01 including gear compressive S/N curve is as follows:

$$Sac = \frac{C_2}{N_F^{\frac{1}{6.8433}}} \tag{13}$$

where, *Sac* is the allowable gear compressive stress (N/mm$^2$), $N_F$ is the No. of cycles, $C_2$ is the coefficient.

### *2.6. Calculation of Gear scoring factor*

Gear scoring factor [4] is based on Gear Handbook of D.W Dudley. To predict scoring failure of differential planetary gears, *PVT* (N·m/s·mm) is as follows: In the case of external gear contact, the actual gear scoring factor of sun gear and pinion gear are,

$$PVT_S = \frac{\pi N_S}{0.08525}\left(\frac{Z_P + Z_S}{Z_P}\right)(A_S - PR_S \times SIN\Phi)^2 \tag{14}$$

$$PVT_P = \frac{\pi N_P}{0.08525}\left(\frac{Z_P + Z_S}{Z_S}\right)(A_P - PR_P \times SIN\Phi)^2 \tag{15}$$

In the case of internal gear contact, the actual gear compressive stresses of pinion gear and ring gear are,

$$PVT_P = \frac{\pi N_P}{0.08525}\left(\frac{Z_R - Z_P}{Z_R}\right)(A_P - PR_P \times SIN\Phi)^2 \tag{16}$$

$$PVT_R = \frac{\pi N_P}{0.08525}\left(\frac{Z_R - Z_P}{Z_R}\right)(PR_R \times SIN\Phi - A_R)^2 \tag{17}$$

where, P is the actual compressive stress ($N/mm^2$), V is the sliding velocity (m/sec), T is the contact length from pitch point to contact point (mm), N is the speed of gears (rpm), Z is the number of gear teeth, Φ is the operating pressure angle.

Gear scoring factor, *PVT* must not exceed 124,019(N/sec·mm) to prevent scoring failure under HRC 60, carburized gears in mineral oil condition.

### 2.7. The results of gear bending and compressive stress analysis

The calculating actual gear bending and compressive stresses of planetary gear system for mixer gearbox and considering allowable gear bending and compressive stresses as well as the findings of the safety factors and verification of the problems of gear strength for the calculated specifications of the planetary gear system in mixer gearbox have been presented in this paper precisely. Figure 5(a) shows the results of gear bending stress analysis and figure 5(b) shows the results of gear compressive stress analysis of planetary gear system consisted of five gears (A to E) in mixer gearbox.

It can be shown that actual gear bending & compressive stresses of the differential planetary gears are under the allowable gear bending and compressive stresses in these S/N curves. Thus, calculation results are set safely and have verified as valid.

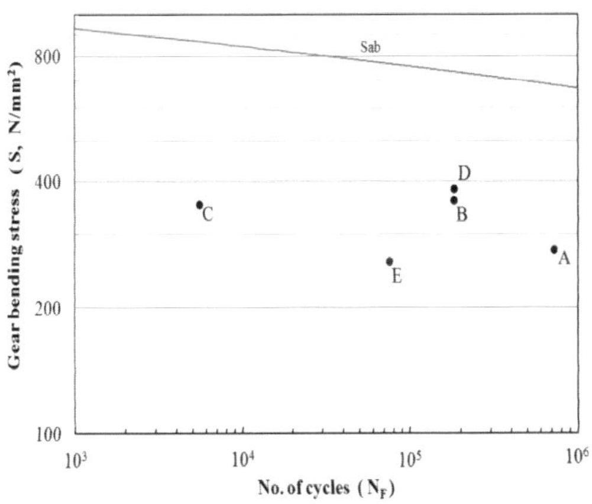


(a) Bending stress

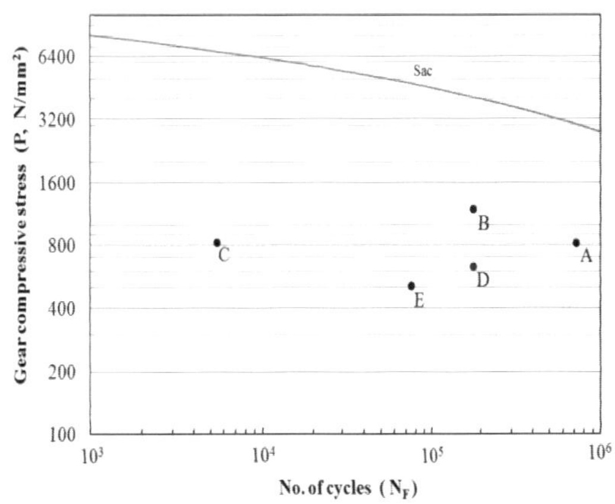


(b) Compressive stress

**Figure 5.** The results of gear stress analysis

### 2.8. The results of Gear Scoring factor analysis

Table 5 shows the results of calculated actual scoring factor, PVT and judge the safety of gear scoring failure considering the limit value, 124,019(N/sec·mm).

Table 5. Calculated actual gear scoring factor.

| | Sun gear | No. 1 pinion gear | | No. 1 ring gear |
|---|---|---|---|---|
| Max. torque(N·m) | 227.6 | 4,380.6 | | 10,012.5 |
| Speed(rpm) | 421.08 | 106.13 | | 3.19 |
| Actual scoring factor(N·m/sec·mm) | 3,031.2 | 5,741.6 | 15,708.0 | 1,085.6 |

(a) Sun gear + No. 1 pinion gear + ring gear

| | No.2 pinion gear | No.2 ring gear |
|---|---|---|
| Max. torque(N·m) | 4,380.6 | 3,579.7 |
| Speed(pm) | 106.13 | 44.00 |
| Actual scoring factor(N·m/sec·mm) | 23,733.7 | 163.7 |

(b) No. 2 pinion gear + No.2 ring gear

## 3. Conclusion

In this study, actual gear bending and compressive stresses of the differential planetary gears using Lewes & Hertz equation and the calculated specifications of differential planetary gears of mixer gearbox for $8m^3$ grade concrete mixer truck have been analyzed and the results with the data of allowable bending and compressive stress from the Stress-No. of cycles curves of gears, based on Gear Handbook of D.W. Dudley and AGMA Standard 218.01 have been evaluated. The outcomes of this study can be summarized as:

(1) The strength of the differential planetary gears and the developed programs have been verified, in respect of the result of gear bending and compressive stress analysis of calculated specifications of differential planetary gears of mixer gear box for $8m^3$ grade concrete mixer truck.

(2) In respect of the result of actual scoring factor analysis of calculated specifications of the differential planetary gears, scoring failure was not predicted.

(3) The developed programs calculating the specifications and analyzing the gear bending and compressive stresses and gear scoring factor of the planetary gear system for gearbox are expected to be effectively utilized. And future research on more excellent planetary gear system of the various reducers for construction machines is expected to be still performed.

## References

[1] Lewes W 1983 *Investigation of Strength of Gear Teeth Proc. Eng. Club,* Philadelphia, 38-55.
[2] Imwalle D E 1972 *Load Equalization in Planetary Gear Systems* ASME publication at the Mechanisms Conference & International Symposium on Gearing and Transmissions, pp 232-8.
[3] Seager D L 1972 *Load Sharing among Planet Gears* SAE publication No. 700178 pp 651-6.
[4] Cunliffe F, Smith J D and Welbourn D B 1980 *Dynamic Tooth Loads in Epicyclic Gears* Transactions of the ASME Journal of Engineering for Industry, pp 578-84.
[5] Castellani G, and Castelli V P 1980 *Rating Gear Strength* ASME Paper No. 80-C2/DET-88 pp 37-43.
[6] Coy J J, Townsend D P and Zaretsky E V 1975 *Dynamic Capacity and Surface Fatigue Life for Spur and Helical Gears* ASME Paper No. 75-Lub-19 pp. 56-73.
[7] Oda S and Koji T 1981 *Bull. JSME* **24(190)** 24-30.
[8] AGMA Standard 218.01 1982 *The Pitting Resistance and Bending Strength of Spur and Helical Involute Gear Teeth* The American Gear Manufacturers Association, pp 14-20.
[9] Dudley D W 1984 *The Handbook of Practical Gear Design*, 2nd Edition, Mcgraw-Hill, pp 1. 27-1.32, 2.1-2.12, 3.1-3.45, 3.78-3.112.
[10] Bae M H, Jang S K and Lee S Y 2009 *Automotive & Continuously Variable Transmission* 2nd Edition Sun Hak publication, pp37-44.

## Acknowledgment

This work was supported by the Ministry of Trade, Industry & Energy (A010600035-Korea), and the authors are gratefully appreciative of the support.

# Experiment Analysis and Modelling of Compaction Behaviour of Ag60Cu30Sn10 Mixed Metal Powders

**Mengcheng Zhou[1], Shangyu Huang[1, 2, *], Wei Liu[1], Yu Lei[1] and Shiwei Yan[1]**

[1]School of Materials Science and Engineering, Wuhan University of Technology, Wu Han 430070, China
[2]State Key Laboratory of Materials Processing and Die & Mould Technology, Huazhong University of Science and Technology, Wu Han 430070, China

**Abstract.** A novel process method combines powder compaction and sintering was employed to fabricate thin sheets of cadmium-free silver based filler metals, the compaction densification behaviour of Ag60Cu30Sn10 mixed metal powders was investigated experimentally. Based on the equivalent density method, the density-dependent Drucker-Prager Cap (DPC) model was introduced to model the powder compaction behaviour. Various experiment procedures were completed to determine the model parameters. The friction coefficients in lubricated and unlubricated die were experimentally determined. The determined material parameters were validated by experiments and numerical simulation of powder compaction process using a user subroutine (USDFLD) in ABAQUS/Standard. The good agreement between the simulated and experimental results indicates that the determined model parameters are able to describe the compaction behaviour of the multicomponent mixed metal powders, which can be further used for process optimization simulations.

## 1. Introduction

Due to the bad plastic workability, cadmium-free silver based brazing filler metals with high proportions of additional elements such as Zn, Sn, In, Ga, are too brittle to be made into thin sheets using conventional manufacturing processes [1, 2]. Recently, a novel process method that combines powder compaction and sintering was employed to fabricate thin sheets of the brittle and fragile cadmium-free silver based filler metals [3, 4]. The green powder compact is produced by cold die compaction of mixed metal powders and then undergoes a sintering process to attain the final thin sheets of filler metals with the desired properties. Since the any defects, cracks and density inhomogeneity generated at compaction stages will persist throughout the sintering process. Thus the properties of the filler metal sheet significantly depend on the quality of green compact which was related to the powder behaviour during compaction process. Therefore, it is of practical significance to investigate the compaction behaviour of mixed metal powders. In this paper, green compacts of filler metals Ag60Cu30Sn10 were fabricated, and the compaction behaviour of the mixed metal powders were investigated.

The finite element modelling (FEM) is an effective approach to investigate the powder behaviour considering the efficiency and cost. Based on simulations, the whole powder compaction process is simulated, and the powder densification behaviours, such as the distributions of stress and density, can be analyzed and predicted. In addition, the process parameters and tooling design optimization could be realized. While a proper constitutive model is vital for modelling the powder compaction process with continuum mechanics. So far, researches on metal powder compaction were mainly focused on densification of pure powders due to the complexity in densification of mixed powders [5]. To describe the metal powder compaction behaviour, the ''classical elastoplasticity'' models were frequently employed, such as Kuhn [6], Green [7], Shima-Oyane [8] models. These models were expanded from the classical von-Mises model, which were generally proposed from uniaxial compression tests of sintered powder compacts. But they fail to capture the shearing phenomenon in powder compact which is significantly important during unloading and ejection process and are not suitable for describing the early stage of compaction [9, 10]. Moreover, most of these models were proposed for pure powder. To model the compaction behaviours of mixed metal powders, mixed yield functions were proposed by Kim [5, 11] with various mixing rules of soft and hard powders. Han [12] used the Voigt mixed-mode model to determine the yield strength and elastic parameters of the Fe and Al composite powders in simulations of the single action die compaction process. However, these models are too complicated and not applicable for multicomponent mixed metal powders in this paper. Since the densification behaviour of mixed metal powders and soil materials have many similarities, such as both of them are mixtures of particles with different mechanical features, material properties varied with densities. Meanwhile, the "soil mechanics" model, modified Drucker-Prager Cap (DPC) model, has been frequently employed to simulate the metal powder compaction [13, 14]. Thus, the modified DPC model was introduced to characterize the densification behaviour of mixed metal powders [3].

In this work, the compaction densification behaviour of Ag60Cu30Sn10 mixed metal powder was experimentally investigated. The density-dependent modified DPC plasticity model as well as a linear elasticity law was employed to model the powder densification behaviour. The model parameters and friction coefficients between the powders and die wall were determined experimentally. Validation experiments and finite element modelling of powder compaction process with a user subroutine (USDFLD) in ABAQUS/Standard were carried out to validate the obtained model parameters.

## 2. Constitutive model

The modified DPC[15] model employed in this work is a continuum mechanics model as shown in figure 1. The model assumes that the material is isotropic and its yield surface consists of three segments: a shear failure surface ($F_s$), a cap yield surface ($F_c$), and a transition region ($F_t$). The yield locus and plastic flow potential are dependent on the hydrostatic pressure stress $p$ and the Mises equivalent stress $q$. The equations of the three surfaces are expressed as

$$F_s = q - p\tan\beta - d = 0 \tag{1}$$

$$F_c = \sqrt{(p-p_a)^2 + \left(\frac{Rq}{1+\alpha-\alpha/\cos\beta}\right)^2} - Rd\left(d + p_a\tan\beta\right) = 0 \tag{2}$$

$$F_t = \sqrt{(p-p_a)^2 + \left[q - \left(1-\frac{\alpha}{\cos\beta}\right)\left(d + p_a\tan\beta\right)\right]^2} - \alpha\left(d + p_a\tan\beta\right) = 0 \tag{3}$$

where $d$ and $\beta$ are the cohesion and friction angle, respectively. $R$ and $\alpha$ are parameters controlling the shape of the cap surface and the transition yield surface. The variable $\alpha$ is typically between 0.01 and 0.05, and is set as 0.02 in this work. $p_a$ is an evolution parameter. $p_b$ is the hydrostatic compression yield stress which defines the position of the cap, and is generally expressed as a nonlinear function of volumetric plastic strain

To fully define the yield surface, five parameters are required: $d$, $\beta$, $R$, $p_a$ and $p_b$. In addition, two elastic parameters, the Young's modulus $E$ and Poisson ratio $\upsilon$, and friction coefficient $\mu$, are required to describe the elastic behaviour of the powder material and the friction between the powders and die wall, respectively. The material parameters were determined experimentally as function of relative density and the calibration methods were described in our previous research [3].

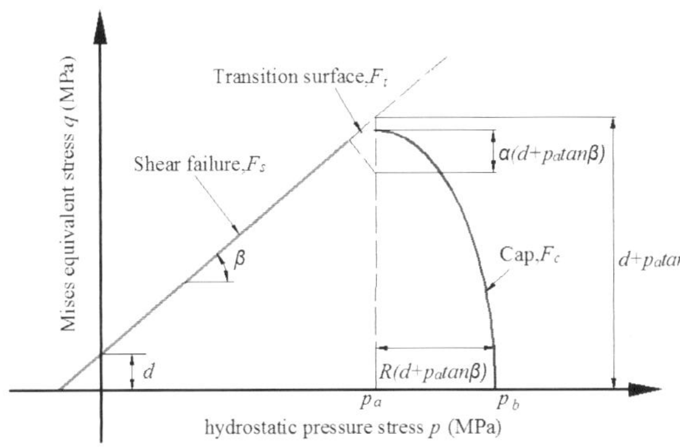


**Figure 1.** Yield surfaces of the modified Drucker-Prager Cap model[15].

**Figure 2.** SEM images of the mixed metal powders.

## 3. Materials and experiments

### *3.1. Materials*

The compositions and physical properties of Ag60Cu30Sn10 mixed powder were listed in table 1. To prepare the mixed metal powder, pure metal powders were weighed by an electronic balance according to their mass fractions. Subsequently, the powders were homogeneously mixed in vacuum using a planetary ball mill to prepare the mixed metal powders. The powder morphology was illustrated by SEM images as shown in figure 2. Based on the equivalent density method [3], the theoretical density of Ag60Cu30Sn10 was calculated as 9.582 g/cm$^3$.

**Table 1.** The composition and physical properties of two mixed metal powders.

| Mixed metal Powders | Element components | Mass fraction (*wt%*) | Theoretical density (*g/cm$^3$*) | Particle sizes (*μm*) | Theoretical density (*g/cm$^3$*) |
|---|---|---|---|---|---|
| Ag60Cu30Sn10 | Ag powder | 60 | 10.53 | 1~45 | 9.582 |
| | Cu powder | 30 | 8.94 | 1~38 | |
| | Sn powder | 10 | 7.28 | 1~58 | |

### *3.2 Experiments*

To investigate the densification behaviour of Ag60Cu30Sn10 mixed metal powders, series uniaxial compaction, diametrical compaction and instrumented die compaction tests were performed by using a WAW-600 (HUALONG, China) material testing system with a capacity of 600 KN. Uniaxial compaction and diametrical compaction tests were performed on samples with different relative densities to measure the axial compressive strength and radial tensile strength of compacts. In uniaxial compaction tests, the specimen diameter and height are 10 mm and 15 mm, respectively, and the compaction speed is 2 mm/min. In diametrical compaction tests, the compaction speed is 0.25 mm/min, the specimen diameter is 16mm and the specimen's thickness is ≤ 4mm. The instrumented die compaction tests were performed using the compaction apparatus as shown in figure 3. The strain-gauge measurement method [3] were employed to measure the radial stress during the powder compaction. The internal diameter, external diameter and height of the die are 10 mm, 20 mm and 50 mm, respectively. In the instrumented die compaction tests, an amount of 5.27 g Ag60Cu30Sn10 powder was manually poured into the die to achieve a uniform packing. The filling height was measured, and the initial filling relative density was calculated as 0.45. Loading and unloading were carried out by controlling the movement of the top punch with a constant load rate 0.2 KN/s to press the powder to different relative densities level, while the lower punch maintains stationary. The axial displacements of the load cylinder, the two axial forces and radial stress were recorded simultaneously. Two experiment cases are designed to characterize the friction between powders and die wall. Die

IOP Conf. Series: Materials Science and Engineering **317** (2018) 012011 doi:10.1088/1757-899X/317/1/012011

compaction tests with an unlubricated die, the inner die surface was cleaned by acetone and ethanol. Die compaction tests with a lubricated die, the inner die surface was lubricated using zinc stearate alcohol solution.

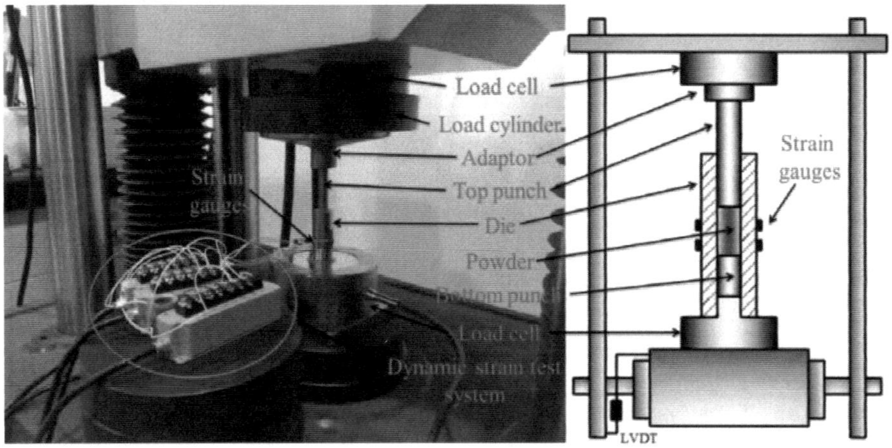


**Figure 3.** Schematic diagram of the instrumented die compaction test apparatus.

## 4. Results and discussion

### *4.1. Determination of the material parameters*

The axial compressive and radial tensile strengths of powder compacts were measured by the diametrical compression and uniaxial compression tests, from which the cohesion $d$ and friction angle $\beta$ were determined as functions of relative density shown in figure 4. The cap surface parameters $R$, $p_a$ and $p_b$, elastic parameters $E$ and $\upsilon$, were determined by instrumented die compaction tests which were shown in figures 5 and 6. All the experimentally calibrated material parameters were accurately fitted and extrapolated to low initial relative density regimes (< 0.6) to model the powder compaction behaviour. The friction coefficients $\mu$ between powder and the die wall in lubricated/unlubricated die were determined using the Janssen-Walker theory [16] as shown in Figure 6 (c). It can be observed that in lubricated die compaction, $\mu$ increased with top punch pressure at low pressure regime ($\leq$ 50MPa), and then decreased gradually. But at the high pressure regime (> 125MPa), $\mu$ almost maintained a constant value of 0.08. In unlubricated die compaction, $\mu$ increased with top punch pressure at low pressure regime ($\leq$ 70MPa), and then almost kept a constant value of 0.19.

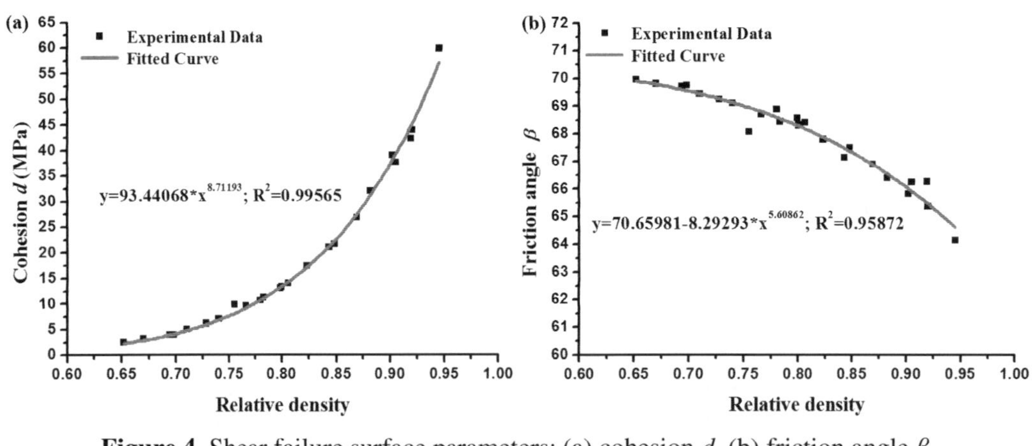


**Figure 4.** Shear failure surface parameters: (a) cohesion $d$, (b) friction angle $\beta$.

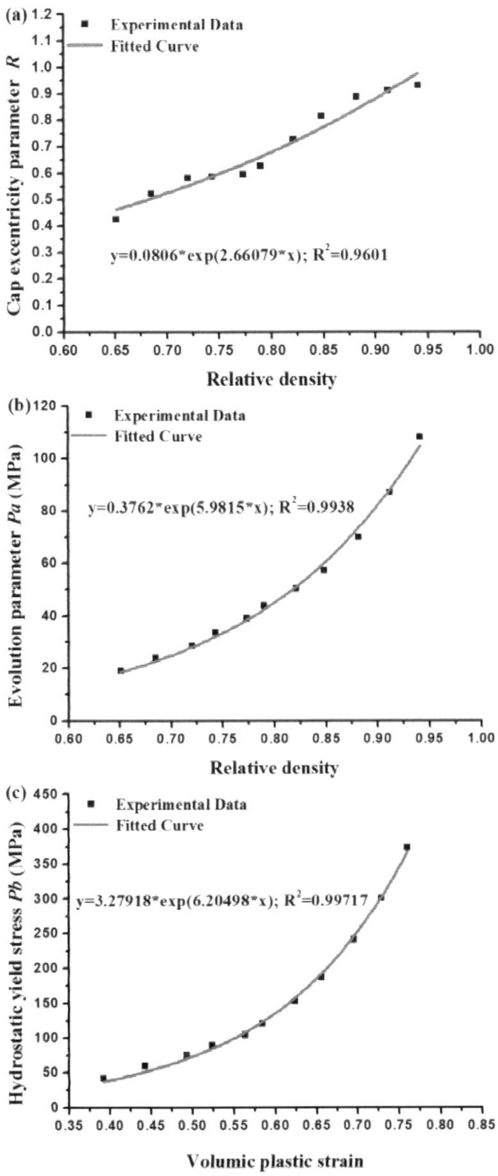


**Figure 5.** Cap surface parameters: (a) $R$, (b) $p_a$, (c) $p_b$.

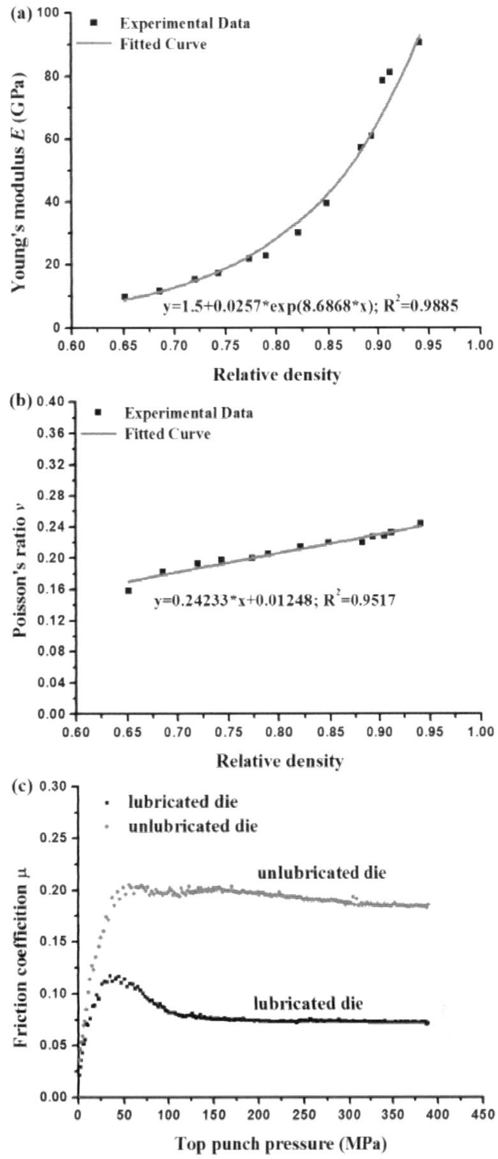


**Figure 6.** Elastic parameters and friction coefficient: (a) $E$, (b) $\upsilon$, (c) $\mu$.

*4.2. Verification of the obtained material parameters*

To validate the material parameters, validation experiments and simulations of powder compaction with lubricated/unlubricated die were performed. The characteristics of powder compacts prepared in the validation experiments, such as initial filling height $H_0$, the top punch displacement $H_1$, initial relative density$\rho_0$, final relative density $\rho_1$ and compaction pressure $P$, are presented in table 2. Since the geometry corresponding to the die compaction was axisymmetric, thus a quarter of 3D finite element model for the die compaction were analysed in ABAQUS/Standard. In simulations, the powder compact with a diameter of 10mm was modelled as a deformable continuum with C3D8R elements. The top/bottom punches and die were modelled as discrete rigid bodies with R3D4 elements. The friction coefficient between powder and tools were defined with a Coulomb friction model and were set as a constant value (table 2). In simulations, the top punch moved downward to press the powder during compaction and then moved upward during unloading, the bottom punch maintained

IOP Conf. Series: Materials Science and Engineering **317** (2018) 012011 doi:10.1088/1757-899X/317/1/012011

stationary. The determined elastic and plastic material parameters in section 4.1 were used in simulations. To characterize the evolution of material parameters with the relative density, a user subroutine USDFLD with a user-defined field for relative density was employed to update the elastic and plastic material parameters of powder during each time step.

**Table 2.** The characteristics of powder compacts in the validation experiments.

| Compacts | $m$ (g) | $H_0$ (mm) | $H_1$ (mm) | $\rho_0$ | $\rho_1$ (experiment) | $\rho_1$ (simulation) | $P$ (MPa) (experiment) | $P$ (MPa) (simulation) |
|---|---|---|---|---|---|---|---|---|
| Lubricated die $\mu$=0.08 | 5.286 | 15.55 | 8.04 | 0.452 | 0.912 | 0.916 | 389 | 389.8 |
| Unlubricated die $\mu$=0.19 | 5.27 | 15.2 | 7.49 | 0.461 | 0.893 | 0.884 | 389.2 | 393.1 |

The simulated results of relative density distributions in powder compacts after decompression were presented in figure 7. Due to the effect of the friction between the die and powders, the relative density is higher at the top corner contact with the die and lower in the bottom corner. It is also observed that because of the higher friction in unlubricated die, the density gradient is more remarkable. Figure 8 shows the comparison between the simulated and experimental loading-unloading curves. The simulated curves all matched well with the experimental data both in the loading and unloading steps. And the simulated results of the final relative density and compaction pressures were almost consisting with the experimental measurements as presented in table 2. The good agreement between the simulated and experimental results indicates that the experimentally determined density-dependent model parameters are able to precisely characterize the compaction behaviour of Ag60Cu30Sn10 mixed metal powders, and can be further used for process optimization simulations.

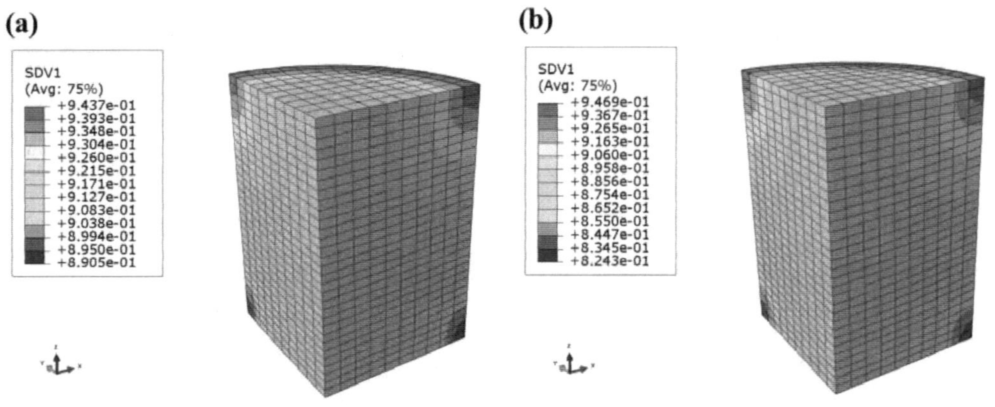


**Figure 7.** Relative density (SDV1) distribution in powder compact after decompression: (a) lubricated die, and (b) unlubricated die.

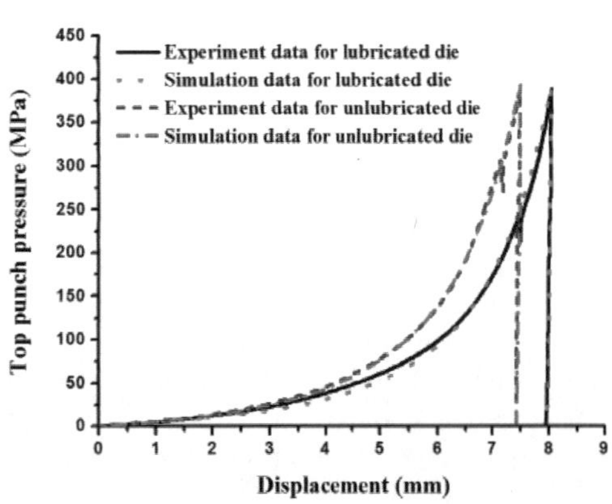


**Figure 8.** Comparison between the simulated and experimental loading-unloading curves.

## 5. Conclusions

The compaction densification behaviour of Ag60Cu30Sn10 mixed metal powders were investigated experimentally. Based on the equivalent density method, the density-dependent modified DPC model was employed to model the powder compaction behaviour. All the elastic and plastic material parameters were experimentally identified and were validated by validation experiments and finite element simulations of powder compaction process in ABAQUS/Standard with a user subroutine (USDFLD). The good agreement between the simulated and experimental results indicates that the determined model parameters can describe the mixed metal powder compaction behaviour, and can be further used for process optimization simulations.

## Acknowledgments

This work was supported by National Natural Science Foundation of China (No.51475345), the Open Fund Project of State Key Laboratory of Materials Processing and Die & Mould Technology, Huazhong University of Science and Technology (No. P2015-01).

## References

[1] Lai Z, Xue S, Lu F, et al. 2009 *China Weld* **18** 33-8.
[2] Weise W, Voelcker A, Kaufmann D, et al. 1996 *Cadmium-free silver alloy brazing solder, method of using said solder, and metal articles brazed with said solder*: US US5531962.
[3] Zhou M, Huang S, Hu J, et al. 2017 *Powder Technol.* **305** 183-96.
[4] Andrieux J, Dezellus O, Bosselet F, et al. 2008 *J. Phase Equilibria Diffus.* **29**1 56-62.
[5] Kim K T, Cho J H 2001 *Int. J. Mech. Sci.* **43** 2929-46.
[6] Kuhn H A, Downey C L 1971 *Int. J. Powder Metall.* **7** 15-25.
[7] Green R J 1972 *Int. J. Mech. Sci.* **14** 215-24.
[8] Shima S, Oyane M 1976 *Int. J. Mech. Sci.* **18** 285-91.
[9] Lee S C, Kim K T 2002 *Int. J. Mech. Sci.* **44** 1295-1308.
[10] Lee S C, Kim K T 2007 *Mater. Sci. Eng.* A **s445-446** 163-9.
[11] Cho J H, Kim K T 2001 *Int. J. Mech. Sci.* **43** 921-33.
[12] Han P, An X Z, Zhang Y X, et al. 2015 *J. Min. Metall.* **51** 163-71.
[13] Zadeh H K, Jeswiet J, Kim I Y 2013 *Int. J. Adv. Manuf. Technol.* **68** 1785-95.
[14] Zhou R, Zhang L H, et al. 2013 *Trans. Nonferrous Met. Soc. China.* **23** 2374-82.
[15] Han L H, Elliott J A, Bentham A C, et al. 2008 *Int. J. Solids Struct.* **45** 3088-106.
[16] Sinka I C, Cunningham J C, Zavaliangos A 2003 *Technol.* **133** 33-43.

# Energy Center Structure Optimization by using Smart Technologies in Process Control System

**Svetlana V Shilkina**[1]

[1]129337, 26, Yaroslavskoye Shosse, Moscow, Russia, Moscow State University of Civil Engineering

**Abstract.** The article deals with practical application of fuzzy logic methods in process control systems. A control object – agroindustrial greenhouse complex, which includes its own energy center – is considered. The paper analyzes object power supply options taking into account connection to external power grids and/or installation of own power generating equipment with various layouts. The main problem of a greenhouse facility basic process is extremely uneven power consumption, which forces to purchase redundant generating equipment idling most of the time, which quite negatively affects project profitability. Energy center structure optimization is largely based on solving the object process control system construction issue. To cut investor's costs it was proposed to optimize power consumption by building an energy-saving production control system based on a fuzzy logic controller. The developed algorithm of automated process control system functioning ensured more even electric and thermal energy consumption, allowed to propose construction of the object energy center with a smaller number of units due to their more even utilization. As a result, it is shown how practical use of microclimate parameters fuzzy control system during object functioning leads to optimization of agroindustrial complex energy facility structure, which contributes to a significant reduction in object construction and operation costs.

## 1. Introduction

Competitive production development efficiency is largely determined by the level of its automation. Building of a new production facility in this days requires a pre-investment feasibility study involving professional experts at all stages and, especially, at front-end engineering design stage. This allows to determine all main parameters of the project at an initial stage with sufficient accuracy, which contributes to significant cost reduction, confirms data reliability for potential creditors and counterparties, significantly shortens designing and construction time, since equipment basic composition and basic technical solutions are identified immediately. In addition, involvement of experts with advanced knowledge of how to use smart control technologies, innovative solutions in construction and structuring of automated process control systems (APCS), ensures significant benefits associated with resource optimization, and significant costs reduction at the stage of built object operation [1, 2].

The article purpose is to consider practical use of a fuzzy logic controller-based control systems to optimize agroindustrial production processes which structurally includes its own energy center. To show possible cut of investor's costs at project feasibility study stage in a specific context of a new greenhouse complex construction with the use of up-to-date smart object control systems allowing to optimize energy center structure.

The experts examined the project to justify agroindustrial greenhouse complex construction investments. Construction of this agricultural enterprise is planned using an "open field" principle, i.e.

IOP Conf. Series: Materials Science and Engineering **317** (2018) 012012 doi:10.1088/1757-899X/317/1/012012

on an isolated empty plot of land, but with possible brining of main utilities in future. The article is not aimed at describing full economic and mathematical model of the project, but only at considering the use of up-to-date control methods, adapting models and algorithms when constructing facility, system ACS. Proposed smart ACS will significantly optimize the resources used, reduce equipment procurement and project energy circuit construction costs [3-5].

## 2. Feasibility study

Several options for object power supply – the greenhouse complex – were estimated as part of the project examination. In particular, the options that take into account possible connection to external power grids, as well as installation of own power generating equipment with various modifications. The estimation showed: one of the main problems is that there are power consumption surges in current process resulting in extremely uneven daily power consumption schedule (figure 1).

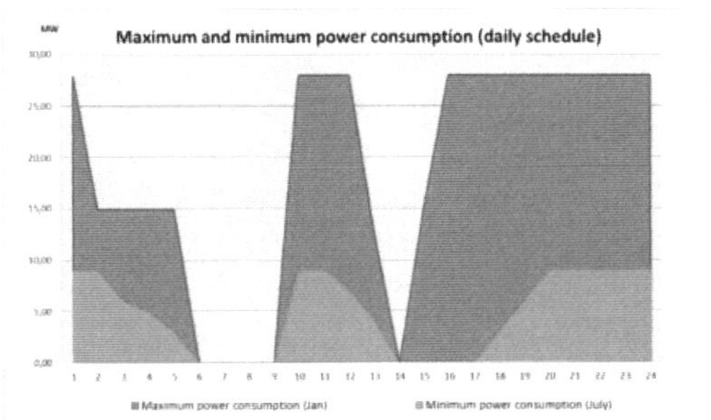


**Figure 1.** Daily power consumption schedule.

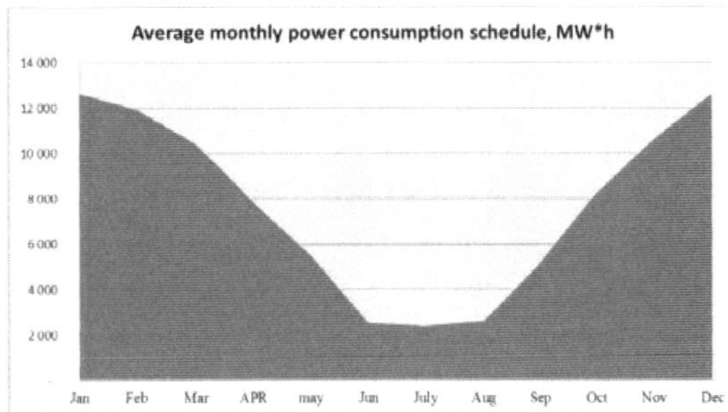


**Figure 2.** Annual power consumption schedule.

Such a daily utilization of the energy center (figure 1) and uneven average monthly consumption (figure 2) showed that operation according to third party consumer power supply schedule is inefficient, and taking into account process connection costs and current power consumption costs the project turns out to be nearly cost-ineffective, since the main costs of the greenhouse facility are energy supply.

Alternatively, building of own gas reciprocating unit-based (GRU) energy center was estimated. Initially, the investor planned to install four GRUs of 8.18 MW each (figure 3, figure 4) based on the object needs. This particular equipment is used due to external reasons, which will not be dwelt on separately herein. However, the use of four GRUs with a total capacity exceeding 32 MW has many negative consequences:

1. Evident power redundancy: low utilization of the fourth unit throughout the year: 79% - the first unit, 55% - the second one, 31% - the third one, 5% - the fourth one (with 8000 hours and above of annual utilization estimated). Energy center peak utilization exceeds 90% (in winter daytime).
2. If current daily utilization of the energy center is kept, operation according to third party power consumption schedule is impossible, which greatly complicates solution of the problem through selling surplus power on the side. A prerequisite is availability of consumers in local power consumption area with clear schedules of daily and annual power consumption. Even if such a consumer is available now there is no guarantee of its permanency.
3. The energy center capacity exceeds 25 MW. Selling power from such facilities to the retail market is generally not allowed by a power market regulator, while power sale price in the wholesale market is 2-3 times lower than the retail one. To work at retail price, it is necessary to reduce the installed capacity or to unify the greenhouses and energy center into a single process enterprise and a separate legal entity that can sell surplus power in the retail market. This is accompanied by significant legal risks and dependence on the power market regulator in the future.

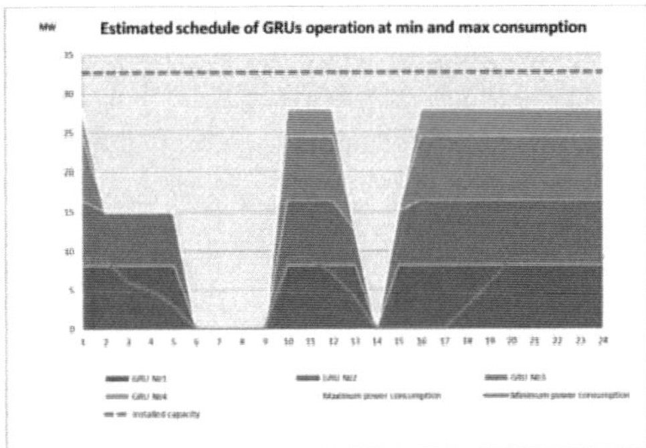


**Figure 3.** Four GRUs operation schedule.

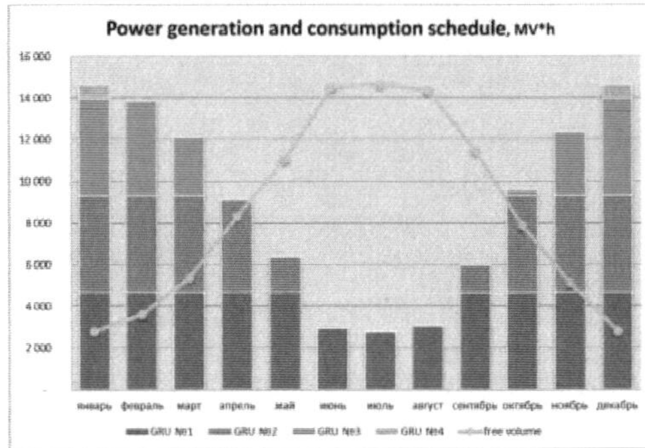


**Figure 4.** Power generation and consumption schedule.

A financial model was developed for the investor to allow a comparative analysis of various options for energy hub arrangement. The options providing for installation of two, three and four GRUs were estimated with peak loads covered by purchasing missing capacities in the power market and selling surplus power. For the four GRUs option, several sub-options with and without connection to external power grids and possible partial selling of surpluses were estimated (table 1).

Conclusions of preliminary assessment of the various options:

1. From the point of view of energy self-sufficiency and safety against game rule changes and price fluctuations, the most preferable option is operation in “power island” mode, where no complex power selling schemes are required, possibility of autonomous operation with no external power supply infrastructure is ensured.
2. The least efficient options are construction of energy center with four GRUs and selling surplus power to the market, or refuse to sell (the fourth unit is backup).
3. In any case EBITDA calculated on the basis of reduction of cost for power purchasing due to GRUs generation is between 200 and 400 million rubles depending on the power hub configuration.

**Table 1.** Comparative analysis of the energy hub arrangement options.

| Name | Basic option* 4 GRUs | Option 1 2 GRUs | Option 2 3 GRUs | Option 3 4 GRUs | Option 4 4 GRUs |
|---|---|---|---|---|---|
| Installed capacity, MW | 32.7 | 16.4 | 24.5 | 32.7 | 32.7 |
| Construction cost, million rubles | 2408.80 | 1284.00 | 1607.60 | 1931.30 | 2090.50 |
| Power required for process connection / power distribution scheme, MW | 27.86 | 18.57 | 9.3 | - | 9.3 |
| Process connection cost / power distribution scheme, million rubles | 477.5 | 318.3 | 159.2 | - | 159.2 |
| Power purchase on the market, million rubles per year | - | 144 | 30.6 | - | - |
| Power sale to third party consumers, million rubles per year | 163.1 | - | - | - | 121.1 |
| Need to optimize power consumption schedule | - | - | - | yes | yes |
| Payback period, taking into account process connection, years | 6.6 | 5.2 | 5.3 | 6.4 | 6 |

From the point of view of economic factors combination and confidence in power supply, the investor was advised:

1. To optimize greenhouses power load schedule (peaks decrease, dips mitigation) by change of a greenhouse facility process control system. To do this, it is advised to involve automated process control systems construction experts to develop equipment functioning optimal algorithms. The goal is to optimize the system operation, taking into account refusal to use the fourth unit with redistribution of the remaining units utilization.
2. Development of a legally reliable organizational scheme for working in the retail power market for the three GRUs energy center option.
3. Various technical solutions based on thermal energy consumption schedule analysis, after comparison with electrical energy consumption schedule. Including the use of carbon dioxide from the units exhaust gas to increase greenhouses performance up to 25%.

## 3. Fuzzy logic control system

As a result of the above recommendations, studies to develop a fuzzy algorithms-based control system with automatic modification of the process system to reduce energy costs by optimizing power consumption were carried out [5].

For ACS experts a requirements specification was drawn up. It provided for development of the fuzzy algorithm and a production rules system on its basis, the implementation of which in real time would ensure a more even consumption of electrical and thermal energy, which would result in energy costs reduction through the GRUs operation optimization. Fuzzy control system generalized structure can be illustrated by the following chart (figure 5) [1, 3].

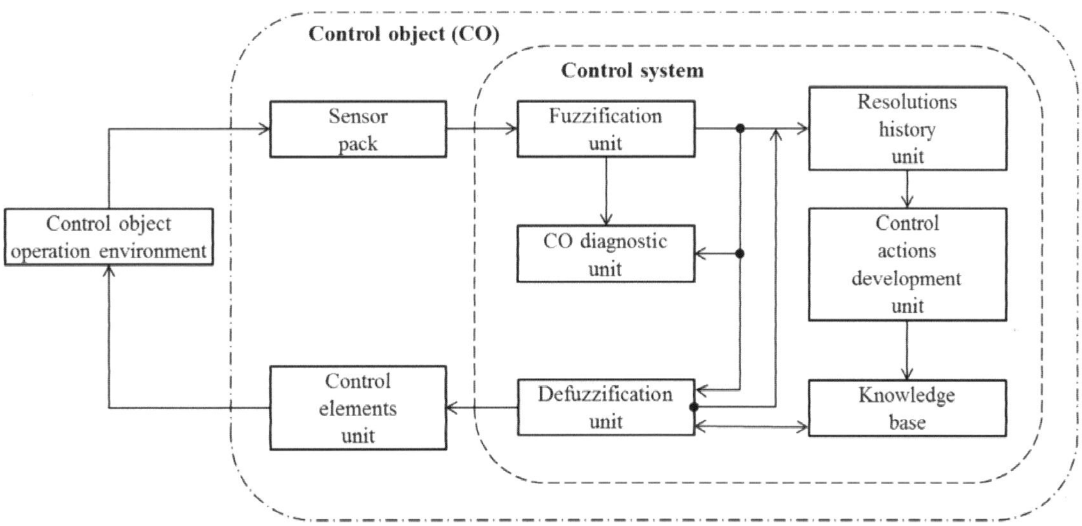


**Figure 5.** Structure chart of the fuzzy logic system.

Development of the fuzzy control system provides for the following issues to be solved [1, 3, 5]:

- presentation of the information used for control in terms of fuzzy logic (fuzzification task);
- framing of a fuzzy inference rule base taking into account process engineers' expert knowledge and control actions determination algorithm (control actions development task taking into account resolutions history);
- description of control in terms of fuzzy logic to get physical control actions in the system (defuzzification task).

Resolution of the above issues is related to the need to carry out studies and to develop appropriate mathematical models applicable to processes control [3, 5, 6].

At the initial stage of fuzzification in respect to greenhouse complex process parameters control, a general list of regulated parameters, linguistic variables and terms structure to describe the control object state, conditions for its functioning and control actions were determined taking into account the results of process features analysis. At this stage a complete list of parameters that can be regulated by the developed control system is determined. Main power-consuming parameters were chosen:

- temperature inside the greenhouse (especially at the soil level);
- greenhouse illumination;
- humidity of soil and air inside the greenhouse;
- external perturbing factors (environmental impact).

Membership functions of information and control signal physical values in linguistic variable terms were determined taking into account degree of interrelation between these components and expert estimates. After solving the problem of fuzzification, linguistic variables and their terms were stated to describe process microclimate parameters condition and to control them, and functions of membership connecting process physical parameters with linguistic variables were obtained.

Implementation of the algorithm next stage allowed to develop a rule base for describing a control algorithm of the fuzzy controller on the basis of experimental data and experience gained by process experts. The obtained rules determine microclimate parameters combination selection inside the object based on the environmental parameter values and effect of each parameter on all the others.

The solution to the next task of fuzzy inference rules development is to determine a resulting function of membership for control actions using the rule base and fuzzy operations. The resulting rule base unit contains a set of linguistic rules reflecting system operation algorithm.

During fuzzy inference, the fuzzy controller determines the control actions membership function value through temperature, humidity and illumination control channel in terms of fuzzy control. To determine a control actions physical value, defuzzification is performed for control actions logical values. Defuzzification is associated with conversion of the control membership function for each control channel into physical value of control action on the actuators [3, 5].

Based on the results of the object main parameters study, a system of fuzzy control of the object process microclimate parameters is proposed. General structure of the greenhouse complex microclimate parameters fuzzy control system is shown in figure 6.

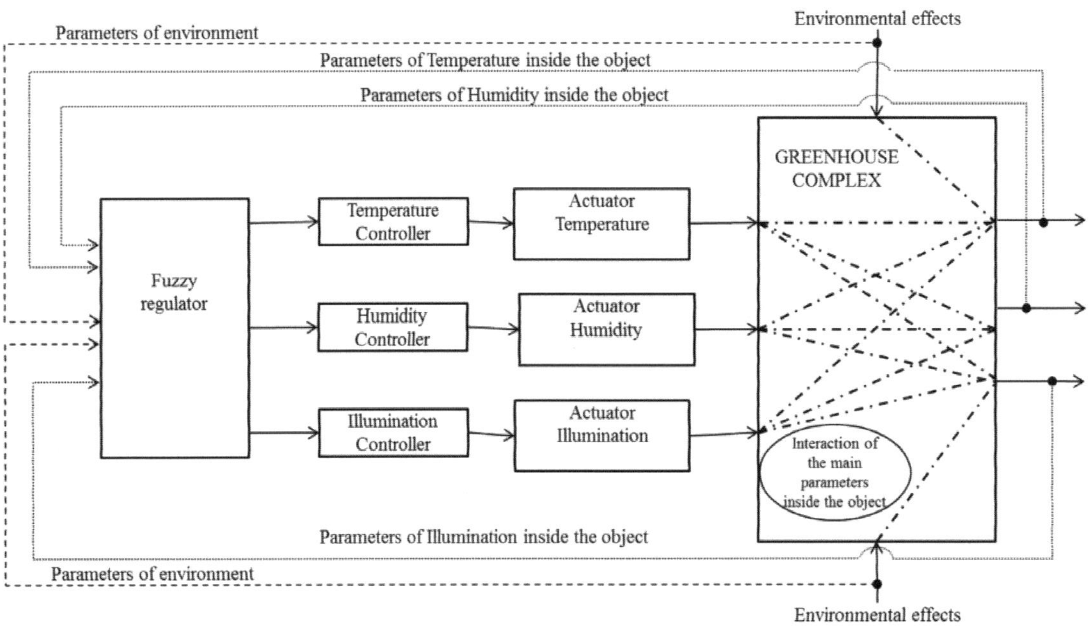


**Figure 6.** Greenhouse complex microclimate parameters fuzzy control system.

## 4. Conclusion Findings

According to the estimations, introduction of the proposed control system will significantly improve the situation regarding GRUs uneven utilization during winter season, which in turn will allow us to speak with a high degree of confidence about possible operation of the energy center with three GRUs (option 2) and to optimize its structure using smart technologies. Optimum control of the energy center operation will result in peak load decrease in winter season to a level sufficient for stable operation of the control object in these conditions (figure 7).

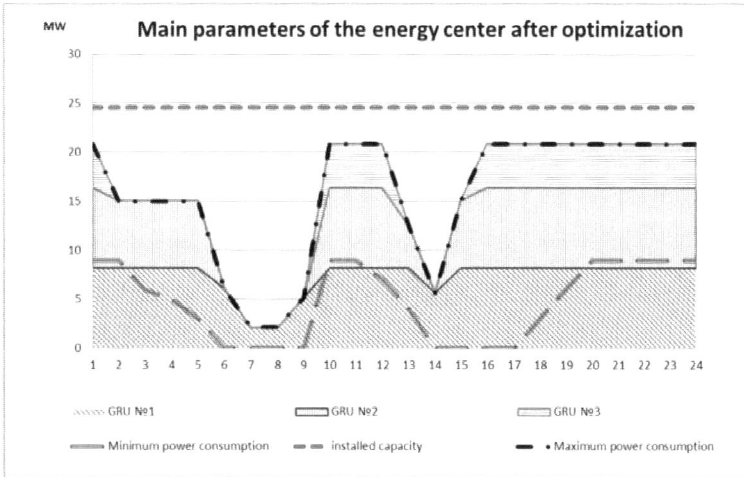


**Figure 7.** Schedule of power generation and consumption after optimization of the object control with the use of the fuzzy control system.

The figure shows that the peak loads are reduced, power consumption dips are smoothed.

Thus, development of the greenhouse complex processes smart control system based on fuzzy logic, taking into account best practice employed by process engineers in the form of the knowledge base, results in real optimization of the energy center structure and, accordingly, the level of expenditure.

**Reference**

[1] Vasilyev V I, 2009 *Smart control systems. Theory and practice: Teaching aid.* Radiotekhnika, 387.

[2] Aliyev R A 1991 *Production control with fuzzy source information*, Energoatomizdat.

[3] Peshko M S, Shkapov P M 2015 *Adaptive system intelligent control of process parameters in food industry* Omsk scientific bulletin. Series of devices, machines and technologies. **2(140)** 118-22.

[4] Shepherd A B, Batty W J 2003 *Build. Serv. Eng. Res. Tech.* **24** 35-45.

[5] Kolyazov K A, Kayashev A I 2010 *Fuzzy algorithm-based energy-saving control system* Energy efficiency and energy security at industrial enterprises and housing and public utilities: third all-Russian scientific and practical workshop, Ufa: Academy of Sciences of the Republic of Bashkortostan; Gilem, 119-23.

[6] Kolyazov K A, Vorobyova A V, Shiyanova N I 2007 *Ind. Autom.* **7** 55-7.

IOP Conf. Series: Materials Science and Engineering **317** (2018) 012013 doi:10.1088/1757-899X/317/1/012013

# An Engineering Method of Civil Jet Requirements Validation Based on Requirements Project Principle

**Yue Wang[1], Dan Gao[1], Xuming Mao[1]**

[1]Shanghai Aircraft Design and Research Institute, No. 5188, Jinke Road, Pudong New Area, Shanghai, China

**Abstract.** A method of requirements validation is developed and defined to meet the needs of civil jet requirements validation in product development. Based on requirements project principle, this method will not affect the conventional design elements, and can effectively connect the requirements with design. It realizes the modern civil jet development concept, which is "requirement is the origin, design is the basis". So far, the method has been successfully applied in civil jet aircraft development in China. Taking takeoff field length as an example, the validation process and the validation method of the requirements are detailed introduced in the study, with the hope of providing the experiences to other civil jet product design.

## 1. Introduction

With more and more advanced technologies applied in the advanced civil jet development, the system integration plays more and more important role in the industry. Due to this, a new design principle, called Requirements Project, is formed and studied, and finally achieves great development. The core concept of it is Requirements and Requirements Management, which makes it different from the conventional design principle.

The civil jet of China developed bumpily and discontinuously, and many design principles are transplanted from military aircraft design system. The gap still exists between China and other countries in civil jet manufacture, especially in the design concept.

After more than twenty years' efforts and development, the advanced design principles are introduced into Chinese design working, especially in the regional civil jet and single aisle large civil jet. Based on advanced requirements project design principle, a civil jet requirements validation method and process for aircraft concept design phase are defined. They have been applied successfully in design working, and the application prospect is wide in the future.

## 2. The advanced civil jet requirements project principle

The core factors of requirements project principle are requirements and their management. The requirements are identifiable design elements, which can be validated and verified with the method of simulations and experiments. In the whole product lifecycle, requirements management shall identify, transfer, break down, manage and control the requirements. In the management process, the requirements are traceable, complete, unique and coincident [1], see figure 1.

Requirements are the basis of the aircraft and its systems design working, giving the design working direction. If the requirements are not complete, correct or coincident, the product will not reach the design requirements and objects. Therefore, the definition of top level requirements is important, which is the origin of the product, see figure 1 with red dashed zone. This paper would focus on this part, and propose a new method of requirements validation.

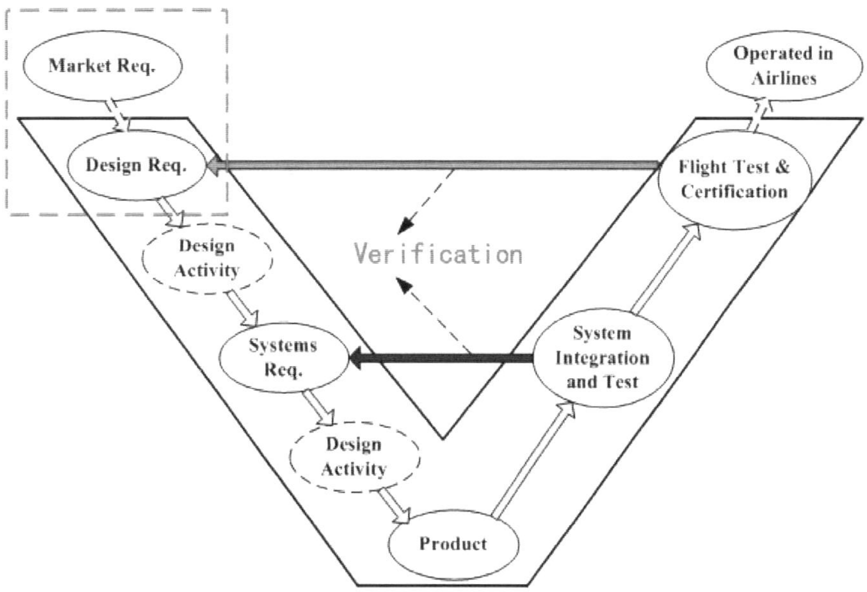


**Figure 1.** Requirements and Management Frame [1].

## 3. A method of civil jet requirements validation based on requirements project principle

The design principle based on requirements project has been applied successfully in many industries, such as auto industry. However, due to the development state of the civil jet industry and the limited cognitive of the principle in China, it is lately that the principle is introduced into the civil jet industry of China.

Comparing with the conventional design principle, the requirements project has its own advantages, for instance, the design is requirement-based, the design input and output is clear, and the design flow management can be easily controlled. Although the new principle is applied, the conventional design principles should not be abandoned completely. It is necessary to combine two principles effectively to achieve "1+1 ⩾ 2" effect.

Base on the study on the design principle put up by the foreign advanced civil jet manufacturers and the analysis of the design principle of China, three rules for building new design system are proposed as follows:

(1) The origin design elements which are the basis of product should not be abandoned;

(2) Increasing the design elements facing customer's requirements to direct and lead the design work;

(3) Defining the input and output of every design element to make management easier.

A new design method is drawn from the experience and summed up, which concentrates on top level design requirements validation phase. See figure 2 for details.

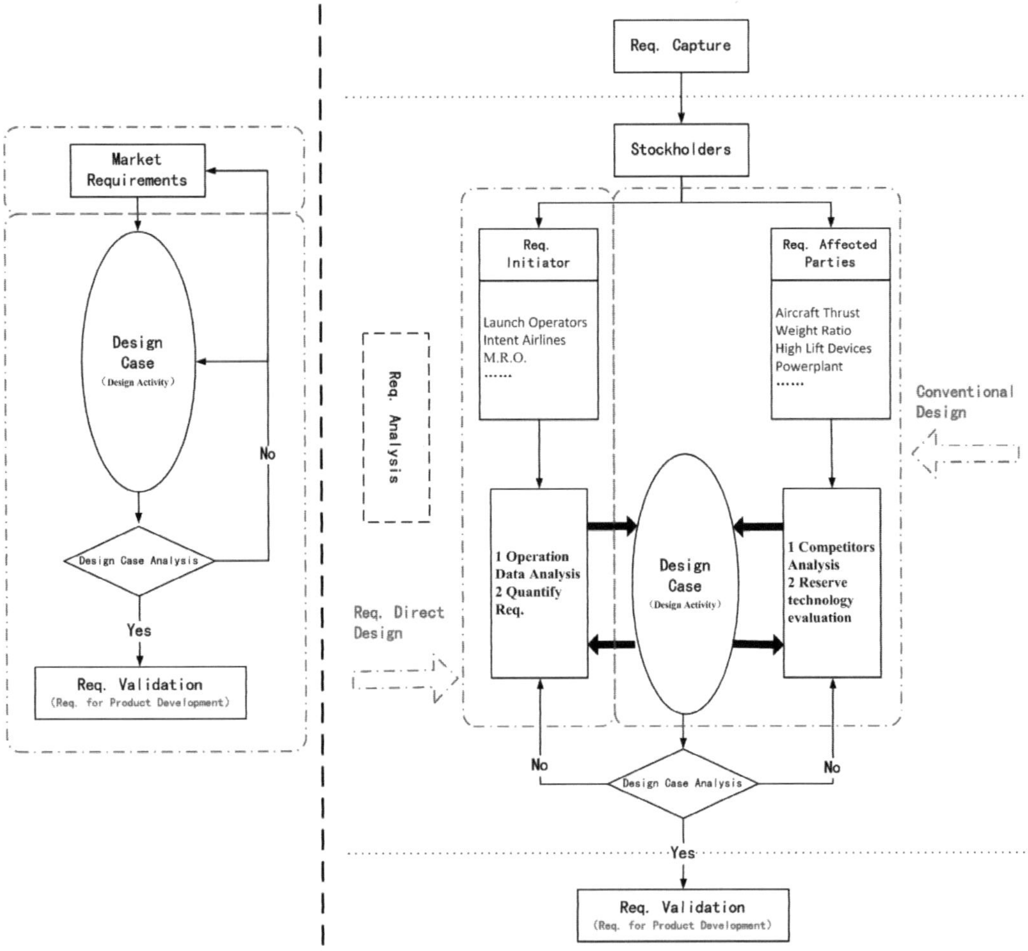


A) Conventional Design Method    B) Design Method Based on Req. Project Principle

**Figure 2.** Design Methods Frame.

Figure 2 shows the primary difference between the two design principles is that the design elements facing customers is now sinking into design level. On one hand the importance of market has not decreased, on the other hand the design elements have been completed, and the design work meets the requirements of market directly.

There are three steps in the whole requirements validation process, which are requirements capture, requirements analysis and requirements validation.

Requirements capture is the origin of product development, which directly facing customers, designers, manufacturers, customer service staff, and salesmen.etc. Because of the characteristics of market, more attentions are usually paid to the external characteristics and performances of the product, and the internal characteristics are usually ignored, such as product construction and work principle. So the requirements in the requirements capture phase are not the real requirements. These requirements are need to be confirmed. They just reflect a trend lack of completeness, correctness and feasibility. A key phase of product development, called "Requirements Analysis", is necessary to convert and transmit these requirements.

The requirements analysis is a bridge connecting the requirements origin with the product development. Therefore, the analysis to the "Requirements Initiator" is important, see figure 2. On one hand it could reflect the requirements project principle, on the other hand it is the key phase to translate these requirements in a proper way. Besides, in this phase, conventional design elements are the basis of the requirements analysis, the analysis and evaluation of competitors and technical reserve are also the basis of design case drawing-up.

The last step is "Requirements Validation". To connect "Requirements Initiator" with "Requirements Affected Party" effectively, an appropriate platform, the design case of the product, is needed. With the design case, all the design elements can be handled and leveraged on the same platform. The product requirements are defined, and the relationship between the requirements origin and product development is established in the phase.

## 4. Applications in product development

Requirements are the basis of the product development, and their quantities are huge. For easily understood, the requirement of "take off field length" would be introduced as an example.

### *4.1. Take off field length*

According to CCAR Part 25, take off field length is defined as follow:

*4.1.1. Take off distance [2].* For given operational conditions (temperature, pressure altitude, weight, etc.):

a) The takeoff distance on a dry runway is the greater of the following values:

$TOD_{N\text{-}1\ dry}$ = Distance covered from the brake release to a point at which the aircraft is at 35 feet above the takeoff surface, assuming the failure of the critical engine at $V_{EF}$ and recognized at $V_1$.

$1.15TOD_{N\ dry}$ = 115% of the distance covered from brake release to a point at which the aircraft is at 35 feet above the takeoff surface, assuming all engines operating.

b) The takeoff distance on a wet runway is the greater of the following values:

$TOD_{dry}$ = Takeoff distance on a dry runway (see above).

$TOD_{N\text{-}1\ wet}$ = Distance covered from brake release to a point at which the aircraft is at 15 feet above the takeoff surface, ensuring the $V_2$ speed to be achieved before the airplane is 35 feet above the takeoff surface, assuming failure of the critical engine at $V_{EF}$ and recognized at $V_1$.

*4.1.2. Accelerate-stop distance [2].* The accelerate-stop distance on a dry runway is the greater of the following values:

$ASD_{N\text{-}1\ dry}$ = Sum of the distances necessary to:

- Accelerate the airplane with all engines operating to $V_{EF}$
- Accelerate from $V_{EF}$ to $V_1$ assuming the critical engine fails at $V_{EF}$ and the pilot takes the first action to reject the takeoff at $V_1$
- Come to a full stop
- Plus a distance equivalent to 2 seconds at constant $V_1$ speed

$ASD_{N\ dry}$ = Sum of the distances necessary to:

- Accelerate the airplane with all engines operating to $V_1$, assuming the pilot takes the first action to reject the takeoff at $V_1$
- With all engines still operating come to a full stop
- Plus a distance equivalent to 2 seconds at constant $V_1$ speed

a) The accelerate-stop distance on a wet runway is the greater of the following values:

$ASD_{dry}$

$ASD_{N\text{-}1\ wet}$ = same definition as $ASD_{N\text{-}1\ dry}$ except the runway is wet

$ASD_{N\ wet}$ = same definition as $ASD_{N\ dry}$ except the runway is wet

*4.1.3. Balanced field [3-5].* For a given takeoff weight, any increase in $V_1$ leads to a reduction in both $TOD_{N\text{-}1}$. The reason is that all engines acceleration phase is longer with a higher $V_1$ speed, and, consequently, in case of an engine failure occurring at $V_{EF}$, the same $V_2$ speed can be achieved at 35 feet at a shorter distance.

As a result, the following graph providing the takeoff/rejected takeoff distances as a function of $V_1$ can be plotted. This graph clearly shows that a minimum distance is achieved at a particular $V_1$ speed. This speed is called "balanced $V_1$", and the corresponding distance is called "balanced field". See figure 3 for details.

In the design work, balance field is usually chosen for evaluation [3-5].

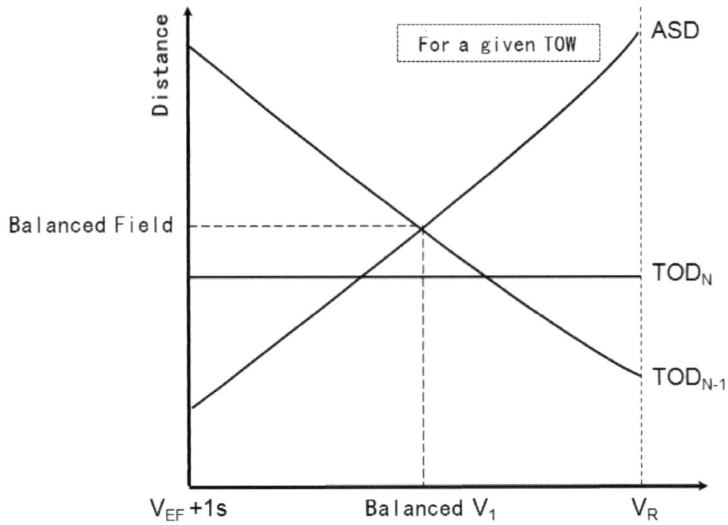


**Figure 3.** Balanced Field Definition [3-5].

*4.2. Requirements Capture*

a) Market investigation: intent airlines, intent leasing aircraft companies

b) Investigation background: the shorter take off field length is, the wider the market of airlines is.

c) Initial requirement: take off field length is no more than A m (MTOW, SL, ISA+15 °C).

Note: The initial requirement is a partly prepared requirement with potential problems, especially when it is analyzed with conventional methods, so more time and more resources are needed to refine it. See requirements analysis for details as follow.

*4.3. Requirements Analysis*

The conventional design method and the new method are both used to analyze the requirements, so it is easy to see the different processes and results between the two methods, and the advantages and benefits of the new method.

*4.3.1. Analysis with conventional method.* According to the conventional method, the initial requirement is directly transmitted to design team and is used in design. After analysis of design case, which reaches the requirement of take off field length, 5 problems are found out as follows:

a) The thrust weight ration of aircraft is more than competitors.

b) The take off field length is shorter than competitors.

c) The take off thrust is close to the maximum capability of engine for baseline, and the stretched aircraft is limited.

d) The thrust of engine is bigger than competitors, so its life becomes shorter and the maintenance cost is higher.

e) For decreasing the thrust, the bigger wing areas and more effective high lift device are applied.

All the problems are the critical for design case. If these problems could not be fixed appropriately, the competitiveness of the product would be affected.

The situation mentioned above is very typical with conventional method. These problems could be found out, and the primary design element causing the problems could be located, but there is no way to solve them effectively. In this condition, the design team can only choose to feed the problems back to the market team. But the market team could not understand the design case, so it is difficult for them to find a right way to solve them. Due to the lack of requirements analysis, large quantity of design time and resources are wasted.

*4.3.2. Analysis with the new method.* In the new method, when initial requirements have been confirmed, the next important step is to classify and confirm the requirements stakeholders. See table I for details.

**Table 1.** The Requirements Stakeholders of TOFL.

| The Requirements Stakeholders of TOFL | |
|---|---|
| Requirements Initiators (Primary) | Launch Operators<br>Intent Airlines<br>Potential Customers |
| Requirements affected parties (Primary) | Aircraft Thrust Weight Ratio<br>High Lift Devices Design<br>Powerplant<br>Stretched Aircraft Design |

After requirements stakeholders classification, the analysis work of requirements initiators and requirements affected parties start in parallel. The analysis work of requirements initiators plays an important role in the requirements analysis period, which is the core of the requirements project, so this chapter would focus on this part, which would help understanding the initial requirements correctly, analyzing the initial requirements in right way, translating the initial requirements into the real requirements for product development, and promoting the product development positively.

a) Analysis of requirements initiators

Main analysis idea is:

- According to the initial requirement of TOFL, the airports have been classified and collected, whose runway length is between A m to A+600 m.
- According to these airports, the airlines with these airports should be classified and collected.
- The airlines should be picked up, which are operated by customers and potential customers. And these airlines characteristics should be analyzed.
- Transmitting this analysis data to requirements affected parties for requirements validation.

After analysis work, the results are that:

- Among the intent customers, there is no airline operating at the airports whose runway length is between A m and A+400 m.
- There are only two potential airlines operating at the airports, but the range of these airlines is relatively short, about a half of design range of aircraft. So there is no limit to aircraft operating in these airlines, because the aircraft take-off weight is lighter than maximum take-off weight.
- The analysis result of requirements initiators is that take off field length is no more than A+400 m (MTOW, SL, ISA+15°C), and the result is transmitted to requirements affected parties.

After above analysis, we can see clearly that the initial requirement is usually not precise, only reflecting the trend of market requirement. It is not proper to be used in product development directly. We should evaluate the initial requirement objectively, dig the data behind the initial requirement deeply, trade off them, and quantify the result.

It is necessary to point out that the conventional design method is lack of analysis of requirements initiators, which induces unsmooth connection between market and design.

b) Analysis of requirements affected parties

The analysis of requirement affected parties is the same as the conventional design work. The primary work is competitor's analysis and reserve technology evaluation. After the analysis of requirements initiators, we could find that the new requirement of TOFL falls into a reasonable range, which is resulted from requirements affected parties. Therefore, design case development turns easier.

*4.4. Requirements Validation*

After the result of requirements analysis is brought into the preliminary design case, the problems which usually come into in conventional design method is no longer exist. The whole design case is balance and feasible. So the validated requirement of TOFL is no more than A+400 m (MTOW, SL, ISA+15 °C).

## 5. Conclusions

For meeting the needs of civil jet requirements validation in product development, a method of requirements validation is developed and defined, which is based on requirements project principle. And the following results have also been achieved:

a) The new method makes the connection more smoothly between market and design.

b) In the new method, the conventional design elements and requirements elements are reset and recombined effectively, which leads to maximum effectiveness.

c) This new method has been successfully applied in civil jet aircraft development of China.

## References

[1] SAE. ARP4754A, 2010 *Guidelines for Development of Civil Aircraft and Systems.*

[2] CAAC. CCAR PART 25, *AIRWORTHINESS STANDARDS: TRANSPORT CATEGORY AIRPLANES.*

[3] Chen B S, Li Y J, Wang Z Y 2005 *Aircraft Design Manual, volume 5, Civil Aircraft General Design.* Beijing: Aviation Industry Press.

[4] Egbert T 1976 *Synthesis of subsonic airplane design.* Netherlands: Delft University Press.

[5] Denis H 2000 *Aircraft Conceptual Design Synthesis.* Great Britain: Antony Rowe Limited, Chippenham, Wiltshire, UK.

IOP Conf. Series: Materials Science and Engineering **317** (2018) 012014 doi:10.1088/1757-899X/317/1/012014

# Inhibition of Photocatalytic Activity of Basic Blue-41 by ZnO Modified Surface with Amino Silane

**S Limsapapkasiphon[1], W Sirisaksoontorn[1] and A Songsasen[1]**

[1]Department of Chemistry, Faculty of Science, Kasetsart University, 50 Ngamwongwan Road, Lad Yao, Chatuchak, Bangkok 10900, Thailand

**Abstract.** The reduction of the photo catalytic efficiency of ZnO can be achieved by modifying its surface with amino silane, which synthesized through condensation reaction under basic condition. The pH of solution was varied from 8 to 14 during the synthesis and was found that pH 12 was the most suitable pH for the preparation. All of ZMAS were characterized by Elemental Analysis which showed the highest percentage of nitrogen at 3.1064% and IR technique which indicated the Si-O-Zn bond at about 1000 $cm^{-1}$. The photodegradation property of ZMAS prepared at pH 8-12 toward basic blue 41 was retarded when compared with the unmodified ZnO. Effect of mole ratio of ZnO:APTES (1:0.1, 1:0.5, 1:1, and 1:2) in the preparation of ZMAS was investigated. The photodegration activity of ZMAS prepared at mole ratio of ZnO:APTES as 1:0.5 to 1:2 toward basic blue 41 was retarded when compared with the unmodified ZnO. The coating of amino silane on ZnO surface did not have much effect on the band gap energy of modified ZnO. The absorption edge of ZMAS was only slightly shifted from 392 to 397 nm.

## 1. Introduction

The first synthesis dye has been discovered since 1856 by William Henry Perkin. At present more than 10,000 synthesis dyes are available in the market. The textile industry usually uses these dyes to create products. In order to use these dyes, the Textile factories need to improve the quality of the dyes to be durable for various environments such as: sunray, water, temperature etc. When these dyes are dumped into natural water sources, they will pollute and destroy the environment [1]. As a result, more awareness of the effects began to emerge, but the demand for dyes was not exhausted. The natural dyes are another way to reduce the problem because it is a green chemistry. Natural dyes are more environment friendly than synthetic dyes. However, the big problem of to use the natural dyes in the textile industry is the operation cost because it requires a lot of land to produce and they are not durable to the environments such as sunray, water, and temperature [2].

To retard the photodegradation of dyes by UV ray from the sunlight, many kind of UV absorber such as UVA-1, UVA-2, and UVA-3 were integrated into the dyes. In this research, we want to improve the light resistant property of dyes by using UV absorber and light stabilizer. Zinc Oxide can be used to absorb UV radiation with a wide band gap of 3.4 eV. In order to keep the UV absorber properties but prevent the enhancing photocatalytic degradation property, the surface of Zinc Oxide will be modified by amino silane [3]. Amino silane reacts easily on the surface of Zinc Oxide by condensation reaction and the Nitrogen on amino silane chain will trap radicals to prevent the photodegradation of dyes.

## 2. Experiment and characterization

### *2.1. Synthesis of modified surface ZnO (ZMAS)*

*2.1.1. Effect of pH on the synthesis of ZnO modified surface by amino silane.* The synthesis of ZnO modified surface by (3-Aminopropyl) triethoxysilane (APTES) was prepared following this process: 2 g of ZnO was stirred in 50 ml of KOH solution for 1 hour (pH was varied from 8-14). Then 2.878 ml (mole ratio of ZnO:APTES was fixed as 1:0.5) of APTES was added and stirred for 24 hours at 60 $^{o}$C on sand bath. The excess APTES was washed with distilled water. Finally, the white precipitate was dried under vacuum.

*2.1.2. Effect of mole ratio of ZnO:APTES on the synthesis of ZnO modified surface by amino silane.* The synthesized of ZnO modified surface by APTES (ZMAS) was performed as mentioned 2.1.1. but the pH was fixed at pH 12. The mole ratio of ZnO:APTES was varied as 1:0.1, 1:0.5, 1:1, and 1:2, respectively.

### *2.2. Characterization*

The functional group of APTES was analyzed by Fourier Transform Infrared Spectroscopy (FT-IR, Bruker EQUINOX-55). The quantity of APTES on surface of ZnO was recorded by Elemental Analysis (EA, LECO CHNS-932 and VTF-900). The UV absorption property of ZMAS was analyzed by UV-visible Diffuse Reflectance Spectroscopy (UV-DRS, JASCO V-670).

### *2.3. Photocatalytic study of ZnO modified surface by amino silane*

The photodegradation experiment was carried out under UV lamp. Initially, 0.01 g of the ZnO modified surface by APTES (ZMAS) was added into 100 ml of 10 ppm BB41 solution. The degradation reaction was operated under dark condition for half an hour and then operated under UV lamp. The system was first left in the dark in order to attain the absorption equilibrium before the degradation. In this condition, the solution was continually stirred under UV lamp. To monitor the reaction, the sample was collected by a syringe in order to measure the amount of initial (before the addition of the ZMAS) and remaining concentration of BB41 in each period of time. For avoidance of the ZMAS contamination in the collected sample, stirrer was switched off 10 minutes before collecting the sample. Then all of the collected samples were centrifuged for 5 minutes to reduce the dispersion of the fine powdered ZMAS. After that, the centrifugation was transferred into a test tube. Finally, their concentrations were measured by a JASCO 7800 UV-Vis spectrophotometer at wavelength 610 nm.

## 3. Result and discussion

### *3.1. Effect of pH on the synthesis ZnO modified surface by amino silane*

The hydrated metal oxides can be presented as MOH on the surface, when Zinc oxide was soaked in a basic condition. Then silanol group, which hydrolyzed by the nucleophilic attack of the oxygen contained in water on silicon atom according to the equation 1, will react with the OH group on the metal surface to form ZMAS. However, it was found that the concentration of hydroxide ion at pH 8 and pH 10 were not high enough to change all the surface of Zinc oxide into OH group before the coupling reaction with APTES [4]. Therefore pH 12 and pH 14 were more suitable for the preparation of ZMAS in this experiment. The mechanism of oxalation condensation is stated in equation 2 and 3 respectively.

$$\text{\textasciitilde Si-OR} + H_2O \rightarrow \text{\textasciitilde Si-OH} + ROH \quad (1)$$

$$\text{Zn-OH} + \text{OH-Si\textasciitilde} \rightarrow \text{Zn-OH-Si-OH} \quad (2)$$

$$\text{Zn-OH-Si-OH} \rightarrow \text{Zn-O-Si} + H_2O \quad (3)$$

The oxalation condensation occurs from lone pair electrons of oxygen, which act as nucleophile, attack to metal atom through the nucleophilic addition process. Therefore, the metals must have empty orbitals for obtain electrons from nucleophile. The Silicon atom has empty 3d orbitals. So, the hydroxo group of Zinc is attacking to Si atom followed by $H_2O$ elimination to form Zn-O-Si bond (as shown in equations 2 and 3 respectively).

### *3.1.1. Characterization*

**Table 1.** Percentage of CHN and IR spectra of ZMAS prepared at various pH when the mole ratio of ZnO:APTES was fixed at 1:0.5.

| | Elemental Analysis | | | Fourier Transform Infrared Spectroscopy | | |
|---|---|---|---|---|---|---|
| | Carbon (%) | Hydrogen (%) | Nitrogen (%) | N-H stretching | C-H stretching | Si-O-Zn stretching |
| **ZnO** | 0 | 0 | 0 | - | - | - |
| **ZMAS-pH 8** | 7.1848 | 2.4147 | 2.4271 | 3412, 3381 | 2924 | 987 |
| **ZMAS-pH 10** | 7.3647 | 2.3120 | 2.5366 | 3417, 3252 | 2925 | 995 |
| **ZMAS-pH 12** | 8.8993 | 2.4691 | 3.1064 | 3428, 3265 | 2925 | 1006 |
| **ZMAS-pH 14** | 0.9734 | 0.1542 | -0.1544 | - | - | - |

The modifying process of ZMAS (pH was varied from 8-12) was successful which confirmed by the percentage of nitrogen in the products as shown in table 1. The amount of nitrogen was related to the amount of silane group on the surface of ZnO. From IR spectra, the Si-O-Zn stretching vibration peak was found at about 1000 $cm^{-1}$ [5]. At pH 14, most of silanol group are deprotonated and react with itself through the condensation reaction as shown in equation 4 and 5;

$$\sim Si\text{-}OH + OH^- \rightarrow \sim Si\text{-}O^- + H_2O \quad (4)$$

$$\sim Si\text{-}O^- + HO\text{-}Si\sim \rightarrow \sim Si\text{-}O\text{-}Si\sim + H_2O \quad (5)$$

As shown in figure 1, the photodegration property of ZMAS prepared at pH 8-12 toward basic blue 41 was retarded when compared with the commercial ZnO semiconductor but not the ZMAS prepared at pH 14. This indicated that the amino silane can prohibit the catalytic efficiency of ZnO for photodegradation of dyes. There are two possible explanations for the reduction of catalytic efficiency of ZnO when modified by amino silane. Firstly, the silanol group on the surface of ZnO is react with nearby silanol group through condensation reaction to form the polymer of Si-O-Si bond (figure 2). The hydrocarbon chains on silane will prevent water from approaching the ZnO (hydrophobic effect), so the hydroxyl radical is not occurred from water splitting process. Secondly, the radical trapping, nitrogen on amino silane can trap radical, which generated by water splitting proceed. The mechanism is similar to HALS as shown in figure 3. In this work the preparation condition at pH 12 was preferred due to the highest percentage of nitrogen in the product.

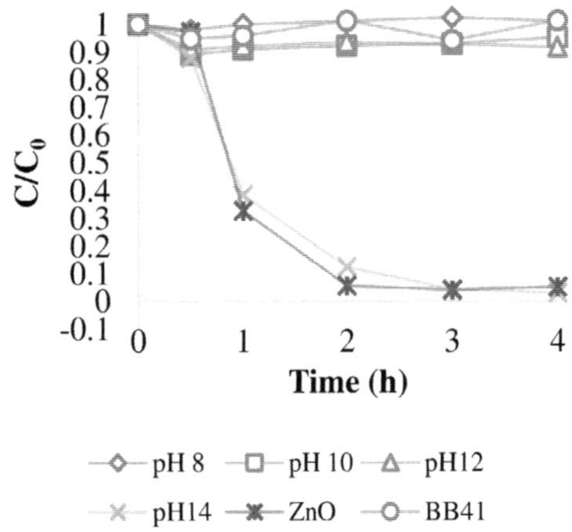


**Figure 1.** The photodegradation of BB41 under UV irradiation for 4 hours by ZMAS prepared at various pH.

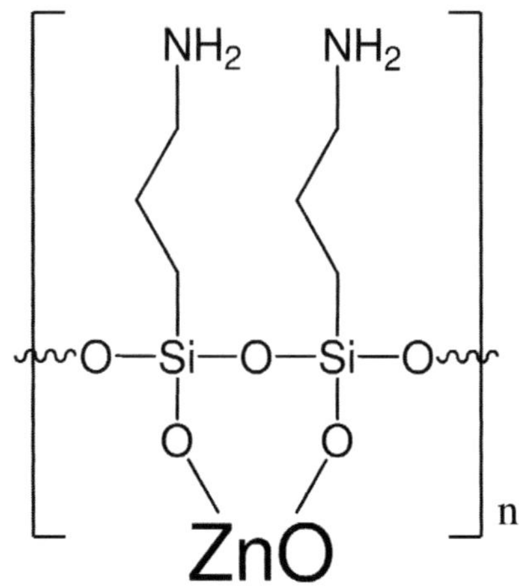


**Figure 2.** Condensation reaction to form the polymer of Si-O-Si bond at the surface of ZnO.

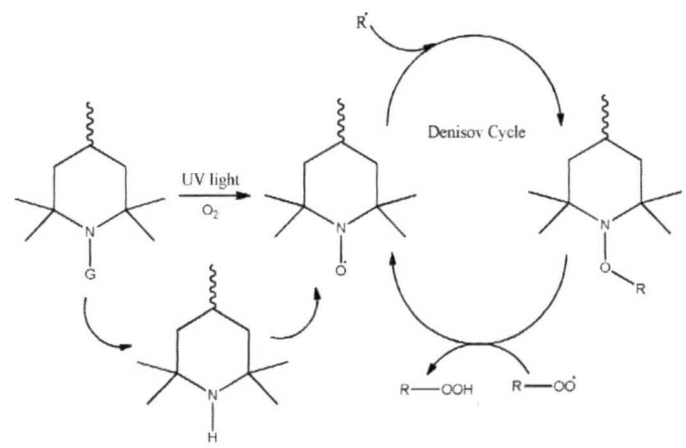


**Figure 3.** Possible mechanism of HALS activation and radical scavenging.

*3.2 Effect of ZnO:APTES mole ratio on the synthesis of ZMAS*

The synthesis of ZMAS when the mole ratio of ZnO:APTES were 1:0.1 and 1:0.5 was successful, which did not leave sticky solid on the surface of the reaction vessel, unlike the mole ratio of ZnO:APTES at 1:1 and 1:2. The sticky solid may occurred from the reaction of the excess APTES that can form Si-O-Si bond at the surface of the glass vessel.

*3.2.1. Characterization*

**Table 2.** Percentage of CHN and IR spectra of ZMAS prepared at various mole ratio of ZnO:APTES at pH 12

| Mole ratio of ZnO:APTES | Elemental Analysis | | | Fourier Transform Infrared Spectroscopy | | |
|---|---|---|---|---|---|---|
| | Carbon (%) | Hydrogen (%) | Nitrogen (%) | N-H stretching | C-H stretching | Si-O-Zn stretching |
| **1 : 0** | 0 | 0 | 0 | - | - | - |
| **1 : 0.1** | 1.1413 | 0.16855 | -004144 | - | - | 1019[a] |
| **1 : 0.5** | 8.8993 | 2.4691 | 3.1064 | 3428, 3265 | 2925 | 1006 |
| **1 : 1** | 9.4540 | 2.6487 | 3.5290 | 3268 | 2929 | 991 |
| **1 : 2** | 11.687 | 3.3014 | 4.3634 | 3308 | 2930 | 994 |

[a]very weak intensity.

The results from the elemental analysis as shown in table 2 show the percentage of nitrongen in the ZMAS was related to amount of silane which came from the increasing concentration of APTES. The IR spectra of All ZMAS products also showed the Si-O-Zn stretching at about 1000 $cm^{-1}$ and only when the mole ratio of ZnO:APTES equal was 1:0.1 the very weak intensity peak was observed. As shown in Figure 4, the photodegration property of ZMAS prepared at mole ratio of ZnO:APTES various from 1:0.5 to 1:2 with BB41 was retarded when compared with the commercial ZnO but not the ZMAS prepared at mole ratio of ZnO:APTES equal as 1:0.1. As shown in Figure 5, the coating of amino silane on ZnO surface did not have much effect on the band gap of ZnO. The absorption edge of ZMAS was only slightly shifted from 392 to 397 nm. Therefore, ZMAS still has UV absorber property.

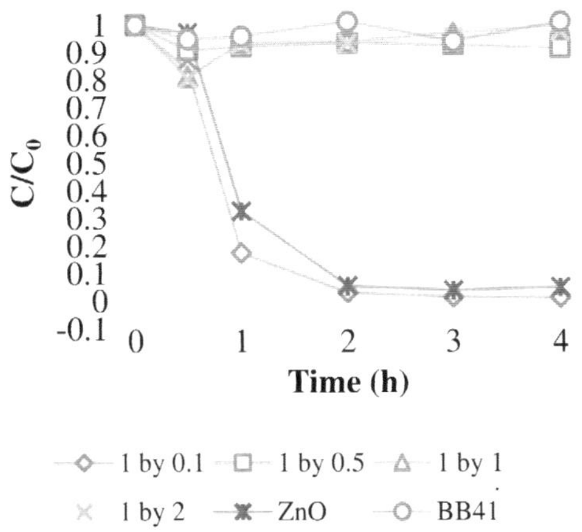


Figure 4. The photodegradation of BB41 under UV irradiation for 4 hours by ZMAS prepared at various mole ratio of ZnO:APTES.

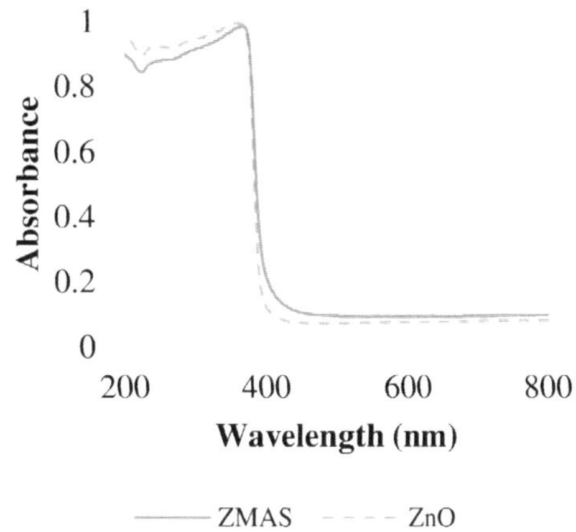


Figure 5. DRS- UV-visible spectra of the ZMAS and ZnO.

## 4. Conclusion

The suitable condition to prepare ZnO modified surface by amino silane (ZMAS) is at pH 12 with the mole ratio of ZnO:APTES as 1:0.5. The prepared ZMAS can still absorb UV light and can retard the photocatalytic efficiency of ZnO towards BB41. The prepared ZMAS may possible to be used as dye stabilizer for the textile industry.

## 5. References

[1] Insaf A, Yasmine S, Ines J, Francisco P and Wissem M 2016 *J. Dev. Drugs* **5**.
[2] Woo K K, Ngou F H,Ngo L S, Soong W K and Tang P Y 2011 *Am. J. Food Technol.* **6(2)** 140-8.
[3] Grasset F, Saito N, Li D, Park D, Sakaguchi I, Ohashi N, Haneda H, Roisnel T, Mornet S and Duguet E 2003 *Food Chem.* **360** 298-311.
[4] Kosmulski M 2009 *J. Colloid Interface Sci.* **337** 439-48.
[5] Farzi G A, Tayebee R and Naghibinasab S 2015 *Int. J. Nano Dimens.* **6(1)** 67-75.

## Acknowledgment

We would like to thank the Department of Chemistry, Faculty of Science, Kasetsart University and the SAST for all research facilities. We also thank the Graduate school, Kasetsart University and Faculty of Science, Kasetsart University for supporting the expense to present this work in ICEAT 2017.

IOP Conf. Series: Materials Science and Engineering **317** (2018) 012015 doi:10.1088/1757-899X/317/1/012015

# Optimization of Thermal Object Nonlinear Control Systems by Energy Efficiency Criterion.

**Vladimir A. Velichkin[1], Vladimir A. Zavyalov[2]**

[1]Cand. Tech. Sciences, National Research Moscow State University of Civil Engineering, Russia
[2]Dr. Tech. Sciences, prof., National Research Moscow State University of Civil Engineering, Russia

**Abstract.** This article presents the results of thermal object functioning control analysis (heat exchanger, dryer, heat treatment chamber, etc.). The results were used to determine a mathematical model of the generalized thermal control object. The appropriate optimality criterion was chosen to make the control more energy-efficient. The mathematical programming task was formulated based on the chosen optimality criterion, control object mathematical model and technological constraints. The "maximum energy efficiency" criterion helped avoid solving a system of nonlinear differential equations and solve the formulated problem of mathematical programming in an analytical way. It should be noted that in the case under review the search for optimal control and optimal trajectory reduces to solving an algebraic system of equations. In addition, it is shown that the optimal trajectory does not depend on the dynamic characteristics of the control object.

## 1. Introduction

Heat treatment is the most energy-intensive stage in the production of construction materials and products [1, 2]. A process should be automated so we can control it effectively. The quality of automatic control systems (ACS) applied will determine to a wide extent how successful automation will be. In turn, the quality of ACS is determined by the assessment of its quality, i.e. by the optimality criterion adopted at the design stage.

Today, there are many optimality criteria developed through practice or based on the "minimal action" principles of theoretical mechanics. In this regard, linear and quadratic optimality criteria are distinguished [2-5]. The application of linear criteria requires a numerical solution of the optimal control problem. Using integral quadratic criteria allows us to design regulators in an analytical way [4].

The authors have developed an integral quadratic energy efficiency criterion, the mathematical expression of which includes controllable and control variables with coefficients that are parameters of the control object mathematical model. This approach helps determine the control device parameters, optimal control of the optimal control system and motion trajectory in an analytical way.

## 2. Research methods

When designing an energy-efficient control system, the following Modern Control Theory methods were applied [3].

The optimal control problem was solved with respect to thermal control objects.

In our case, the optimal control problem can be formulated as a mathematical programming problem

$\max\ J = \int_0^T \{[AX + B(x)U]' \cdot [AX + B(x)U]\} d\tau$ with constraints $\frac{dX}{d\tau} = AX + B(x)U$;

$$X(0) = 0; X(T) = X_T. \tag{1}$$

The auxiliary functional takes the following form

$$H = \frac{[AX + B(x)U]' \cdot [AX + B(x)U]}{2} + \Lambda \cdot [AX + B(x)U]. \tag{2}$$

The optimal control and trajectory can be calculated from the following Euler-Lagrange differential equations

$$\frac{\partial H}{\partial U} = 0 = \frac{\partial [AX + B(x)U]'}{\partial U} \cdot [AX + B(x)U] + \frac{\partial [AX + B(x)U]'}{\partial U} \cdot \Lambda; \tag{3}$$

$$-\frac{\partial H}{\partial X} = \frac{d\Lambda}{d\tau} = -\frac{\partial [AX + B(x)U]'}{\partial X} \cdot [AX + B(x)U] + \frac{\partial [AX + B(x)U]'}{\partial X} \cdot \Lambda; \tag{4}$$

$$\frac{\partial H}{\partial \Lambda} = \frac{dX}{d\tau} = AX + B(x)U. \tag{5}$$

After certain algebraic manipulations, the system of equations (3)–(5) takes the following form

$$B'(x) \cdot [AX + B(x)U + \Lambda] = 0; \tag{6}$$

$$\frac{d\Lambda}{d\tau} = -\frac{\partial [AX + B(x)U]'}{\partial X} \cdot [AX + B(x)U + \Lambda]; \tag{7}$$

$$\frac{dX}{d\tau} = AX + B(x)U \tag{8}$$

According to equations (6) and (7) it follows that $\frac{d\Lambda}{d\tau} = 0$, since

$$[AX + B(x)U + \Lambda] = 0.$$

Then it can be argued that $\Lambda = const$.

According to equations (8) and (9), we can obtain the dependence of the speed of the object (1) moving along the optimal trajectory on the value $\Lambda$ $\frac{dX}{d\tau} = -\Lambda$.

Hence, we can conclude that the object (1) should move along the optimal trajectory at a constant speed $\frac{dX}{d\tau} = const$. Since the initial $X(0)$ and final $X(T)$ states of the control object are given, the optimal speed can be calculated from the formula

$$\frac{dX}{d\tau} = \frac{[X_T - X_0]}{T} \tag{9}$$

and the optimal trajectory $\rightarrow X^0(\tau) = X_0 + \frac{\tau \cdot [X_T - X_0]}{T}$.

The optimal control can be determined from equations (8) and (9)

$$U^0(X) = [B(x)]^{[-1]} \cdot \left\{ \frac{[X_T - X_0]}{T} - AX \right\} \tag{10}$$

or

$$U(\tau) = [B(x)]^{[-1]} \cdot \left\{ [I - A \cdot \tau] \cdot \frac{[X_T - X_0]}{T} - AX_0 \right\}$$

The results of analytical work can be illustrated by a practical example. Let the control object is a dryer, in which the drying process is controlled by the temperature of the drying space environment. As is known, evaporation happens when heat is supplied to material in the drying process. To maintain a dryer's internal environment temperature at the desired level, the common practice is to utilize industrial regulators in order to implement positional, proportional and proportional-integral regulation laws. The above laws are intended only to ensure the specified accuracy of regulation. The efficiency of the use of thermal energy in the calculation of systems utilizing such regulators is not taken into account. To compare the energy efficiency of a traditional system with the optimal one by criterion (2), it is necessary to determine the mathematical model of the dryer. The calculated model of the dryer and the structural diagram of heat flows can be represented as in figure 1.

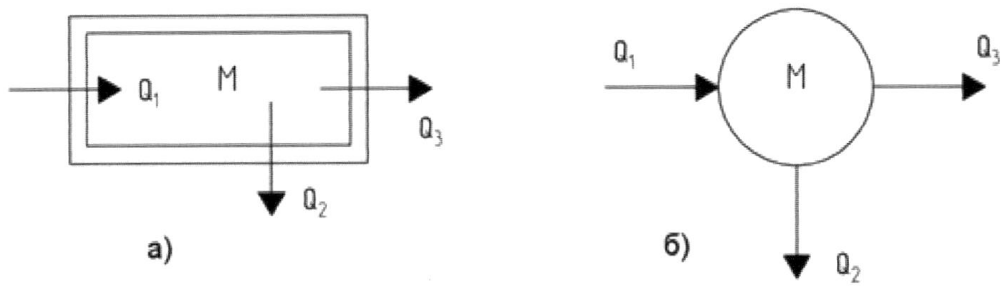


**Figure 1.** a) calculated model of the dryer; b) structural diagram of heat flows.

The following notations are used here: $Q_1 = cGt_1$ - heat flow supplied by the heat-transfer agent, $Q_2 = cGt_1$ - heat flow leaving with the spent heat-transfer agent, $Q_3 = FK(t_1 - t_{oc})$; $M = cmt_1$ - heat accumulation in the dryer's internal environment; $c$- heat-transfer agent specific heat capacity; $G$- heat-transfer agent flow rate; $t$- coolant temperature; $t_1$- temperature of the dryer's internal environment; F - dryer inner surface area; $m$ = dryer's internal environment mass; $K$ - coefficient of heat transfer from the dryer's internal environment to the external environment; $t_{oc}$- external environment temperature.

According to the structural diagram (figure 1(b)), the heat balance can be represented by a differential equation

$$\frac{dM}{d\tau} = Q_1 - Q_2 - Q_3$$

or

$$cm \cdot \frac{dt_1}{d\tau} = cGt - cGt_1 - FKt_1 - Fkt_{oc} \tag{11}$$

If the constant temperature ($t_{oc}$) of the environment is taken as a reference point, then equation (11) will be as follows (at the constant temperature ($t$) of the heat-transfer agent):

$$\frac{dx}{d\tau} = ax + (ex + f)u, \tag{12}$$

where: $x = t_1 - t_{oc}$ ; $a = -\frac{FK}{cm}$; $u = G$; $e = -\frac{1}{m}$; $f = \frac{t}{m}$.

If we put $(ex + f) = b(x)$, then equation (15) takes the following form

$\frac{dx}{d\tau} = ax + b(x)u$, which is analogous to equation (1).

In this example, the criterion (2) is represented by the functional

$$J = \int_0^T [ax + b(x)u]^2 d\tau \tag{13}$$

Then the optimal equation problem can be formulated as follows

$$\min J = \int_0^T [ax + b(x)u]^2 d\tau \tag{14}$$

with the following constraints: $\frac{dx}{d\tau} = ax + b(x)u$; $x(0) = x_0$; $x(T) = x_T$.

The auxiliary functional is determined by the expression

$$H = \frac{[ax + b(x)u]^2}{2} + [ax + b(x)u]\Lambda. \tag{15}$$

The problem (17) can be tackled by Pontryagin's method

$$\frac{\partial H}{\partial u} = 0 = b(x)[ax + b(x)u]; \tag{16}$$

$$\frac{\partial H}{\partial x} = -\frac{d\Lambda}{d\tau} = \left\{\frac{\partial[ax + b(x)u]}{\partial x}\right\} \cdot [ax + b(x)u + a\Lambda] \tag{17}$$

$$\frac{\partial H}{\partial \Lambda} = \frac{\partial x}{\partial \tau} = ax + b(x)u. \tag{18}$$

After certain algebraic manipulations, the equation (16) takes the following form

$$b(x) \cdot [ax + b(x)u + \Lambda] = 0. \tag{19}$$

Since $b(x) \neq 0$, equations (17) and (18) can be written down as follows

$$[ax + b(x)u] + \Lambda = 0; \tag{20}$$

$$[ax + b(x)u] - \frac{dx}{d\tau} = 0. \tag{21}$$

According to equations (20) and (21) it follows that $\frac{dx}{d\tau} = -\Lambda$.

The equation (17) can be replaced by the expression

$$-\frac{d\Lambda}{d\tau} = \left\{\frac{\partial[ax + b(x)u]}{\partial x}\right\} \cdot 0. \tag{22}$$

We can conclude from the equation (22), that

$$\frac{d\Lambda}{d\tau}=0. \tag{23}$$

Then the conjugate variable vector has constant coordinates

$$\Lambda = const \tag{24}$$

and, consequently, the vector

$$\frac{dx}{d\tau}=const. \tag{25}$$

Based on the obtained results (24) and (25), as well as constraints of the original problem (14), it is possible to calculate the analytic expressions of the optimal trajectory

$$\left(\frac{dx}{d\tau}\right)^0=\frac{\left[x_T-x_0\right]}{T}; \tag{26}$$

$$x^0(\tau)=x_0+\frac{\tau\left[x_T-x_0\right]}{T}; \tag{27}$$

and control

$$u^0(x)=\frac{\left[x_T-x_0\right]}{\left[T\cdot b(x)\right]}-\frac{a}{b(x)}x; \tag{28}$$

$$u^0(\tau)=\frac{\left[\left(x_T-x_0\right)\left(\frac{1-a\tau}{T}-ax_0\right)\right]}{\left[e\left(x_0+\frac{\tau\left[x_T-x_0\right]}{T}\right)+f\right]}. \tag{29}$$

Let $x_0=0$ ; $x_T=1$; $a=-1$; $b(x)=(ex+f)$; $e=-1$; $f=1$; $T=4$.

Then the behavior of the optimal system in the transition from the state $x_0=0$ to $x_T=1$ can be illustrated by the following diagrams (see figure 2) obtained on the basis of the following expressions:

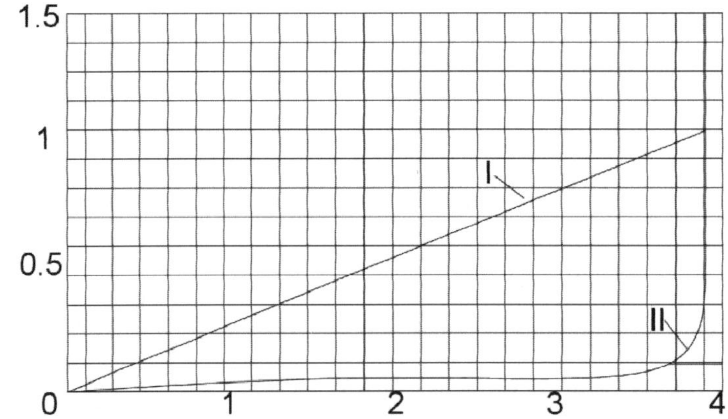


**Figure 2.** Optimal trajectory (I) and control (II) diagrams.

$$x^0(\tau)=\frac{\tau}{4}; \tag{30}$$

$$u^0(\tau) = \frac{(1+\tau)}{(4-\tau)}. \quad (31)$$

In the example above, the most unfavorable initial data were chosen, where at $\tau \to 4$ and $u(4) \to \infty$.

Therefore, the target $x_T$ cannot be achieved within a given time period, because of an infinitely big control. Hence, the problem (14) is solved taking into account the constraints on control $\lfloor u^0(\tau) \le u_{\max} \rfloor$. Thus, the problem (14) is solved in two steps: first, we should solve the problem analytically without taking into account the constraints on control, and then change the duration of the control interval or boundary conditions to achieve the required accuracy of the value $x_T$.

In practice, the probability of such ratios is very small, since the value $f >> e$, i.e. $f/e = t$. Generally, the temperature of heat-transfer agents used in the drying of construction materials exceeds 100°C.

## 3. Conclusion

The design methodology developed simplifies the procedure for designing energy-efficient one-dimensional and multi-dimensional systems, because there is no need to solve systems of non-linear differential equations.

Since the mathematical expression of the proposed energy efficiency criterion includes controllable and control variables with coefficients that are parameters of the control object mathematical model, the problem of synthesizing the optimal automatic control system reduces to identifying the control object.

It should be noted that the optimal trajectory does not depend on the parameters of the mathematical model of the control object.

The practicality of the proposed method is proven by the above example of solving the optimization problem.

## Reference

[1] Marsov V I, Slavutskiy V A 2015 *Automatic control of technological processes at construction material sites* Stroyizdat, pp. 288.
[2] Zavyalov V A, Pushkaryov S M 2002 *Energ.* **6**.
[3] Chaki F 1975 *Modern management theory* Mir, pp. 424.
[4] Egorov S V 2014 *Elements of identification and optimization of controlled systems.* MEI, pp. 224.
[5] Zavyalov V A, Krylov E N 2008 *Mech. Constr.* **3** 28-31.

# Ultra-low Temperature Curable Conductive Silver Adhesive with different Resin Matrix

**Xingli Zhou[1], Likun Wang[1], Qingwei Liao[1]*, Chao Yan[1, 2], Xing Li[1] , Lei Qin[1]**

[1]Research Center of Sensor Technology, Beijing Information Science & Technology University, Beijing 100192, China
[2]Tianjin Key Laboratory of Optoelectronic Detection Technology and System, School of Electronics and Information Engineering Tianjin Polytechnic University, Tianjin 300387, China

**Abstract.** The ultra-low temperature curable conductive silver adhesive with curing temperature less than 100 °C needed urgently for the surface conductive treatment of piezoelectric composite material due to the low thermal resistance of composite material and low adhesion strength of adhesive. An ultra-low temperature curable conductive adhesive with high adhesion strength was obtained for the applications of piezoelectric composite material. The microstructure, conductive properties and adhesive properties with different resin matrix were investigated. The conductive adhesive with AG-80 as the resin matrix has the shorter curing time (20min), lower curing temperature (90°C) and higher adhesion strength (7.6MPa). The resistivity of AG-80 sample has the lower value (2.13 × 10-4Ω·cm) than the 618 sample (4.44× 10-4Ω·cm).

## 1. Introduction

The surface conductive treatment is important to the dielectrics when they used in the devices. Piezoelectric composite materials are widely used in transducers for acoustics, medical and other fields [1-4]. During the preparation of the transducer, the adhesion strength of the electrode layers for the piezoelectric composite transducers is a difficult research point due to the polymer components in the composite material cannot bear high temperature, requiring low curing temperature and short curing time conductive adhesive which also shows strong conductive ability and adhesion strength. Conductive silver adhesive had been widely used in the field of IT and materialsl [5-12]. In recent decades, with the rapid development of semiconductor devices, conductive silver adhesive developed rapidly. M Bouguettaya [10] reported the Intrinsic Conducting Polymer (ICP) used to prepare conductive adhesives, providing excellent electrical and thermal properties for carbon black and polyaniline. X S Li [12] prepared silver nanowires as conductive adhesive for the conductive filler by alcohothermal method, and it has good electrical and mechanical properties. However, currently reported conductive silver adhesive showing the curing temperature of 150 ~ 300 °C [13-20]. It is not low enough for the piezoelectric composite material. In this paper, to meet the application of piezoelectric composite materials, the ultra-low temperature (90°C) fast curing (30min) with high adhesion strength conductive silver adhesive were studied. 618 and AG-80 were chosen to be the resin matrix.

## 2. Experimental section

The resin matrix solution is firstly prepared by Epoxy Resin (AG-80), curing agent Ethylenediamine (EDA), Dibutyl Phthalate (DBP, C16H22O4, Analytical Pure), Carboxylated- terminated liquid acrylonitrile rubber (CTBN, Analytical Pure), Silane Coupling Agent (SCA-403, Analytical Pure) and accelerator Diethylene Glycol Gutyl Ether (DGBE, Analytical Pure) in proportion of 100:160:10:10:10:10. Then add acetone (Analytical Pure), prepared resin matrix solution, and nano-silver (100 nm flaky pellets) in proportion of 10:30:60 to obtain the final ultra-low temperature curable conductive silver adhesive. The powder X-ray diffraction analysis was carried out at the 4B9A beamline at the Beijing synchrotron radiation facility (BSRF) with an X-ray wave length of 0.15406nm. The surface morphological was observed using a Scanning Electron Microscope (SEM: EVO-18, ZEISS, Germany). The component analysis was carried out by Energy Dispersive Spectrometer (EDS: EVO-18, ZEISS, Germany). The thermal analysis was carried out by thermogravimetric analysis (SDT Q600, TA, USA) and differential scanning calorimetry (DSC: SDT Q600, TA, USA).

## 3. Results and discussions

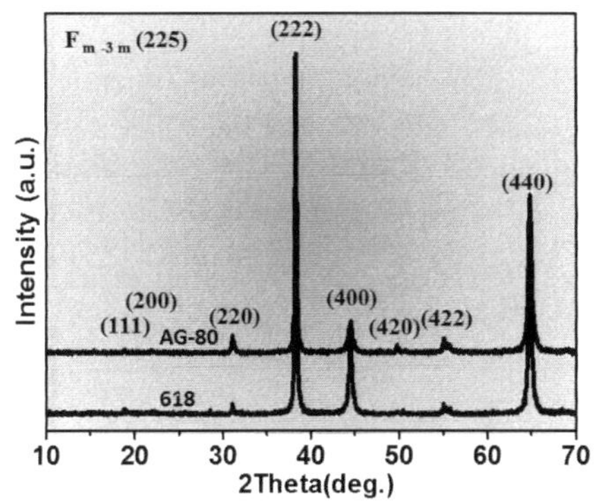


**Figure 1.** The synchrotron radiation X-Ray Powder Diffraction patterns of cured ultra-low curable conductive silver adhesive with resin matrix of 618 and AG-80, respectively.

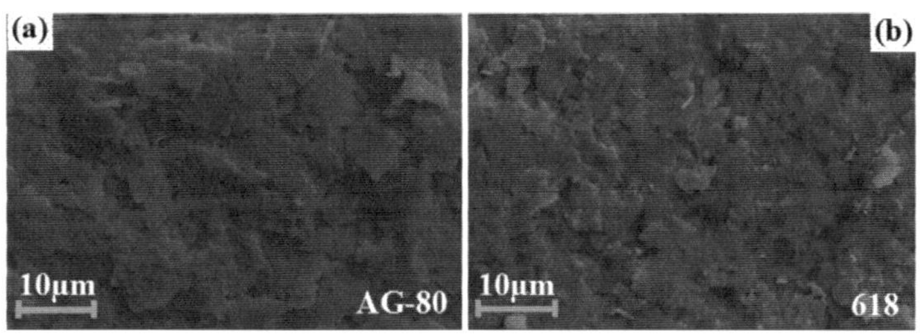


**Figure 2.** The SEM image of cured ultra-low temperature curable conductive adhesive with resin matrix of AG-80 and 618, respectively.

Figure 1 shows the synchrotron radiation X-Ray Powder Diffraction patterns of cured ultra-low curable conductive silver adhesive with resin matrix of 618 and AG-80, respectively. Both two samples show the single Ag phase with space group of Fm -3 m (225) (PDF No. 42-0479). Figure 2 shows the SEM image of cured ultra-low temperature curable conductive adhesive with resin matrix of AG-80 and 618, respectively. It can be seen that after the curing, the conductive surface shows good sintered representation. The whole surface of sample formed a good conductive path, in other words, the silver flake grain arranged neatly, and connected tightly. The sample with 618 as resin matrix has relative neater arrangement. As the characteristics of the large specific surface area, high surface free energy, high activity, various internal crystal defects in nano-silver particles, the condensation reaction of organic, and the thermal movement of silver particles made adjacent nano-silver particles move and gather during heat-curing. The conductive adhesive formed larger structure by crystal boundary migration and edge blending of crystal particles.

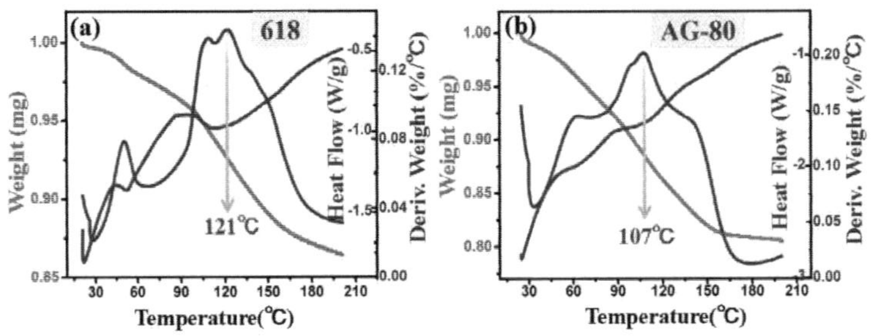


**Figure 3.** The thermal analysis results of the conductive silver adhesive with resin matrix of 618 and AG-80, respectively.

Figure 3 shows the thermal analysis (TGA, DTG, DSC) results of the conductive silver adhesive with resin matrix of 618 and AG-80, respectively. Cured products of AG-80 shows weight loss significantly in the 70 °C to 150 °C, and gradually slowed down after 150 °C until the weight loss rate reached 17%, and then the curing reaction tends to complete. Cured products of 618 shows weight loss significantly in the range of 100 °C to 175 °C. Until the weight loss rate reached 13%, the curing tends to complete. This is due to the epoxy groups of AG-80 is more than 618, and the activity is much higher than 618. The reaction releases a lot of heat, which can make sure fast curing at low-temperature. The weight loss rate of cured product of AG-80 and 618 reached the maximum at 107 °C and 121°C, respectively. The initial increase of heat release indicates the beginning of crosslinking of the resin. With the increase of the heating rate, the exothermic amplitude is also increasing. The initial decomposition temperature of cured product of AG-80 and 618 are 50 °C and 70°C. In addition, it can be seen that thermal stability of 618 epoxy resin is poor.

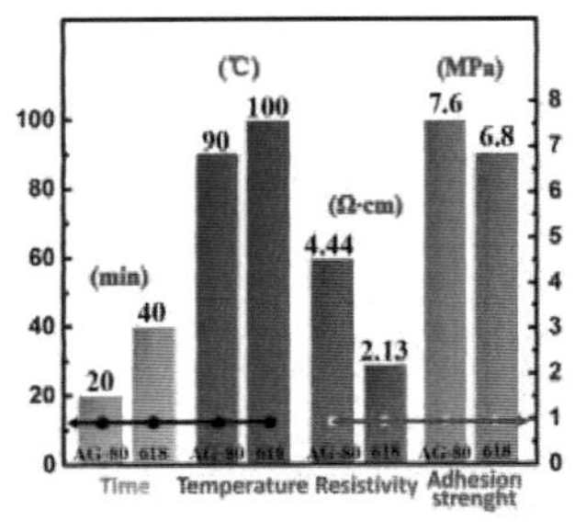


**Figure 4.** The process parameters, conductive property and adhesion property of the conductive adhesive with resin matrix of 618 and AG-80, respectively.

Figure 4 shows the process parameters, conductive property, and adhesion strength of cured ultra-low curable conductive silver adhesive with resin matrix of 618 and AG-80, respectively. It can be found that the conductive adhesive with AG-80 as the resin matrix has the shorter curing time, lower curing temperature and higher adhesion strength. As described above, it is due to the difference of epoxy group between 618 and AG-80. The resistivity testing following "ASTM D257-2007" standard. It can be seen from figure 5 that the resistivity of 618 sample has the lower value 2.13 × 10-4Ω cm than the AG-80 sample (4.44× 10-4Ω cm). It may because the better micromorphology of 618.

## 4. Conclusions

In summary, an ultra-low temperature curable conductive adhesive with different resin matrix (AG-80 and 618) were investigated for surface conductive treatment of piezoelectric composite material. Both two samples show the single Ag phase with space group of $F_{m\ -3\ m\ (225)}$ (PDF No. 42-0479), and the surface with conductive treatment formed a good conductive path in all direction. The conductive adhesive with AG-80 as the resin matrix has the shorter curing time, lower curing temperature and higher adhesion strength. The resistivity of 618 sample has the lower value 2.13 × 10-4Ω cm than the AG-80 sample (4.44× 10-4Ω cm).

## References

[1] Jang C, Sharifi M, Palmese, G R, Abrams C F 2014 *Polymer*, **55(16)** 3859-68.

[2] Lan T, Kaviratna P D, Pinnavaia T J 1995 *Chem. Mater.* **7(11)** 2144-50.
[3] Sang J L, Sohn H, Hong J W 2010 *J. Nondestruct. Evaluat.* **29(2)** 75-91.
[4] Sato S, Yamada K, Inagaki N 2006 *Med. Biol. Eng. Comp.*, **44(5)** 353-62.
[5] Pettersen S R, Redford K, Njagi J, Kristiansen H, Helland S, Kalland E, Goia D, Zhang Z, He J 2017 J. Electr. Mater. 1-11.
[6] Rane S B, Seth T, Phatak G J, Amalnerkar D P, Das B K 2003 Mater. Lett. **57(20)** 3096-100.
[7] Zhao S Y, Li X, Mei Y H, Lu G Q 2016 J. Electr. Mater. **45(11)** 1-11.
[8] Peng P, Hu A, Gerlich A P, Zou G, Liu L, Zhou Y N 2015 *Acs. Appl. Mater. Interf.* **7(23)** 12597.
[9] Kwon J, Cho, H, Eom H, Lee H, Suh Y D, Moon H, Shin J, Hong S, Ko S H 2016 *Acs. Appl. Mater. Interf.* **8(18)** 11575.
[10] Bouguettaya M, Vedie N, Chevrot C 1999 *Synth. Met.* **102(1-3)** 1428-31.
[11] Marcq F, Demont P, Monfraix P, Peigney A, Laurent C, Falat T, Courtade F, Jamin T 2011 *Microelectron. Reliabil.* **7** 1230-4.
[12] Li X S, Xiang X Z, Wang L, Bai X J 2016 *Rare Met.* 1-5.
[13] Morris J E 2017 *Polymers*, **3(1)** 191-7.
[14] Nakayama K, Nagai A, Iida N 2007 *Mater. Trans.* **48(3)** 594-9.
[15] He S, Zhang X, Yang B, Xu X, Chen H, Zhou C 2017 *Chinese Phys. B,* **26(7)** 426-32.
[16] Park S D, Yoo M J, Kang N K, Park J C, Lim J K, Kim D K 2004 *Macromol. Res.* **12(4)** 391-8.
[17] Lu C A, Lin P, Lin H C, Wang S F 2006 *Japanese J. Appl. Phys.* **45(9A)** 6987-92.
[18] Bai J G, Zhang Z Z, Calata J N, Lu G Q 2006 *IEEE Trans. Comp. Pack. Tech.* **29(3)** 589-93.
[19] Gural P S, Jenniskens P 2007 *Japanese J. Appl. Phys.* **46(1)** 251-5.
[20] Heckmann T, Souvignet T, Naccache D 2017 *Dig. Investigat.* **21** 53-64.

**Acknowledgements**
This work was supported by the National Natural Science Foundation of China (No. 11604363, 61471047, 61671068, 61204008, 11505211), and the Beijing College Innovation Capability Promotion Plan of Beijing Municipal Institutions (No. TJSHG201510772015). The project is also supported by The Importation and Development of High-Caliber Talents Project of Beijing Municipal Institutions (No. CIT&TCD201504053).

# Effect of Diluent on Ultra-low Temperature Curable Conductive Silver Adhesive

**Xingli Zhou[1], Likun Wang[1], Qingwei Liao[1]*, Chao Yan[1], Haibo Du[1], Lei Qin**

[1]Research Center of Sensor Technology, Beijing Information Science & Technology University, Beijing 100192, China

Abstract. The ultra-low temperature curable conductive silver adhesive needed urgently for the surface conductive treatment of piezoelectric composite material. The effect of diluent acetone on ultra-low temperature curable conductive silver adhesive were investigated for surface conductive treatment of piezoelectric composite material. In order to improve the operability and extend the life of the conductive adhesive, the diluent was added to dissolve and disperse conductive adhesive. With the increase of the content of diluent, the volume resistivity of conductive adhesive decreased at first and then increased, and the shear strength increased at first and then decreased. When the acetone content is 10%, the silver flaky bonded together, arranged the neatest, the smallest gap, the most closely connected, the surface can form a complete conductive network, and the volume resistivity is 2.37 × 10-4Ω · cm, the shear strength is 5.13MPa.

## 1. Introduction

Piezoelectric composite materials are widely used in transducers for acoustics, medical and many other fields [1-3]. The adhesion strength of the electrode layers for the piezoelectric composite transducers is a difficult research point due to the polymer components in the composite material cannot bear high temperature during the preparation of the transducer, requiring low curing temperature and short curing time conductive adhesive which also shows strong conductive ability and adhesion strength. Conductive silver adhesive had been widely used and rapid development in the field of IT and materials [4-10]. X. S. Li [10] prepared silver nanowires as conductive adhesive for the conductive filler by alcohothermal method, and it has good electrical and mechanical properties. However, currently reported conductive silver adhesive showing the curing temperature of 150 ~ 300 °C [11-17], which is a too much higher than the temperature suitable for the surface conductive treatment of piezoelectric composite material. In this paper, we prepared ultra-low temperature fast curing with high adhesion strength conductive silver adhesive, and the effect of diluent on the properties were investigated.

## 2. Experimental section

The resin matrix solution is firstly prepared by Epoxy Resin (AG-80), curing agent Ethylenediamine (EDA), Dibutyl Phthalate (DBP, C16H22O4, Analytical Pure), Carboxylated- terminated liquid acrylonitrile rubber (CTBN, Analytical Pure), Silane (Analytical Pure), Silane Coupling Agent (SCA-403, Analytical Pure) and accelerator Diethylene Glycol Gutyl Ether (DGBE, Analytical Pure) in proportion of 100:160:10:10:10:10. Then add acetone (Analytical Pure), prepared resin matrix solution, and nano-silver (100 nm flaky pellets) in proportion of (5~18):30:60 to obtain the final ultra-low temperature curable conductive silver adhesive. The surface morphological was observed using a Scanning Electron Microscope (SEM: EVO-18, ZEISS, Germany).

IOP Conf. Series: Materials Science and Engineering **317** (2018) 012017 doi:10.1088/1757-899X/317/1/012017

## 3. Results and discussions

Figure 1 shows the diagram of the key preparing process to prepare the ultra-low curing temperature conductive silver adhesive. The ultra-low temperature curable conductive silver adhesive is prepared by two-step method. The first step is to prepare resin matrix solution and the second step is to prepare final conductive silver adhesive. The epoxy resin, curing agent, and additives were mixed in proportion of 100:150:10. And then stir the mixture was at room temperature to obtain the corresponding solution, which would then be deformed for 8 to 10 minutes. The obtained resin matrix solution would be mixed with silver powder and acetone solution. Then, stir the slurry by a constant temperature magnetic stirrer. The final conductive adhesive was obtained by vacuum treatment of stirred homogeneity mixture slurry. The prepared conductive silver adhesive was coated evenly on the surface of the piezoelectric composite material which was then putted into the vacuum box to be cured at 90 °C for 30 minutes with different dose of diluent.

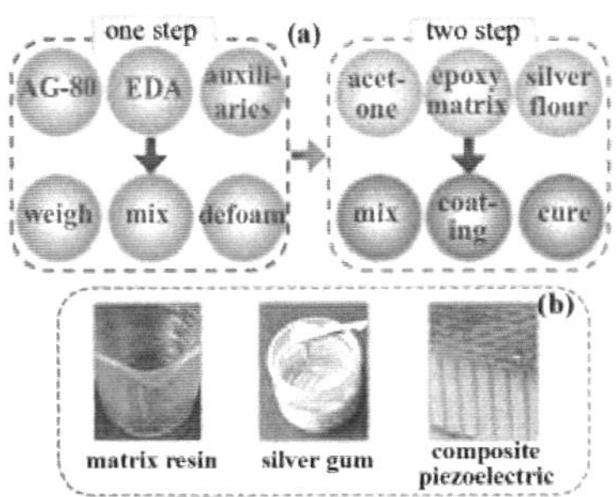


**Figure 1.** The diagram of the key preparing process to prepare the ultra-low curing temperature conductive silver adhesive.

Figure 2 (a) and (b) shows the molecular structure of acetone and AG-80, respectively. Figure 2 (c) shows the chemical and physical properties of acetone and AG-80, respectively. The viscosity of Ag-80 increase from 2000mPa·s to 3000~5000mPa·s when the temperature increase to 50°C. In order to improve the operability and extend the life of the conductive adhesive, the diluent was added to dissolve and disperse conductive adhesive, and reduce the viscosity of conductive adhesive, improve the coating ability and fluidity. Acetone is an inactive diluent, does not participate in the curing reaction, and will evaporate during the curing process, so that increasing the shrinkage of resin.

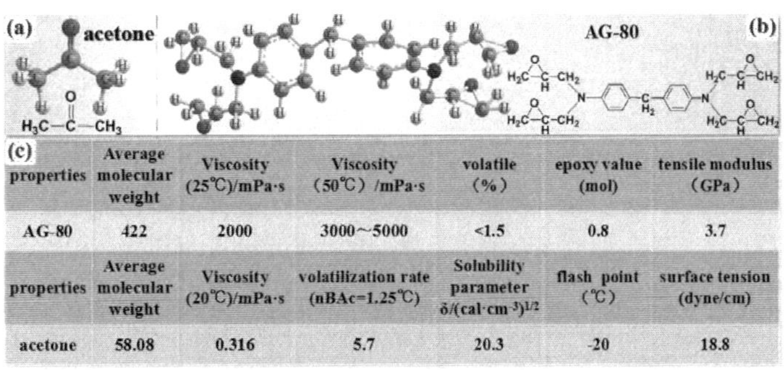


(c)

| properties | Average molecular weight | Viscosity (25℃)/mPa·s | Viscosity (50℃) /mPa·s | volatile (%) | epoxy value (mol) | tensile modulus (GPa) |
|---|---|---|---|---|---|---|
| AG-80 | 422 | 2000 | 3000~5000 | <1.5 | 0.8 | 3.7 |
| properties | Average molecular weight | Viscosity (20℃)/mPa·s | volatilization rate (nBAc=1.25℃) | Solubility parameter $\delta/(cal \cdot cm^{-3})^{1/2}$ | flash point (℃) | surface tension (dyne/cm) |
| acetone | 58.08 | 0.316 | 5.7 | 20.3 | -20 | 18.8 |

**Figure 2.** The chemical structural formula, characteristics contrast of AG-80 and 618. (a) and (b) shows the molecular structure of acetone and AG-80, respectively; (c) shows the chemical and physical properties of acetone and AG-80, respectively.

Figure 3 shows the volume resistivity and shear strength of conductive adhesive with different content of the diluent. It can be seen from the figure that with the increase of the content of diluent, the volume resistivity of conductive adhesive decreased at first and then increased, and the shear strength increased at first and then decreased. When the acetone content is 10%, the volume resistivity is $2.37 \times 10^{-4}\Omega \cdot cm$ and the shear strength is 5.13MPa. When the acetone content is 8%, the volume resistivity is $8.06 \times 10^{-4}\Omega \cdot cm$ and the shear strength is 6.54MPa. This is due to the addition of a small amount of diluent can improve the shrinkage of the resin, making the contact between the silver powder is better, improving the conductivity of the conductive adhesive, and can increase the cohesive energy density of the resin system and enhance the mechanical properties of the resin system. However, the added too much diluent will affect the curing reaction, resulting in the decrease of contact of conductive adhesive particles, reduce the conductivity. At the same time, the resin is over crosslinked, resulting the resin system becomes brittle, the rigidity is weakened and the shear strength is reduced.

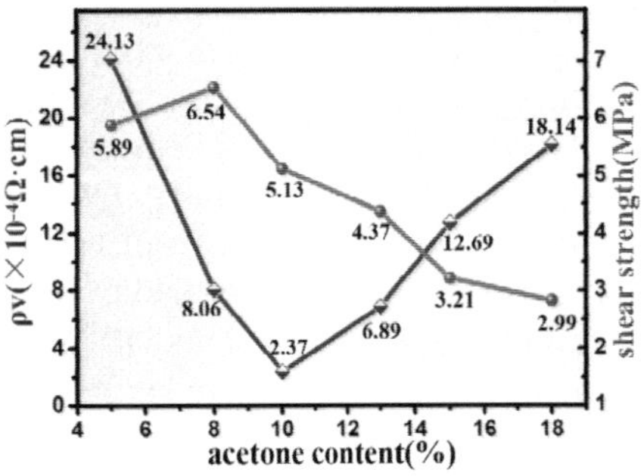


**Figure 3.** The volume resistivity and shear strength of conductive adhesive with different content of the diluent.

Figure 4 shows the SEM images of conductive silver adhesive with different content of diluent. It can be seen that when the diluent content of 10%, the silver flaky bonded together, arranged the neatest, the smallest gap, the most closely connected, and the surface can form a complete conductive network. This is due to adding diluent can reduce the resin viscosity, enhance the resin shrinkage, so that closer contact between the silver. When the amount of diluent is too small, the matrix resin is too thick, resulting in silver powder is not open, and forming a large number of porous microstructure. Furthermore, the addition of diluent affects the curing reaction to a certain extent. When the amount of diluent is too much, the conductive silver adhesive cannot be a good curing, affecting the uniform distribution of silver powder, reduce the conductivity. This also explains why the conductive resist has the smallest volume resistivity at a diluent content of 10%.

**Figure 4.** The SEM images of conductive silver adhesive with different content of diluent.

## 4. Conclusions

In summary, the effect of diluent acetone on ultra-low temperature curable conductive silver adhesive were investigated for surface conductive treatment of piezoelectric composite material. In order to improve the operability and extend the life of the conductive adhesive, the diluent was added to dissolve and disperse conductive adhesive. With the increase of the content of diluent, the volume resistivity of conductive adhesive decreased at first and then increased, and the shear strength increased at first and then decreased. When the acetone content is 10%, the silver flaky bonded together, arranged the neatest, the smallest gap, the most closely connected, the surface can form a complete conductive network, and the volume resistivity is $2.37 \times 10^{-4}\Omega \cdot cm$, the shear strength is 5.13MPa.

## References

[1] Jang C, Sharifi M, Palmese, G R, Abrams C F 2014 *Polymer*, **55(16)** 3859-68.
[2] Lan T, Kaviratna P D, Pinnavaia T J 1995 *Chem. Mater.* **7(11)** 2144-50.
[3] Sang J L, Sohn H, Hong J W 2010 *J. Nondestruct. Evaluat.* **29(2)** 75-91.

[4] Pettersen S R, Redford K, Njagi J, Kristiansen H, Helland S, Kalland E, Goia D, Zhang Z, He J 2017 J. Electr. Mater. 1-11.
[5] Rane S B, Seth T, Phatak G J, Amalnerkar D P, Das B K 2003 Mater. Lett. **57(20)** 3096-100.
[6] Zhao S Y, Li X, Mei Y H, Lu G Q 2016 J. Electr. Mater. **45(11)** 1-11.
[7] Peng P, Hu A, Gerlich A P, Zou G, Liu L, Zhou Y N 2015 *Acs. Appl. Mater. Interf.* **7(23)** 12597.
[8] Bouguettaya M, Vedie N, Chevrot C 1999 *Synth. Met.* **102(1-3)** 1428-31.
[9] Marcq F, Demont P, Monfraix P, Peigney A, Laurent C, Falat T, Courtade F, Jamin T 2011 *Microelectron. Reliabil.* **7** 1230-4.
[10] Li X S, Xiang X Z, Wang L, Bai X J 2016 *Rare Met.* 1-5.
[11] Morris J E 2017 *Polymers*, **3(1)** 191-7.
[12] Nakayama K, Nagai A, Iida N 2007 *Mater. Trans.* **48(3)** 594-9.
[13] He S, Zhang X, Yang B, Xu X, Chen H, Zhou C 2017 *Chinese Phys. B,* **26(7)** 426-32.
[14] Park S D, Yoo M J, Kang N K, Park J C, Lim J K, Kim D K 2004 *Macromol. Res.* **12(4)** 391-8.
[15] Lu C A, Lin P, Lin H C, Wang S F 2006 *Japanese J. Appl. Phys.* **45(9A)** 6987-92.
[16] Bai J G, Zhang Z Z, Calata J N, Lu G Q 2006 *IEEE Trans. Comp. Pack. Tech.* **29(3)** 589-93.
[17] Gural P S, Jenniskens P 2007 *Japanese J. Appl. Phys.* **46(1)** 251-5.

**Acknowledgements**
This work was supported by the National Natural Science Foundation of China (No. 11604363, 61471047, 61671068), and the Beijing College Innovation Capability Promotion Plan of Beijing Municipal Institutions (No. TJSHG201510772015). The project is also supported by The Importation and Development of High-Caliber Talents Project of Beijing Municipal Institutions (No. CIT&TCD201504053).

IOP Conf. Series: Materials Science and Engineering **317** (2018) 012018 doi:10.1088/1757-899X/317/1/012018

# Basic Snow Pressure Calculation

**Shouzhi Hao[1*], Jian Su[1]**

[1]Dalian Ocean University, Dalian, Liaoning, China

**Abstract.** As extreme weather rising in recent years, the damage of large steel structures caused by weather is frequent in China. How to consider the effect of wind and snow loads on the structure in structural design has become the focus of attention in engineering field. In this paper, based on the serious snow disasters in recent years and comparative analysis of some scholars, influence factors and the value of the snow load, the probability model are described.

## 1. Introduction
In 2007, Liaoning Province suffered the biggest snow disaster since the meteorological record, which caused great economic losses to the economy and heavy casualties of Liaoning Province. In contrast, similar incidents had occurred in Tibet in 1956, most parts of northern China in 1977, the southern region of Xinjiang in 1983 and the snow disaster in southern China in 2008.Through investigating and analyzing the common characteristics of such events are caused by excessive snow load, so the related problems about snow load in structure design should be concerned widely [1, 2].

## 2. Data processing of relevant materials
The snow pressure data can be obtained from the local meteorological bureau's weather statistics for the past years and the result is the closest to the local actual situation. It can also be calculated based on basic snow pressure value in the current load code in China. On the basis of the old code, the present load code for the design of concrete building structure has added to the basic snow pressure data in recent years, the basic snow pressure value in some areas has been increased, which accords with the actual situation of frequent accidents caused by abnormal snow load in recent years. In addition, in the absence of snow pressure record, the following two methods can be used to calculate the basic snow pressure:

(1) The relationship between the snow depth $h$, density $\rho$ and the basic snow pressure $s$ can be calculated by the following formula:

$$s=h\rho g. \quad (1)$$

In the formula: $h$ —The snow depth, vertical depth from the surface of the snow to the ground(m).
$\rho$ —The density of snow($t/m^3$), the specification in the appendix to all regions of China has carried on the provisions.
$g$ —Acceleration of gravity,$9.8m/s^2$.

(2) The snow pressure is calculated indirectly by converting precipitation into snow depth. The snow pressure is calculated indirectly by converting precipitation into snow depth. The daily average temperature is 3.1°C as the critical temperature value, the rainfall can be used as the snowfall data when the temperature is less than or equal to 3.1°C, ensuring snowfall rate is 98.5% and the rainfall data into snow depth with 1:10 conversion ratio.

## 3. The calculation method of annual maximum snow pressure

### *3.1. Establishment of snow load probability distribution model*

According to the content of the code of structural design in China, the probabilistic model of live load on the floor, wind load and so on is assumed to be based on the stationary binomial stochastic process, that is to translate the sample data in the process of the stochastic load into a data model of equal time period, so the snow load probability model will use a stationary binomial stochastic process, the steps are as follows [3]:

(1) The building design reference period is T, which is divided into r equal time periods;
(2) The probability of occurrence of snow load is p, and the probability of non-occurrence is q=1-p within any period of time;
(3) In any time period, the probability distribution function of snow load at any time point is equal to the cumulative distribution function of snow load and the amplitude of snow load on each time period is independent and identically distributed;
(4) The probability of the occurrence of snow load and the amplitude of snow load in different time periods are independent of each other.

The probability distribution of the maximum annual snow pressure is based on the probability distribution of the extreme value type I according to the specification. The distribution function is:

$$F(x)=\exp\{-\exp[-\alpha(x-u)]\} \quad (\alpha>0, -\infty<u<\infty) \tag{2}$$

In the formula: x—The annual maximum snow sample
u—The positional parameters of the distribution, the value of the sample distribution;
$\alpha$—The size parameter of distribution;
$\sigma$—The standard deviation of the sample;
$\mu$—The average of the samples.

The maximum snow pressure with a return period of R is determined as follows:

$$x_R=u-\frac{1}{\alpha}\ln\left[\ln\left(\frac{R}{R-1}\right)\right] \tag{3}$$

### *3.2. Parameter estimation of α and μ in extreme value distribution*

The methods of estimating the parameters of the extreme value type I probability distribution is the moment method, the maximum likelihood method and the Gumbel method [4-6].

(1) The moment method

Moment estimation method is to estimate the parameters of the whole numerical characteristics by using the moments of samples. It is intuitionistic and practical in mathematical calculation. The newest load specification in China uses this method to estimate the parameters of α and μ, the formula is as follows:

$$\alpha=\frac{1.28255}{\sigma} \tag{4}$$

$$u=\mu-\frac{0.57722}{\alpha} \tag{5}$$

(2) The maximum likelihood method

According to the sample, constructs the likelihood function. Each group of samples is brought into the likelihood function, and then the maximum likelihood estimator is obtained as the parameter of the model. The parameter estimation formula is:

$$\sum_{i=1}^{n}\exp[-\alpha x_i+1/(n-1)+(-\alpha/n)\sum_{i=1}^{n}x_i]-n=0 \tag{6}$$

$$\mu=1/[\alpha(n-1)]+1/n\sum_{i=1}^{n}x_i \tag{7}$$

In the formula: $x_i$ —The i-th observed value of the sample

This method is widely used in various parameter estimations and has a good fit, especially for the case when the sample size is small.

(3) The Gumbel method

Gumbel method is the combination of empirical guarantee function and the actual parameters to estimate the parameters. The annual maximum snow assumed to be an ordered sequence: $x_1 \leq (\geq)\ x_2 \leq (\geq)\ x_3 \leq (\geq) \cdots \leq (\geq)\ x_n$ then the experience guaranteed function is:

$$F(x_i)=\frac{i}{n+1} \tag{8}$$

The probability of equal or large (small) to the i-th value is about i / (n+1). when n samples of max (small) values are arranged from large (small) to small (large) is about F ($x_i$) = F^($x_i$). from the law of large numbers, we know that when the number of samples is large enough, the empirical guarantee function will approximate the population distribution function. Similar to the moment method, we can get this formula:

$$\alpha=\frac{\sigma(y)}{\sigma(x)} \tag{9}$$

$$\mu=E(x)-\frac{E(y)}{\alpha} \tag{10}$$

### *3.3. Evaluation of the results of parameter estimation*

It is usually used to fit the test index of Kolmogorov to test the parameters of the parameter estimation. The formula is as follows:

$$K_f=D_n\sqrt{n} \tag{11}$$

In the formula: n—The size of a sample

$D_n$—At all points, the maximum value of the difference between the theoretical distribution of the empirical distribution and the hypothesis.

$$D_n=\max\left\{\left|\hat{F}(x_i)-F(x_i)\right|\right\} \tag{12}$$

When the value of the index is smaller, it means that the fitting effect is better.

## 4. Analysis of Calculation Results

Taking Dalian, Shenyang, Dandong, Dalian and Chaoyang four cities in Liaoning province China as examples, the maximum snow pressure statistics are obtained from the data collected through the Internet. The data are plotted in two years as shown in the following figure. The parameters of the parameters estimation are evaluated by the method of moment estimation and Gumbel method, and then use the Kolmogorov fitting test to test the parameters.

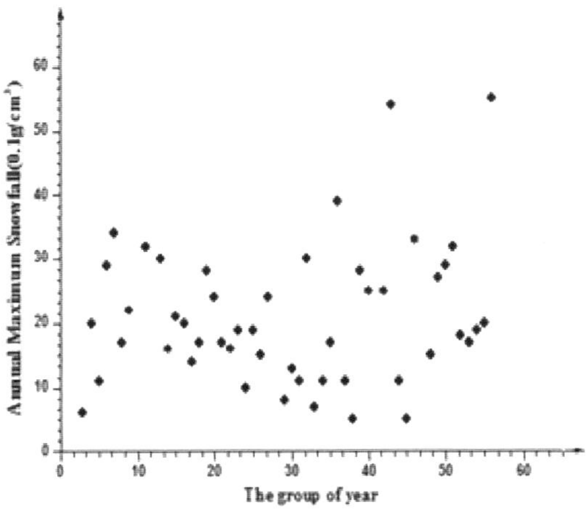


**Figure 1.** The annual maximum snowfall of Shenyang.

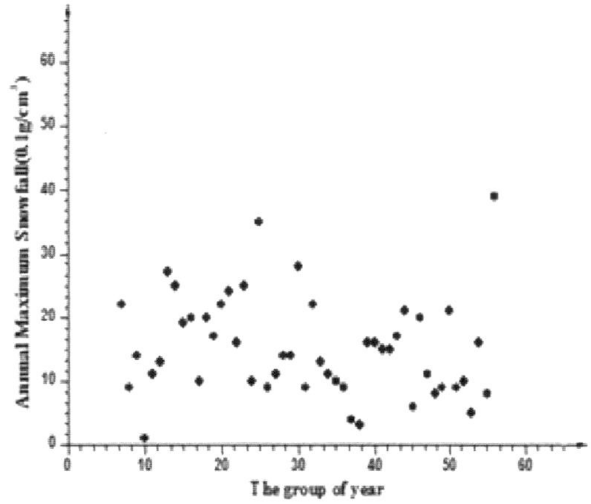


**Figure 2.** The annual maximum snowfall of Dandong.

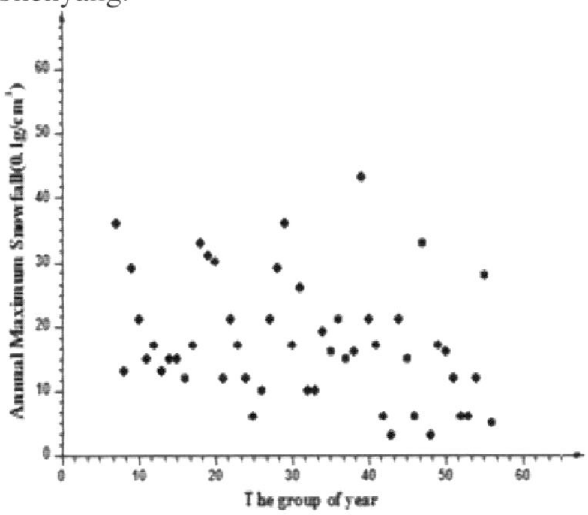


**Figure 3.** The annual maximum snowfall of Dalian.

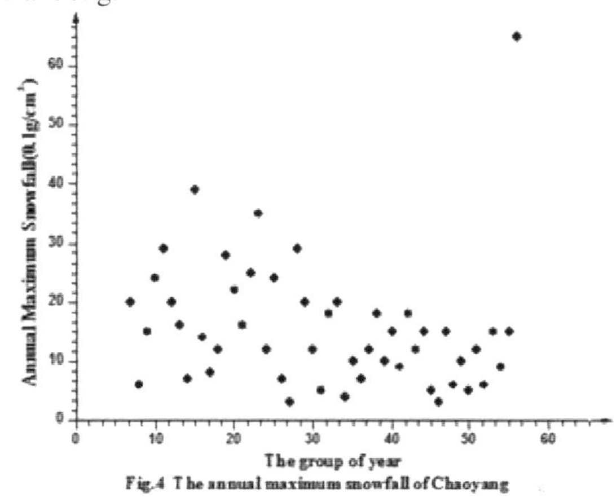


**Figure 4.** The annual maximum snowfall of Chaoyang.

**Table 1.** Comparison of the basic snow pressure calculation results.

| City | The moment estimation method | | The Gumbel method | | The value in the code | | $K_f$ | |
|---|---|---|---|---|---|---|---|---|
| | **n=10** | **n=30** | **n=10** | **n=30** | **n=10** | **n=30** | **Moment** | **Gumbel** |
| **Shenyang** | 0.45 | 0.51 | 0.41 | 0.50 | 0.40 | 0.50 | 0.543 | 0.451 |
| **Dandong** | 0.27 | 0.31 | 0.29 | 0.34 | 0.35 | 0.49 | 0.787 | 0.623 |
| **Dalian** | 0.32 | 0.34 | 0.33 | 0.37 | 0.40 | 0.57 | 0.583 | 0.453 |
| **Chaoyang** | 0.32 | 0.34 | 0.33 | 0.32 | 0.40 | 0.50 | 1.291 | 1.211 |

The calculation results in the table are obviously larger than the specified value in the load specification. The reason is that the average value and the standard deviation of the samples are not correctly treated for unrecorded values and abnormal values.

## 5. Conclusion

1. It can be used to estimate the parameters directly in the calculation and also suitable for most cities in Liaoning Province, Otherwise, some regions can be calculated by convenient moment estimation method.

2. Analysis on the calculation results of other cities in Liaoning province is not listed in the paper found that most of the cities' basic snow pressure value after the revision of the new building load code, it can be applied to the engineering practice.

3. For the default values and outliers in the samples, the final calculation results will have a great impact because of the influence of local temperature and altitude on the snow depth during the basic snow pressure statistics lead the accumulation of errors. How to deal with such problems still needs studying.

## References

[1] Huang H H 2007 *Meteorolog.* **33** 101-6.
[2] Zhang Y N 2010 *J. South China Univ. Tech. (Nat. Sci. Ed.).* **38** 108-12.
[3] Yang Q W, Gu L 1986 *Time series analysis and dynamic data modeling*, Beijing Institute of Industry Press, Beijing.
[4] Duan Z D, Zhou D C 2004 *J. Harbin Inst. Tech.* **36** 1605-9.
[5] Duan Z D, Ou J P 2002 *China Civil Eng. J.* **35** 11-6.
[6] Li H N, Wang Y 2009 *J. Nat. Disast.* **18** 15-26.

IOP Conf. Series: Materials Science and Engineering **317** (2018) 012019 doi:10.1088/1757-899X/317/1/012019

# Mathematical modeling of a process the rolling delivery

**Mikhail A. Stepanov, Andrey A. Korolev**

PhD. of Tech. Sci., Assistant Professor, Moscow State University of Civil Engineering
Postgraduate student, Moscow State University of Civil Engineering

**Abstract.** An adduced analysis of the scientific researches in a domain of the rolling equipments, also research of properties the working material. A one of perspective direction of scientific research this is mathematical modeling. That is broadly used in many scientific disciplines and especially at the technical, applied sciences. With the aid of mathematical modeling it can be study of physical properties of the researching objects and systems. A research of the rolling delivery and transporting devices realized with the aid of a construction of mathematical model of appropriate process. To be described the basic principles and conditions of a construction of mathematical models of the real objects. For example to be consider a construction of mathematical model the rolling delivery device. For a construction that is model used system of the equations, which consist of: Lagrange's equation of a motion, describing of the law conservation of energy of a mechanical system, and the Navier – Stokes equations, which characterize of the flow of a continuous non-compressed fluid. A construction of mathematical model the rolling deliver to let determined of a total energy of device, and therefore to got the dependence upon the power of drive to a gap between of rolls. A corroborate the hypothesis about laminar the flow of a material into the rolling gap of deliver.

## 1. Introduction

The most important condition of development the modern industry, including of the building domain, this is the rise of intensification manufacture. This is can achieved with the aid of apply of the high-productivity technological equipments.

The delivery and transportation of a material this is a one of basic operation of technological process to manufacture of the building articles. The modern equipments the delivery of a material to realize by next mechanisms and devices: band conveyer, screw, rolling, plate conveyer, scraper conveyer, pneumatic, hydraulic and other kinds [1, 2].

The most widespread type of delivery and transporting device this is the rolling mechanism. It is to be different by simple construction, high reliability, minimum of power consumption.

The scientific researches of the rolling devices were considered in the works [3-5]. Was a study the basic technical and technological parameters this is type of equipments. Was determined the forces on rolls and torques, the power of drive and productivity of devices.

The properties of different kinds the working material, which to pass through rolls, were study in the works [6-8]. Was considered viscous-elastic and viscous-plastic properties of material, was determined rheological characteristics, was compiled equations of the flow, was determined effective viscosity.

On this basis it can be done a conclusion about necessity of a construction of mathematical models the rolling devices on purpose to study their physical properties.

## 2. The basic principles of a construction of mathematical model

A one of the method of scientific researches different objects to serve of a construction of mathematical model. Mathematical modeling a very most important article of research for everything special and applied disciplines. A special meaning that is has in physics and technical sciences [9-12].

In order to construct of mathematical model of a real object (at present case of mechanical system – the rolling deliver) necessary [13-16]:

1. formulate of hypothesis, that is to determine of the laws by which to work the researching object, and to determine the forces acting on the working parts, also to determine of the basic (fundamental) physical laws, which necessary to apply for present research;
2. work out a equations, which to describe with the aid of mathematics the researching process or object, that is direct a construction of mathematical model;
3. solution of mathematical equations.

The real object of research can have several different of mathematical models. This is connected with possibility research of different properties a system, also the choice of the type model. Moreover, a research of complex the real object with the aid of different mathematical models to can substantially to expand of the knowledge about occurring process and raised precision of research.

Also not exclude the use of the same mathematical model (similar differential equations) for a research of different the real object. At that to appear possibility of a substitution the researching objects on another. Therefore, cannot constructed again of mathematical model, and correctly choice and apply already well-known models.

The most important the requirement, which to produced of mathematical model, this is adequacy [15, 17], that is correct accordance of the real object. There under to be understood:

1. correct qualitative description of the properties the researching object;
2. correct quantitative description this is the properties.

A quantitative prediction not always demonstrated enough precisely. First of all, this is noticeably at social and biological sciences. Sometimes them necessary neglected and at technical sciences because of considerably to complicate research of object.

An initial stage of a construction any mathematical model this is to receipt the most complete conception about the modeling object. In the case of mechanical system, at present case of the rolling deliver, to determine the basic parts, from which consist the researching device, their joint work and action, the forces acting on the working parts and on a work material.

The next stage of a construction of mathematical model this is formulated of mathematical problem. The problem can be concrete, that is determination of any parameter the researching system, or to express of the indefinite form, that is research any process, occurring inside system, or a conduct of system in general and so on.

Formulate of mathematical problem for a research of the rolling devices at the indefinite form: «Optimization of a process the rolling delivery of viscous-plastic materials».

At present example to choice as parameter optimizations of power of drive the rolling device. Further to study in detail of mathematical model this is parameter can be specified and changed.

## 3. The construction of mathematical model the rolling device

As the initial system of the equations of mathematical model the rolling deliver of viscous-plastic the building materials, for example the clay for ceramic articles, to accept Lagrange's equation of a motion for the rolling device and the Navier – Stokes equations of the flow non-compressed fluid for a material. During a construction of mathematical model the flow of a material considered as laminar.

In accordance with principle of the least action every mechanical system to move likes this as integral of an action equal:

$$S = \int_{t_1}^{t_2} L(r,v,t)dt \tag{1}$$

where $L(r,v,t)$ - function of Lagrange with specified, accordingly, the coordinates of points system $r$, velocity of points system $v$ at determinate the moment of the time $t$.

Integrated function (1), and have a number of transformations, to get Lagrange's equation of a motion:

$$\frac{d}{dt}\cdot\frac{\partial L}{\partial v}-\frac{\partial L}{\partial r}=0 \tag{2}$$

Or another mode:

$$\frac{d}{dt}\cdot\frac{\partial L}{\partial v}=\frac{\partial L}{\partial r} \tag{3}$$

The Navier – Stokes equations for volumetric the flow of non-compressed viscous fluid to have three components [18, 19], which in general kind can be written down the following mode:

$$\rho\cdot\left(\frac{\partial v}{\partial t}+\sum_{i=1}^{3}v_i\frac{\partial v}{\partial x_i}\right)=-grad\ p+\mu\cdot\left(\sum_{i=1}^{3}\frac{\partial^2 v}{\partial x_i^2}\right)+F \tag{4}$$

where $\rho$ - density of a material, $kg/m^3$;
$v$ - the velocity, m/s;
$grad\ p$ - gradient of the pressure, Pa;
$\mu$ - viscosity of a material, $Pa\cdot s$;
$F$ - the external force, N.

In order that the flow of non-compressed viscous fluid in a gap the rolling device to be continuously, necessarily to observe of a condition non-tearing of the stream. That is to determine of equation Saint-Venant, expressing the principle of mass conservation in the cases setting of the motion, and has the following mode:

$$\frac{\partial v_x}{\partial x}+\frac{\partial v_y}{\partial y}+\frac{\partial v_z}{\partial z}=0 \tag{5}$$

where $v_x$, $v_y$, $v_z$ - the components of velocity, m/s.

Now necessary to solve this is system of equations, in order that to find of parameters, which to be choice as optimization, at present case this is the power of drive the rolling device.

## 4. The solution of mathematical model

The distribution of viscous non-compressed fluid in a gap between by rolls to be occur into two direction: on axis X, as showed on figure 1, that is on length of the stream, and on axis Y, that is on width of a sheet (on figure 1 not showed). The sizes on axis Z, that is on thickness of a sheet, to be substantially less and them in further calculations can be neglect.

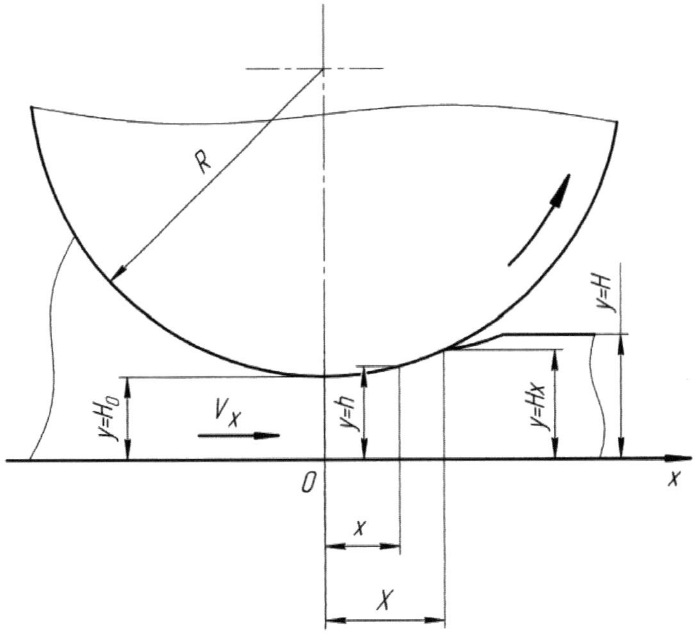


**Figure 1.** The flow of viscous fluid into a gap of rolls.

Thereupon that a function of Lagrange (equation 1) this is a differential equation of a motion, that is to be the law conservation of energy [17]:

$$L = \frac{m \cdot v^2}{2} - U(r) = E \tag{6}$$

where m – the mass, kg;
$v$ – the velocity, m/s;
$U(r)$ – the potential energy of system, Joule;
E – a total energy of system, Joule.

In the case of flat-parallel the flow of viscous fluid the equation of a total energy to accept the following mode:

$$E = \frac{m \cdot \left( v_x^2 + v_y^2 \right)}{2} - U(x, y) \tag{7}$$

where $v_x$, $v_y$ – the components of velocity on axis X and Y, m/s.

The Navier – Stokes equations of a motion of continuously non-compressed fluid also simplify in connection with introduce into allowing, and to be contain the two components, which to accept the following mode:

$$\rho \cdot \left( \frac{\partial v_x}{dt} + v_x \cdot \frac{\partial v_x}{\partial x} + v_y \cdot \frac{\partial v_x}{\partial y} \right) = -\frac{\partial p}{\partial x} + \mu \cdot \left( \frac{\partial^2 v_x}{\partial x^2} + \frac{\partial^2 v_x}{\partial y^2} \right) + F_x \tag{8}$$

$$\rho \cdot \left( \frac{\partial v_y}{dt} + v_x \cdot \frac{\partial v_y}{\partial x} + v_y \cdot \frac{\partial v_y}{\partial y} \right) = -\frac{\partial p}{\partial y} + \mu \cdot \left( \frac{\partial^2 v_y}{\partial x^2} + \frac{\partial^2 v_y}{\partial y^2} \right) + F_y \tag{9}$$

where $\rho$ - density, $kg/m^3$;
$v_x$, $v_y$ – the components of velocity on axis X and Y, m/s;
$p$ - the pressure, Pa;
$\mu$ - viscosity of a material, $Pa \cdot s$;
$F_x$, $F_y$ – the external forces on axis X and Y, N.

Accordingly changed the equation of the non-tearing of Saint-Venant and to accept the following mode:

$$\frac{\partial v_x}{\partial x} + \frac{\partial v_y}{\partial y} = 0 \quad (10)$$

where $v_x$, $v_y$ – the components of velocity on axis X and Y, m/s.

The solution of the equation (7) to give a chance to got the most precisely of a total energy of system the rolling deliver, and therefore the most precisely to calculate the power of drive the rolling device. A system of equations (8 – 10) to characterize of flat-parallel the flow of viscous fluid into a gap of rolls delivery device.

In the work [4] adduced the values of experimental researches the rolling device for high-plastic viscous materials with plastic's number more than 25%. Was a study delivery of pieces for different the gaps between of rolls. Was a got the factual values of the power of drive device. A comparison of the calculating and experimental values (figure 2) showed insignificant divergence of the results (on the average not more than 5%).

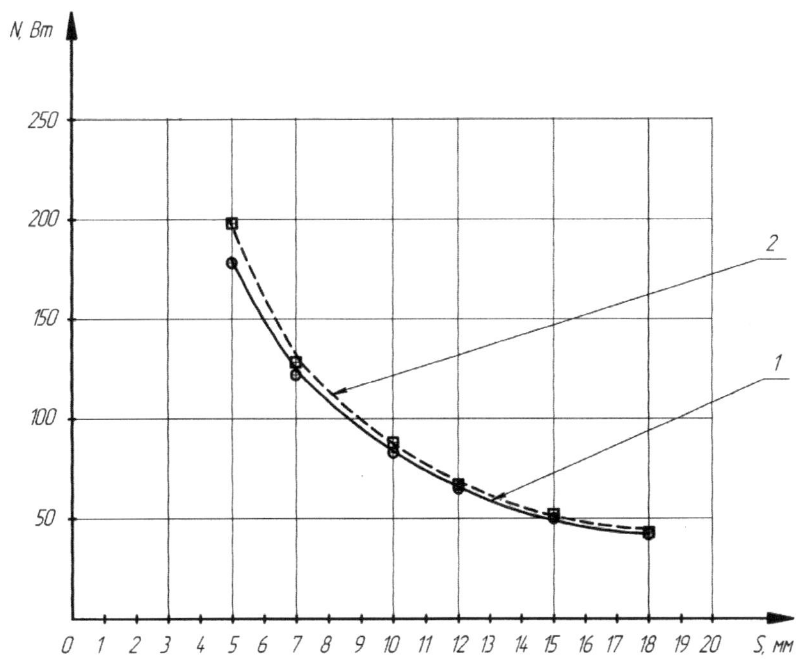


**Figure 2.** The diagram of dependence upon power to a gap between of rolls:
1. – factual power; 2. – calculated power.

IOP Conf. Series: Materials Science and Engineering **317** (2018) 012019 doi:10.1088/1757-899X/317/1/012019

## 5. Conclusion

- A construction of mathematical model to let considered everything factors, which influence on experiment, will choice of the optimum variant and to minimize of quantity the researching factors;
- A construction of mathematical model to let determined of dependence upon power of drive the rolling device to a gap between of rolls, which to be express by exponential function;
- Mathematical model to corroborate the hypothesis about laminar the flow of a material into the rolling gap of deliver, which enough precisely described with the aid of the Navier – Stokes equations;
- A construction of mathematical model the rolling deliver to let determined of a total energy of device and therefore the most precisely to calculate the power of drive and to reduce of energy costs.

## References

[1] Uvarov V A, Karpachyov D V 2008 *Machines for technological transporting of the building materials and articles* Belgorod: Publishing house of BGTU, pp. 156.
[2] Uvarov V A, Stepanov M A, Koshkaryov E K 2013 *Machines for technological transporting of the building materials and articles* MGSU, pp. 216.
[3] Lukyanov N A, Stepanov M A, Korolev A A 2014 *Mech. Constr.* **1** 8–11.
[4] Lukyanov N A, Stepanov M A, Korolev A A 2014 *Mech. Constr.* **12** 24–7.
[5] Shrubchenko I V, Hurtasenko A V, Sharapov R R, Duyun T A, Shchetinin N A 2015 Mod. Appl. Sci. **9(1)** 195.
[6] Lukyanov N A, Stepanov M A, Korolev A A 2013 *Vestnik MGSU*, **5** 43–8.
[7] Gridchin A M, Yadykina V V, Trautvain A I, Sharapov R R, Zhukova A A 2014 Res. J. Appl. Sci. **9(12)** 1053–8.
[8] Sharapov R R, Kapyrin P D, Lozovaya S Yu, Yadykina V V, Agarkov A M 2016 *Research deducting efficiency of the inertial hub with adjustable parameters* MATEC web conf 5-th International scientific conference «Integration, partnership and innovation in construction science and edication», **86** 7.
[9] Andreev V K 2015 *Mathematical models of continuum mechanics* Lan, pp. 231.
[10] Arnold V I 2000 *Mathematical methods of classic mechanics* Editorial URSS, pp. 408.
[11] Blekhman I I, Myshkis A D, Panovko Ya G 2007 *Applied mathematics: theme, logic, specific of approach With examples from mechanics* Publishing house LKI, pp. 376.
[12] Meyer W J 2004 *Concepts of mathematical modeling* – N Y – London: Dover publications, pp. 448.
[13] Zeldovich Ya B, Myshkis A D 2014 *The elements of applied mathematics* LENAND, pp. 592.
[14] Myshkis A D 2007 *Applied mathematics for the engineers* Phys Math Lit, pp. 688.
[15] Myshkis A D 2014 *The elements of theory mathematical models* LENAND, pp. 200.
[16] Tykhonov A N, Samarsky A A 1999*The equations of mathematical physics* Publishing house of MGU, pp. 798.
[17] Fedotkin I M 2015 *Mathematical modeling of technological processes* Book-house «LIBROCOM», pp. 416.
[18] Temam R, Miranville A 2005 *Mathematical modeling in continuum mechanics* Cambridge: Cambridge University Press, pp. 320.
[19] Turenko A V 2004 *A calculation of clay-reprocessing equipments and the presses of plastic forming for a manufacture of ceramic the building articles* MGSU, pp. 114.

# Investigation of the Stability of a Two-Span Bridge with the use of a High-Precision Laser Displacement Sensors

**O Poddaeva[1], P Churin[2], A Fedosova[3] and S Truhanov[4]**

[1]Head of Laboratory, Moscow State University of Civil Engineering (MGSU), 129337, 26, Yaroslavskoye Shosse, Moscow, Russia
[2]Senior Researcher, Moscow State University of Civil Engineering (MGSU), 129337, 26, Yaroslavskoye Shosse, Moscow, Russia
[3]Senior Researcher, Moscow State University of Civil Engineering (MGSU), 129337, 26, Yaroslavskoye Shosse, Moscow, Russia
[4]Assistant professor, Moscow State University of Civil Engineering (MGSU), 129337, 26, Yaroslavskoye Shosse, Moscow, Russia

**Abstract.** Studies of aerodynamics of bridge structures are an actual problem. Such attention is paid to the study of wind influence on bridge structures not at all by chance; a large number of cases of loss of stability of such structures are known under the influence of wind up to their complete destruction. The development of non-contact systems of measuring equipment allows solving this problem with a high level of accuracy and reliability. This article presents the results of experimental studies of wind impact on a two-span bridge using specialized measuring system based on high-precision laser displacement sensors.

## 1. Introduction

Nowadays bridge building's calculation of wind loads on the territory of Russian Federation is regulated by SP 35.13330.2011 Bridges and pipes.

SP 35.13330.2011 Bridges and pipes contains all basic rules of calculations of wind loads on bridge constructions including to take into account the dynamic component of the wind load, to which bridge structures are most susceptible. In a part of wind loads SP 35.13330.2011 "Bridges and pipes" contains many references on calculation methods from SP 20.13330.2016 "Loads and impacts" and using for classic buildings. Aerodynamic checking of stability consists from SP 35.13330.2011 "Bridges and pipes": critical speed, responding for dangerous aeroelastic phenomena, gotten from aerodynamic model test results or determined by calculation must be bigger than maximum wind speed possible in bridge's area but less than 1,5 times.

In that case the information for calculation of wind loads on span structure of the bridge specified in normative documents are clearly insufficient (in SP "Loads and impacts" aerodynamic coefficients for typical cross sections are missed, data of the Strouhal number are given only for single rectangular cross sections; in SP "Bridges and pipes" in appendix N, only the coefficient of drag for the parts and elements of span structures of bridges is given, without taking into account the features of the shape of the elements).

That is why, according to point 5.48 SP "Bridges and pipes" for hanging and cable-stayed bridges, as well as steel beam bridges with spans of more than 100 m, it is necessary to check for aerodynamic stability and spatial rigidity. For structures with dynamic characteristics substantially different from those of the constructed bridges, in addition to analytical calculations, appropriate studies should be carried out on the models.

Such attention is paid to the study of wind influence on bridge structures not at all by chance; a large number of cases of loss of stability of such structures are known under the influence of wind up to their complete destruction. In our country, first, a story with the Volgograd "dancing bridge" (2010) [1, 2], the case with the Takoma bridge in the USA is widely known [3, 4].

## 2. Test method

The methods of numerical modeling of the wind effect on bridge structures are currently poorly developed, this is primarily due to the complexity of the mathematical solution of the connected problem, the modeling of the complex turbulent structure of the wind flow around the object under study, taking into account the dynamic similarity of this object (mass inertial and frequency similarity) [5, 6].

For this reason, we will focus on experimental research in a wind tunnel.

Currently, three main types of models and their corresponding experimental research methods are distinguished:

• Fully aeroelastic models. In this case, all the dynamic and mass-inertial characteristics of the object or its individual element are modeled. Such models are used in the testing of bridge structures at the construction stage, as well as in the testing of complete models of bridge structures. The model allows reproducing all the vibrational modes of a real object, which allows obtaining the most complete and reliable results [7, 8].

• Sectional (shut-off) models. Geometrically similar models, which are part of an elongated structure. The model is made as rigid as possible, the dynamic characteristics are reproduced by using spring suspensions in specialized stands [9, 10].

The most common method is the testing of shut-off models of span structures, since this method is optimal in terms of completeness of the results obtained, labor intensity and economic feasibility. The basic requirements for sectional models and directly to the technology of testing are well known and described in the scientific and technical literature.

Of the main features it is necessary to note the presence of a specialized test bench designed for such tests, with the presence of spring suspensions for modeling the calculated values of the natural frequencies of oscillations of the span structure and the level of its damping, as well as the seats for fastening high-precision force-moment sensors for measuring the lift, resistance and torque around the longitudinal axis [11, 12].

## 3. Wind speed

One of the most interesting types of bridge structures are structures with two parallel-unconnected span structures, located at a minimum distance (up to 5 meters). Such constructions include the Crimean bridge under construction, the Volgograd Bridge was designed, but only one span was built, which is one of the reasons for the occurrence of aerodynamic instability. In addition to the characteristic features of the other bridge structures of the phenomena of aerodynamic instability (vortex resonance, flutter, galloping), the appearance of a buffeting effect is possible here. Buffing is one of the types of self-oscillations, representing forced oscillations of the entire structure or its parts caused by the periodic disruption of turbulent vortices from the structural elements located in front of them when they are flowing. In this case, a breakdown of turbulent vortices from a frontal span structure located along the stream is characteristic.

The object under study is two parallel metal beam girders 150 m long (figure 1).

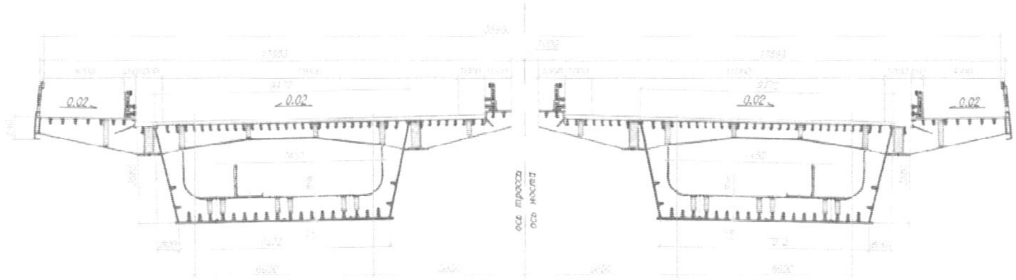

**Figure 1.** Cross-section of the bridge.

The canvas of the bridge is at a level of 10 m from the ground, for this altitude the standard speed value is equal to

$$U_0 = 19{,}5\ m/s \tag{1}$$

The calculated value of the speed taking into account the safety factor 1.4 will be

$$U_{calc} = U_0 \cdot \sqrt{1.4} = 23\ m/s \tag{2}$$

According to SP Bridges and pipes, the critical speed for estimating the flutter of the bridge is determined by the formula:

$$U_{flut} = 1.5 \cdot U_{calc} = 34{,}5 m/s \tag{3}$$

## 4. Making a layout

Taking into account the dimensions of the working part of the wind tunnel (4 mx 2.5 m), the scale of the layout of the model 1:70 was chosen as optimal from the clutter conditions. As a material for manufacturing the model, aluminum sheet is chosen to provide high accuracy of geometric similarity, lightness and high rigidity.

The model consists of two unrelated fragments of span structures of the bridge. Their length is 1714 mm, the width of each span is 254.4 mm, and the total width is 523 mm (taking into account the gap equal to 14.3 mm).

If the section model has the required mass and inertia characteristics, which are determined by the similarity laws, then the actual operation of the span structure (hereinafter - PS) will be similar and can be directly deduced from the behaviour of the PS model in the wind tunnel.

The design of the model in combination with the selected material provides not only a geometric similarity, but also the correct distribution of masses corresponding to the natural object. Thus, the geometric places of the centers of mass of the model and the natural object also correspond [6-8].

All frequency characteristics of span structures are modeled by specialized springs of the measuring stand.

Recommended values: frequency F1 = 0.542 Hz for bending vibrations and F2 = 1.003 Hz for torsional vibrations were used as reference values in this study. The level of damping of the model is measured by conducting a test and is characterized by the value of the logarithmic decrement $\delta$. In this case, $\delta = 0.02$, which corresponds to the standard value of the logarithmic decrement for steel structures.

In accordance with the proportion in which the dimensions of the model and the bridge are related, the following table defines the scale of wind speed, dimensions and frequencies.

IOP Conf. Series: Materials Science and Engineering **317** (2018) 012020 doi:10.1088/1757-899X/317/1/012020

**Table 1.** Relations for the model and the real object.

| Waveform | Real object | Model | Speed scale U* |
|---|---|---|---|
| Bending | 0,542 Hz | 9,8 Hz | 3,87 |
| Torque | 1,003 Hz | 18,4 Hz | 3,81 |

## 5. Conducting research

When carrying out dynamic tests, the model is fixed on spring suspensions in a specialized stand (figure 2). The characteristics of spring suspensions and the corresponding flow velocity scale are determined at the design stage of the model. Movements of the model are measured using 4 optical sensors, which operate on the principle of triangulation of the laser beam. The measurement does not affect the free movement of the suspended model, bending and torsion are measured in the same way by adding or reducing the number of sensor signals.

**Figure 2.** Model of a span structure on spring suspensions.

During the research we used laser displacement sensors LAS-T-500 with the following characteristics:

- range of displacement measurements: 100 ... 600 mm;
- minimum resolution: 0.03 mm;

The speed of the wind is measured with a Pitot tube connected to a high-sensitivity pressure sensor.

All motion sensor signals first pass a low-pass filter, then are digitized with a 16-bit analog-to-digital converter at a rate of 1000 values per second. The statistical calculation of the recorded data makes it possible to obtain mean values, rms, maximum, minimum, etc. for all measurement channels. The displacement is measured relative to the zero level at a wind speed of 0 m/s.

In figure 3 - 4 shows the results of the research, for the most unfavorable angle (+3°) from the point of view of aerodynamics. Stability measurement is presented as a plot of the PS displacement in full scale, depending on the average wind speed also in full scale. Individual graphs are performed for vertical displacement (bending vibrations) and angular displacements (torsional oscillations).

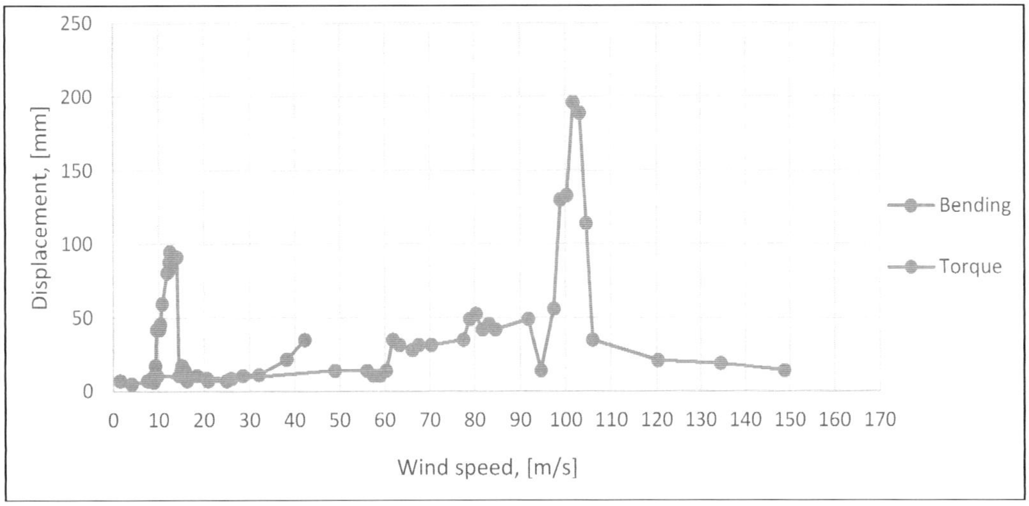


**Figure 3.** Dependence of the oscillation amplitude of the span structure of the bridge on wind speed, flow direction α = +3° (span on the side of the oncoming stream).

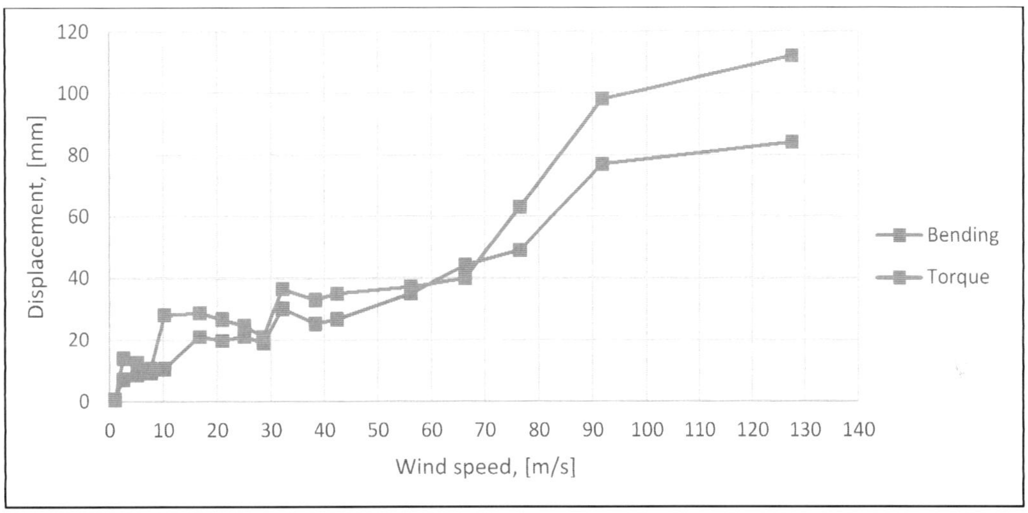


**Figure 4.** Dependence of the oscillation amplitude of the span structure of the bridge on wind speed, flow direction α = +3° (span located in the track).

## 6. Conclusions

Thus, the dynamic tests of the model showed good stability of the span structures at wind speeds below the critical flutter velocity for all flow directions. However, with the flow direction α = + 3°, the model underwent oscillations with the effect of the vortex resonance effect and the maximum amplitude of the vertical oscillations equal to the maximum permissible values: 100 mm - at a flow velocity of 11 to 15 m/s for the span located on the side of the oncoming stream, the span structure located in the track is not subject to aerodynamic instability phenomena.

## Reference

[1] Huber P and Brelser M 2013 *Control of DecN Vibrations of Volgograd Bridge with Semi-Active Tuned Mass Damper System* Proc. of the 13th World Conf. on Seismic Isolation, Energy Dissipation and Active Vibration Control of Structures.

[2] Gazzola F 2013. *Electron. J. Differ. Equ.* **211** 1-47.

[3] Matsumoto M, Shirato H, Yagi T, Shijo R, Eguchi A and Tamaki H 2003 *J. Wind Eng. Ind. Aerodyn.* **91(12)** 1547-57.
[4] Green D and Unruh W G 2006 *J. Phys.* **74(8)** 706-16.
[5] Arioli G and Gazzola F 2015 *Appl. Math. Mod.* **39(2)** 901-12.
[6] Cai C S and Chen S R 2004 *J. Wind Eng. Ind. Aerodyn.* **92(7)** 579-607.
[7] Scanlan R H 1987 *J. Eng. Mech.* **113(4)** 555-75.
[8] Chen X and Kareem A 2003 *J. Eng. Mech.* **129(8)** 885-95.
[9] Yang Y, Zhou R, Ge Y, Mohotti D and Mendis P 2015 J. Wind Eng. Ind. Aerodyn. **145** 196-208.
[10] Zhang Z, Ge Y and Chen Z 2014 *Wind Struct.* 19(3).
[11] Churin P and Poddaeva O I 2014 *Appl. Mech. Mater.* **477** 817-21.
[12] Fedosova A, Churin P and Ageev N 2017 *Computational and experimental study of probability of galloping and torsional divergence at span structure MATEC* Web of Conferences **117.**

**Acknowledgments**
The work was financial supported by the Ministry of Education and Science of the Russian Federation within the framework of the state #7.6075.2017/BCh, Project «Investigation of the phenomena of aerodynamic instability of building structures in aeroelastic statement, including the development of an innovative methodology for analysing meteorologica data to refine the parameters of the wind load».
All tests were carried out using research equipment of The Head Regional Shared Research Facilities of the Moscow State University of Civil Engineering.

# An Aggregated Method for Determining Railway Defects and Obstacle Parameters

**Daniil Loktev[1], Alexey Loktev[2], Roman Stepanov[1], Viktor Pevzner[3], Kanat Alenov[4]**

[1]Department of Structural and Theoretical Mechanics, National Research Moscow State Construction University, Moscow, 129337, Russia
[2]Department of Transport Construction, Russian University of Transport, Moscow, 125190, Russia
[3]Department "Complex issues of traffic safety", VNIIZHT, Moscow, 129626, Russia
[4]Kyzylorda State University m.Korkyt Ata, Kyzylorda, 120014, Kazakhstan

**Abstract.** The method of combining algorithms of image blur analysis and stereo vision to determine the distance to objects (including external defects of railway tracks) and the speed of moving objects-obstacles is proposed. To estimate the deviation of the distance depending on the blur a statistical approach, logarithmic, exponential and linear standard functions are used. The statistical approach includes a method of estimating least squares and the method of least modules. The accuracy of determining the distance to the object, its speed and direction of movement is obtained. The paper develops a method of determining distances to objects by analyzing a series of images and assessment of depth using defocusing using its aggregation with stereoscopic vision. This method is based on a physical effect of dependence on the determined distance to the object on the obtained image from the focal length or aperture of the lens. In the calculation of the blur spot diameter it is assumed that blur occurs at the point equally in all directions. According to the proposed approach, it is possible to determine the distance to the studied object and its blur by analyzing a series of images obtained using the video detector with different settings. The article proposes and scientifically substantiates new and improved existing methods for detecting the parameters of static and moving objects of control, and also compares the results of the use of various methods and the results of experiments. It is shown that the aggregate method gives the best approximation to the real distances.

## 1. Existing Approaches

The stream of digital images may be used to determine the distance to a real-world object that manages the intellectual environment by using natural and intuitive communication of the information.

In modern computer vision systems ToIE (technology of intelligent environments) there are several ways to obtain additional information about the behavior and state of tracked objects: multisensory approach (stereovision) [1, 2], construction of image perspective [3], the use of fixed camera and additional lighting of the object [4], the use of methods of wave analysis [5-7], as well as a special video detector calibration [8-10]. Each of these methods could be used in practice depending on the task [11, 12]. Among the many methods for determining the parameters of behavior and state of a tracked object, the most popular are techniques of stereo vision; determination of blur for a given motion of the imaging detector; definition of blurring for different colors included in the Full HD image of the object. Each method has its advantages and disadvantages, most often associated with technical and software opportunities. The paper develops a method of determining distances to objects by analyzing a series of images and assessment of depth using defocusing [7-9] using its aggregation with stereoscopic vision. This method is based on a physical effect of dependence on the determined distance to the object on the obtained image from the focal length or aperture of the lens. With regard to the problem of determining the external defects of the railway, the distance and speed of objects on the railway, a system using the presented method on track geometry car can be mounted (figure 1). For developed systems, it is proposed to use ontology-based interfaces [6], whose advantages are presented in the paper [7].

**Figure 1.** View from the track geometry car.

## 2. Method implementation

When the video detector focuses on the object at a certain distance, other objects that are closer or further the focal point, form a blur circle depending on the distance on the plane of the image. Blur of the image of the object can be different nature and be determined by motion of the object or detector, boundaries of the object on the image, an aggregate state of the object, as well as various settings of the video detector (focal length, shutter speed, and aperture). As the defining the relations between the main parameters of the video detector and the location of the object in the exposition the following expression is used [8]:

$$1/f = 1/D_{ob} + 1/D_{fip}, \tag{1}$$

where $f$ – the focal length; $D_{ob}$ – distance from a given point of the object to the lens of video detector; $D_{fip}$ – the distance from the center of the lens to a focused image of the object.

In the calculation of the blur spot diameter it is assumed that blur occurs at the point equally in all directions [17]. According to the proposed approach, it is possible to determine the distance to the studied object and its blur by analyzing a series of images obtained using the video detector with different settings. Figure 2 (blue line) represents the averaged dependence of the blur from the distance to the object when using three video detectors and the distance to the object in the focus is 1.2 m. For more accurate estimates of the definition of geometric parameters of behavior and the state of the studied object a statistical approach to define specific parameters and to evaluate their accuracy is used. Object's blur is calculated as an average value for all measurements of blur $\sigma_i$ of the object at points of its boundary:

$$\sigma = \frac{1}{n}\sum_{i=1}^{n}\sigma_i, \tag{2}$$

where $n$ — the number of measurements.

IOP Conf. Series: Materials Science and Engineering **317** (2018) 012021 doi:10.1088/1757-899X/317/1/012021

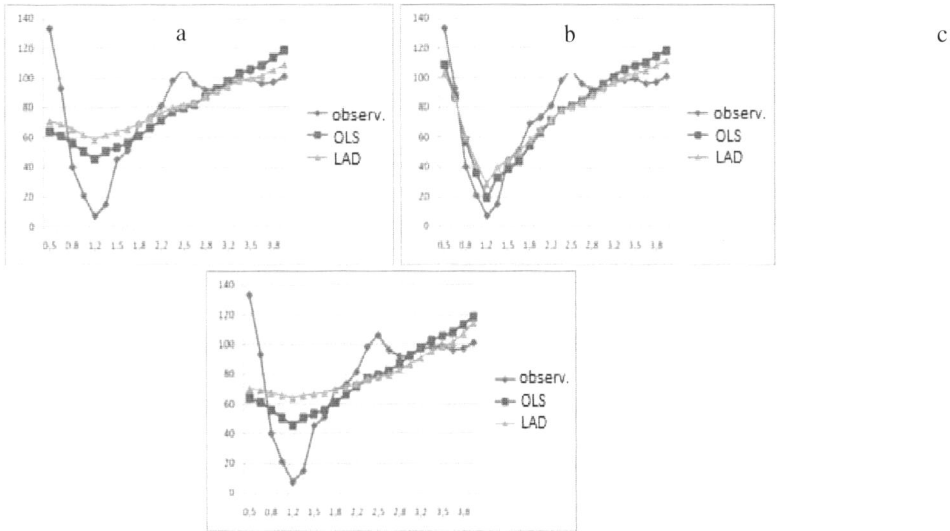


**Figure 2.** The dependence of the blur from the distance to the object for the standard functions:

a) linear; b) logarithmic; c) exponential.

The statistical approach to evaluate the deviation the dependence of the distance and the blur from the different types of standard functions (logarithmic, exponential, linear) (figure 2) is used. The statistical approach includes the method of least squares (OLS) and the method of least absolute deviations (LAD), as well as Bayesian estimates, for which it is necessary to minimize the risk for different loss functions (quadratic, rectangular, linear) [11,13].

$X = (X_1, \ldots, X_M)^T$ – object location vector (distance to object) and $\theta_1, \theta_2, \ldots, \theta_n$ – results of measurements of the average blur of an object. Each observation can be represented as a sum:

$$\theta_i = \eta_i(X) + \Delta_i, \tag{3}$$

where $\eta_i(X)$ – the known function, relative to the calculated estimate of the distribution; $\Delta_i$ – the measurement error.

To estimate the vector of the parameters of the distance to the object is proposed to use a class of statistical estimates M – estimates (that is estimates that provide the minimum sum of any of the functions from data):

$$X(\theta) = \arg\min_{x}\left(\sum_{i=1}^{n} L(\theta_i - \eta_i(X))\right), \tag{4}$$

where $L(x)$ – the valuation function.

For least-squares estimation the function $L = z^2$ is used, by the method of least modules the function $L = |z|$ is used. According to the analysis of these estimates the function $\eta_i(X)$ closest to observations is chosen. The function L(x) is chosen in such a way as to ensure the desired estimation properties (unbiasedness, consistency and efficiency) under conditions when data is taken from a known distribution, and the sufficient resistance to deviations from this distribution. The effectiveness of estimates relative to each other can be found by comparing the variances of estimates: the evaluation, whose dispersion is the smallest (i.e. less variance) is more effective. For sufficient resistance to deviations from the given function, it is necessary to build robust (sustainable) processing algorithms, that is, algorithms that are highly effective under conditions where the characteristics of the error distributions vary within the given classes.

There are two options for obtaining stable estimates of observations $\theta_i$ (3): a variation of a known function $\eta_i(X)$ with unknown error distribution $\Delta_i$ and using standard deviation of observations $\theta_i$; the use of the standard deviation for various functions of distributions of observation errors.

Using the methods of least squares and smallest modules, we estimate the unknown parameters of the chosen function $\eta_i(X)$. By substituting the functions of the linear form $\eta_i(X) = a + bx_i$, the estimated parameters $\hat{a}$ and $\hat{b}$ will be calculated according to the formulas:

$$\hat{b} = \left(\overline{x\theta} - \overline{x}\overline{\theta}\right) \Big/ \left(\overline{x^2} - \overline{x}^2\right); \quad \hat{a} = \overline{\theta} - \hat{b}\overline{x}, \tag{5}$$

where $\overline{x}$, $\overline{\theta}$, $\overline{x\theta}$ - the mathematical expectation of the distance to the object, the mathematical expectation of the measured observable values and their multiplication accordingly.

As a function $\eta_i$ (X) we consider linear, exponential, and logarithmic functions (figure 3):

$$\eta_i(X) = a + b\left|x - x_0\right|, \quad \eta_i(X) = a + b\left|\ln(x - x_0)\right|, \quad \eta_i(X) = a + b\exp\left(\left|cx - x_0\right|\right) \tag{6}$$

Assume that, based on the sample of the measurement results $\theta_1$, $\theta_2$, ...$\theta_n$, estimates are obtained: $X_1 = \theta_1^*$ (the OLS method) and $X_2 = \theta_2^*$ (the LAD method). In order to the final evaluation was unbiased and effective, one should choose the valuation method of processing observations that will give the least variation of estimates. Then the algorithm of final evaluation will be as follows:

$$X = W_1X_1 + W_2X_2; \quad W_i = \begin{Bmatrix} 1, if D_i \le D_j \\ 0, if D_i > D_j \end{Bmatrix}; \quad i = 1,2; j = 1,2; j \ne i, \tag{7}$$

where $D_i$, $D_j$ – selective variances of estimates by the OLS and the LAD algorithms.

## 3. An integrated approach based on the algorithms of stereovision and image blur

Simultaneously with the analysis of defocused images, it is also possible to use stereovision, then we can talk about an integrated approach to determining the characteristics of the objects. In this case, the features of two methods are taken into account, the parameters of methods are compared for further use in a single algorithm.

In the study of aggregation methods of stereovision and evaluation of the blur, it is necessary to determine the boundaries within which the methods should be used, to estimate the error of these methods and also to consider the possibility of replacing one method by another by introducing refinement coefficients. If the stereo vision system is configured accurately, then the adding of blur analysis method expands the limits of applicability of the method. If there are obstacles for one of the cameras or the scopes don't intersect in the system of stereo vision, the method of blur analysis also expands the applicability of the aggregate method.

Otherwise, blur analysis can be used to determine the refinement coefficient k from formula (7) by applying OLS. For this, it is necessary to consider the distance where it is possible to apply the blur analysis method, that is, where the error of this method is not more than the threshold of 5% (for example, when using cameras with 50 mm focal length the distance up to 3 m is considered), and at the same time it is necessary to memorize the difference of the object's location in the coordinate system of the left and right cameras using the method of stereo vision. The distances are calculated by formulas:

$$r_{stereo} = k \cdot f \cdot d / (\Delta x \cdot Sx); \quad r_{blur} = \eta(\sigma), \tag{8}$$

where $\eta(\sigma)$ – the function obtained from (4) with minimization of errors (from the results of the experiments $\eta(\sigma) = e^{|(\sigma-a)/b|} + 0.2$).

So, we get:

$$\sum_N \left(r_{blur} - r_{stereo}\right)^2 \rightarrow min \tag{9}$$

Taking into account the above formula, we obtain for N observations in the range of distances where the use of the blur analysis method is most effective (up to 3 m or up to 6.5 m – given the camera's refocusing [8]):

$$\sum_{N}\left(\eta(\sigma)-k\cdot f\cdot d/\left(\Delta x\cdot Sx\right)\right)^{2}\rightarrow min \tag{10}$$

Thus, we find the refinement coefficient *k*. In determining the distances at which the use of the blur analysis is impossible due to the large error, one can use the method of stereo vision with the use of the refinement coefficient *k*. The use of the aggregate method helps to increase the measured distance up to 10 m with an error of less than 5%. The accuracy of the proposed approach at such distances higher than separately applying of the blur analysis or method of stereovision without a refinement coefficient.

## 4. Results of experiments

The variances of errors and the mathematical expectation for each of the taken as known functions (6-9) are shown in figure 3. It is seen that the variance of the errors from function, the parameters of which are estimated by OLS will be less than the variance of the errors by LAD, that is, the estimate of the least squares is more stable. Also, the estimation of errors with the use of least squares is unbiased, while the mathematical expectation when using LAD is not zero, that indicates the bias of error estimates.

To smooth out possible "emissions", we apply the Kalman filter to the initial observations and analyze estimations of OLS and LAD for the three functions discussed above after application of the filter.

The Kalman filter algorithm is based on induction and in the simplest case, when using a linear function as a theoretical function, it can be represented in the form of an iterative formula:

$$\theta_{k+1}^{opt}=K\cdot\theta_{k+1}+(1-K)\cdot(\theta_{k}^{opt}+u_{k}),\quad u_{k}=\eta(x_{k+1})-\eta(x_{k}), \tag{11}$$

where $0\leq K\leq 1$ – Kalman coefficient; $\theta_k^{opt}$ and $\theta_{k+1}^{opt}$ – the current and the following filtered values of observations; $\theta_{k+1}$ – the following observation; $u_k$ – a known value responsible for the evolution of the system, $\eta(x_k)$ and $\eta(x_{k+1})$ – the current and the following values of the theoretical function $\eta$.

For logarithmic and exponential functions, the equation will change due to the nonlinearity of the increments:

$$\theta_{k+1}^{opt}=K\cdot\theta_{k+1}+(1-K)\cdot(\theta_{k}^{opt}\cdot u_{k}),\quad u_{k}=\eta(x_{k+1})/\eta(x_{k}), \tag{12}$$

As the initial filtered observation the first observation is taken, i.e. $\theta_1^{opt}=\theta_1$. Using Kalman filter it is possible to reduce the error variance by appropriately selecting the coefficient K in the filter. Using the method of least squares and the method of least modules we will estimate the parameters of the function $\eta_i$ (X) in different dependencies of the observations, passed through the Kalman filter (figure 3).

Thus, using the Kalman filter and taking into account the variation of its coefficients in the analysis of the obtained functions the error distributions have a smaller variance, than without the use of a Kalman filter, that is, the evaluation is more efficient, hence it can be concluded that the Kalman filter provides an opportunity to obtain a more accurate theoretical function. Estimate with the least variance is the OLS estimation for the theoretical function (7) with parameters *a* = 10.696, *b* = 93.122 using the Kalman filter with coefficient K=0.1.

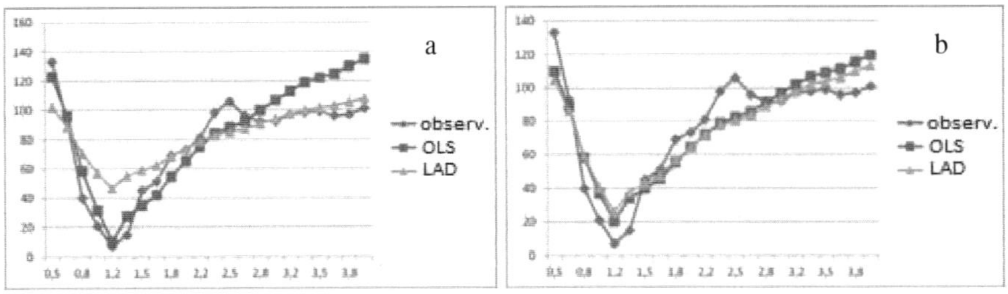


**Figure 3.** Dependence of the variance of the distance for a logarithmic model with the coefficient Kalman: a) – with coefficient K=0.1; b) – with coefficient K=0.7.

## 5. Results of using the combined method

The limits of applicability of the method depend on the selected focal length of the camera, exposure and pixel size of CCD matrix camera, stereobase. For the focal length of 50 mm and stereobase – 0.05 m, the limits of applicability of the developed method are presented in figure 4.

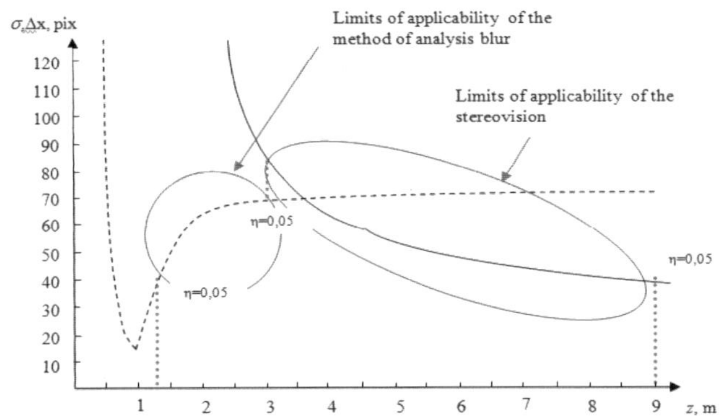


**Figure 4.** Limits of applicability of aggregated method.

**Figure 5.** Detection of external railway defects.

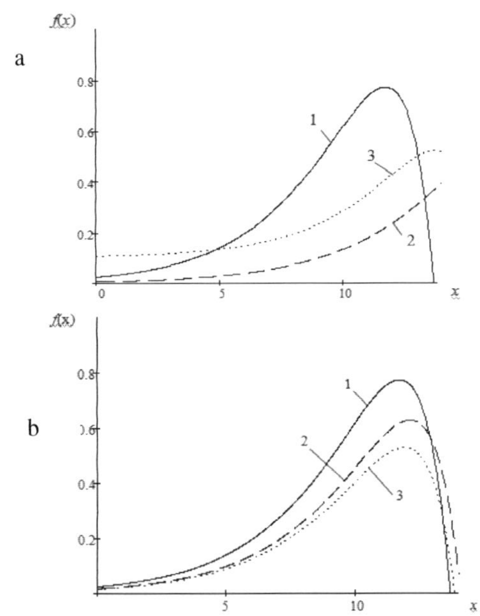


**Figure 6.** Dependence of accuracy of the distance determination to an object for different methods (a), for different focal lengths of video cameras (b).

The operation of the system under real conditions for detection of railway defects is presented in figure 5.

Figure 6a shows that using the combined method (curve 1) the accuracy of the measurement of the distance to the object is substantially increased in comparison with the methods based on stereo vision (curve 2) and the estimation the blurring of the image of the identified object (curve 3). The actual distance to the object at the time of photographing is 11.8 м. In figure 6b, a comparative analysis of the accuracy of determining the distance to the object with a change in the focal length: curve 1 corresponds to the focusing point at 11.0 m, curve 2 – 11.5 m, curve 3 – 10.5 m.

## 6. Conclusion

Studies have shown that for determination of parameters of the function which most accurately describes the dependence of the distance to the object from the blur, the optimal method is the method of least squares. The article proposes and scientifically substantiates new and improved existing methods for detecting the parameters of static and moving objects of control, and also compares the results of the use of various methods and the results of experiments. It is shown that the aggregate method gives the best approximation to the real distances. In general, the method proposed in this paper for determining the geometric characteristics of an object from the analysis of its image and the algorithm of estimation the values obtained with its help can be used in complex monitoring, diagnostic and control systems.

## References

[1] Kraft H, Frey J, Moeller T, Albrecht M, Grothof M, Schink B, Hess H, Buxbaum B 2004 *3D-camera of high 3D-frame rate, depth-resolution and background light elimination based on improved PMD (photonic mixer device)-technologies* OPTO 2004, AMA Fachverband, Nürnberg, 45-9.

[2] Kuhnert K D, Langer M, Stommel M, Kolb A 2007 *Dynamic 3D-Vision* Vision Systems: Applications. 311-34.

[3] Akimov D, Vatolin D, Smirnov M 2011 *Single-Image Depth Map Estimation Using Blur Information.* Proc. of the 21st GraphiCon International Conference on Computer Graphics and Vision. 112-6.

[4] Loktev A A, Loktev D A 2008 *Tech. Phys. Lett.* **34(11)** 960-3.

[5] Subbarao M, Yuan T, Tyan J K 1997 *Integration of Defocus and Focus Analysis with Stereo for 3D Shape Recovery* Proc. of SPIE Conference on Three Dimensional Imaging and Laser-Based Systems for Metrology and Inspection III, V. 3204, Pittsburgh PA, 214-9.

[6] Loktev A A, Loktev D A 2015 *Contemp. Eng. Sc.* **8(11)** 467–5.

[7] Lin H Y, Chang C H. 2006 *Opt. Eng.* **45(12)** 1-12.

[8] Levin A, Fergus R, Durand Fr, Freeman W T 2007 *ACM Trans. Graph.* **26(3)** 124-32.

[9] Cremers D, Soatto S 2005 *Int. J. Comp. Vis.* **62(3)** 249-65.

[10] Alfimtsev A N, Loktev D A, Loktev A A 2012 *Vestnik MGSU,* **11** 242–52.

[11] Jiwani M A, Dandare S N 2013 *Int. J. Sc. Res. Publ.* **3(6)** 1-6.

[12] Loktev A A, Loktev D A 2015 *Contemp. Eng. Sc.* **8(17)** 789–97.

# Amorphous Silica Micro Powder Additive Influence on Tensile Strength of One-Ply Particle Board

A V Pitukhin[1], G N Kolesnikov[2], N G Panov[3], S B Vasilyev[4]

[1]Professor of Petrozavodsk State University Petrozavodsk, Russia
[2]Professor of Petrozavodsk State University Petrozavodsk, Russia
[3]Docent of Petrozavodsk State University Petrozavodsk, Russia
[4]Professor of Petrozavodsk State University Petrozavodsk, Russia

**Abstract.** The methods and results of experimental investigation on the additive influence of amorphous silica micro powder when mixed in the glue for one-ply particle board are presented in the article. Wooden particles of coniferous and hardwood species as well as glue solution based on carbamide-formaldehyde resin were used for boards manufacturing. The amorphous silica micro powder contained particles on the average 8 μm by the size and specific surface 120...400 m2/g was used in experiment. The samples were tested to determine their physical-mechanical properties. It was found that 1 % amorphous silica micro powder additive increases the breaking point of one-ply particle board under tensile stress by 143 %.

## 1. Introduction

On the basis of the previous investigations and literature data, the hypothesis of shungite nanopowder influence on physical-mechanical properties of wood particle board (PB) was formulated. The first test of this hypothesis was done in the course of semi-industrial experiments for one-ply PB production with the shungite nanopowder use. The test results [1-3] proved that shungite nanopowder added in glue solution considerably increased the strength under static tensile. The investigation has shown that the schungite nanopowder when mixed in the glue increases the physical-mechanical properties of a particle board. However, it should be noted that the cost of nanostructured powders of shungit is currently high. Also it remains unclear which of the components of shungite substance has the greatest effect on the strength of the glue and particle board as a whole. In our study shungit consists about 30 % of carbon and 60 % of silica. Silica is a more available and cheap material. Therefore, it was planned to investigate the effect of silica addition on physical-mechanical properties of particle boards.

Here we present the methods and results of experimental investigation on the additive influence of amorphous silica micro powder (ASMP) when mixed in the glue for one-ply PB based on carbamide-formaldehyde resin (CFR). It should be mentioned that solutions were made both with ASMP and without ASMP additives. The ASMP additives content was from 0.10 to 1.00 % of absolutely dry CFR mass. The purpose of the investigation was to determine the breaking tensile point. The results obtained were used for the analysis of the change of physical-mechanical properties of one-ply PB, influenced by ASMP additive mixed into glue solution based on CFR.

## 2. Investigation materials and methods

The wooden particles to manufacture the boards were produced out of mixture of coniferous (fur, pine) wood and hardwood (birch, aspen). The mixture composition was as follows: coniferous particles mass share made 60 %, and hardwood particles share made 40 %. The porosity of particle mixture made 80 %, and the density made 309 $kg/m^3$. The particle size was such that during

IOP Conf. Series: Materials Science and Engineering **317** (2018) 012022 doi:10.1088/1757-899X/317/1/012022

stratification they have passed through a sieve with holes of 3 mm diameter. The wooden particles moisture content made 4 %.

Under experimental conditions the CFR was used with the following physical-chemical resin property indices: dry residue mass share – 68 %, gelatinization time at 100 °C – 68 seconds, conditional viscosity (nuzzle 4 mm) – 80 seconds, hydrogen ions concentration at 20° C – 8.1, refraction coefficient at 20° C – 1.471, ultimate miscibility resin with water – 1:4.5.

The reference density of the board was 850 kg/m$^3$. The content of absolutely dry resin in the PB was 10 % of absolutely dry wood mass.

The average particle size of ASMP – 8 microns. The nanoporous structure of surfaces of the particles provided the specific surface 120...400 m$^2$/g.

Five types of glue solutions for boards manufacturing were composed. The first one (the basic glue solution) was prepared of 66 % water CFR solution and 20 % water hardening agent $NH_4Cl$. The content of hardening made 1 % of resin solution mass. Four other glue solutions were mixtures of basic one and ASMP additive. The ASMP addition was as follows: 0.10 %, 0.25 %, 0.50 %, 1.00 % of absolutely dry resin mass.

The sample PBs were manufactured with the use of laboratory equipment: a mixer and a hot press. The resinification of wooden particle was conducted in a laboratory mixture. The packages for boards shaping were made manually. The cold sub-pressing of packages was performed before pressing in hot press. The hot pressing was carried out at 190° C, pressure 2.5 MPa, the joining velocity of hot press boards was 3 mm/min. The package exposure in press after board joining was 2 minutes. After extracting a ready board out of the press, it was subjected to conditioning. In the course of the experiment one-ply particle boards with the thickness 16 mm, by dimension 280□280 mm were produced.

The samples for experimental investigation of their physical-mechanical properties were made out of these boards. To determine one and the same index the samples were cut out of different board parts. The sample length difference was not more than 0.5 mm. The sample thickness difference was not more than 0.2 mm. The external sample surfaces were smooth. Also there were no edge chips and corner crumbles. Before testing the samples were conditioned during 24 hours at 20 ± 2 °C and relative air humidity 65 ± 5 %.

The tensile strength perpendicular to the board face was found with the use of the samples with 50 × 50 mm dimension. Special blocks were glued to the sample faces with the help of melting glue. The testing block was put into pinchers of the testing device, and then the distance between them was being increased with constant speed 10 ± 1 mm/min up to breaking point. Maximum loading was registered with arrow indicator with 1% precision.

## 3. Result and discussion

According to the investigation results it is established that the breaking point under tensile stress of one-ply PB made with the use of the glue solution on the basis of CFR and 1.00 % ASMP additive from absolutely dry resin mass exceeds the above-mentioned control sample value by 143.4 % (see Table 1).

**Table 1.** Experimental results for breaking point under tensile stress.

| Tensile strength | ASMP content (x), % | | | | |
|---|---|---|---|---|---|
| | 0.00 | 0.10 | 0.25 | 0.50 | 1.00 |
| *Mean value, MPa (y)* | 0.99 | 1.01 | 1.23 | 1.33 | 1.42 |
| *STD, MPa* | 0.02 | 0.24 | 0.19 | 0.12 | 0.11 |

Thus, one can conclude that the ASMP additive increases the one-ply PB breaking point under tensile stress. It was also found that test data show a direct relationship between the breaking point value under tensile stress and ASMP content (figure 1).

Earlier researchers have found, that nano and micro powder of amorphous silica when mixed in the glue change its viscosity [4, 5]. This leads to an improvement of adhesive properties of glue composition and increases PB strength. From the other hand it was established that silica additive can build a reinforcing nanostructure mesh inside gelatinized glue [1]. This structure is the main cause of strength increase. Therefore, under certain conditions, an increase of shungite and silica additives does not increase the strength of the PB.

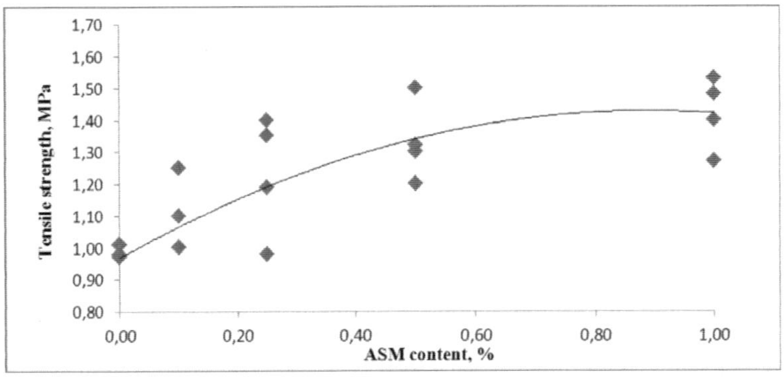


**Figure 1.** Polynomial approximation of the breaking point value under tensile stress of ASMP content.

## 4. Conclusions

The findings of our experimental investigation are as follows:

1. It has been established that the breaking point under tensile strength of one-ply PB made with the use of the glue solution on the basis of CFR can be increased by ASMP additive;

2. It was established that within the extremum was reached when ASMP additive content was 1.0 % of absolutely dry resin mass.

3. In the studied circumstances the correlation between ASMP additive content and breaking point under bending strength may be described by the equation $y = -0{,}5855x^2 + 1{,}0374x + 0{,}9675$, where y is the breaking strength under bending, MPa; x is the ASMP content, %.

## References

[1] Pitukhin A V, Vasilyev S B, Panov N G, Kolesnikov G N 2014 Austrian J. Tech. Nat. Sci. **7-8** 22-4.

[2] Panov N G, Pitukhin A V, Rozhkov S S, Tsvetkov V E, Sanayev V G, Firulina O V 2012 *Wood chipboards of carbamide-formaldehyde resin modified with nanosized shungite* Proceedings of Moscow Forest State University. **2** 135-8.

[3] Panov N, Rozhkov S, Pitukhin A 2011 *Water Resistance increase in three-ply particleboards based on carbamide-formaldehyde resin resulting from injection of nanostructured shungite filler into adhesive*, Proceedings of Petrozavodsk State University. **8** 88-93.

[4] Zhang Z et al. 2016 *Self-formation of elastomer network assisted by nano-silicon dioxide particles: A simple and efficient route toward polymer nanocomposites with simultaneous improved toughness and stiffness*, Chem. Eng. J. **285** 439–48.

[5] Pitukhin A V, Vasilyev S B, Panov N G, Kolesnikov G N 2016 Micro and Nano Powder Additive Influence on Bending Strenght of One-ply Particle Board Technology Ltd, 82–5.

# Corrosion Effects on the Fatigue Crack Propagation of Giga-Grade Steel and its Heat Affected Zone in pH Buffer Solutions for Automotive Application

**H S Lee**[1]

[1]Department of Automotive Management Engineering, Joongbu University, Goyang-si, Gyeonggi-do, 10279, South Korea

**Abstract.** Corrosion fatigue crack propagation test was conducted of giga-grade steel and its heat affected zone in pH buffer solutions, and the results were compared with model predictions. Pure corrosion effect on fatigue crack propagation, particularly, in corrosive environment was evaluated by means of the modified Forman equation. As shown in results, the average corrosion rate determined from the ratio of pure corrosion induced crack length to entire crack length under a cycle load were 0.11 and 0.37 for base metal and heat affected zone, respectively, with load ratio of 0.5, frequency of 0.5 and pH 10.0 environment. These results demonstrate new interpretation methodology for corrosion fatigue crack propagation enabling the pure corrosion effects on the behavior to be determined.

## 1. Introduction

Recently, the application of ultra-high strength steel sheet has been increasing in the automobile industry to improve the fuel efficiency by reducing weight of the vehicle body [1]. It is very important to analyse the characteristics of corrosion of giga-grade steels developed for this purpose, and to find ways to apply them to actual vehicles. The corrosion resistance characteristic of steel is caused by a thin layer of oxides in particular chromium and iron oxides, and despite a thickness of a few nanometers, this oxide layer is so strong that it effectively isolates the steel from the environment 2-3. Once a layer has formed, the metal has become passivated, and the oxidation or rusting rate would slow down to less than 0.05 mm per year. Unfortunately, this kind of ideal process does not always take place; the oxide layer has been damaged without re-passivation, and the result has been shown serious corrosion [2]. Since many study and chemical applications have been conducted, most of the cases were done in NaCl or acid environments. However, the other cases in chemical industries and power plants deal with alkaline solutions and use giga-grade steels in a broad area of products including transfer pipe, heat exchanger, container, etc. [3]. In this study, the fatigue crack propagation (FCP) and corrosion fatigue crack propagation (CFCP) were investigated by experimentally in air together with pH 7.0 of $KH_2PO_4(1\%)+Na_2HPO_4(1\%)+H_2O(98\%)$ and pH 10.0 of $Na_2CO_3(1\%)+Na_2B_4O_7.10H_2O(1\%)+H_2O(98\%)$ environments, respectively. And then, a methodology to evaluate the corrosion effect on fatigue crack propagation in corrosive environment has been established based on the modified Forman equation by using the experimental da/dN–$\triangle$ K relations and average corrosion rate. The mechanisms of corrosion fatigue crack propagations, together with the pure corrosion effect on the crack propagation thus can be determined by the newly developed scheme.

## 2. Chemical reaction in corrosive environment

The chemical reactions of giga-grade steel with pH buffer solutions were investigated. Buffer solution of pH 7 is chemically reacted as $KH_2PO_4 + H_2O \rightarrow H_3PO_4 + KOH$ with $PO_4^{3-} + H_2O \rightarrow HPO_4^{2-} + OH^-$. In case of pH 10, it is reacted as $Na_2B_4O_7 + H_2O \rightarrow H_3BO_3 + NaOH$ with $CO_3^{2-} + H_2O \rightarrow HCO_3^- + OH^-$. Both of the processes produce $OH^-$, and the $OH^-$ ions reacted with $Fe^{2+}$ ions from metal dissolution processes at anodic sites. The formation of anodic sites by disruption of the protective passive film on the metal surface leads to the metal dissolution. The anodic iron dissolution reaction is presented by $Fe \rightarrow Fe^{2+} + 2e$. This is balanced by the cathodic reaction of oxygen on the adjacent surface as $O^2 + H_2O + 2e \rightarrow 2OH^-$. Due to the continuing Fe dissolution, an excess of positive ions $Fe^{2+}$ is accumulated in the anodic area, which are necessary for the growth of pitting. In addition, a presence of $Na^+$ ions affects to the acceleration of anodic dissolution especially at the tip of the crack produced by pitting as an anode despite of the surface in non-pitting region as a cathode because the $Na^+$ ions act as a catalyst in chemical ionization of $Fe(OH)_2$ as $Fe(OH)_2 + 2Na^+ \rightarrow Fe^{2+} + 2NaOH$. Consequently, the larger and deeper pits which formed on the surface of the specimens exposed in pH buffer solutions. And the presence of $OH^-$ ions affects to the hydrolysis as $Fe^{2+} + 2OH^- \rightarrow Fe(OH)_2$ and also to the rust formation as $4Fe(OH)_2 + 2H_2O + O_2 \rightarrow 4Fe(OH)_3$.

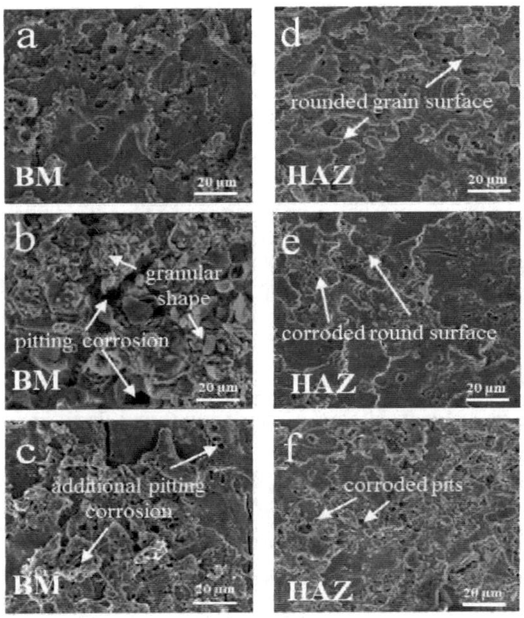


**Figure 1.** SEM observations of giga-grade steel immersed in buffer solutions of pH 7 and pH 10 for 7 days; (a) BM in air, (b) BM in pH 7.0, (c) BM in pH 10.0, (d) HAZ in air, (e) HAZ in pH 7.0, (f) HAZ in pH 10.0.

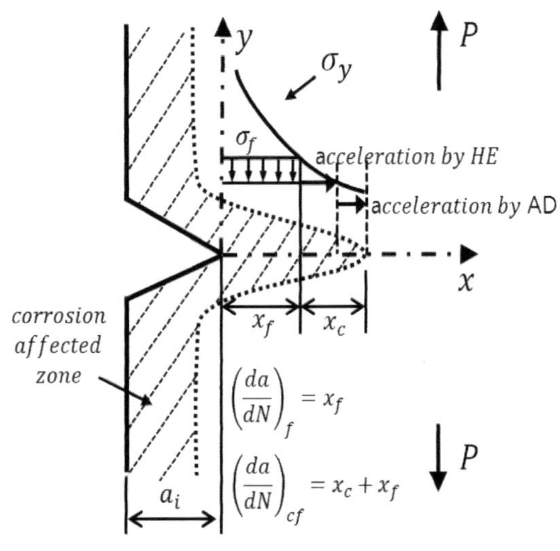


**Figure 2.** Schematic diagram showing the corrosion fatigue crack propagation process [4].

Based on these chemical reaction mechanisms, base material (BM) and heat affected zone (HAZ) of giga-grade steel were prepared for experiment. Before commencement of the corrosion fatigue test, specimens of giga-grade steel were in buffer solutions of pH 7 and pH 10 for 7 days resulting in pre-chemical reactions and the microstructures obtained at this stage are presented in figure 1. Figures 1a, 1b and 1c show the corroded surface of BM in air, pH 7 and pH 10 environments, respectively. Pitting corrosions were presented with rust formed granular shape on the surface of specimen in pH 7

IOP Conf. Series: Materials Science and Engineering **317** (2018) 012023 doi:10.1088/1757-899X/317/1/012023

environment and additional pitting corrosions were evolved on the granular shape surface. Figures 1d, 1e and 1f show the corroded surface of HAZ in air, pH 7 and pH 10 environments, respectively. HAZ had more rounded grain surface in air environment compare to the BM in figure 2d. The round grain surface was corroded in pH 7 in figure 1e and severely corroded pits were evolved in pH 10 environment in figure 1f.

## 3. Corrosion fatigue crack propagation

It is kwon that the corrosion fatigue crack propagation (CFCP) is caused by the combination of corrosion damage and relative slip of the metal at the crack tip under cyclic loading. This process is controlled by the three major mechanisms: relative slip of the metal at the crack tip, the bare surface corrosion of the fractured material elements and the crack closure. The corrosion in the CFCP process may occur mainly on the bare surface after the material elements are fractured [4]. In aqueous environment, two possible mechanisms are associated with the corrosion damage on the CFCP process: anodic dissolution (AD) at the crack tip, and hydrogen penetration by AD into the material elements resulting in hydrogen embrittlement (HE). In consequence, both AD and HE lead out the acceleration of crack propagation where it is not the case of air environment [5]. As shown in figure 2, as a state of pre-cracked length of $a_i$ in the specimen, a material element exposing crack tip may be fractured by applied load, P when the level of stress at the crack tip reaches failure strength of the material. Only considering the stress effect without any corrosion such as FCP in a cyclic loading hysteresis, following processes occur: crack opening by increasing load - crack propagation due to fracture of the material element at crack tip - crack closing by decreasing load - crack opening [6]. However, in a corrosive environment, CFCP may occur by the processes: crack opening by increasing load - longer crack propagation due to decrease of failure strength by HE - additional crack propagation with blunting at the crack tip by AD - crack closing by decreasing load - crack opening [4]. Herein, the rate of fatigue crack propagation (FCP) is given by $\left(da/dN\right)_f = x_f$, where $x_f$ is the length of crack propagation in a loading cycle. Meanwhile, the CFCP will be accelerated by the two different mechanisms, HE and AD in the corrosive environment as mentioned before. Total amount of the CFCP length in a cyclic load therefore gives the CFCP rate as:

$$\left(\frac{da}{dN}\right)_{cf} = x_c + x_f \tag{1}$$

In here, $x_c$ is the length of additional crack propagation in a cycle due to the corrosion effect. Although Wang's CFCP model [4] is pretty rare in corrosion fatigue study, the equation has still some limitation on fitting the rapid propagation region III such as the changing slope according to varying load conditions due to the fixed exponent constant, in Paris model. Also, the effect of load frequency was not considered in the equation properly. Therefore, an adequate model which can describe both the FCP and the CFCP behaviours was required to reveal the significance of the parameters related to environmental effects. In order to identify the pure corrosion effect on fatigue crack propagation in corrosive environment, the average corrosion rate ($C_a$) during one loading period $T$ is considered as follows

$$C_a = \frac{1}{T}\frac{x_c}{x_c + x_f} = f\frac{x_c}{x_c + x_f} \tag{2}$$

And then, the $x_c$ term can be rearranged as $x_c = C_a x_f / \left(f - C_a\right)$.

Generally, fatigue crack propagation behaviour of a material has been described by the FCP rate, $da/dN$ against the stress intensity factor range (SIFR), $\Delta K$. Among the several models, the modified Forman equation enables the FCP behaviour to be modeled within the full range of crack propagation such as an initial crack propagation of threshold region (Region I), stable crack growth (Region II), and fast fracture region with rapid-unstable crack growth (Region III). The material's characteristics of FCP behaviour by considering the modified Forman equation give the formulation:

$$\left(\frac{da}{dN}\right)_{fat} = x_f = \frac{M\left(\Delta K - \Delta K_{th}\right)^N}{\left(1-R\right)K_{crit} - \Delta K} \tag{3}$$

where, two parameters, $M$ and $N$ and $\Delta K_{th}$ value at a certain level of SIFR were selected as the characteristics of the FCP behaviour. $M$ and $N$ were indicating the intercept of $da/dN$ -axis and the slope of the curve in linear region, and the subscript *th* and *crit* indicates threshold value and critical value of SIFR, respectively.

Substituting (2) and (3) into equation (1), a CFCP model for determination of corrosion effect can be expressed:

$$\left(\frac{da}{dN}\right)_{cf} = x_c + x_f = \frac{C_a x_f}{f - C_a} + x_f = \left(1 + \frac{C_a}{f - C_a}\right)\left[\frac{M\left(\Delta K - \Delta K_{th}\right)^N}{\left(1-R\right)K_{crit} - \Delta K}\right] \tag{4}$$

In here, $x_f$ and $C_a$ in $x_c$ term can be obtained from the results of air condition, and corresponding pH and loading conditions, respectively. By comparing the FCP and CFCP behaviours of giga-grade steel in air and pH buffer solutions such as pH 7.0 and pH 10.0, and consequently the corrosion effects in alkaline environment therefore can be analysed.

**Table 1.** Thermal cycles for HAZ simulation.

| Method | Preheat temp. (°C) | Peak temp. (°C) | Heating rate (°C/s) | Cooling rate (°C/s) | Cycles and heat input |
|---|---|---|---|---|---|
| GTAW | 25 | 410 | 258 | 5 | 1 cycle (16kJ/cm) |

Experiments were conducted by using the giga-grade steel as a base metal (BM) and heat affected zone (HAZ) to investigate FCP and CFCP of the materials. Table 1 shows the welding cycles to develop the HAZ with relevant temperature cycle which was calculated by the Rosenthal equation for a thick plate [7] as follows:

$$T_t - T_0 = \frac{q/v}{2\pi\lambda t}\exp\left(-\frac{r^2}{4bt}\right) \tag{5}$$

where $T_t$ is the temperature (K), $T_0$ is the pre-heat temperature, $q/v$ is the heat input (kJ/cm), $\lambda$ is the thermal conductivity of giga-grade steel (J/msK), $t$ is the time, $b$ is the thermal diffusivity of giga-grade steel ($m^2$/s), and $r$ is the radial distance from the heat source. In this work, the HAZ was simulated for single-pass welding by using gas Tungsten arc welding (GTAW). A three-point single edge notched bending (3-SENB) fixture and specimens were used in the FCP and CFCP test (ISO 11782-2). A 6-mm notch was produced on every specimen, followed by formation of a pre-crack at the tip of the notch under cyclic fatigue load with load ratios of 0.1, 0.3, 0.5 and load frequencies of 0.1, 0.3, 0.5. FCP and CFCP tests were conducted at room temperature with a 50 KN capacity universal testing machine. As mentioned previously, three levels of the load ratio (*R*) and the frequency (*f*), respectively, 0.1, 0.3 and 0.5, were considered. In addition, CFCP test was conducted by using the 7 days immersed specimens in the same environment condition. In alkaline environment, the corrosion resistance of material is generally reduced by the hydrogen embrittlement. The hydrogen embrittlement is due to the chemical absorption and molecular dissociation which results in the formation of atomic hydrogen, $H^+$. The atomic hydrogen migrates to pitting root regions of high stress concentration. The absorption of the hydrogen ions into the pitting root as a crack tip, through a dislocation motion [8] or attracting by the hydrostatic stress gradients [9], promotes embrittlement of the crack tip material. Microstructures of BM and HAZ in air and pH buffer solutions are shown in figure 3. Compare to the air condition in figures 3a and 3d, pitting corrosion evolution aspects are shown in figures 3b and 3c for BM and figures 3e and 3f for HAZ. The corroded pits were produced with narrow hole-shape in pH 7.0, and enlarged corroded pits were generated in pH 10.0. Overall, faster corrosion evolution occurred in HAZ than BM. Corroded pits occurred mainly in the presence of hydrogen and oxide contained solutions. Hydrogen and oxide ions facilitated a local breakdown of the passive layer, especially imperfections in the surface. The root of the corroded pit played an additional notch effect, so stress concentration was arisen in the near of the root tip.

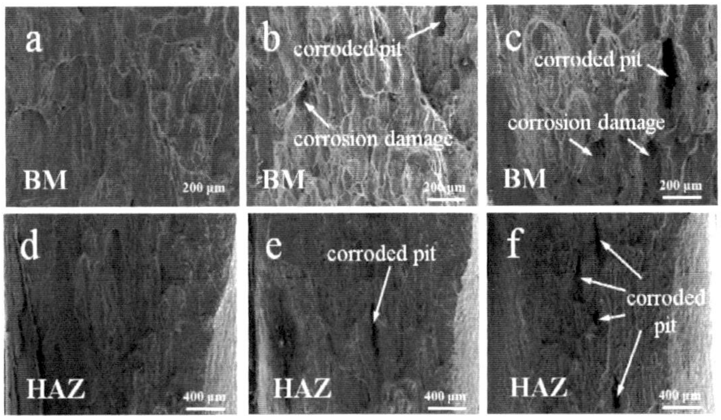


**Figure 3.** SEM observations of giga-grade steel at fractured section after FCP and CFCP tests; (a) BM in air, (b) BM in pH 7.0, (c) BM in pH 10.0, (d) HAZ in air, (e) HAZ in pH 7.0, (f) HAZ in pH 10.0.

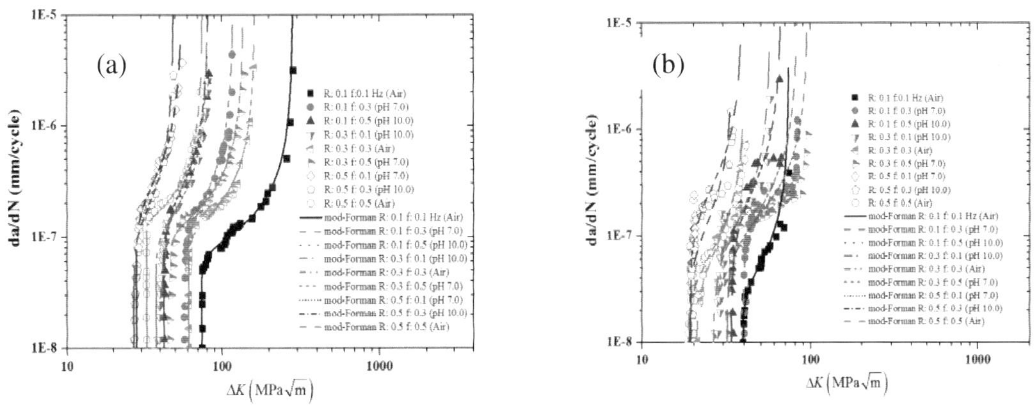


**Figure 4.** Fatigue and corrosion fatigue crack propagation behaviors of giga-grade grade steel under several environment-loading conditions; (a) BM in air, pH 7.0 and pH 10.0, (b) HAZ in air, pH 7.0 and pH 10.0.

In order to obtain the fatigue parameters of M and N and $\Delta K_{th}$ value in a CFCP model of (4), da/dN and SIFR relations of BM and HAZ were examined and the FCP and CFCP rates for the SIFR are shown in figure 4. As can be seen, the modified Forman model demonstrates the whole region of FCP and CFCP behaviours, specifically, regions I, II and III. CFCP rates of the pH buffer solutions were more rapid than the FCPRs in air. And also threshold SIFR values of pH buffer solutions are lower than in air environment.

This is caused by the pitting corrosion cracking which occurs in a corrosive environment due to the stress concentration with the corroded shape and thus a shorter crack propagation path. By using the newly suggested scheme, the average corrosion rates (Ca) in CFCP of BM and HAZ were calculated and the results are expressed in figure 5. As shown in results, more severe corrosion effects were examined in pH 10.0 with higher Ca values than pH 7.0 for both BM and HAZ. In addition, rapid increasing of Ca was found in HAZ while the monotonous increment in BM. It reveals that HAZ is more sensitive to corrosion environment than BM about three times in R=f=0.5 and indicating that thermal damage on the surface of HAZ makes the corrosion solution easy to penetrate. In maximum value of R=f=0.5, more than 30% of the entire crack length is due to the pure corrosion in HAZ of pH 10.0 condition.

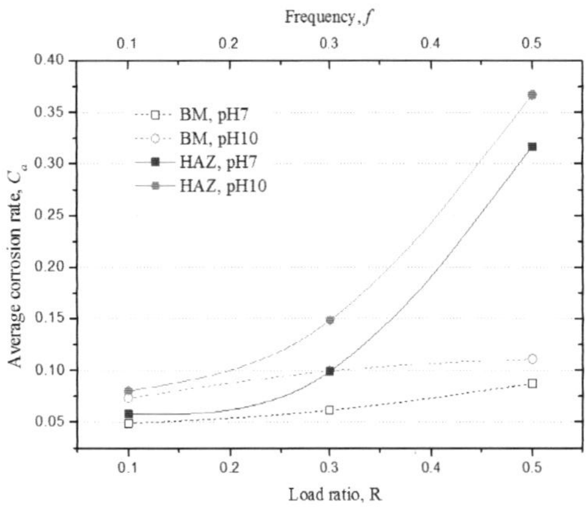


**Figure 5.** Average corrosion rate against the load ratio with frequency of R=*f*=0.1, R=*f*=0.3 and R=*f*=0.5.

## 4. Conclusions

Fatigue and corrosion fatigue crack propagation behaviours of giga-grade steel and its HAZ were investigated under several cyclic loads in alkaline solutions. Not only the fatigue crack propagation in such conditions but the corrosion effects on the behaviour were evaluated in terms of the average corrosion rate with modified Forman equation. The new CFCP scheme enabled the determination of length ratio of crack due to the pure corrosion to the entire crack length during CFCP. The HAZ of giga-grade steel showed higher value of the average corrosion rate than the BM about three times in maximum. The calculated maximum crack propagation length ratio against the total crack propagation length due to the pure corrosion during CFCP process was about 0.37 in HAZ of pH 10.0 and R=*f*=0.5. Based on the chemical reaction analysis in corrosion environment, experimentally measured CFPC behaviours together with model predictions present new interpretation methodology to determine the crack propagation length by pure corrosion effects on the fatigue crack propagation in alkaline solutions.

## References

[1] Kim H, Choi H, Lim J and Cho Y 2011 *Proc. Spring Conf. on The Korean Society of Automotive Engineers,* p 1877.
[2] Chen G, Duquette D 1992 *Metal. Trans. A.* **23** 1563.
[3] Salih S, El-Masri A and Baraka A 2001 *J. Mater. Sci.* **36** 2547.
[4] Wang R 2008 *Int. J. Fatigue* **30** 1376.
[5] Ruiz J and Elices M 1996 *Corros. Sci.* **38** 1815.
[6] Zheng X and Hirt M 1983 *Eng. Fract. Mech.* **18** 965.
[7] Easterling K 1983 *Introduction to the Physical Metallurgy of Welding* (Butterworth).
[8] Henaff G, Petit J and Bouchet B 1992 *Int. J. Fatigue* **14** 211.
[9] Ellyin F and Wu J 1994 *Acta Metall. Mater.* **42** 2709.

IOP Conf. Series: Materials Science and Engineering **317** (2018) 012024 doi:10.1088/1757-899X/317/1/012024

# Discussion on Application of Space Materials and Technological Innovation in Dynamic Fashion Show

**Meilin Huo[1, 2], Chul Soo Kim[3], and Wenhan Zhao[1]**

[1]Art College Northeast Dianli University Jilin, China
[2]Marine Convergence Design Pukyong National University Busan, Korea
[3]Department of Industrial Design Pukyong National University Busan, Korea

**Abstract.** In modern dynamic fashion show, designers often use the latest ideas and technology, and spend their energy in stage effect and overall environment to make audience's watching a fashion show like an audio-visual feast. With rapid development of China's science and technology, it has become a design trend to strengthen the relationship between new ideas, new trends and technology in modern art. With emergence of new technology, new methods and new materials, designers for dynamic fashion show stage art can choose the materials with an increasingly large scope. Generation of new technology has also made designers constantly innovate the stage space design means, and made the stage space design innovated constantly on the original basis of experiences. The dynamic clothing display space is on design of clothing display space, layout, platform decoration style, platform models, performing colors, light arrangement, platform background, etc.

## 1. Introduction

Since beginning of the 21st century, with improvement of people's living and improvement of aesthetic, the general audience concern about the artistic innovation of dynamic fashion show space while appreciating dynamic fashion show. The suggestion is to strengthen combination of science and technology with new ideas and new trends in the modern art and to pay attention to the appearance of new technologies, new methods and new materials, so as to more widely use new technology and new materials for better innovation and constant innovation of fashion show space design on the original basis of experiences.

## 2. Meaning of dynamic fashion show space design

From the perspective of environmental art, space is the place where people behave. Dynamic fashion show space refers to the space and layout, stage decoration style, platform model, performance colors, light arrangement, stage background and other factors for fashion show. In modern dynamic fashion show, designers often use the latest ideas and technology, and spend energy in stage effect and overall environment. Due to development of science and technology in China, concept breakthrough occurs in fashion show, and the stage artists create wonderful "dots" on each part to guide viewers feel the artistic space visual effect. This is combination of art and technology design. In general, the dynamic display space of clothing refers to the space that can be designed in the area of stage art design and the surrounding area of the stage besides the audience [1].

The stage art design of dynamic fashion show originates from theatrical performance, which is different from other arts. The display makes stage art arrangement according to the theme and style. In 1914, the Chicago Garment Manufacturing Association held a large public fashion show at the New Medina Temple in Chicago, which used a runway for the first time. In 1917, the United States began to use films as the backdrop for fashion shows. The mode to use full stage lighting and advanced acoustics in fashion show was firstly adopted by the Unde Creger Fashion Store in the 1970s. He changed the sort of quiet and contemplative atmosphere. The vagaries of lights with strong beat music were integrated with the graceful poise of clothing models, which fully expressed the art effect and functional characteristics of fashion.

## 3. Application form of stage space materials for dynamic fashion show

### *3.1. Scientific and technological static space materials*

In the context of environmental art, static space refers to the space in which people observe the surrounding environment with the same perspective in a static state. In dynamic fashion show, static space refers to all the static space excepting models' dynamic display while the audience stays in the stationary state in the same perspective. In addition to this, the space also includes static stages and the surrounding environment composed of stage type, stage surface, background and property. In stage art design of dynamic fashion show, stage type, stage surface, background and property are the important elements, and also basic conditions of static state space. Appropriate technology innovation of the basic conditions is good for space showing in dynamic fashion show [2].

Te stage for dynamic fashion show is generally called "T" platform. The "T" platform has a simple shape with a wide vision. Its advantage is to fully show the audience clothing and to make it easy to see the color, fabrics and style of clothes. Good viewing effect can be obtained. The width of the conventional extension part to the T platform is from 0.5 to 1.5 m with a height around 1 m and a length from 3 to 15 m [3]. Due to design of the length and width, the fashion show stage makes a smaller distance between the subject and object, makes each viewer sit at the best position as possible as it can, and makes its audience appreciate clothing as appreciating art works. In order to improve the overall effect of performance, many types of stages such as converse T type, square shape stage, U shape stage, S shape stage and Z shape stage appears [4]. Due to change of materials, the materials change from traditional wood to glass, metal, textile, paper, chemical materials PVC and so on. In some performances, designers design the original plane stage into slope surface or use various steps to create sense of layers for better appreciation.

The common form of stage background is board type. There are also some different modeling backgrounds. According to difference of materials, stage background can be divided into hard background, soft background and comprehensive background. Hard background is the background made of hard materials. The common is of the straight plate style. There are also some hard backgrounds that use large pieces of field to make models. Hard backgrounds are usually used in T Stages with runways. At present, on application of science and technology in clothing stage art design, the form mostly used is the use of large LED screens. High flow with large projection is mostly used on the background. Using this background would make the scene appear large and spectacular. Soft background refers to the background made of soft materials (transparent, semi-translucent and opaque). It is usually composed of an awning and a backdrop. Soft background is suitable for casting light and projection screens, most of which are used in performing at grand theatres. The comprehensive background is made with combination of hard background and soft background. To increase the artistic atmosphere of the performance, the comprehensive background can be set if space allows. This construction of the background is complicated and expensive. Usually, it is used in large performances. The tone of the background should be coordinated with the stage. In order to highlight clothing in a better way, pure plain color is for best expression of fashion show.

**Figure 1.** Stage background is the use of LED large screen.

The stage design of fashion show also includes design of props. Although it does not play a major role, it can reach good environmental effect in a better way. The right choice and use of props can enhance the artistic and ornamental feature of the costumes performance, and also increase the performance space of models [5]. In the 2014 Autumn/Winter Paris Fashion Week, the stage space design by CHANEL has a unique style. Its main feature lies in design of props. It made the scene into a supermarket, and the whole place was full of shopping fun with red, blue and yellow bars longitudinal on the bright white walls, perfectly fitting the beautiful red and green shopping carts. There are more than 500 kinds of products on the shelves (totally thousands of items). Models were shuttling between the neat shelves, which made the audience staring. The whole performance was creative and innovative, and this special performance venue and special environmental props kept the viewers' aftertaste.

**Figure 2.** Fall and winter 2014 Paris Fashion Week Chanel stage space design.

IOP Conf. Series: Materials Science and Engineering **317** (2018) 012024 doi:10.1088/1757-899X/317/1/012024

*3.2. Digitization of dynamic space material*

The same space between dynamic space and the static space is that the viewing angle doesn't change and only the surrounding environment changes in the way from one space to another with dynamic visual images. The stage type, stage surface, background and props mentioned on the static space above can also change to be dynamic. With the support of digital technology, the formation of dynamic space is possible [6].

**Figure 3.** Dynamic space of digital technology.

Generally, the dynamic stage is called as a movement stage. The movement platform is a performance stage for dynamic fashion show. Only part is of movement. The movement is divided into translation, lifting and rotating. This movement is of the stage surface. Another movement stage is used in the shopping mall. This makes the escalator in the mall as an innovative dynamic and layer stage and makes the mall for customers to buy things as a show place for models. In 2015, Dior held a men's fashion release meeting in Guangzhou. It used a rotating movable stage and a T stage around the circular stage. Models can do dynamic display on the static T stage, and can also make static display on the dynamic revolving stage, which not only improved the level of performance, but also enhanced performance effect.

The movable background refers to that the background board can move according to design requirements during performance. According to the form of the background board movement, this can be divided into open type, reciprocating type, tilting type, rotary type and so on. Inspiration of dynamic design is from opening and closing doors and windows. With different open and closing styles, different effects can be seen. In general, this kind of background gives people sense of mystery and novelty. It is often associated with what is the situation when the background board is opened. The open-type and reciprocating-type background boards are commonly used. It is simple to operate them compared with the reverse and rotary types. Though the inverse and rotary backgrounds are complex, good effect and high levels can be reached. For warm and wonderful scenes, projector or LED can be use to show landscape sights, outfit effect videos, subtitles and some images relevant to the topic on the background, to combine the movable background with advanced science and technology for good the scene atmosphere and better effect.

**Figure 4.** The movable background combines advanced science and technology.

Today it is worth mentioning that the digitalized light era has arrived. With intervention of multimedia, light and stage art design are gradually integrated. Stage light illuminates the costumes and facial expressions of models. The more important is that lighting engineers use lighting technology to strengthen the artistic effect on the stage and to attract the audience to the model performance area. Light is a kind of art language. It has the directivity and adornment feature in fashion show. The most important is that it has transformability and stage lighting is the soul of "stage". Lighting engineers can make full use of light to display the connotation of the works by costume designers. Therefore, many kinds of light effect would happen if the lighting design is listed in the dynamic performance space and if with many factors such as using computer-aided color changing lights, laser and blinking in a fashion show. According to its motive force, complicated visual effects can be manufactured to strengthen fashion show stage art designing.

**Figure 5.** Digital stage lighting.

### *3.3. Diversification of suspended space materials*

The aesthetic neuroses of modern people have become dull and numb under the bombardment of commercial advertisements. The aesthetic feeling of audience reaches worse effect. Deep impression can hardly be reached. The suspended space design for dynamic fashion show makes dynamic fashion show innovative with a new product display and brand information transmission. The suspended space refers to a space where suspended things are in the air, the base is not equipped with supporting structure and there is a sense of suspension. Use of suspension happened in recent years of promotion performance, modeling contesting and literary performance. It strengthens feeling of atmosphere, and also breaks through the traditional design of stage surface and background up to the extensional environment to the stage. Suspension space design is only for indoor. It firstly considers the height of the indoor, not too high or too short, because a large height would cause long suspension with the feeling of top-heavy and a small height would cause the feeling of depression [7]. The smallest height should be over the height of models. The second is that the range of suspension can be selected on the upper part of the stage and the runway, but it can't affect the model's presentation and costume appearance. Attention should also be paid to the weight of the hanging object. This determines quality of the hanging material. It is not too heavy and is drape. The hanging things can be cloth curtain, yarn, rope, etc. The layout of suspension space can decorate the stage with a good effect. It will give the audience different artistic conception and enrich the models' exhibition methods.

## 4. Conclusion

Today, dynamic fashion show is not pure operation. Instead, it is an artistic creation process with the real sense. Dynamic fashion show space design show people many kinds of expression. It brings people not only visual enjoyment, but also the elaboration of stage art designers' new ideas and different cultural backgrounds, resulting in shocking reaction. The space design for dynamic fashion show in the future will be integrated with various elements of modern science and technology to show the stage effect. It will be the artistic presentation with science, technology, digitization, humanization and diversity.

## Reference

[1] Duan P, Zhang Y 2014 *Art Tech.* **02**
[2] Lv J M 2001 *Mod. Electr. Tech.* **11**.
[3] Zhu H L, Huo M L 2009 *Film Literat.* **14**.
[4] Zhu H L 2009 *Fashion Show Planning and Choreography*. Beijing: China Textile Publishing House.
[5] (British) Trevor·R·Griffith 2000 *Stage Art*. Beijing: China Textile Publishing House.
[6] Li L 2016 *Design and Realization of Digital Demonstration in Representation of Artistic Stage Effects* Beijing University of Technology.
[7] Li T S 2007 *Space Design on Environmental Art Title* Guangxi: Guangxi Art Publishing House.

IOP Conf. Series: Materials Science and Engineering **317** (2018) 012025 doi:10.1088/1757-899X/317/1/012025

# Synthesis, characterization and photocatalytic performance of chemically exfoliated $MoS_2$

S V Prabhakar Vattikuti[1*] and Jaesool Shim[1]

[1]School of Mechanical Engineering, Yeungnam University, Gyeongsan, South Korea, 712-749

**Abstract.** Two-dimensional (2D) layered structure transition metal dichalcogenides (TMDs) has gained huge attention and importance for photocatalytic energy conversion because of their unique properties. Molybdenum disulfide ($MoS_2$) nanosheets were synthesized via one-pot method and exfoliated in (dimethylformamide) DMF solution. Subsequent exfoliated $MoS_2$ nanosheets (e-$MoS_2$) were used as photocatalysts for degradation of Rhodamine B (RhB) pollutant under solar light irradiation. The e-$MoS_2$ nanosheets exhibited excellent photocatalytic activity than that of pristine $MoS_2$, owing to high specific surface area with enormous active sites and light absorption capacity. In addition, e-$MoS_2$ demonstrated remarkable photocatalytic stability.

## 1. Introduction

Photocatalysis process can convert solar energy into chemical energy for wide range of applications in environmental control technologies. The development of visible-driven photocatalysts has received significant attention in recent years [1]. Generally, traditional photocatalysts for example $TiO_2$ have an inherent drawback in that they only active under UV-light irradiation [1, 2]. Therefore, the design of efficient visible light driven photocatalysts is highly desirable.

Molybdenum disulfide ($MoS_2$) has been reported as a distinguished precious metal free photocatalysts for degradation of organic pollutants based on its unique layered features, cost effectiveness and solar energy utilization [3]. In a previous study, we successfully synthesized layered $MoS_2$ and discovered that this precious metal free photocatalyst demonstrated efficient photocatalytic activity for the removal of organic pollutants from aqueous solution under visible light irradiation [3]. Unfortunately, $MoS_2$ alone does not show excellent photocatalytic performance due to recombination of charge carriers. Recently, E. Parzinger et al. [4] developed few layered $MoS_2$ via micromechanical exfoliation from bulk crystal and used as photocatalyst for photocatalytic hydrogen ($H_2$) evolution. B. Han et al. [5] reported intercalated exfoliated $MoS_2$ is used as co-catalyst for photocatalytic hydrogen evolution from water splitting. Bai et al. [6] demonstrated chemically exfoliated metallic $MoS_2$ nanosheets as cocatalyst for improved the photocatalytic activity of commercial $TiO_2$ nanocrystals. However, these synthetic procedure comprising of various sequence of steps and complexity to synthesis.

Therefore, it is necessary to fabricate a new facile method for exfoliation of $MoS_2$ that could be produce an ultra-thin 2D nanosheets at the large scale and low cost for its potential application in various fields.

In this work, we developed a simple and facile chemical route to exfoliate the bulk layered $MoS_2$ materials into few layered e-$MoS_2$ nanosheets. These e-$MoS_2$ can scale-up with ease. The e-$MoS_2$ nanosheets were employed as photocatalyst for degradation of RhB pollutant. The e-$MoS_2$ nanosheets exhibited excellent photocatalytic performance compared to as-synthesizes $MoS_2$ ones. Moreover, the used photocatalyst is successfully recovered and reused three times and observed no loss of photocatalytic activity.

## 2. Experimental details

### *2.1. Synthesis of $MoS_2$ and e-$MoS_2$ nanosheets*

In a typical synthesis procedure, 1.5 mol of ammonium hepta molybdate tetrahydrate and 60 mmol of $C_2H_5NS$ were dissolved in 50 mL of ethanol and 25 mL deionized water under constant stirring. Then, 5 mol/L HCl was used to adjust the pH of the mixed solution to 2–2.8 and the solution was stirred for 30 min at room temperature. Then the above solution was transferred into a 100-mL Teflon-lined stainless steel autoclave, sealed, and maintained at 180°C for 24 h. After being cooled, the product was obtained by centrifuging, washed repeatedly with deionized water and ethanol three times, and dried in a vacuum oven at 120°C for 4h. To prepare few-layer e-$MoS_2$, a sonication method was employed. Typically, 50 mg of the as-synthesized $MoS_2$ and 100 mL dimethylformamide were added to a beaker, which was submerged in a 200-W ultra-sonication machine for 1 h. Then, final precipitates were collected by centrifugation (6000 rpm) and washed with ethanol and dried at 70°C in vacuum oven.

### *2.2. Characterization*

Powder X-ray diffraction (XRD) patterns of the as-prepared photocatalysts were obtained using a Shimadzu LabX X-ray diffractometer (XRD 6100 model) using Cu Kα radiation (λ= 0.154 nm) at 40 kV and 30 mA. X-ray photoelectron spectroscopy (XPS) analysis was conducted using a Thermo Scientific instrument and K-alpha surface analysis. $N_2$ adsorption-desorption isotherms were recorded using a Micromeritics ASAP 2420 surface area analyser at the temperature of liquid $N_2$. Before gas adsorption, all photocatalysts were degassed for 2 h at 180°C. To determine the surface morphology transmission electron microscopy (TEM) at an accelerating voltage of 300 kV. High-resolution TEM (HRTEM) studies were performed using a Hitachi H-7000 and a Tecnai G2 F 20 s-twin TEM.

### *2.3. Photocatalytic activity test*

Typically, the photocatalytic activity of the samples was tested at the natural pH of an RhB organic dye solution (10 ppm in concentration). The test was performed under visible-light irradiation with a MAX-300 Xe lamp (λ> 400 nm). In a typical run, a mixture comprising 50 mg of various photocatalysts was suspended in 100 ml of RhB aqueous solution and stirred for 30 min in order to achieve an adsorption-desorption equilibrium mixture. During light irradiation, 5 ml of the reaction mixture was sampled at 5-min intervals with a syringe. The collected samples were examined with a UV-Vis-NIR Cary 500 spectrophotometer. The photocatalysts were then separated from the degraded solution by centrifuging (5000 rpm), washed with ethanol to fully take away the residual dye moieties then again washed with water and reused for consequent runs.

## 3. Results and discussion

The powder XRD patterns of the e-$MoS_2$ nanosheets is shown in figure 1a. The diffraction peaks of e-$MoS_2$ is matched with JCPDS card: 751539, with 2θ =14.12°, 28.42°, 32.91°,39.53°, 49.41°, 55.44°, 58.62°, and 69.47 corresponding to (002), (004), (100), (103), (105), (106), (110), and (200) planes of hexagonal $MoS_2$ materials and were confirmed by previous reports [3]. The XRD results represented that the obtained materials have pure phase, and no characteristic peaks were observed from impurities. To investigate the chemical compositions and nature of chemical bonding of constituent elements are presented in e-$MoS_2$, X-ray photoelectron spectroscopy (XPS) was carried out, resulted are shown in figure 1b-d. The existence of core levels of Mo and S elements confirms by survey, results reflecting the high purity of e-$MoS_2$ as shown in figure 1b. XPS pattern of Mo 3d with characteristic peaks at binding energies of 229.34 and 232.85 eV, which are ascribed to the spin orbit of Mo $3d_{5/2}$ and Mo $3d_{3/2}$ of the $Mo_2^+$ chemical state, respectively (figure 1c). The two peaks at binding energies of 161.18 and 162.34 eV are ascribed to the S $2p_{3/2}$ and S $2p_{1/2}$ of the $S_2^-$ chemical state [3], respectively (figure 1d).

figure 2 shows the morphology of e-$MoS_2$ nanosheets was investigated by scanning electron microscopy (SEM) and high resolution transition electron microscopy (HRTEM). figure 2a show exfoliated MoS2 are distributed randomly. To further confirm the purity of e-$MoS_2$, SEM EDX study was carried out. It is obvious from figure 2b, that the entire sample comprising of Mo and S elements, no other impurities were detected. Figure 2 c shows HRTEM image of the pristine $MoS_2$. From the HRTEM images (figure 2 c-e), it is clear that e-$MoS_2$ grew in sheet like morphology having very dense and uniform structure with an average length in few nanometers to submicron. SAED pattern of e-$MoS_2$ represents the existing of pure phase grew along with diffraction planes of (100). (103), (105), and (110).

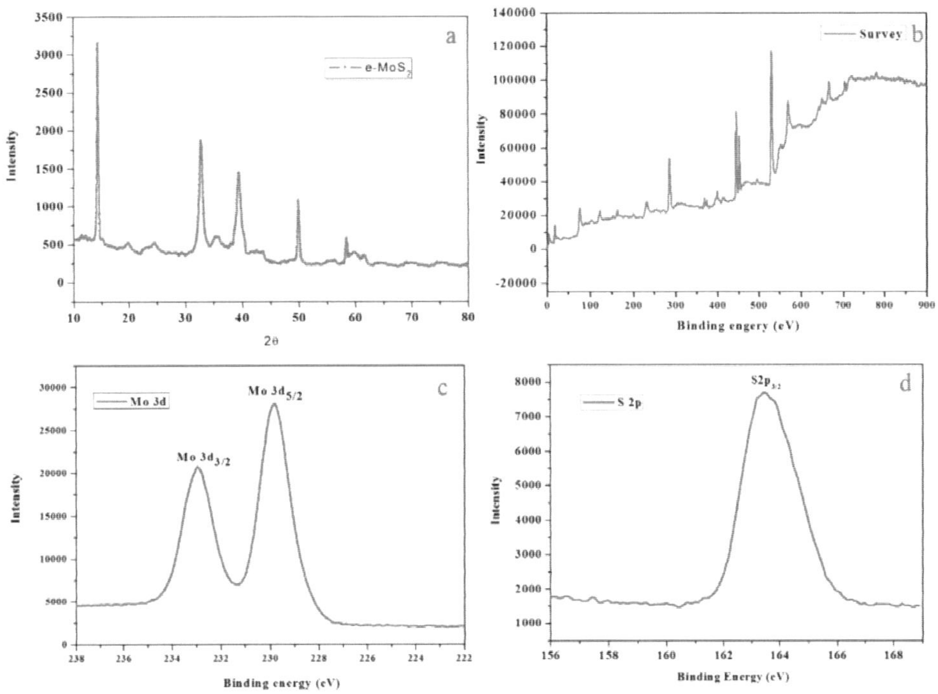


**Figure 1.** (a) XRD pattern of e-$MoS_2$, (b) survey spectra of e-$MoS_2$, (c) XPS spectra of Mo 3d and (d) XPS spectra of S2p element.

The photocatalytic activity of the pristine $MoS_2$ and e-$MoS_2$ were studied with RhB degradation under simulated solar light illustration. Figure 3a displays the sequential changes of the UV–vis absorption spectra of RhB solution treated with simulated solar light as a function of time, which were obtained in the presence of e- $MoS_2$. There is a strong adsorption peak at 554 nm that decreases gradually with as the irradiation time increases from initial to clear ones at 60 min, which reflects the pink color of the initial RhB solutions gradually faded during the process of photodegradation.

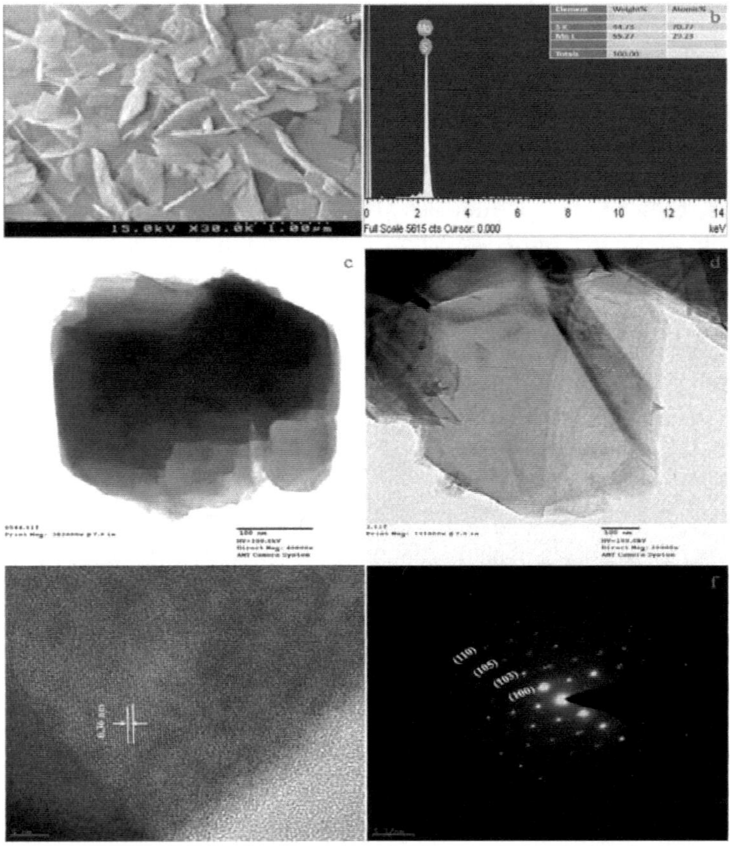

**Figure 2.** (a) SEM image of e-$MoS_2$, (b) SEM-EDX mapping, HRTEM image of (c) pristine $MoS_2$, and (d, e) e-$MoS_2$, and SAED pattern of e-$MoS_2$.

Figure 3b shows the photodegrading efficiency expressed as changes in the RhB organic dye relative concentration ($C/C_0$, C-final concentration, $C_0$-initial concentration), which was obtained as a function of irradiation time with various catalyst under simulated solar light irradiation. A blank experiment was carried out in the absence of photocatalyst and demonstrated no change in the RhB concentration within 60 min of reaction under simulated solar light irradiation. The e-$MoS_2$ can greatly enhance the photocatalytic activity under simulated solar light irradiation. Under simulated solar light irradiation, both photocatalysts exhibited remarkable photodegradation performance. The highest photocatalytic activity was obtained over e-$MoS_2$, with which more than 99.94% of the RhB was degraded within 60 min, owing to more active sites for RhB adsorption and degradation and enhance the photodegradation [3]. These results designate that the e-$MoS_2$ play an important role in the adsorption and photocatalytic degradation process of RhB. Figure 3c displays the kinetic plot of ln ($C/C_0$) with respect to time, which followed the first-order kinetics. To examine the photostability and reusability of the photocatalyst, three successive recycling tests were accomplished for the degradation of RhB under simulated solar light irradiation, as shown in figure 3d. After three cycles, the e-$MoS_2$ demonstrated no significant loss in activity and the decomposition efficiency was consistent. Thus, the e-$MoS_2$ are very stable with prolonged irradiation time and significantly enhanced the visible light photocatalytic activity. This result denotes that e-$MoS_2$ photocatalyst have high stability and can be reusable.

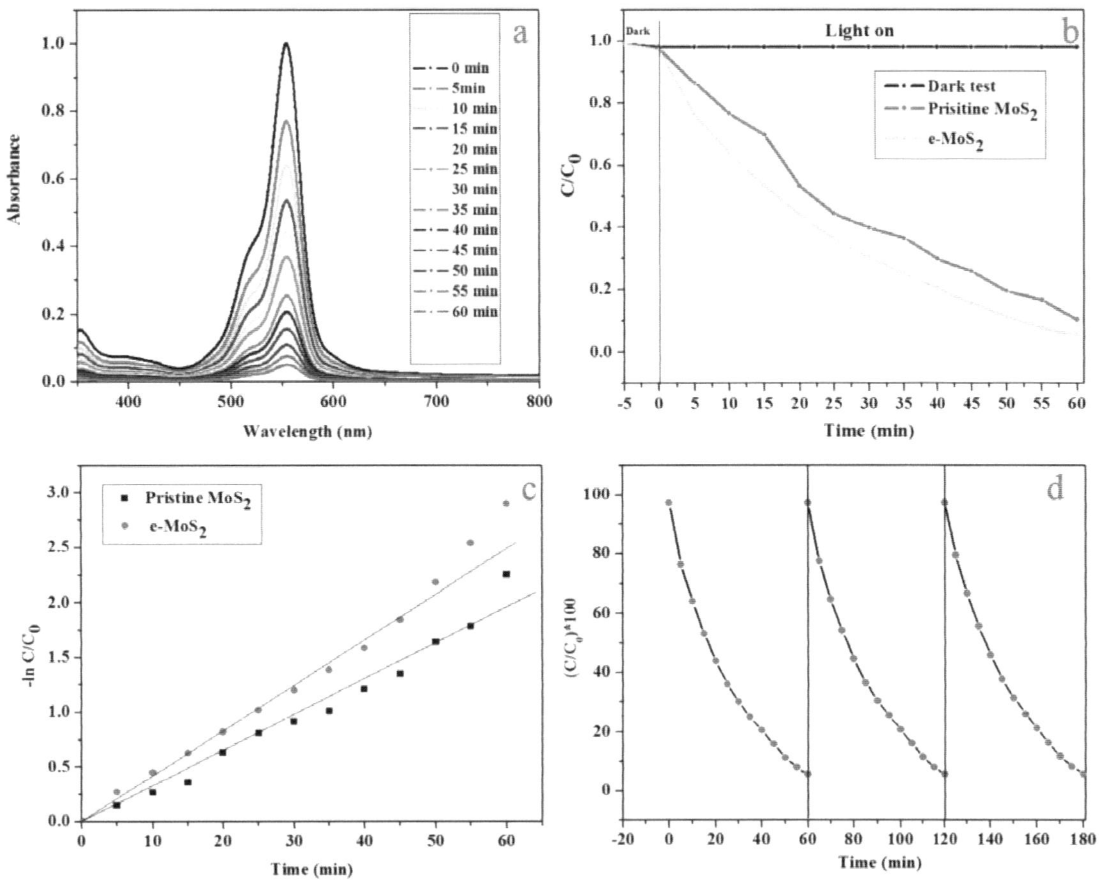


**Figure 3.** (a) UV-Vis absorption spectra changes of RhB aqueous solution in the presence of e-$MoS_2$, (b) Photocatalytic activities, and (c) Kinetic linear simulation of RhB photocatalytic degradation of pristine $MoS_2$ and e-$MoS_2$ photocatalysts, and (d) Recycling test results of the e-$MoS_2$ photocatalyst for the degradation of RhB under simulated solar light irradiation.

## 4. Conclusions

$MoS_2$ nanosheets were synthesized via a hydrothermal method and then, chemically exfoliated $MoS_2$ into few layered nanosheets in presence of DMF. The comparative photocatalytic studies were performed over pristine $MoS_2$ and exfoliated $MoS_2$ (e-$MoS_2$) for photodegradation of RhB pollutant under simulated solar light irradiation. The e-$MoS_2$ nanosheets showed greater photocatalytic degradation of RhB under simulated solar light irradiation compared to pristine one. In addition, the photocatalyst activity demonstrated good stability during the photoreactions with no significant deactivation observed after three test runs with the recycled catalyst. The simplicity this entire synthetic process could facilitate numerous significant applications for exfoliation of other inorganic materials.

## References

[1] Martin D J, Liu G, Moniz S J A, Bi Y, Beale A M, Ye J and Tang J 2015 *Chem. Soc. Rev.* **44** 7808.
[2] Dong H, Zeng G, Tang L, Fan C Z, Zhang C, He X and He Y 2015 *Water Research* **79** 128.
[3] Vattikuti S V P, Byon C and Reddy C V 2015 *Superlattice. Microstruct.* **85** 124.
[4] Parzinger E, Miller B, Blaschke B, Garrido J A, Ager J W, Holleitner A and Wurstbauer U 2015 *ACS nano*, **9 (11)** 11302
[5] Han B and Hu Y H 2016 *Energy Sci. Eng.* **4(5)** 285.
[6] Bai S, Xiaoyi L W, Junteng C and Xiong D Y 2015 *Nano Res.* **8(1)** 175.

IOP Conf. Series: Materials Science and Engineering **317** (2018) 012026 doi:10.1088/1757-899X/317/1/012026

# Block-Module Electric Machines of Alternating Current

**I Zabora**[1]

[1]Department of automation and electrical supply, Moscow State University of Civil Engineering, Moscow, Russia

**Abstract.** The paper deals with electric machines having active zone based on uniform elements. It presents data on disk-type asynchronous electric motors with short-circuited rotors, where active elements are made by integrated technique that forms modular elements. Photolithography, spraying, stamping of windings, pressing of core and combined methods are utilized as the basic technological approaches of production. The constructions and features of operation for new electric machine – compatible electric machines-transformers are considered. Induction motors are intended for operation in hermetic plants with extreme conditions surrounding gas, steam-to-gas and liquid environment at a high temperature (to several hundred of degrees).

## 1. Introduction

Flat disk asynchronous electric motors with short-circuit rotor are manufactured in some countries and occupy a certain niche for many mechanisms. Small axial dimensions, open active slot-tooth area offers solutions for design problems of integrating motors without use of any intermediate mechanisms, as well as afford implementation of advanced technologies for making windings and magnetic cores [1-4]. They include: printed wiring and photolithography, vacuum sputtering, stamping of windings, pressing of core etc. that allows to avoid coiling operations from making winding. This is of special importance for microminiature motors [5, 6]. The author of the paper has patented some of constructive and technological solutions [3, 7, 8] described below.

High-speed (8000 rpm) disk induction motors can serve one of examples of embedding a three-high bloomer construction small-section rolling mill (figure 1).

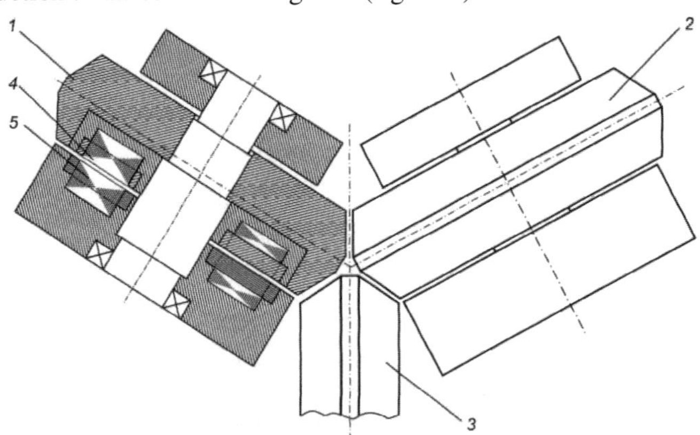


**Figure 1.** The construction arrangement of a three-high bloomer with embedded disk induction motors. ***1, 2*** – auxiliary rolls with embedded rotors; ***3*** – basic roll, rotated by the basic power electric motor; ***4*** – disk rotor with a short-circuited winding; ***5*** – disk stator with a stamped-welded winding.

Such solution gives 1.5-2 times increase in roll speed and, at the same time, reduces the dimensions of the stand.

In addition, the motor is cooled by the technological water in the rolling mill, what increases of current density in the windings [6].

## 2. General dimensions of modular block motors and induction distribution

Properties of disk motors substantially depends on the relation outer ($D_{out}$) and inner ($D_{in}$) diameters of active parts: $a = D_{out}/D_{in}$.

The diameters equation is:

$$(D_{out}+D_{in})^2(D_{out}-D_{in}) = V_A,$$

where

$$V_a = \frac{48{,}8P'10^5}{\alpha_\delta k_f k_{W1} AB_\delta n_1}, \text{ m}^3$$

– active volume of machine, defined from machine constant by initial data and electromagnetic loads; $P'$ means electromagnetic power; $AB_\delta$ – electromagnetic loads; $n_1$ – rotation frequency; $\alpha_\delta$ – coefficient of the pole zone; $k_f$ – factor of form of induction curve; $k_{W1}$ – winding factor.

In view of relationship between outer and inner diameter, equation becomes:

$$(aD_{out}+D_{in})^2(aD_{out}-D_{in}) = V_{A.}$$

This expression defines of inner diameter $D_{in}$ – dependence for given initial data and electromagnetic loads:

$$D_{in} = \sqrt[3]{\frac{1}{(a+1)^2(a-1)}}\sqrt[3]{V_a} = k_{in}\sqrt[3]{V_a}.$$

Dependencies of outer $D_{out}$ and inner diameters values, as well as of active length and are expressed as:

$$D_{out} = \sqrt[3]{\frac{1}{(a+1)^2(a-1)}}\sqrt[3]{V_a} = k_{out}\sqrt[3]{V_a};$$

$$D_{mid} = 0{,}5\sqrt[3]{\frac{(a+1)}{(a-1)}}\sqrt[3]{V_a} = k_{mid}\sqrt[3]{V_a};$$

$$l_d = 0{,}5\sqrt[3]{\left(\frac{a-1}{a+1}\right)^2}\sqrt{V_a} = k_l\sqrt[3]{V_a},$$

where $k_{out}$, $k_{mid}$ and $k_l$ are respectively, coefficients of outer and inner diameters, and of active length.

Electromagnetic moment that accounts for uneven distribution of induction $B_\delta(R)$ along the radius $R$:

$$M=C_M\int_{R_{in}}^{R_{out}} B_\delta(R)RdR = C_M R_{mid} l_0\left(B_{\delta mid} + B'_\delta\right),$$

where $R$ means a current radius, $C_M$ means a proportionality coefficient, $R_{mid}$ – a middle radius, $l_0$ – an active length, $B_{\delta mid}$ – an induction at the middle diameter, and $B'_\delta$ – an additional induction that may be both positive and negative.

The distribution of magnetic induction in a working gap along the active length may be obtained under the calculation block diagram (figure 2) by method of successive approximations, that breaks the active length into $n_\Delta$ sections ($n_\Delta$ = 5÷10) of $l_{\Delta\delta}$ width and calculation of induction $B_{\delta i}$, average a for each section from its middle radius $R_{mid}$:

$$k'_\mu(R) = \mu_0 F_\Sigma / B_\delta(R) k_\delta(R) 2\delta.$$

In the process, certain values of $B'_\delta(R)$ are preset and calculations made until the $k_\mu(R)$ and $k'_\mu(R)$ values agree with intended degree of accuracy:

$$\left(k_\mu(R) - k'_\mu(R)\right) / k_\mu(R) < \varepsilon = 0{,}01.$$

Magnetomotive forces of magnetic circuit ($F_\Sigma$ = const) is obtained at the middle radius $R_{mid}$ (within the limits of section $l_{\Delta\delta}$) by the value $B_\delta$.

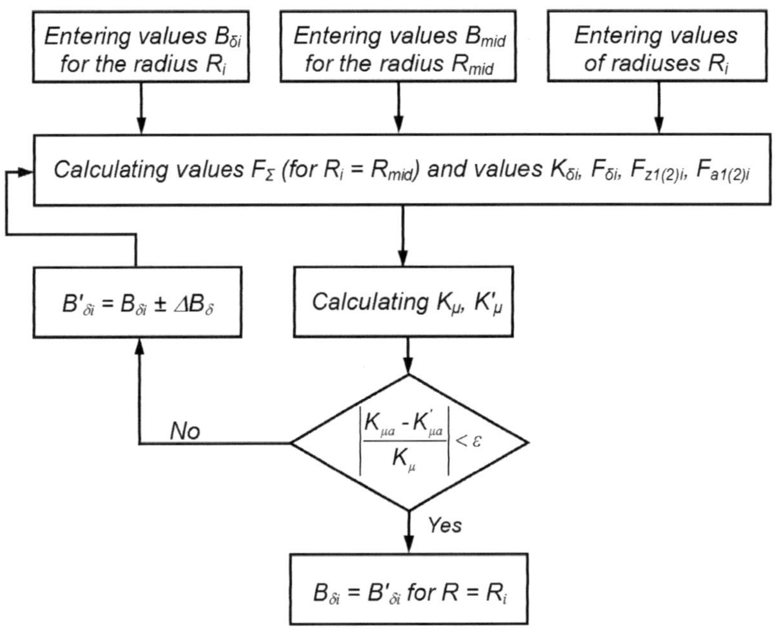


**Figure 2.** The calculation scheme for induction distribution $B_\delta(R)$ in working gap.

## 3. Compatible electric machines-transformers

As the compatible construction: the electric machine -transformer it is possible to examine three probable options:

– cylinder-type electric machine (motor or generator) and transformer;
– end (disc) machines (motor and generator);
– mixed: cylinder-type transformer – disc motor.

In every option, the transformer converts the AC voltage and number of phases.

For the first option, electric machines and transformers can be based on serial produced machines either in single-body or two-body modification that provides for standard manufacturing.

Block-module execution of combined electric motors-transformers allows to solve following problems:

– to reduce voltage on electric motor to 1÷2 Volt and practically to remove grooving insulation and electric conductor insulation;

– to raise electric and thermal loads, to increase fill factor of groove stator winding to unit and, thereby, to lower mass and overall dimensions of electric motors.

This option seems to be the best for the electric motor operating in extreme environment. In all known constructions of hermetic motors made for corrosive environment the stator is placed outside the object (environment) and has a sealed wall (screen) that segregates it from the rotor. This increases the gap between the stator and the rotor and reduces energy indicators.

The report offers block-module execution of the combined electric motors-transformers [5-9] which allow to exclude these negative factors.

When the primary distributed three-phase winding of the transformer ***1*** with $p$ couples of poles gets supply from a three-phase circuit of $f$ frequency, the rotary magnetic fields appears. When the secondary $Z$-phase winding of transformer ($Z$ means the number of windings) crosses bars ***4***, magnetic fields induces $Z$- phase electromotive force with the same amount of poles $2p$. Sealed lead-in rods ***7*** and arc-shaped elements ***10*** provide an electrical connection between the winding and the $nZ$-phase bar winding of the stator; the former passes the currents and creates a step curve of magnetomotive forces (MMF) (figure 4) with less component of higher harmonics for the electric motor.

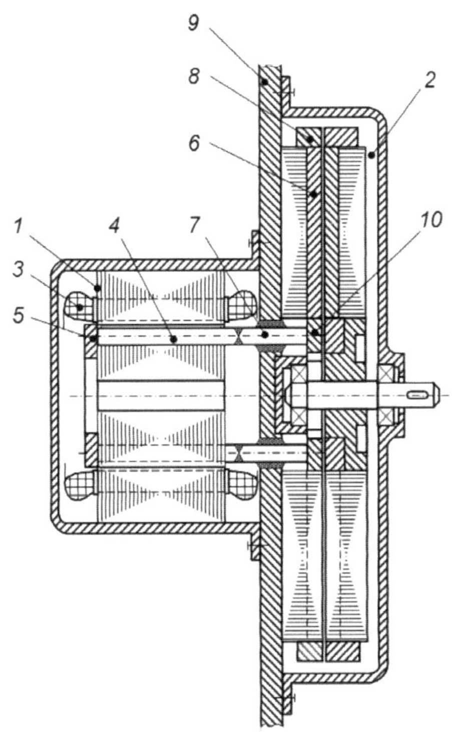


**Figure 3.** Longitudinal section of the electric machine
***1*** – cylindrical three-phase transformer with a rotating magnetic field; ***2*** – disk electric motor; ***3*** – primary winding of a transformer; ***4*** – multiphase (z-phase) bar winding; ***5*** – short circuit ring; ***6*** – rod stator winding; ***7*** – sealed lead-in rods; ***8*** – short circuit ring; ***9*** – leakproof partition; ***10*** – arched conductive segments.

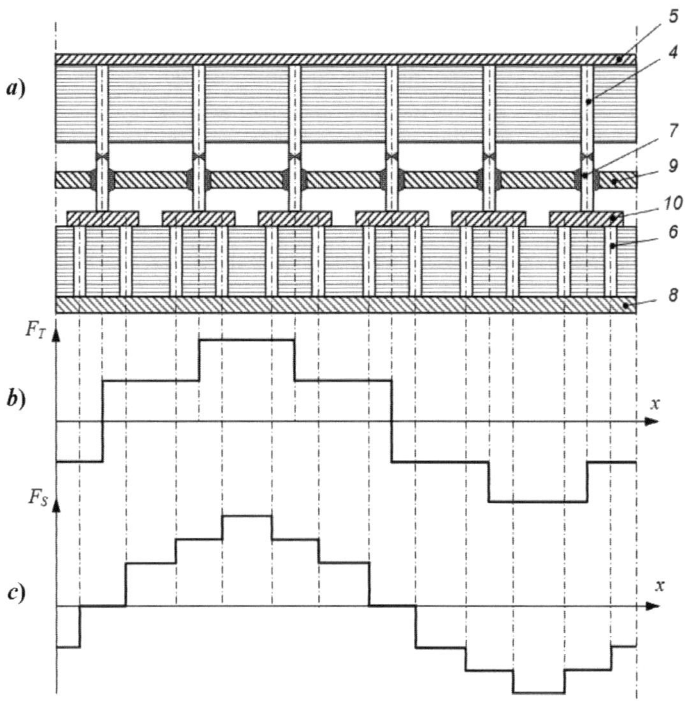


**Figure 4.** Scanning active surfaces of transformer and stator from outside of rods (***a***): distribution magnetomotive forces (MMF) $F_T$ of transformer (***b***) and MMF $Fs$ of electric motor (***c***) on effective diameter.

## 4. Conclusion

The designs and features of operation for new disk induction motors of block-module type are a part of motors-transformer aggregates for extreme condition environments are considered.

The new patented constructive decisions, allowing to raise technical and economic indicators compatible electric machines-transformers at the expense of *n*-multiple division of rod windings in a motor part of aggregate are offered;

Manufacturing of a disk rotor is offered with double-sided stator with Z/2 the radial grooves, shifted on half beard division and is mutual-cut with cores hexagonal section.

## References

[1] Vildanov C, Zabora I, Trutko D 2000 *Elektrotekhn.* **9.**
[2] Zabora I, Stavinskiy A 2000 *UDMTU* **1(367)** 136-140.
[3] Zabora I, Vildanov K et al 2001 RF Patent 2173926 RF Electric motor for hermetic objects *Otkryt. Izobret.* **26.**
[4] Zabora I, Stavinskiy A, Kimstach O, Kazanskiy S 2002 *Elektrotekhn.* **3.**
[5] Vildanov C, Zabora I, Berezkina N 2012 *Russian Electr. Eng.* **83(1)** p 21-6.
[6] Vildanov C, Zabora I, Berezkina N 2012 *Russian Electr. Eng.* **83(3)** p 142-6.
[7] Zabora I, Vildanov K et al 2013 RF Patent 2487454 RF Electric motor-transformer aggregate *Otkryt. Izobret.* **19.**
[8] Zabora I, Vildanov K et al 2014 RF Patent 2507665 RF Electric motor-transformer aggregate *Otkryt. Izobret.* **5.**
[9] Bespalov V, Vildanov K, Zabora I, Chernov R 2016 *Russian Electr. Eng.* **87(10)** 549-53.

# Synthesis of Calcium Phosphate Composite Organogels by Using Emulsion Method for Dentine Occlusion Materials

**C Nopteeranupharp[1], K Akkarachaneeyakorn[1], A Songsasaen[1]**

[1]Faculty of science, department of chemistry, Kasetsart University, 50 Ngamwongwan street, Lat Yao, Chatuchak, Bangkok, Thailand 10900

**Abstract.** Dentinal hypersensitivity (DH) is one of the most human's problems caused by the erosion of enamel. There are many methods and materials to solve this problem. Calcium phosphate is an excellent alternative for curing this symptom because of its osteoconductivity, and biocompatibility properties. The low-cost and low-toxicity calcium phosphate nanogel was fabricated by using emulsion method and characterized by using TEM, EDX, and DLS techniques. The results showed that P123 (poly (ethylene oxide)$_{19}$-*block*-Poly (propylene oxide)$_{69}$-*block*-poly (ethylene oxide)$_{19}$) has played a major role as template and gel formation, SDS was used as a surfactant to form water-in-oil emulsion nanodroplets with circle-like shape. Moreover, the ability of synthesised organogel to occlude the exposed dentine tubules was tested on the model of human's dentine slices. The results showed that calcium phosphate composite organogel can be efficiently occluded on dentine slice, characterized by SEM technique, after 1 day.

## 1. Introduction

Dentinal hypersensitivity (DH) is described as a "sharp pain of short duration which occurs when the stimulus of chemical or thermal origin comes in contact with an exposed dentine surface" [1]. It is mainly caused by the erosion of enamel due to human's habits in daily life such as eating and drinking. Within enamel, there are numerous tiny dentine tubules filled with fluid. When enamel is destroyed, dentine tubules can be easily contacted with the outer stimuli and the fluid in dentine tubules are oscillated [2].

From previous works, many researchers have been continually studying on how to reduce DH symptom and producing various desensitizing agents such as bioactive glass (BG), fluoride, oxalate, and hydroxyapatite (HA) [2]. Calcium phosphate (CaP), especially hydroxyapatite, is one of the most popular materials widely used in biomedical, biomaterial and bioengineering applications because of its biocompatibility, osteoconductivity, non-inflammable, bioactive, biodegradable and non-toxicity properties. Moreover, the methods and conditions of CaP preparation such as sol-gel, microwave, and hydrothermal methods are reported to have an effect on the sizes, morphologies and phases of CaP [3-5].

Herein, calcium phosphate composite organogels were synthesized by the main components of P123 (poly (ethylene oxide)$_{19}$-*block*-Poly (propylene oxide)$_{69}$-*block*-poly (ethylene oxide)$_{19}$), compared with calcium phosphate nanoparticle without the addition of P123 to study the gel formation and the effect of it on the emulsion formation, SDS (sodium dodecyl sulfate) as surfactant, eucalyptus oil as an organic phase, and calcium and phosphate source in aqueous phase using emulsion method. Emulsion methods are known to be low-cost procedures performing under mind condition and can be used to produce small particle sizes (in nanometer to submicron) [6]. Pluronic polymer, poly (ethylene oxide)$_{19}$-*block*-Poly (propylene oxide)$_{69}$-*block*-poly (ethylene oxide)$_{19}$ (P123), has widely used as a template to form gel, and control the size, structure and morphology of particles [7]. In this study, P123 was added in order to form a gel which could facilitate the application of the product on the exposed teeth. Eucalyptol was used as an organic solvent due to its low toxicity. In addition, calcium phosphate nanogels were tested in vitro with human's dentine slices to demonstrate the occlusive performance of fabricated organogel.

## 2. Material and methods

All chemicals were of analytical grade without any further purification. Poly (ethylene oxide)$_{19}$-*block*-Poly (propylene oxide)$_{69}$-*block*-poly (ethylene oxide)$_{19}$ (P123) (M.W. 5800 Da), eucalyptol, sodium dodecyl sulfate (SDS), calcium chloride dihydrate ($CaCl_2 \cdot 2H_2O$) and disodium hydrogen phosphate ($Na_2HPO_4$) were purchased from Sigma Aldrich.

### *2.1. Synthesis of calcium phosphate nanoparticles by using microemulsion method*

Five milititres of eucalyptol was transferred in vial containing SDS ($7.7x10^{-4}$ mol, 0.222 g) and then stirred vigorously overnight. Then, 100 µl of $CaCl_2 \cdot 2H_2O$ ($7.4x10^{-4}$ mol, 0.1095 g) was added into the vial prior to be kept stirring vigorously for further 30 minutes. Two mililitres of $Na_2HPO_4$ aqueous solution ($7.2x10^{-4}$ mol, 0.128 g) was added dropwise into the same vial under vigorous stir. The solution was stirred for another 30 minutes to complete the reaction before being left stand still for approximately 15 minutes to allow the phase separation. Finally, the oil fraction was collected and kept in another vial.

### *2.2. Synthesis of calcium phosphate composite organogel by using microemulsion method*

0.2 g of P123 (M.W. 5800 Da, $3.5x10^{-5}$ mol) was mixed with 5 ml of eucalyptol and kept stirring overnight. SDS ($7.7x10^{-4}$ mol, 0.222 g) was added into the vial before being stirred vigorously overnight once again. Then, 100 µl of $CaCl_2 \cdot 2H_2O$ ($7.4x10^{-4}$ mol, 0.1095 g) was added into the vial and then kept stirring vigorously for another 30 minutes. $Na_2HPO_4$ aqueous solution ($7.2x10^{-4}$ mol, 0.128 g, 2 ml) was added dropwise into the vial under vigorous stir. The solution was stirred for 30 minutes before being aged at room temperature for phase separation (approximately 15 minutes) and the oil fraction was collected and evaporated under vacuum at room temperature to remove solvent.

### *2.3. Characterization*

The morphology and size of synthesized calcium phosphate nanoparticles and the synthesized calcium phosphate composite organogel were characterized by using transmission electron microscope (TEM). Briefly, 20 µl of samples was dispersed into 100 µl of ethanol and then sonicated for 10 minutes before dropped onto carbon-coated copper grids and left to dry in air overnight. Dynamic light scattering (DLS) was used to confirm the size of nanoparticles contained in the emulsion and organogel from TEM. Shortly, 3 ml of samples was added into the cuvette for analysis using Malvern Zetesizer instrument.

### *2.4. Dentine tubule occlusion and mineralization*

The model of dentine tooth slices were kindly supported by the faculty of dentistry, Mahidol University and prepared as previous description [8]. They were kept in 1:1 ethanol: water to prevent the growth of some bacteria. Firstly, each of dentine slices was sonicated in deionized water for 15 minutes and dried in air. Then, the samples of calcium phosphate composite organogel were applied on dentine slices and kept under moisture condition for 5 days. After that, the dentine slices were washed by ethanol extensively in order to remove the excess of gel and oil. Finally, the dentine slice samples were placed on carbon stub and sputter-coated with gold ready to be characterized by scanning electron microscope (SEM) for investigating the occlusion efficiency of samples on tooth.

## 3. Results and discussions

Calcium phosphate composite organogel was synthesized by emulsion method in the presence of P123, SDS, eucalyptol, $CaCl_2$ and $Na_2HPO_4$ as precursors. In the beginning, colorless solution was occurred when P123 was added into eucalyptol (figure 1A). The solution then became turbid after the addition of surfactant, SDS (figure 1B). Calcium and phosphate source in aqueous solution were respectively added into the solution to form the emulsion. Finally, the aqueous phase (lower) and oil phase (upper) were separated (figure 1C), and then the water-in-oil emulsion layer was collected into the new vial (figure 1D). Compared with the calcium phosphate nanoparticles without P123 by emulsion method, the calcium phosphate composite organogel was more obviously viscous. The results confirmed that P123 was performed a major role in gel formation.

**A B C D**

**Figure 1.** Digital photograph showing process of calcium phosphate composite organogel synthesis after the addition of (A) P123 (B) SDS (C) calcium and phosphate aqueous solution and (D) the oil fraction containing calcium phosphate nanocomposite.

The size and morphology of calcium phosphate nanoparticles contained in organogel compared to calcium phosphate nanoparticles prepared in the absence of P123 were characterized by TEM. The image of calcium phosphate nanoparticles (figure 2-left) showed the circle-like shape with approximately 10-60 nm diameters and connected with fibocous structure with 5 nm thickness. Whereas, the TEM image of calcium phosphate nanoparticles contained in organogel (figure 2-middle) showed the circle-like shape with 10-40 nm diameters and the network of calcium phosphate with 5 nm thickness. Moreover, the large particles, approximately 90-110 nm diameters, which presumably from agglomeration of the small particles by themselves were observed. Referred from previous paper [7], the results showed that SDS was the main component to form water-in-oil emulsions and nanofibrous structure, whereas the addition of P123 as template could still stabilize the spherical structure. However, nanofibrous structure might be changed to the thicker network nanofibers. EDX technique was used to characterize the components in calcium phosphate composite organogel. The EDX spectra (figure 2-right) showed the peak of carbon, phosphorous, calcium, oxygen, and sulfur which confirmed that calcium phosphate nanoparticles could be synthesized by emulsion method. To confirm the sizes of calcium phosphate nanoparticles and organogel, DLS technique was used and the result indicated that the major size of calcium phosphate nanoparticles was 80 nm approximately (figure 3), which is relevant in nanometer range, but calcium phosphate organogel could not be measured by DLS technique because the structure of organogel was not indiscrete circle-shape structure.

The ability of synthesized calcium phosphate composite organogel to occlude dentine tubules was tested in vitro with human's dentine slices. The synthesized organogel was applied on the dentine slices, which kept into the humid condition for 1 and 5 days. Dentine slices were washed with ethanol to eliminate the excess of SDS, P123 and other matrixes before being characterized by SEM. Compared to the dentine slice without the application of calcium phosphate organogel (figure 4A), the results showed that the dentine tubules were almost fully occluded after the calcium phosphate after 1 day of organogel application (figure 4B) and almost covered all the surface of dentine slice after 5 days (figure 4C). As the results, it confirmed that the mineralization process on dentine with high stability was occurred. Due to the fact that dentine mainly consists of hydroxyapatite (HAp), it can lead to the biomineralization of synthesized calcium phosphate nanoparticles on the dentine tubules.

IOP Conf. Series: Materials Science and Engineering **317** (2018) 012027 doi:10.1088/1757-899X/317/1/012027

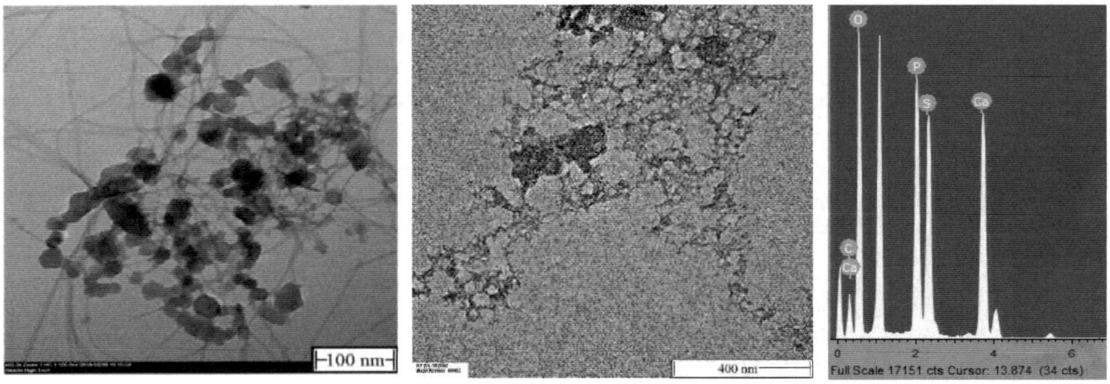


**Figure 2.** TEM images of calcium phosphate nanoparticles without P123 (left), calcium phosphate with P123 (middle), and EDX spectra of calcium phosphate from TEM (right).

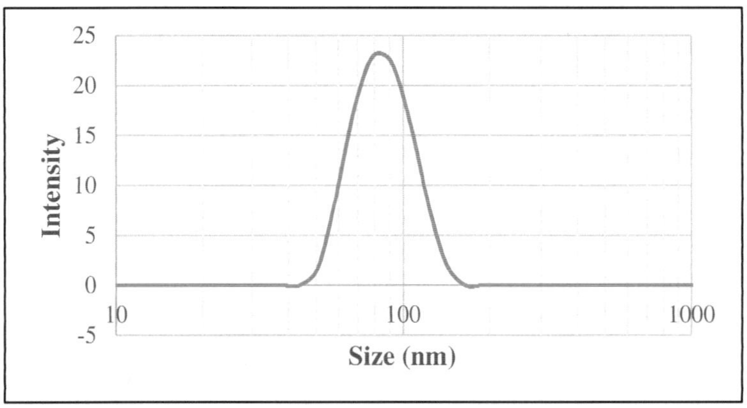


**Figure 3.** DLS peak of calcium phosphate nanoparticles

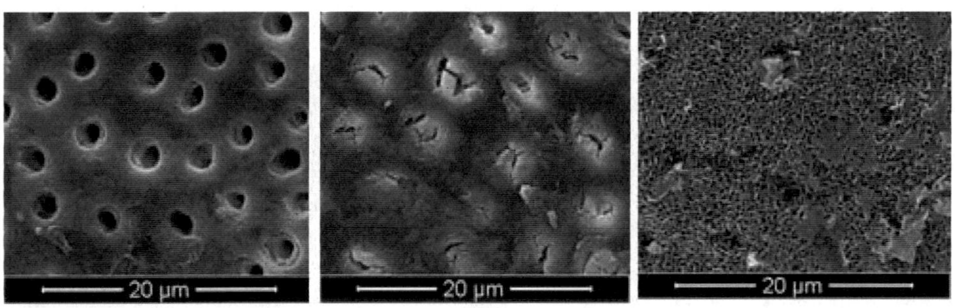


A B C

**Figure 4.** SEM images of dentine slices washed with ethanol (A) before applying calcium phosphate organogel and after applying calcium phosphate organogel for (B) 1 day and (C) 5 days, showing the occlusion of the dentine tubules with synthesized calcium phosphate.

## 4. Conclusion

In conclusion, calcium phosphate composite organogel can be synthesized by using emulsion method, which used eucalyptol as low-toxicity organic phase, P123 as a template for calcium phosphate and gel formation, and SDS as surfactant to form water-in-oil emulsion. This method is easy to synthesize, low-cost, and require mild condition. TEM and DLS techniques showed that size and morphology of organogel was circle-like in nanosize. Moreover, EDX spectra confirmed the component of calcium and phosphate in organogel. For the dentine occlusion efficiency, calcium phosphate organogel was efficiently occluded the dentine tubules in vitro within 1 day. In the future, organogel will be improved for finding the optimum condition to synthesize in order to apply for dentine occlusion.

## References

[1] Farooq I, Moheet I A and AlShwaimi E 2015 *Arch. Oral Biol.* **60(9)** 1246-53.
[2] Miglani S, Aggarwal V and Ahuja B 2010 *J. Conservat. Dentist. JCD* **13(4)** 218-24.
[3] Bohner M et al 2013 *Biomatter.* **3(2)** e25103.
[4] Lin K, Wu C and Chang J 2014 *Acta Biomater.* **10(10)** 4071-102.
[5] Okada M and Matsumoto T 2015 *Japanese Dent. Sci. Rev.* **51(4)** 85-95.
[6] Muñoz-Espí R, Weiss C K and Landfester K 2012 Curr. Opin. Coll. Interf. Sci. **17(4)** 212-24.
[7] Akkarachaneeyakorn K et al 2014 *Chem. Mater.* **26(20)** 5965-72.
[8] Fletcher J et al 2011 Cryst. Eng. Comm. **13(11)** 3692-7.

## Acknowledgement

The authors would like to thank graduate school, department of chemistry, faculty of science at Kasetsart University (Thailand), and Center for Innovation in Chemistry (PERCH-CIC), department of Chemistry at Mahidol University (Thailand) to support funds for all experimental..

IOP Conf. Series: Materials Science and Engineering **317** (2018) 012028 doi:10.1088/1757-899X/317/1/012028

# Impact Toughness and Dynamic Compression of Ultrafine-Grained Titanium Grade 4

**Ivan V Smirnov[1, 2] and Alexander Y Konstantinov[1]**

[1]Research Institute for Mechanics, Lobachevsky State University of Nizhni Novgorod, 23 Prospekt Gagarina, Nizhny Novgorod, 603950, Russia
[2]Laboratory for Mechanics of Advanced Bulk Nanomaterials for Innovative Engineering Applications, Saint Petersburg University, 7/9 Universitetskaya nab., Saint Petersburg, 199034, Russia

**Abstract.** This work presents the study of an influence of ultrafine-grained (UFG) structure on impact toughness and dynamic compression of commercially pure titanium Grade 4. The UFG structure was provided by means of equal-channel angular pressing according to the Conform scheme (ECAP-Conform, Ufa) with subsequent heat treatment. Impact toughness tests on samples with U-shaped notch were carried out using a drop weight impact machine. Dynamic compression test of cylindrical samples were carried out on a setup with the Split-Hopkinson pressure bar (SHPB-20) by the Kolsky method. It was found that the impact toughness of the UFG titanium is 15% greater than that of its coarse-grained (CG) counterpart. However, fracture process of the CG material requires 1.5 times more energy. The ECAP treatment significantly increased yield strength of the material. At the same time, an increase of compression strain rate by 6 orders resulted in an increase of yield strength of the CG material by 20%, while yield strength of the UFG titanium remained at the level of a quasi-static load. The UFG material showed a well-expressed strain hardening behaviour for all strain rates.

## 1. Introduction

Ultrafine-grained (UFG) titanium have attracted significant research attention in the last few decades due to its superior mechanical properties, such as high strength and yield stress, as well as good endurance limit [1,2]. However, one lack that prevents its engineering application is a decreased ductility.

A lot of research is being done to increase the ductility of UFG and nanocrystalline titanium. For example, in Ufa State Aviation Technical University a new approach to enhance the strength and ductility of metallic materials was developed by controlling the grain boundary structure that can be considered as grain boundary engineering of UFG metallic materials [3]. The effective approach in ductility enhancement of UFG titanium is formation of a high fraction of high-angle boundaries in the UFG structure with providing recovery processes of non-equilibrium boundaries [4, 5]. Such approach is possible via combined techniques that include equal channel angular pressing (ECAP), subsequent thermomechanical treatment and annealing [5].

Most studies on the mechanical behaviour of UFG titanium are focused on the phenomena occurring under quasi-static loading, whereas research on dynamic response and fracture of the material under high strain rates and pulse loads are rare. Nevertheless, dynamic characteristics are critically important for evaluating the potential application of ultrafine-grained materials.

The purpose of this work is to begin developing data on dynamic response and dynamic fracture parameters of ultrafine-grained titanium Grade 4, which has not only high static strength, but also good ductility.

## 2. Materials and experimental procedures

### *2.1. Materials*

In this work, the states of commercially pure titanium Grade 4 with a coarse-grained (CG) and ultrafine-grained (UFG) structure are considered. The chemical composition of the material, in wt.%, was 0.05C, 0.15Fe, 0.05N, 0.007H, 0.36O, residuals<0.3, and Ti as the balance.

The initial material with a coarse-grained structure was produced by Dynamet Company (USA) in the form of hot-rolled rods with Ø12 mm.

The ultrafine-grained structure was produced in Ufa State Aviation Technical University using equal-channel angular pressing (ECAP) combined with the Conform process. The installation was with an angle of 120° in the ECAP abutment. The rod was processed through a total of six passes using route BC at a temperature of 200 °C. Before the ECAP, the rod was annealed at 680 °C for 1 hour. After the ECAP, the rod was annealed at 250 °C for 1 hour and then drawn with a die diameter of 12 mm. This combination of processing regimes makes it possible to produce UFG titanium with high strength and the greatest ductility, and also to ensure uniformity of mechanical properties along the volume of a rod. A more detailed description of the processing regimes and the corresponding properties of the produced titanium, as well as its structure, can be found in [6, 7].

The characteristic parameters of the materials are given in Table 1.

**Table 1**. Characteristic parameters of titanium Grade 4 before (CG) and after (UFG) ECAP-Conform.

| Structure/ Parameter | CG | UFG |
|---|---|---|
| Average grain size $d$ (μm) | 11 | 0.7 |
| Microhardness HV | 243 | 317 |
| Tensile strength $\sigma_t$ (MPa) | 685 | 1004 |
| Tensile yield strength $\sigma_{0.2}$ (MPa) | 514 | 991 |

### *2.2. Impact toughness tests*

The impact toughness tests were carried out using a drop weight impact test system (Instron CEAST 9350). The Charpy scheme was applied in according to the standard GOST 9454 [8]. The impactor speed at the time of impact was 5 m/s that corresponded to the impact energy ~68 J. Load and energy absorption diagrams were determined by the system of sensors and computer control.

Samples had a length of 55 mm, a height of 8 (working 6) mm and a width of 2 mm and also a U-shaped notch with a curvature radius of 1 mm. The samples were cut from the rod using an electro erosion machine, so that the length of the sample was located along the longitudinal direction of the rod.

### *2.3. Dynamic compression tests*

Tests on dynamic compression of cylindrical samples were performed on a setup with the Split-Hopkinson pressure bar (SHPB) by the Kolsky method [9]. A mathematical model of SHPB is a system of three bars: two infinitely strong and infinitely long thin bars and a "soft" short bar (sample) between them (Figure 1).

An impactor *1* generates a one-dimensional elastic compression wave in an incident bar *2*. This wave propagates along the bar with a speed of c. In view of a difference in an acoustic stiffness of the materials of the bar *2* and sample *4*, the wave is split so that part of it is reflected back, and the other part passes through the sample into the transmitter bar *5*. In this process, the sample can undergo elastic-plastic deformation, while the bars deform only elastically.

IOP Conf. Series: Materials Science and Engineering **317** (2018) 012028 doi:10.1088/1757-899X/317/1/012028

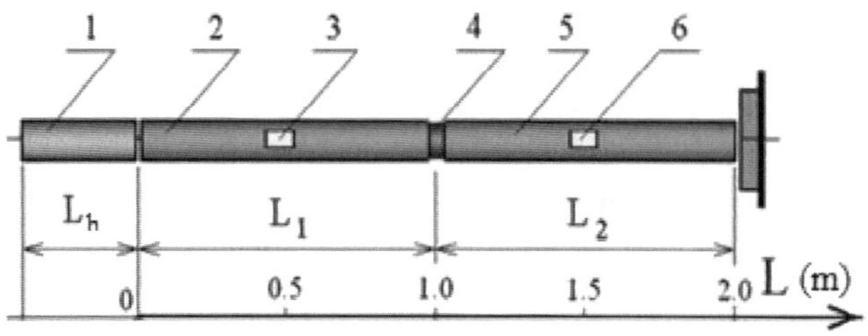


**Figure 1.** A typical circuit of the Split-Hopkinson pressure bar [10]. (Explanations are given in the text.)

The pulses of deformation of the incident bar (incident pulse and reflected pulse) and the transmitter bar (transmitted pulse) were measured using strain gauges *3* and *6* glued to the lateral surfaces of the bars. The signals from the strain gauges were captured on a multichannel computer measuring system. Formulae for the calculation of the mechanical characteristics by the pulses of deformation of the measuring bars in case of the dynamic compression were proposed by Kolsky. A more detailed description of the wave theory and computational formulas can be found in [9, 10].

A set of steel measuring bars with a diameter of 20 mm was used for titanium testing. The system was loaded by a steel impactor with a length of 300 mm and a diameter of 20 mm. Dynamic compression was performed for two impactor speeds (and, correspondingly, strain rates) 13 and 22 m/s.

The samples had a length of 4 mm and a diameter of 8 mm. The samples were cut from the rod using an electro erosion machine, so that the cylinder axis was located along the longitudinal direction of the rod.

Quasi-static tests were carried out using standard test machine (Shimadzu AG-50kNX) with a capturing speed of 1 mm/min. A sample deformation was measured by a traverse displacement.

## 3. Results and discussion

### *3.1. Impact toughness tests*

It is well known that a grain reduction can lead to an increase in metal strength. On the other hand, this leads to a decrease in ductility of the material. However, as shown in [5, 7], the correctly selected sequence of processing regimes of severe plastic deformation and heat treatment allow obtaining the UFG material with good ductility.

Figure 2 shows the results of the impact toughness tests of the titanium with the ultrafine-grained and coarse-grained states. The impact toughness of the UFG titanium is 15% greater than the impact toughness of its original coarse-grained analogue. At the same time, the fracture of coarse-grained material requires 1.5 times more energy.

Fracture surfaces have a dimpled character for both material states, Figure 3. The similar dimpled fractography was observed in impact toughness tests of Ti Grade 5 in [11]. Such a fracture surface indicates a localized plastic deformation, which suggests good perspective for increasing the plastic properties of ultrafine-grained titanium.

### *3.2. Dynamic compression tests*

Figure 4 shows the "engineering stress-engineering strain" curves for the compression of the titanium at different strain rates. The curves for the initial grain state of the material are significantly different for its ultrafine-grained state. The yield point (for 0.2% deformation) at slow deformation is 660 MPa for the CG material and 1020 MPa for the UFG material. An increase in the strain rate by 6 orders led to an increase in the yield stress of the CG material to 810 MPa. However, the yield stress for the UFG material remained at the level corresponding to the quasi-static loading.

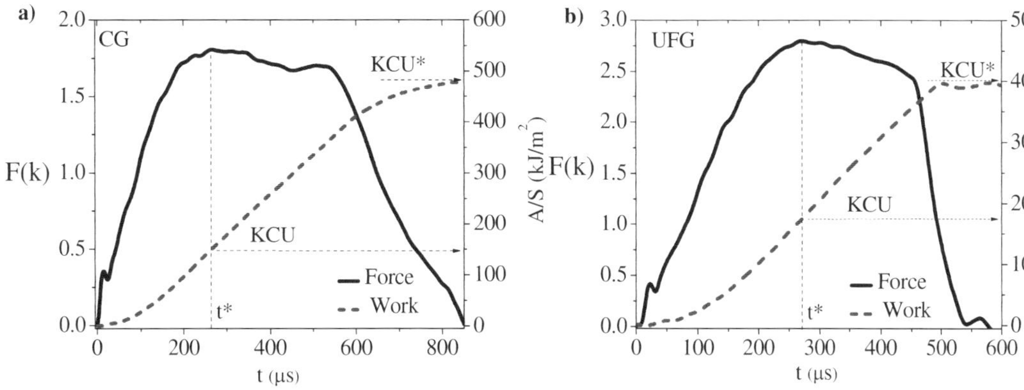


**Figure 2.** Characteristic diagrams in impact toughness tests of titanium. a) coarse-grained state; b) ultrafine-grained state; A is the displacement work; S is the area of a sample cross section in the point of impact; t* is the start time of macro fracture; KCU is the impact toughness; KCU* is the full specific work.

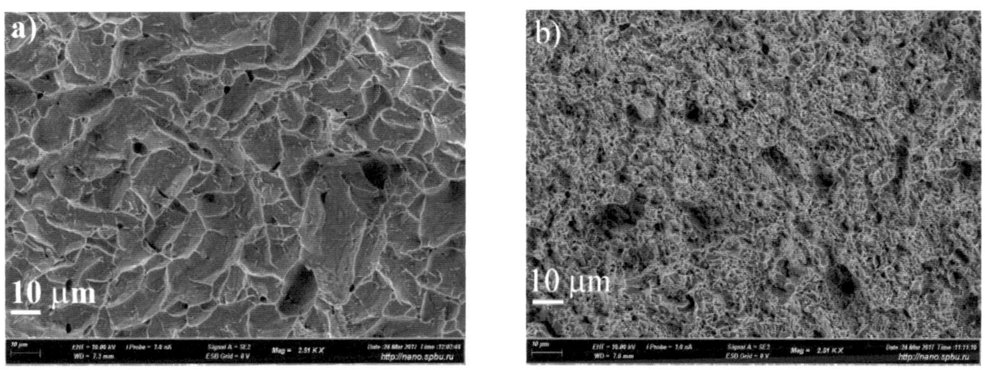


**Figure 3.** Fracture surfaces of titanium Grade 4 after impact toughness tests. a) coarse-grained state; b) ultrafine-grained state.

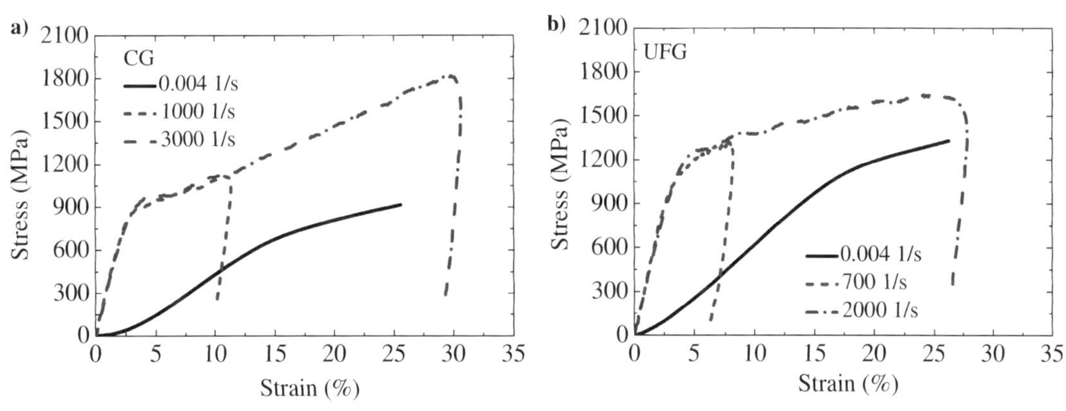


**Figure 4.** Characteristic diagrams in compression tests of titanium for different strain rates. a) coarse-grained state; b) ultrafine-grained state.

Deformation hardening is observed for both states and for all strain rates that is unusual for materials with ultrafine-grained structure. Thus, due to combination modes of ECAP and heat treatment, it was possible to achieve such a structure with large-angle grain boundaries, which promote the mechanisms of grain boundary sliding and contribute to strain hardening during plastic deformation [5], that results in ductility enhancement. However, the ultrafine-grained titanium still showed less ductility than the coarse-grained material at a dynamic load.

## 4. Conclusions

Impact toughness and response on dynamic compression of pure titanium Grade 4 processed by equal-channel angular pressing (ECAP-Conform) were investigated. ECAP processing resulted in an ultrafine-grained (UFG) structure of the material. The impact toughness tests were carried out according to the Charpy scheme for samples with U-notch. The dynamic compression tests were carried out on a setup with the Split-Hopkinson pressure bar at strain rates of up to $10^3$ 1/s. The conclusions are as follows:

1. Impact toughness of UFG titanium processed by optimal regimes and combination of ECAP and heat treatment can be no worse than that of the material prior to processing. In this study the impact toughness of the UFG titanium is 15% greater than that of its coarse-grained (CG) counterpart. Both CG and UFG samples fractured in a ductile manner. However, fracture process of the CG material requires 1.5 times more energy.
2. The UFG titanium showed significantly enhanced yield strength compared to the CG titanium at quasi-static loading. However, an increase in a compression strain rate by 6 orders resulted in an increase in yield strength of the CG material by 20%, while yield strength of the UFG titanium remained at the level of a quasi-static load. Both CG and UFG samples showed strain hardening behavior for all strain rates.

## Acknowledgements

The authors are thankful to Prof. Irina P. Semenova at Ufa State Aviation Technical University for providing the materials. The SEM images were obtained in the Interdisciplinary Center for Nanotechnology at the Research Park of Saint Petersburg University. This paper contains the results of two projects. Mechanical parameters and impact toughness of the materials were investigated within the framework of the Grant of the President of the Russian Federation for young scientists (MK-2587.2017.1). Research on the materials response to dynamic compression was carried out within the grant of the Russian Science Foundation (Grant № 15-19-10032).

## References

[1] Popov A A, Pyshmintsev I Yu, Demakov S L, Illarionov A G, Lowe T C and Valiev R Z 1997. *Scripta Mater.* **37** 1089.
[2] Sergueeva A V, Stolyarov V V, Valiev R Z and Mukherjee A K 2001. *Scripta Mater.* **45** 747.
[3] Valiev R Z 2008. *Materials Science Forum* **584-586** 22.
[4] Valiev R Z, Sergueeva A V and Mukherjee A K 2003. *Scripta Mater.* **49** 669.
[5] Semenova I, Salimgareeva G, Da Costa G, Lefebvre W and Valiev R 2010. *Adv. Eng. Mater* 12 803.
[6] Polyakov A V, Semenova I P and Raab G I 2012. *Rev. Adv. Mater. Sci.* **31** 78.
[7] Polyakov A V, Semenova I P, Valiev R Z, Huang Y and Langdon T G 2013. *MRS communications* **3** 249.
[8] GOST9454 1978. Metals. Method for testing the impact strength at low, room and high temperature.
[9] Kolsky H 1963. Stress Waves in Solids (NY: Dover Publications) p 224.
[10] Bragov A M and Lomunov A K 1995. *Int. J. Impact Eng.* **16**(2) 321.
[11] Semenova I P, Polyakov A V, Polyakova V V, Grishina Yu F, Huang Yi, Valiev R Z and Langdon T G 2017. *Mater. Sci. Eng. A* **696** 166.

IOP Conf. Series: Materials Science and Engineering **317** (2018) 012029 doi:10.1088/1757-899X/317/1/012029

# Dynamics of Longitudinal Impact in the Variable Cross-Section Rods

**R Stepanov[1], D Romenskyi [1], S Tsarenko[2]**

[1] Department of Structural and Theoretical Mechanics, National Research University Moscow State University of Civil Engineering, 26 Yaroslavskoye Shosse, Moscow, 129337, Russia
[2] Department of Strength of Materials, Donetsk National Technical University, 58 st. Artema, Donetsk, 83001, Ukraine

**Abstract.** Dynamics of longitudinal impact in rods of variable cross-section is considered. Rods of various configurations are used as elements of power pulse systems. There is no single method to the construction of a mathematical model of longitudinal impact on rods. The creation of a general method for constructing a mathematical model of longitudinal impact for rods of variable cross-section is the goal of the article. An elastic rod is considered with a cross-sectional area varying in powers of law from the longitudinal coordinate. The solution of the wave equation is obtained using the Fourier method. Special functions are introduced on the basis of recurrence relations for Bessel functions for solving boundary value problems. The expression for the square of the norm is obtained taking into account the orthogonality property of the eigen functions with weight. For example, the impact of an inelastic mass along the wide end of a conical rod is considered. The expressions for the displacements, forces and stresses of the rod sections are obtained for the cases of sudden velocity communication and the application of force. The proposed mathematical model makes it possible to carry out investigations of the stress-strain state in rods of variable and constant cross-section for various conditions of dynamic effects.

## 1. Introduction

The research of dynamic processes is carried out using the elastic rod model for a wide range of objects, such as: elements of drilling equipment [1-3], buildings and structures [4], power impulse systems [5, 6], etc. Especially urgent is the problem of longitudinal impact for power pulse systems, which are used in machines that perform various technological processes: punching, forging, destruction of rocks, concrete coatings, piling, etc. The problem of increasing the productivity of impact machines includes not only increasing the power, but also increasing the efficiency of energy transfer to the process area. The latter is achieved, among other things, on the basis of studying the process of formation of deformation waves by strikers of various geometries and searching for strings structures that create deformation waves with rational parameters [5].

At considering the various models of longitudinal collision of bodies [5, 7], the wave model of Saint-Venant's shock is taken as the basis, since it most fully reflects the real dynamic processes in colliding bodies, and for its practical implementation the d'Alembert method is used. The exact solution of the wave model is given by the Fourier method [8].

At solving the equation of longitudinal oscillations of variable cross-section rods, in the case of impact, arise certain mathematical difficulties, for example, the given equation is an equation with variable coefficients, the orthogonality of eigen functions, etc. Therefore, for solving the such problems are using various simplifying hypotheses. Thus, in [9] systems with distributed parameters are replaced by single-mass systems (Cox's theory) or various refinements [10], or are using an approximation of the dynamic deformations forms by static [11, 12], and in paper [13] is used the method of averaging variable coefficients. In [7, 14] are considered the models of longitudinal impact of various shapes rods with the using of their surface approximation by successively conjugate cylindrical sections. Using the operational calculus in work [3], is carried out an analysis of the using effectiveness of application of elements of drilling equipment (picks, drill rods) with different configurations, and for each of the schemes under consideration, a custom mathematical model is being constructed. The longitudinal impact of rods with conical and hyperbolic forms is considered in [15, 16], solutions of the wave equation for which are obtained in elementary functions, and in [17] dynamical processes of longitudinal impact in hollow rods of conical shape are studied using Bessel functions.

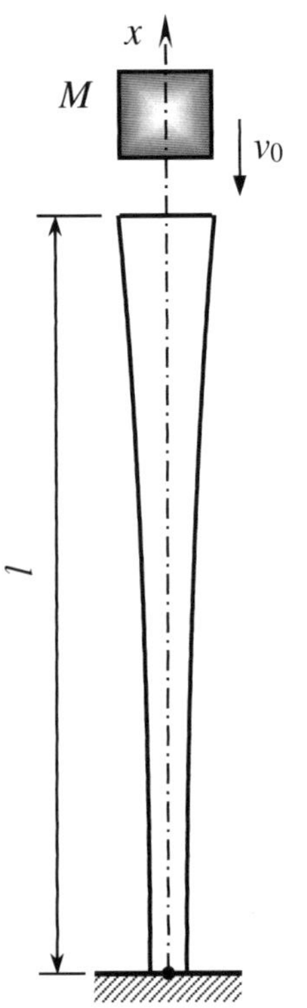


**Figure 1.** Scheme of longitudinal impact of a complex configuration rod.

Thus, based on the fact that in the theory of calculations, at the present time, there is no single approach to constructing a mathematical model of the longitudinal impact of complex configuration rods, the search and development of new solutions to longitudinal impact problems is an actual scientific and practical problem.

## 2. Development of a mathematical model

As a mathematical model of the object under consideration, we take an elastic rod of length l, the distributed mass m and the cross-sectional area F, which varies according to the exponent law from the longitudinal coordinate x:

$$m=\gamma F\ ;\ F=F_2 z^{\mu}\ ;\ z=(1-k)\frac{x}{l}+k\ ,\ \ k=\frac{h_1}{h_2}\ \ (0<k<1) \tag{1}$$

when $\gamma$ – material density, $F_2$ – the cross-sectional area of the larger base of the rod, $h_1$ и $h_2$ – the parameters of the cross dimensions of the upper and lower sections are determined by the geometry of the rod cross-section of the rod (the radius of the cross-sections for conical structures), the exponent $\mu$ depends on the rod configuration, for example, for conical tubes $\mu = 1$, and for a conic rod of solid cross section $\mu = 2$ (figure 1).

Equation of longitudinal displacements $u(x,t)$, taking into account the accepted notation, will have the view [17].

$$z^{\mu}\frac{\partial^2 u}{\partial z^2}+\mu z^{\mu-1}\frac{\partial u}{\partial z}=\frac{\gamma l^2}{E(1-k)^2}z^{\mu}\frac{\partial^2 u}{\partial t^2}+\frac{l^2}{EF_2(1-k)^2}p(z,t), \tag{2}$$

when $E$ – elasticity modulus, $p(x,t)$ – external load.

Let us find the general solution of equation (2) at $p(z,t)\equiv 0$ (the case of free oscillations).

Separating the variables, obtain an equation for the eigen functions whose solution has the view [18].

$$Z_n(z)=z^{\nu}\left(C_1 J_{\nu}(\lambda_n z)+C_2 Y_{\nu}(\lambda_n z)\right) \tag{3}$$

when

$$\nu=\frac{1-\mu}{2}\ ;\ \lambda_n=\frac{\omega_n l}{(1-k)}\sqrt{\frac{\gamma}{E}}\ ,$$

$\omega_n$ – oscillation eigen frequencies, $J_{\nu}(z)$ и $Y_{\nu}(z)$ – Bessel functions.

For the convenience of solving problems with different boundary conditions, we introduce the following notation for functions:

$$A_n(z)=\frac{\pi}{2}\lambda_n k\left(J_\nu(\lambda_n z)Y_{\nu-1}(\lambda_n k)-Y_\nu(\lambda_n z)J_{\nu-1}(\lambda_n k)\right);$$

$$B_n(z)=\frac{\pi}{2}\lambda_n k\left(Y_\nu(\lambda_n z)J_\nu(\lambda_n k)-J_\nu(\lambda_n z)Y_\nu(\lambda_n k)\right);$$

$$C_n(z)=\frac{\pi}{2}\lambda_n k\left(J_{\nu-1}(\lambda_n z)Y_{\nu-1}(\lambda_n k)-Y_{\nu-1}(\lambda_n z)J_{\nu-1}(\lambda_n k)\right);$$

$$D_n(z)=\frac{\pi}{2}\lambda_n k\left(Y_{\nu-1}(\lambda_n z)J_\nu(\lambda_n k)-J_{\nu-1}(\lambda_n z)Y_\nu(\lambda_n k)\right).$$

In point $z=k$ the presented functions take the values:

$$A_n(k)=D_n(k)=1;\; B_n(k)=C_n(k)=0.$$

Using the adopted dependencies, arbitrary constants in (3) can be expressed in terms of displacement $u_0$ and force $N_0$ in the zero section, then the expressions for displacements and longitudinal forces for an arbitrary waveform can be represented as:

$$u_n(z)=Z_n=z^\nu\left(u_0k^{-\nu}A_n(z)+\frac{N_0 l}{EF_2(1-k)}\frac{k^{\nu-1}}{\lambda_n}B_n(z)\right); \tag{4}$$

$$N_n(z)=\frac{EF_2(1-k)}{l}z^{1-2\nu}Z_n'=z^{1-\nu}\left(u_0\frac{EF_2(1-k)}{l}k^{-\nu}\lambda_n C_n(z)+N_0k^{\nu-1}D_n(z)\right). \tag{5}$$

On the basis of the reciprocity theorem [19], at the absence of concentrated masses, the eigen functions will be orthogonal with weight $\rho(z)=z^{1-2\nu}$. To find the square of the eigen functions norm, we first proceed in the same way as in [17] for eigen functions with different indices.

$$\left(\lambda_n^2-\lambda_m^2\right)\int_k^1 z^{1-2\nu}Z_mZ_n dz=Z_m\left(z^{1-2\nu}Z_n'\right)-Z_n\left(z^{1-2\nu}Z_m'\right)\Big|_k^1.$$

Passing to the limit at $m\to n$, are getting

$$2\lambda_n\int_k^1 z^{1-2\nu}Z_n^2(z)dz=\frac{\partial Z_n}{\partial\lambda_n}\left(z^{1-2\nu}Z_n'\right)-Z_n\frac{\partial\left(z^{1-2\nu}Z_n'\right)}{\partial\lambda_n}\Bigg|_k^1.$$

Having determined the partial derivatives by $\lambda_n$, a square of the eigen functions norm find in the form

$$\Delta_n^2=\frac{1}{2}\left(z^{2-2\nu}Z_n^2+\frac{z^{2\nu}}{\lambda_n^2}\left(z^{1-2\nu}Z_n'\right)^2-\frac{2\nu-1}{\lambda_n^2}\left(z^{1-2\nu}Z_n'\right)Z_n\right)\Bigg|_k^1. \tag{6}$$

Now consider the case of forced oscillations, this requires finding a solution $u_2(z,t)$ of non-linear equation

$$\left(z^{1-2\nu}u'\right)'-\beta z^{1-2\nu}\ddot{u}=\frac{\beta}{\gamma F_2}p(z,t),\quad \beta=\frac{\gamma l^2}{E(1-k)^2} \tag{7}$$

The equation solution (7) can be represented as a row by eigen functions

$$u_2(z,t)=\sum_{n=1}^{\infty} w_n(t)Z_n(z).$$

Substituting this row into (7) and taking into account relation $\lambda_n^2 = \beta\omega_n^2$, are getting

$$\sum_{n=1}^{\infty}\left(\ddot{w}_n + \omega_n^2 w_n\right)z^{1-2\nu}Z_n = -\frac{p(z,t)}{\gamma F_2} \quad (8)$$

Applying the Fourier method to equation (8), taking into account the orthogonality property of the eigen functions with weight, obtain the equation with respect to the coefficients $w_n$

$$\ddot{w}_n + \omega_n^2 w_n = -\frac{1-k}{\gamma F_2 l \Delta_n^2}\int_k^1 p(z,t)Z_n(z)dz \quad (9)$$

As an example, consider the scheme of mass impact M on the upper end of the rod, while the mass is held for some time by the rod, and its lower end is rigidly limited (figure 1), such a circuit can simulate a stress-strain state in waveguides of power pulse systems or in piles in the process of their driving. For the adopted scheme, the movement of the bottom end will be $u_0 = 0$, the boundary condition on the second end of the rod

$$N(l,t) = -M\ddot{u}(l,t) \quad (10)$$

initial conditions

$$u(x,0) = 0\ ;\ \dot{u}(x,0) = -v_0 e(x-l) \quad (11)$$

when $e(x)$ – a unit function.

The equation for displacements and longitudinal forces for an arbitrary shape is obtained in the form

$$u_n(z) = \frac{N_0 l}{EF_2(1-k)}Z_n = \frac{N_0 l}{EF_2(1-k)}\frac{k^{\nu-1}}{\lambda_n}z^{\nu}B_n(z);$$

$$N_n(z) = N_0 z^{1-2\nu}Z_n' = N_0 z^{1-\nu}k^{\nu-1}D_n(z).$$

From the boundary condition (10) obtain the equation for finding the eigen values

$$D_n(1) - \zeta\frac{1-k^{2-2\nu}}{2(1-\nu)}\lambda_n B_n(1) = 0 \quad (12)$$

here $\zeta = M / M_0$, $M_0$ – the road mass.

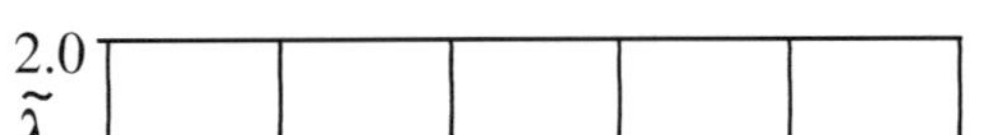


**Figure 2.** Depending $\tilde{\lambda}_1$ from the value of the relative cross dimension parameter *k*.

Taking into account the fact, that in some approximate methods of dynamic calculation [9, 10] is using the value of the first eigen frequency, we estimate the effect of the parameter on the values of the first eigen value of (12)

$$\omega_1 = \frac{\tilde{\lambda}_1}{l}\sqrt{\frac{E}{\gamma}},\ \tilde{\lambda}_n = \lambda_n(1-k),$$

the frequencies of the rods of constant cross section are determined in the same way [8]. Numerical studies of dependence $\tilde{\lambda}_1$ on the parameter value $k$ for conic hollow roads (ν=0) and solid (ν=-0,5) sections, without taking into account the concentrated mass ($\zeta = 0$), are presented at figure

2. It can be seen from the graph that for the case $k \approx 1$ the value $\tilde{\lambda}_1$ coincides with the known value $\tilde{\lambda}_1 = \pi/2 \approx 1{,}571$ [8] (is indicated by the dot-dash line on the graph figure 2).

Given the presence of the concentrated mass on the upper end of the rod (figure 1), the eigen functions will be orthogonal with the weight

$\rho(z) = z^{1-2\nu} + \zeta\frac{1-k^{2-2\nu}}{2(1-\nu)}\delta(z-1)$, where $\delta(z)$ – Dirac's delta function. Then from the relation (6) the square of the eigen functions norm will be determined by the formula

$$\Delta_n^2 = \frac{1}{2\lambda_n^2}\left(\left(1+\zeta\frac{1-k^{2-2\nu}}{2(1-\nu)}\left(\zeta\frac{1-k^{2-2\nu}}{2(1-\nu)}\lambda_n^2 - 2\nu + 2\right)\right)k^{2-2\nu}B_n^2(1) - k^{2\nu}\right).$$

First, we find the movements and forces from the the speed reporting. From the first initial condition (11) it follows that

$$u_v(z,t) = \sum_{n=1}^{\infty}\Phi_n Z_n(z)\sin\omega_n t \tag{13}$$

If the second initial condition (11) is satisfied, then in expression (13) the coefficients $\Phi_n$ will take the view

$$\Phi_n = -\frac{v_0 l}{c}\zeta\frac{\left(1-k^{2-2\nu}\right)k^{\nu-1}B_n(1)}{4(1-\nu)(1-k)\lambda_n^2\Delta_n^2},$$

where $c = \sqrt{E/\gamma}$ - longitudinal wave velocity of the rod [8].

Dependences for displacements, forces and stresses are obtained in the view

$$u_v(z,t) = -\frac{v_0 l}{c}\zeta\frac{\left(1-k^{2-2\nu}\right)k^{2\nu-2}}{4(1-\nu)(1-k)}\sum_{n=1}^{\infty}\frac{B_n(1)}{\lambda_n^3\Delta_n^2}z^{\nu}B_n(z)\sin\omega_n t \tag{14}$$

$$N_v(z,t) = -\frac{v_0}{c}EF_2\zeta\frac{\left(1-k^{2-2\nu}\right)k^{2\nu-2}}{4(1-\nu)}\sum_{n=1}^{\infty}\frac{B_n(1)}{\lambda_n^2\Delta_n^2}z^{1-\nu}D_n(z)\sin\omega_n t \tag{15}$$

$$\sigma_v(z,t) = -\frac{v_0}{c}E\zeta\frac{\left(1-k^{2-2\nu}\right)k^{2\nu-2}}{4(1-\nu)}\sum_{n=1}^{\infty}\frac{B_n(1)}{\lambda_n^2\Delta_n^2}z^{\nu}D_n(z)\sin\omega_n t \tag{16}$$

Taking into account to the sudden weight application $Mg$ (9) will have the view

$$\ddot{w}_n + \omega_n^2 w_n = -\frac{Mg(1-k)}{\gamma F_2 l\Delta_n^2}Z_n(1),$$

and their solution

$$w_n(t) = -u_2\frac{2\nu k^{\nu-1}B_n(1)}{\Delta_n^2\lambda_n^3\left(1-k^{2\nu}\right)}(1-\cos\omega_n t);\ u_2 = \frac{Mgl\left(1-k^{2\nu}\right)}{2\nu(1-k)EF_2}.$$

Thus, in the final form obtain the equations of displacements, forces and stresses from the application of the load

$$u_p(z,t) = -u_2\frac{2\nu k^{2\nu-2}}{\left(1-k^{2\nu}\right)}\sum_{n=1}^{\infty}\frac{B_n(1)}{\Delta_n^2\lambda_n^4}z^{\nu}B_n(z)(1-\cos\omega_n t) \tag{17}$$

$$N_p(z,t) = -Mgk^{2\nu-2}\sum_{n=1}^{\infty}\frac{B_n(1)}{\Delta_n^2\lambda_n^3}z^{1-\nu}D_n(z)(1-\cos\omega_n t) \tag{18}$$

$$\sigma_p(z,t) = -\frac{Mg}{F_2} k^{2\nu-2} \sum_{n=1}^{\infty} \frac{B_n(1)}{\Delta_n^2 \lambda_n^3} z^{\nu} D_n(z)(1-\cos\omega_n t) \qquad (19)$$

Introduce the notation for dimensionless quantities:

$\tau = \frac{t}{l}\sqrt{\frac{E}{\gamma}}$ – dimensionless time, $\tilde{\sigma}_v(z,t) = \frac{\sigma_v(z,t)c}{v_0 E}$ ; $\tilde{\sigma}_p(z,t) = \frac{\sigma_p(z,t)F_2}{Mg}$ – relative stresses, $\tilde{u}_p(z,t) = \frac{u_p(z,t)}{u_2}$ – relative displacement.

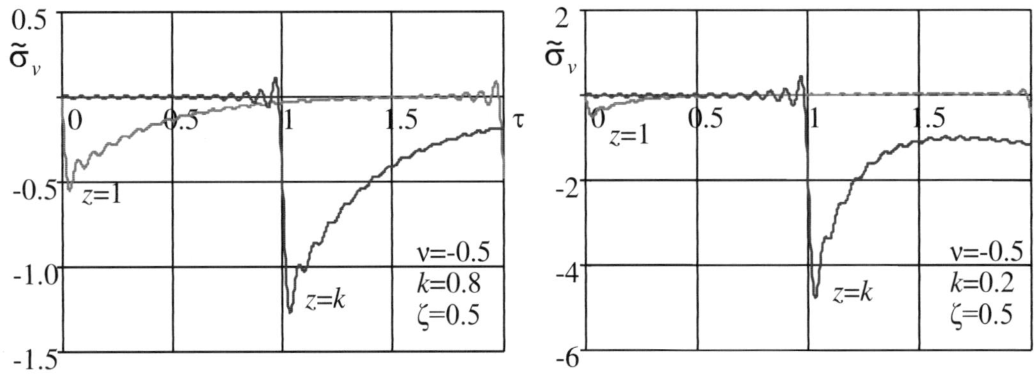


**Figure 3.** The relative magnitude of the stresses in the end sections of the rods of different configurations from the velocity reporting.

At figure 3 are shown the dependencies of stresses $\tilde{\sigma}_v$ from time $\tau$, caused by communication to the upper end of the speed rod from the falling load ($\zeta = 0{,}5$) in the shock end ($z = 1$) and at the lower end ($z = k$), for conical shaped rods of solid section ($\nu = -0{,}5$),with the ratio of the radiuses of the upper and lower sections, respectively: $k = 0{,}8$ и $k = 0{,}2$. Analysis of stresses and stresses for various parameters of the scheme showed that for rods with a decrease in the value of the relative cross dimension *K*, the stresses at the lower end increase, but not in proportion to the decrease in the cross-sectional area, but to a much lesser extent, i.e. the force decreases. Thus, for waveguides of power pulse systems, it is expedient to use rods of constant cross section with sharpening only in the area of the working part, using of rods with a decreasing cross section will lead to a decrease in the efficiency of transferring the force pulse to the working zone.

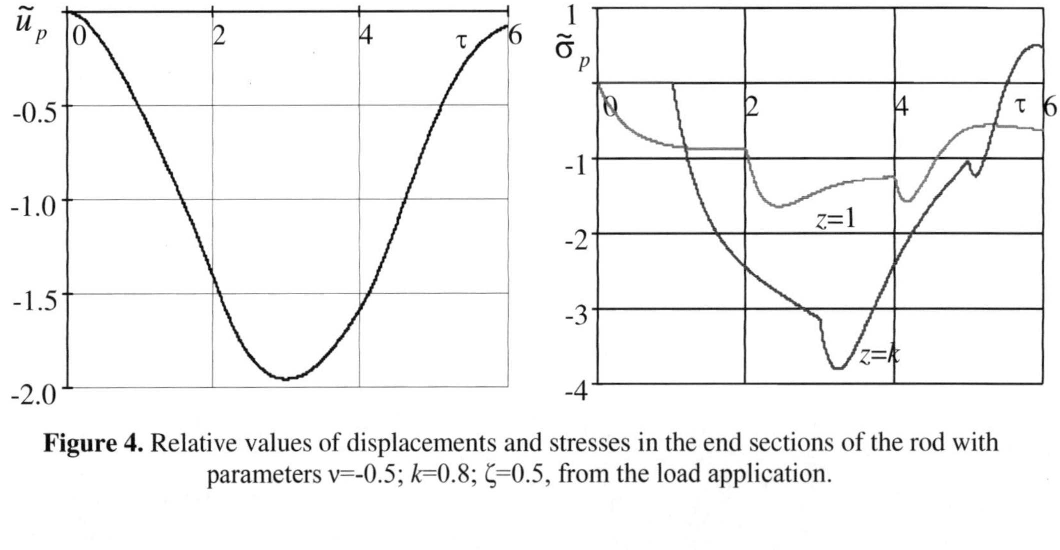


**Figure 4.** Relative values of displacements and stresses in the end sections of the rod with parameters ν=-0.5; *k*=0.8; ζ=0.5, from the load application.

At figure 4. are presented the relative displacements $\tilde{u}_p$ and stresses $\tilde{\sigma}_p$ depending on time $\tau$, caused by a sudden application to the upper end of the power shaft, for a conical shaped rod of solid section ( $\nu = -0{,}5$ ), with the ratio of the radiuses of the upper and lower sections $k = 0{,}8$. The study showed that different parameters of the scheme do not significantly affect the values of the maximum relative displacements and forces, the stresses in the lower section increase in proportion to the reduction in the cross-sectional area. Those, in systems where the transfer of the pulse is mainly due to the application of the load, and not the speed, in this case the configuration of the rod will not affect the transmission efficiency of the power pulse.

## 3. Conclusions

The proposed mathematical model of the rod longitudinal impact with complex configuration allows both direct investigation of the stress-strain state and, in contrast to numerical methods and means of object modeling, the analytical solution makes it possible to estimate the degree of influence of various model parameters on the unknown quantities. The presented algorithm of calculation can be put in the basis of CAD for imitating modeling of complex mechanical systems. It should also be noted that all the parameters of the model are determined and at the value $\nu = 0{,}5$ that corresponds to the rods of the constant section, so can use the unified approach to the construction of rod models with various configurations

## 4. References

[1] Gulyaev V, Lugovoi P and Borshch E 2013 *Int. Appl. Mech* **49(3)** 114–24.
[2] Ulitin G. 2000 *Int. Appl. Mech* **36(10)** 1380–4.
[3] Shaposhnikov I I 2011 *Mechanical engineering* **21** 95–107.
[4] Anil K. Chopra 2014 *Dynamics of Structures: Theory and Applications to Earthquake Engineering* Prentice Hall **4** 992.
[5] Alimov O, Manzhosov V and Yerem'yants V 1985 *Impact. The spread of the strain waves in the shock systems* Moscow **4** 357.
[6] Beksalov Ye, Manzhosov V and Slepukhin V 2010 *Journal UlGTU* **3(51)** 24–8.
[7] Manzhosov V and Slepukhin V 2011 *Simulation of longitudinal impact in rod systems of non-uniform structure* Ulyanovsk 208.
[8] Birgera I and Panovko Ya 1968 *Strength, stability, vibrations. Handbook in three volumes* Moscow **3** 567.
[9] Sarkisov G 1971 *Calculations of drilling and casing columns* Moscow 206.
[10] Panovko Ya 1977 *Introduction to the theory of mechanical shock* Moscow 232.
[11] Artobolevskiy I, Bogolyubov A, Bolotin V and et al 1978 *Oscillations of linear systems* Moscow **1** 352.
[12] Biderman V 1972 *Applied theory of mechanical oscillations* Moscow 416.
[13] Ulitin G and Tsarenko S 2016 *Journal YUUrGU* **8(1)** 43–8.
[14] Timofeyev Ye and Zhukov I 2016 *Modern mechanical engineering. Science and education* **5** 540–9.
[15] Ulítín G and Tsarenko S 2014 *Strength of materials and theory of structures* **93** 56–63.
[16] Ulítín G and Tsarenko S 2014 *Vibrations in technics and technologies* **3(75)** 37–42.
[17] Ulitin G and Tsarenko S 2015 *Int. Appl. Mech* **51(1)** 102–7.
[18] Kamke E 1977 *Differentialgleichungen: Lösungsmethoden und Lösungen* Leipzig: Teubner 636.
[19] Kiselev V 1964 *Structural mechanics. Special course. Dynamics and stability of structures* Moscow 332.

# Study on a Biometric Authentication Model based on ECG using a Fuzzy Neural Network

**Ho J. Kim[1] and Joon S. Lim[1]**

Department of Computer Engineering, Gachon University, Seongnam-si, Republic of Korea

**Abstract.** Traditional authentication methods use numbers or graphic passwords and thus involve the risk of loss or theft. Various studies are underway regarding biometric authentication because it uses the unique biometric data of a human being. Biometric authentication technology using ECG from biometric data involves signals that record electrical stimuli from the heart. It is difficult to manipulate and is advantageous in that it enables unrestrained measurements from sensors that are attached to the skin. This study is on biometric authentication methods using the neural network with weighted fuzzy membership functions (NEWFM). In the biometric authentication process, normalization and the ensemble average is applied during preprocessing, characteristics are extracted using Haar-wavelets, and a registration process called "training" is performed in the fuzzy neural network. In the experiment, biometric authentication was performed on 73 subjects in the Physionet Database. 10-40 ECG waveforms were tested for use in the registration process, and 15 ECG waveforms were deemed the appropriate number for registering ECG waveforms. 1 ECG waveforms were used during the authentication stage to conduct the biometric authentication test. Upon testing the proposed biometric authentication method based on 73 subjects from the Physionet Database, the TAR was 98.32% and FAR was 5.84%.

## 1. Introduction

Due to rapid advancements in IT technology, issues regarding privacy violation prevention and data protection have become increasingly important and the demand for biometric authentication is on the rise. Existing authentication technology is reliant on passwords or objects, which involves the risk of exposure, forgetting, loss, or theft. However, the more advanced biometric authentication technology uses an individual's unique data to verify identify. Unique data is generally categorized into action-based features and physical features. Action-based features include voice, signature, or walk, while physical characteristics include fingerprints, iris, face, and ECG. Of these, fingerprints, voice, iris, and face are used as the main mobile biometric authentication features [1-3]. Fingerprint authentication has been used for over 100 years in various application fields. A fingerprint is formed when the sweat glands form grooved into the palm of the hand. They appear around 13-19 weeks while in the womb and their shape never changes. However, fingerprints may become damaged and there is still the risk that an attacker may steal finger the fingerprint. Voice authentication recognizes the user through an individual's unique speech patterns. These speech types are behavioural and are formed as a result of physiological factors. Voice authentication is the most natural authentication method since the system's equipment does not require contact with the user's body. However, the disadvantage of the voice authentication system is that there is no reliability. There are dangers of manipulation where someone may imitate the voice of someone who is already authenticated in the voice authentication system database. Moreover, a person's voice is inconsistent according to their present heath conditions there may be overlapping noise, hence its authentication rate suffers compared to other biometric authentication methods. Therefore, its drawback is that it is not suitable for quiet indoor environments compared to public areas. The iris is known to be different even among identical twins and never changes throughout a person's lifetime as long as there is no major external trauma or illness. The patterns of the iris are formed within 6 months after birth and are determined at around 2-3 years old. The shape of the iris on the left and right differs for every individual. The advantage of iris authentication is that the target can be detected with a camera even at a distance of about 1m. However, there is the issue of privacy violations because the camera films an individual's face. Face authentication is a method of authenticating the shape of an individual's face, and is most commonly used when a person is verifying another person's identity. Face authentication involves assessing whether or not a person is who they say they are by comparing the present image with a face image in the database. To authenticate a face, it is important for the device that will process extracted features to have quick operation processing speeds. Moreover, data will change significantly depending on disguises, aging, hair length, direction of light, or expression, thus there is the disadvantage of decreased reliability in fields that required higher security. While there are many advantages and disadvantages to commonly used authentication methods, ECG biometric authentication involves unique biometric features from the electrical physiological factors of the heart, location and size of the heart, physical conditions, etc. ECG biometric authentication is advantageous in that it does not require any particular actions or positions and enables unrestrained measurements in the mobile environment, and is difficult to duplicate or forge since it involves biometric signals that are recorded from a person's beating heart [4, 5].

The first study on ECG biometric authentication was proposed by Lena Biel [6] in 2001. Henceforth, biometric authentication algorithms using ECG have been introduced by T. W. Shen [7], Steven A. Israel [8], Konstantinos N, Plataniotis [9], and Gerd Wubbbeler [10]. The characteristics of ECG biometric authentication usage thus far have mainly developed through the reference point analysis method and structural analysis method. Although the reference point analysis method was widely researched in the beginning, the trend of research is moving toward the structural analysis method because since the former method was found to directly reflect detection authentication errors as is [4]. The structural analysis method extracts features by categorizing signals into single cardiac impulse units from all ECG signals. Individual authentication algorithm studies are being conducted using various categorization methods such as the neural network, SVM, KNN, etc. as classifiers [11-14]. There are also research based on the number of ECG data channels and ECG measurement status. Although individual authentication using multichannel ECG signals achieves a near 100% accuracy, this accuracy decreases when the number of the channel is reduced [15, 16]. Biometric authentication technology using ECG is highly likely to become commercialized like the now ubiquitous fingerprint authentication technology. With the recent rise mobile device dependency, there is need for biometric authentication technology with increased convenience while maintaining a certain level of accuracy and reliability regarding personal information protection and authentication.

This paper used the structural analysis method, which extracts features using the structure of ECG waveforms, for effective biometric authentication using the low computing power of the mobile environment. Further, a biometric authentication test was conducted using the ECG of 73 subjects in the Physionet Database, which is a 1-lead signal that minimizes the number of electrodes for simple measurements in the mobile environment [23]. Haar-wavelet transformation was used as the structural analysis method for extracting ECG signal features, and the features were extracted using the wavelet coefficients a4, d3, and d4. A study that authenticates an individual's waveforms by training in the neural network with weighted fuzzy membership functions (NEWFM) based on the extracted features was proposed [17, 18].

Section 2 is regarding the structure and training of the neural network with weighted fuzzy membership functions, which is a related study, and Section 3 includes a detailed explanation of the proposed ECG biometric authentication method. Next, Section 4 is regarding the experiment and results for a fair performance evaluation, and Section 5 is the conclusion.

IOP Conf. Series: Materials Science and Engineering **317** (2018) 012030 doi:10.1088/1757-899X/317/1/012030

## 2. Proposed ECG Biometric Authentication Model

Figure 1 is a diagram of the biometric authentication model proposed in this study. The proposed biometric authentication model is divided into preprocessing, biometric feature extraction, and biometric authentication. During the preprocessing stage, the sampling rate and amplitude are normalized, and signals are extracted in ECG waveform units, then the ensemble average is applied to obtain a stable feature value even from a changing heart rate [24]. Noise removal and ECG waveform division can be performed simultaneously by extracting features using wavelet transformation during the biometric feature extraction process, and a strong resolution is achieved for the factors of each feature in the ECG signal. Biometric features are extracted through Haar-wavelet transformation coefficients a4, d3, and d4, and the extracted features are entered into the neural network with weighted fuzzy membership functions for use in order to categorize ECG waveforms in the proposed biometric authentication model.

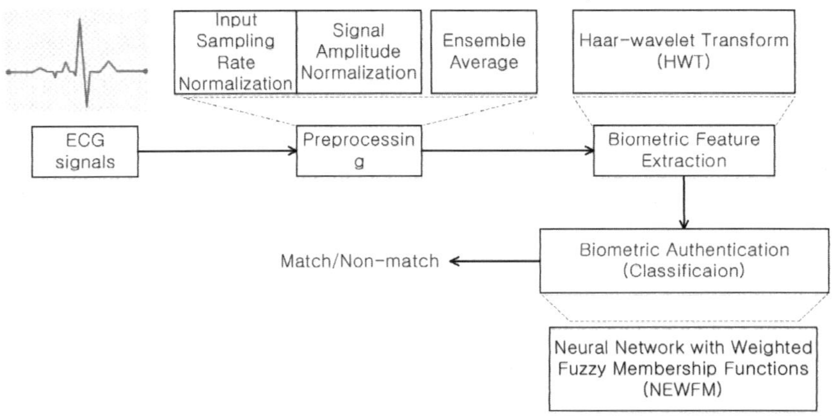


**Figure 1.** Structure of the proposed ECG biometric authentication model.

### 2.1. Preprocessing

The individual ECG signals that are entered are expressed through $E$ in Formula (1). $r$ in Formula (1) is the number of samples in individual ECG signals that were entered, and a ECG signal length of 2,000 seconds is used for input.

$$E = (e_1, e_2, e_3, e_4, \ldots, e_n) \quad (1)$$

ECG signals have varying sample rates depending on the measurement instrument. To experiment on signals that were obtained from different instruments, there must first be a process of converting the signals of various input sampling rates into an average sampling rate signal. Further, the intensity of heartbeats among subjects of ECG measurement will differ according to whether or not they exercise, take medication, etc. Therefore, there must also be a process of normalizing the ECG signal amplitude. preprocessing is performed by calculating an individual's average ECG waveform and applying the ensemble average of all ECG waveforms to calculate the average space of average ECG waveform samples and align ECG waveforms [24].

*2.1.1. Input Frequency Sampling Rate Normalization.* Since the ECG sampling rate varies according to the ECG measurement instrument, there must first be a normalization process into an ECG signal of a standard sampling rate with a unified sampling rate of diverse ECGs. The sampling rate normalization process can be divided into up-sampling normalization and down-sampling normalization. Up-sampling normalization converts signals with low sampling rates into signals with

IOP Conf. Series: Materials Science and Engineering **317** (2018) 012030 doi:10.1088/1757-899X/317/1/012030

high sampling rates while down-sampling normalization converts signals with high sampling rates into signals with low sampling rates process 1 shows the up-sampling normalization process through pseudocodes and process 2 shows the down-sampling normalization process. The initial sampling rate is received as a variable in the input_sample_rate and the standard sampling rate that is desired for output is received as a variable in standard_sample_rate. The two variables are divided to calculate the rate, then ECG samples are added or deleted to match this rate and the sampling rate is normalized and stored in the *sig*.

```
Input: E
Input: Input_sample_rate
Input: Standard_sample_rate
rate=
     Input_sample_rate/Standard_sample_rate
for i=0 to n
    index=i+1
    num_1=round(index*rate+0.5, 1)
    num_2=num_1+1
    sig[i]=(e[num_1]*rate* ndex)/num_2
end for
```

**Process 1.** Up Sampling Normalization.

```
Input: E
Input: Input_sample_rate
Input: Standard_sample_rate
rate=
     Input_sample_rate/Standard_sample_rate
len=floor(length(E)/rate, 1)
sig[len]=0
for i to len
    index=floor(i*rate, 1)
        num_1=integerthrow(i*rate)
        sig[i]=e[index]
    if num_1==0
        i=i+1
        sig[i]=e[i]
    end if
end for
```

**Process 2.** Down Sampling Normalization.

*2.1.2. ECG signal amplitude normalization.* To normalize the intensity of the ECG signal that is entered, the maximum value of the input ECG signal is showed through $E_{max}$ in Formula (2) and the minimum value of the input ECG signal is shown through $E_{min}$ in Formula (3). *r* in Formula (2) and (3) is the length of the input ECG signal.

$$E_{max} = (e_1, e_2, e_3, e_4, \ldots, e_n) \quad (2)$$

$$E_{min} = (e_1, e_2, e_3, e_4, \ldots, e_n) \quad (3)$$

The variable that will be stored after normalizing the amplitude is defined through $E_{NA}$ in Formula (4). Formula (5) shows the process of normalization between 0 and 1 by using the maximum and minimum value of ECG signals.

$$E_{NA} = (a_1, a_2, a_3, a_4, \ldots, a_n) \quad (4)$$

$$a_i = \frac{e_i - E_{min}}{E_{max} - E_{min}}, where i = 1,2,3,4,\ldots,r \quad (5)$$

Process 3 shows the process of normalization between 0 and 1 as a (pseudocode). This process receives the ECG signal that will undergo amplitude normalization at $E_{NA}$ , then the value that is obtained by normalizing amplitude using the maximum and minimum values of each ECG signal is stored in $E_{NA}$.

*2.1.3. Ensemble average.* Although the P-wave and T-wave of ECG signals show significant changes on the time axis, the QRS complex is an unchanging feature on the time axis. In order to obtain a stable feature value in a state where heart rate changes, this study used an alignment method that emphasizes the QRS complex. The Pan-Tompkins QRS complex detection algorithm was used to detect the location of R-peaks in the ECG signal [21]. R in Formula (6) stores the location value of R-peaks that are detected through the QRS complex detection algorithm, and m is the number of pulse beats that are found in individual ECG signals. One pulse beat can be defined as one ECG waveform. Formula (7) is shown as a 2-dimensional vector by applying the ensemble average of one ECG waveform [24]. k of Formula (7) is the ensemble average of one ECG waveform, and m is the number of ECG waveforms that are found in the ECG signal.

$$R = (r_1, r_2, r_3, r_4, \dots, r_m) \tag{6}$$

$$E_{AL} = \begin{pmatrix} l_{11} & l_{12} & \cdots & l_{1k} \\ l_{21} & l_{22} & \cdots & l_{2k} \\ \vdots & \vdots & & \vdots \\ l_{m1} & l_{m2} & \cdots & l_{mk} \end{pmatrix} \tag{7}$$

The P-wave of all ECG waveforms is an atrial depolarization that occurs when an electrical stimulation from the sinus node contracts the atrium, and the time required is between 0.05 to 0.12 seconds. The T-wave is the ventricular repolarization that shows period of recovery after the atrium was contracted. The time required is between 0.1 to 0.25 seconds. Therefore, 40 samples and 80 samples were extracted from the left and right of the R-peak respectively. These 120 samples were configured as one ECG biometric feature. Figure 2 shows the ECG signal before applying the ensemble average as a dotted line and the ECG signal after applying the ensemble average as a solid line.

Process 4 is the process of applying the ensemble average. The ECG signal that has undergone the normalization process in the previous stage in $E_{AL}$ is received as input. The location value of the R-peak that was detected through the Pan-Tompkins QRS complex detection algorithm is stored in peak[r]. This process involves applying the ensemble average that calculates the average space of an ECG signal's waveform and re-aligning it based on the R-peak to store it into $E_{AL}$ once again.

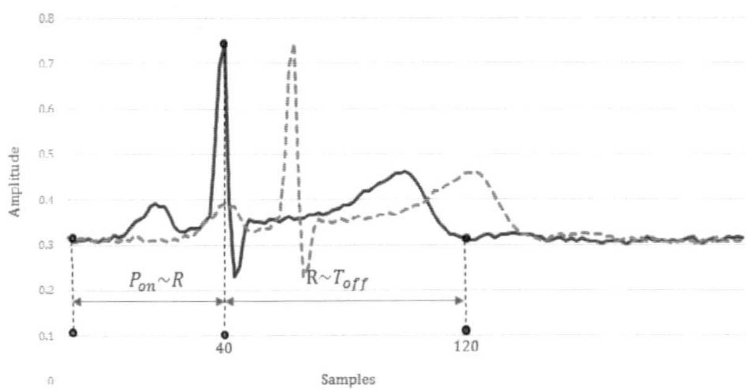


**Figure 2.** ECG waveform before applying ensemble average (dotted line) and ECG waveform after applying ensemble average (solid line)

Input: $E_{NA}$
[1]: $E_{max}$=max($E_{NA}$)
[2]: $E_{min}$=min($E_{NA}$)
[3]: for $i$ = 0 to $r$
[4]: $E_{NA}[i]$=($E_{NA}[i]$-$E_{min}$)/ ($E_{max}$-$E_{min}$)
[5]: end for

Input: $E_{AL}$
Input: *peak[r]*
[1]: *sig_num*=[], *sig*=[]
[2]: for $i$ = 0 to $n$
[3]: *before_loc=peak[i]*-40
[4]: *after_loc=peak[i]*+80
[5] *sig_num*= transpose($E_{AL}$*[before_loc:after_loc]*)
[6]: $E_{NA}$.append(*sig_num*)
[7]: end for

**Process 3**. Amplitude Normalization **Process 4.** Ensemble Average

*2.2. Feature extraction using haar-wavelet transform*

Wavelet transformation is a multi-resolution system that converts input sampling frequencies into other sampling frequencies to process signals in different frequency bands. Wavelet transformation is categorized into the scale coefficient (*a*) and transition element (*d*) from the wavelet generating function ($\psi$), and can be shown through Formula (8).

$$\psi(a,d)(x) = \frac{1}{\sqrt{a}}\int_{-\infty}^{\infty}\psi\left(\frac{t-d}{a}\right)dt \tag{8}$$

The scaling coefficient (*a*) and transition element (*d*) in Formula (8) are converted for the wavelet transformation of ECG signals. This can be categorized into high frequency signal $d_2^j[x(n)]$ and low frequency signal $a_2^j[x(n)]$, and ECG signal is shown through $d_2^j[x(n)]+a_2^j[x(n)]$ [20].

Formula (9) and (10) show that the number of samples becomes divided in two according to a decrease in Level j. Formula (9) is a high frequency detail signal. The input signal's high frequency component is divided through Level *j* according to the transition element (*d*) of the wavelet function. Formula (10) is the frequency component's approximation signal and divides the frequency component though scale factor (*a*) of the input signal by Level *j*.

$$d_{2^j}[x(n)] = \sum_{k=1} d_{2^j}(k)\psi_{2^j}(n-2^j k) \tag{9}$$

$$a_{2^j}[x(n)] = \sum_{k=1} a_{2^j}(k)\psi_{2^j}(n-2^j k) \tag{10}$$

In this paper, wavelet features were extracting using the Haar-wavelet transform (HWT), which is the most simple and easy to use. Figure 3 shows the multi-resolution analysis process of dividing input signals into high frequency portions and low frequency portions, then dividing the low frequency portion again using the same method. The wavelet coefficients in Level (1-2) are influenced by frequency interference, and the wavelet coefficients in Level (3-4) are influenced by base line conversions. *x[n]* in figure 3 shows the input ECG signals. *g[n]* refers to the detail signal and *h[n]* refers to approximation. ↓2 means that the number of the transformation coefficient during the down-sampling process gets reduces in half. The wavelet coefficients that are extracted as features of biometric authentication included 7 samples from a4, 15 samples from d3, and 7 samples from d4, hence a total of 29 wavelet coefficients were converted into on ECG biometric feature. Figure 4 is a diagram of the wave coefficients a4, d3, and d4 of MIT-BIH Normal Sinus Rhythm Record No. 16265.

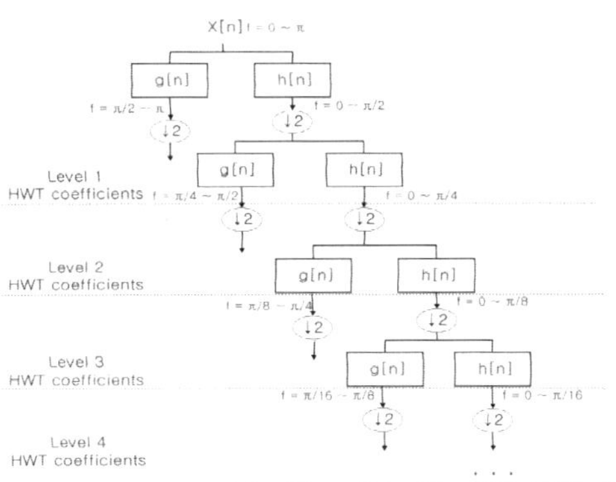

**Figure 3.** Division of scale Level 4 ECG signals t hrough Haar-wavelets.

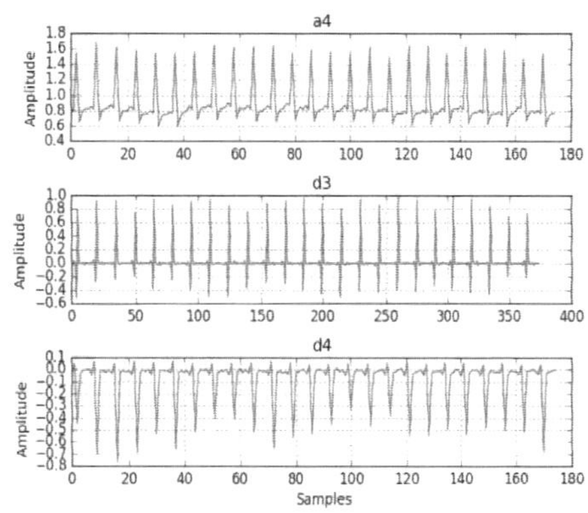

**Figure 4.** a4, d3, and d4 signals of MIT-BIH Normal Sinus Rhythm Record No. 16265.

*2.3. Training in the neural network with weighted fuzzy membership functions*

The neural network weighted fuzzy membership functions (NEWFM) is a supervised fuzzy neural network that categorizes classes using the boundary sum of weighted fuzzy membership functions that were trained from input. [Figure 5] is a diagram of the structure of the neural network with weighted fuzzy membership functions. The neural network with weighted fuzzy membership functions consists of the input layer, hyperbox layer, and output layer. The input layer has N nodes and each node is connected to one input feature. The number of nodes in the hyperbox layer is the number obtained by multiplying the number of nodes in the input layer and the number of nodes in the output layer. The nodes in the hyperbox layer are bundled as the hyperbox group, and the number of hypperboxes is identical to the number of nodes in the output layer. Each node is connected to one hyperbox. The weighted fuzzy membership functions have three linguistic variables that are large, medium, or small, and the data that is entered through the results of the weighted fuzzy membership function is assessed through the boundary sum of this variable [17, 18].

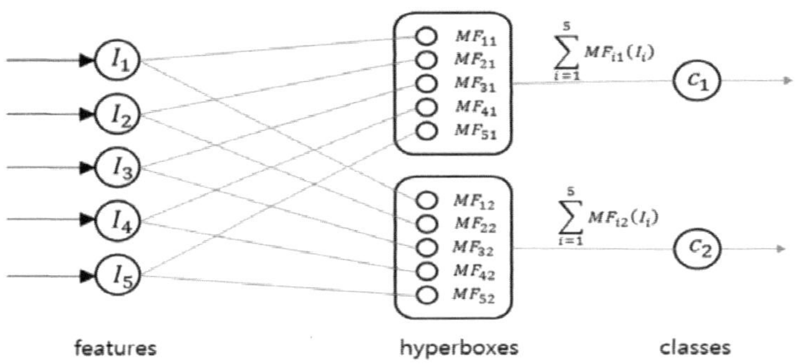


**Figure 5.** Structure of the neural network with weighted fuzzy memb ership functions (NEWFM)

To apply biometric authentication using ECG to the NEWFM, each ECG biometric feature that was extracted through Haar-wavelets was matched to one input node to compose authenticated class nodes and unauthenticated class nodes of the output layer. The process of training ECG waveforms from the user and those that are not from the user are registered in the NEWFM, and authentication is when an individual's waveforms are categorized through the results of the trained weighted fuzzy membership functions.

## 3. Experiment and results

In order to make multicategory classifications through a secondary categorization algorithm that differentiates Class 1 and Class 2, the neural network with weighted fuzzy membership functions uses the One Against One[19] evaluation strategy to select the classifier with a high Sugeno value.

The TAR (true acceptance rate) and FAR (false acceptance rate) indices from Formula (11) and (12) were used to evaluate the experimental results of ECG biometric authentication.

$$TAR = \frac{TN}{FN+TN} \times 100 \tag{11}$$

$$FAR = \frac{FP}{TP+FP} \times 100 \tag{12}$$

Here, TP (true positive) shows the number of times a person in the database was authentication, TN (true negative) shows the number of times the information of a person who is not in the database was denied, FN (false negative) shows the number of times a person who is in the database was denied, and FP (false positive) shows the number of times the database of another person who is not in the database was authenticated [22]. The false acceptance rate (FAR), in which the actual values differs from expected values), serves as an important factor of evaluation in the biometric authentication algorithm. FAR is when another user's ECG waveforms are authenticated as the user's ECG waveforms. Table 1 shows the results of the ECG biometric authentication using the above performance index.

This study used ECG data from the Physionet Database [29], which can be compared with other similar studies such as [25-28]. The Physionet Database consists of the European ST-T Database, MIT-BIH Normal Sinus Rhythm Database, MIT-BIH Arrhythmia Database, and QT Database. Of these databases, 73 records were used to verify the biometric authentications. The length of ECG records in the Physionet Database varied from 30 minutes up to 24 hours. These signals were analyzed by doctors and are used in various research projects in medical and electronic fields.

2,000 seconds of continuous ECG signals from a random start point in each ECG record was extracted as one ECG section. When the sampling rate and amplitude are normalized for each ECG section and the ensemble average is applied in ECG waveform units, an ECG waveform sample composed of 120 samples and a 2-dimensional vector with m ECG waveforms are generated in one ECG waveform. The Haar-wavelet transform (HWT) is performed on the ECG waveform sample to extract wavelet coefficients a4, d3, and d4 as features, and the wavelet coefficient composed of 29 samples is extracted as the ECT biometric feature. The wavelet coefficient that is extracted as the final feature is entered into the NEWFM biometric authentication model to perform ECG biometric authentication.

The proposed biometric authentication model extracted 10, 15, 20, 25, 30, 35, and 40 random ECG biometric features from each record for use in training. The remaining unused ECG biometric features were used to test authentication. One ECG biometric feature is the feature from one ECG waveform, and refers to the length of one heartbeat. One heartbeat refers to the length between ECG signal point $P_{on}$ to $T_{off}$. Figure 6 shows the results of experimental training using 10 to 40 ECG biometric features from 73 subjects from the Physionet Database.

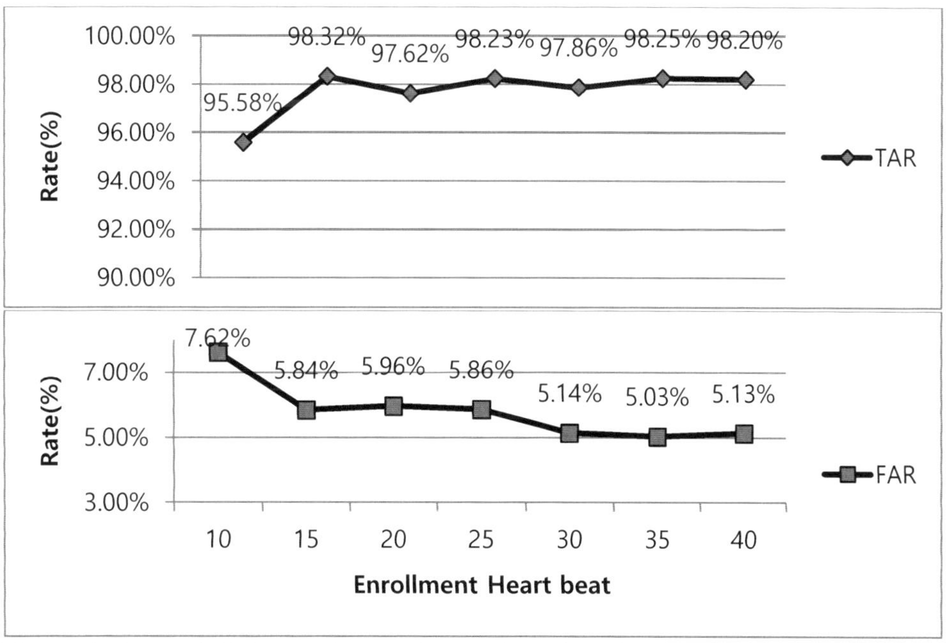


**Figure 6.** TAR and FAR according to changes in registration time from 10 to 40 seconds.

Upon testing the biometric authentication algorithm in the neural network with weighted fuzzy membership functions according to changes in the number of ECG biometric features that are used for authentication, the TAR was 98.32% and FAR was 5.84% for 15 seconds. The registration time was short while having a stable authentication rate among all the test results. Therefore, 15 ECG biometric features, meaning 15 ECG waveforms, were registered in the NEWFM, and an individual's waveform was authenticated using one ECG waveform. Table 1 compares the results from other ECG biometric authentication.

**Table 1.** Results of ECG biometric authentication.

| Algorithm | Subjects | TAR | FAR | Authent. lenght | Enroll lenght |
|---|---|---|---|---|---|
| Singh et al.[26] | 73 | 82.00% | 7.00% | 1/2 of record | 1/2 of record |
| Wubbeler et al.[8] | 74 | 97.00% | 3.00% | 10 s. | 10 s. |
| singh et al. [30] | 73 | 95.55% | 3.00% | 10 heart beat | 10 heart beat |
| Juan et al.[28] | 73 | 84.93% | 1.29% | 4 s. | 30 .s |
| Proposed Algorithm | 73 | 98.32% | 5.84% | 1 heart beat | 15 heart beat |

## 4. Conclusion

This study tested the ECG of 73 subjects from the Physionet Database [23]. After the normalization of the sampling rate and amplitude from continuous ECG signals and categorizing each ECG waveform based on R-peaks, Haar-wavelet transform (HWT), which can extract features without detecting a reference point, was used to extract ECG biometric features. The Haar-wavelet transform coefficients a4, d3, and d4 were selected as one ECG biometric feature. Biometric authentication was performed on the selected ECG biometric features using the classifiers of the neural network with weighted fuzzy membership functions (NEWFM) [17, 18]. 15 ECG waveforms were used for training NEWFM classifiers. TAR was 98.32% and FAR was 5.84% upon testing with one ECG waveform. This shows that registering just 15 ECG waveforms is sufficient, and more diverse authentication waveform quantities should be tested for authentication in the future. The results of this study were compared against four existing algorithm results. The proposed algorithm had a shorter training time and testing time, and was just as reliable as other algorithms. Various other features will be added in future studies, and feature selection will be used to remove unnecessary features to improve the TAR and FAR and achieve an even shorter registration time.

## References

[1] Arantes M, Ide A N and Saito J H 2003 *A system for fingerprint minutiae classification and recognition* Neural Information Processing, 2002. ICONIP '02. Proceedings of the 9th International Conference on.

[2] Weaver A C 2006 *IEEE Comp. Soc.* **39** 96-7.

[3] Frischholz R W and Dieckmann U 2000 *IEEE Comp. Soc.* **33** 64-8.

[4] Sidek K, Sufi F, Khalil I and AI-Shammary D 2010 *An efficient method of biometric matching using interpolated ECG data* Biomedical Engineering and Sciences (IECBES), 2010 IEEE EMBS Conference on.

[5] Wang Y, Agrafioti F, Hatzinakos D and Plataniotis K N 2007 *EURASIP J. Adv. Sign. Proc.* doi: https://doi.org/10.1155/2008/148658.

[6] Biel L, Pettersson O, Philipson L and Wide P 2001 *IEEE Trans. Instrum. Measur.* **50** 808-12.

[7] Shen T and Tompkins W 2005 *Biometric Statistical Study of One-Lead ECG Features and Body Mass Index (BMI)* Conf Proc IEEE Eng. Med. Biol. Soc. doi: 10.1109/IEMBS.2005.1616629.

[8] Israel S A, Irvine J M, Cheng A, Wiederhold M D, Wiederhold B K 2005 *Patt. Recogn.* **38** 133-42.

[9] Wübbeler G, Stavridis M, Kreiseler D, Bousseljot R and Elster C 2007 *Patt. Recogn. Lett.* **28** 1172-17.

[10] Adrian D, Chan C, Senior Member, 2008 *IEEE Trans. Instr. Measur.* **57** doi:10.1109/TIM.2007.909996.

[11] Lourenço A, Silva H, Fred A 2012 *ECG-based biometrics: A real time classification approach* Machine Learning for Signal Processing (MLSP), 2012 IEEE International Workshop on

[12] Zhao Z and Yang L 2011 *ECG identification based on Matching Pursuit* Biomedical Engineering and Informatics (BMEI), 2011 4th International Conference on.

[13] Jongshill L, Youngjoon C, and Inyoung K 2012 *J. Appl. Math.* **2012** 12.
[14] Ikenna O, Po-Hsiang L, Alan D K, Joseph A O, Erik J S and John W R 2012 *IEEE Biometr. Compend. IEEE RFIC Virt. J.* **7** 1812-24.
[15] Biometric Measurements and Systems for Security and Medical Applications. *IEEE WORKSHOP. 2014.* (BIOMS 2014) pp. 99.
[16] Zhao Z, Yang L, Chen D and Luo Y 2013 *A human ECG identification system based on ensemble empirical mode decomposition* US National Library of Medicine National Institutes of Health.
[17] Lim J S, Wang D H, Kim Y S and Gupta G 2005 *Neurocomput.* **69** 969-74.
[18] Lim J S 2009 *Syst. IEEE Trans. Neural Networks* **20(3)** 522-27.
[19] Knerr S, Personnaz L and Dreyfus G 1990 *Neurocomput.* **68** 41-50.
[20] Yoon S J, Kim G J and Chang P S 2012 *J. Korea Inform. Sci. Soc.* 101-87.
[21] Pan J and Tompkins W J 1985 *IEEE Trans. Biomed. Eng.* 230-236, doi:10.1109/TBME.1985.325532.
[22] Kim T S and Kim H D 2005 *Lect. Note Comput. Sci.* **3682** 284-90.
[23] A. L. Goldberger, L. A. N. Amaral, L. Glass, J. M. Hausdorff, P. C. H. Ivanov, R. G .Mark, J. E. Mietus, G. B .Moody, C-K. Peng, H. E. Stanley 2000 *Circulation* **101(23)** e215-e220 Circulation Electronic Pages; http://circ.ahajournals.org/content/101/23/e215.full.
[24] Lim S H, Min K G, Lee J H, Jang D P and Kim I Y 2014 *Depart. Biomed. Eng.* ***35** 42-9.*
[25] Wübbeler G, Stavridis M, Kreiseler D, Bousseljot R D, and Elster C 2007 *Patt. Recogn. Lett.* **28** 1172-5.
[26] Singh Y N and Singh S K 2012 *J. Inform. Sec.* **3(1)** 39-48.
[27] Singh Y N and Singh S K 2013 *J. Eng.* **2013** 8.
[28] Arteaga-Falconi J S, Osman H AI and Saddik A EI 2015 IEEE Trans. Instrum. Measure. **65(3)** 591-600.
[29] G. B. Moody, R. G. Mark, A. Goldberger 2011 *PhysioNet: Physiologic signals, time series and related open source software for basic, clinical, and applied research,* Proceedings of the 33rd Annual International Conference of the IEEE Engineering in Medicine and Biology Society, EMBC, doi:hdl.handle.net/1721.1/76352.
[30] SinghMehtaNitin S amd Lingayat S 2007 Development of entropy based algorithm for cardiac beat detection in 12-lead electrocardiogram Signal Processing 87 pp. 3190-3201 doi: https://doi.org/10.1016/j.sigpro.2007.06.009

**Acknowledgement**
This research was supported by Basic Science Research Program through the National Research Foundation of Korea (NRF).
This research was supported by the MSIT (Ministry of Science and ICT), Korea, under the ITRC (Information Technology Research Center) support program (IITP-2017-0-01630) supervised by the IITP (Institute for Information & communications Technology Promotion).

IOP Conf. Series: Materials Science and Engineering **317** (2018) 012031 doi:10.1088/1757-899X/317/1/012031

# Development of a Portable Torque Wrench Tester

**Y Wang[1], Q Zhang[1], C Gou[1] and D Su[1]**

[1]Metrology Testing Center, China Academy Of Engineering Physics, No.64, Mianshan Road, Mianyang 621900, China

**Abstract.** A portable torque wrench tester (PTWT) with calibration range from 0.5 Nm to 60 Nm has been developed and evaluated for periodic or on-site calibration of setting type torque wrenches, indicating type torque wrenches and hand torque screwdrivers. The PTWT is easy to carry with weight about 10 kg, simple and efficient operation and energy saving with an automatic loading and calibrating system. The relative expanded uncertainty of torque realized by the PTWT was estimated to be 0.8%, with the coverage factor $k$=2. A comparison experiment has been done between the PTWT and a reference torque standard at our laboratory. The consistency between these two devices under the claimed uncertainties was verified.

## 1. Introduction

Hand torque wrenches are necessary for assembling in various fields, such as automotive, ship-building, aerospace, power and other industries. The quality of assembling procedure is highly dependent on the accurate torque control. Accurate and reliable measurement of the torque value will help to improve and enhance product performance. In most cases, it is assumed that a torque wrench will deliver the requested torque value if set correctly. However, torque wrenches have moving parts that will be subject to wear. They are also subject to harsh operating conditions in the field, which elevates the risk of damage [1]. In order to maintain consistent accuracy, calibration or verification of torque wrenches or screwdrivers on a periodic basis would be advisable. However, this is only part of the requirement. Data obtained from the calibration of hundreds of wrenches over several year period indicates that there is still a big probability that wrench failure occurs prior to recalibration periodically, especially those overused. Those disqualified wrenches may cause bolting related failure till recalibration. Consequently, it is reasonable that wrenches themselves need to be on-site verified on a moveable torque wrench tester (TWT) just before assembling. There are many commercial TWTs available for hand torque tools, most of which are designed only for laboratorial conditions. That heavy laboratorial equipment with poor mobility are inconvenient for on-site calibration. Therefore, we developed a portable torque wrench tester (PTWT) for on-site calibration or verification of several commonly used hand torque tools, namely, setting type torque wrenches (STWs), indicating type torque wrenches (ITWs) and hand torque screwdrivers (HTSs). The relative expanded uncertainty of torque realized by the PTWT was evaluated in the 0.5~60 Nm torque range and was estimated to be 0.8%, with a coverage factor, $k$, being equal to 2. The PTWT was compared with a reference torque standard (RTS) at our lab. The calibration results obtained by these two devices were coincided within the claimed uncertainties.

## 2. Establishment of equipment

### *2.1. Equipments*

Light in weight, easy to operate and automatic procedure was what we pursue with this equipment. Two easily detachable torque transducers were adopted to cover the target range from 0.5 Nm to 60 Nm. A servo motor in combine with a planet gear was applied to realize automatic calibration procedure and output a rated capacity of 60 Nm, which free our hands and conserve time. In the design process of the mechanical structure, finite element calculations had been carried out to optimize the clamping structure geometry with the aim to guarantee strength and reliability.

A schematic of the PTWT is shown in figure 1. The main components were installed onto a framework assembled of aluminium profile. A handle on the top and castors with brakes at four corners makes it much easier to move. Motor, planet gear, torque transducer and torsion axis of wrench were connected along the measurement axis. Various square drives were prepared in different sizes to enable torque wrenches of various sizes and shapes to be installed on either of the two torque transducers.

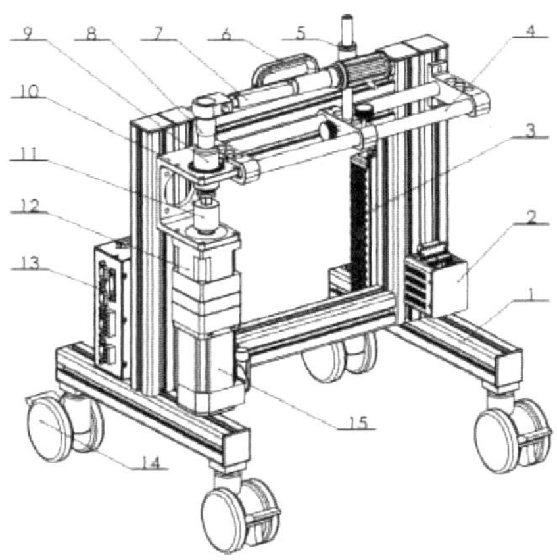


**Figure 1.** Schematic of the PTWT.

1. Profile; 2. Driver; 3. Power supply; 4. Guiders; 5. Rod for loading point; 6. Handle; 7. Torque wrench; 8. Square drive; 9. Torque transducer; 10. Holder; 11. Adapter; 12. Planet gear; 13. PLC device; 14. Castor with brake; 15. Motor.

### *2.2. Calibration procedures*

A PLC device is used to control the movements of the motor and run a calibration procedure via a PC that acts as man-machine interface. Due to diversity of principles and structures, we designed different calibration procedures for each type of wrenches as illustrated in figure 2, in which, 'setting mode' is for STWs, 'indicating mode' for ITWs (includes digital type and analogue type) and 'manual mode' for HTSs or other torque tools need to be manually calibrated. The machine can operate almost fully automatically with the exception that, manual mode being selected, or installing and uninstalling a wrench whenever required, or exchanging a torque transducer during the measurement process.

*2.2.1. Setting mode.* Internal structure of a STW is a combination of leveraged structure and spring. Once the applied torque is large enough to overcome resistance produced by the internal spring, protuberant node of the lever will strike inner wall of the wrench immediately, thus makes a 'click' sound and torque released. Calibration of a STW is the measurement of the peak torque just before releasing to a known torque standard such as a TWT and adjustment to within specific tolerances if necessary. The aim of developing an automatic procedure is to improve efficiency in contrast to the conventional manual procedure. Thus, the loading and unloading process should be as fast as possible. However, loading too fast upon peak point may lead to overloading and thus in return do great damage to the tested wrench. Consequently, a tested wrench is first loaded to 80% of the target torque at a high speed according to JJF 707 [2]. Then, decrease loading speed gradually, and activate peak torque identification simultaneously. Once a peak is detected, note the peak value and unload quickly. Otherwise, continue to load and identify. Note that all the calibration procedure is done automatically by a calibration program implanted in the PLC device in figure 1. The calibration program would remind you to unmount the tested STW to preset the next target torque after all the repetitive tests of one test point is done.

*2.2.2. Indicating mode.* Calibration procedure of ITWs is almost the same as that of STWs before 80% of the loading process, as shown in figure 2b. Instead of peak identification, a tested ITW would be gradually loaded to the target torque and maintained for a few seconds to note the reading of the ITW. Afterwards, load to the next point till 100% of torque range is done. Repeat the above steps till test times is enough.

*2.2.3. Manual mode.* Figure 2c presents calibration procedure of HTSs or other torque tools need to be manually calibrated. Before starting a calibration procedure, set display mode of the tested torque tool to peak mode instead of track mode if the tested tool is an indication one. First, brake the motor axis to prevent the torque transducer from rotating. And then, manually apply load to the torque transducer through the tested torque tool. Activate peak torque identification once the applied load reaches 80% of the target torque. Note reading of the tested tool and unload when a torque peak is identified. Stop the calibration procedure when test times are enough, otherwise move to the next step.

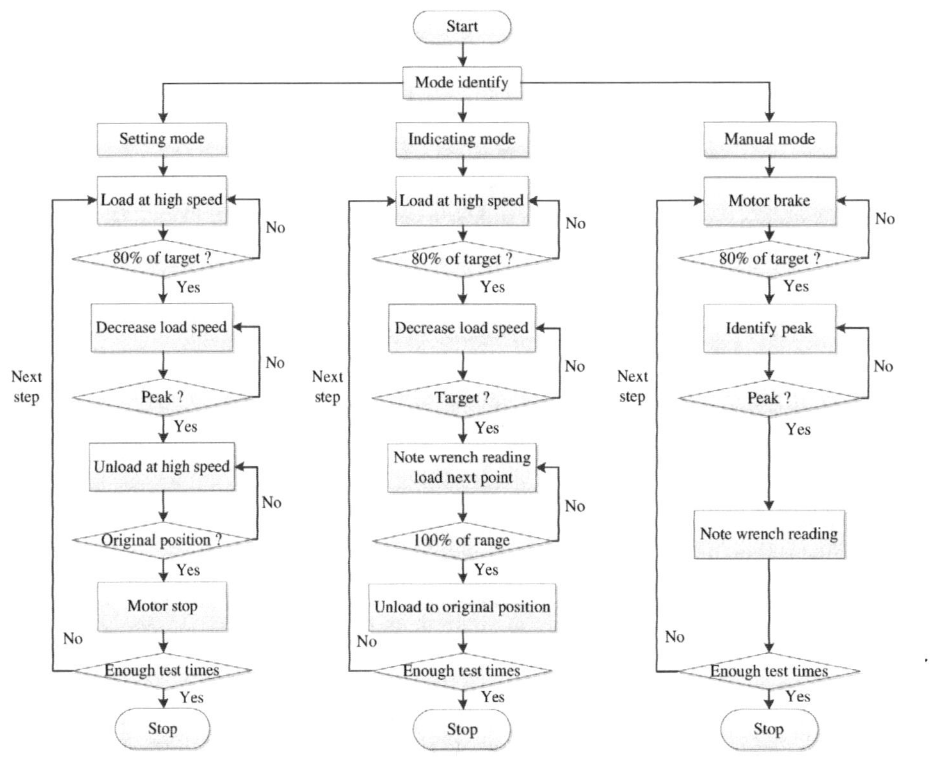


**Figure 2.** Calibration procedure.

a. Setting mode; b. Indicating mode; c. Manual mode.

## 3. Uncertainty evaluation

### *3.1. Uncertainty of the calibration results*

The characteristics of the new PTWT were evaluated according to GJB 2749A [3] whereas uncertainty of calibration was evaluated according to ISO/IEC guide 98-3 [4], and evaluation methods are described as following. The relative expanded uncertainty $U_r$ of calibration results for every measuring point is expressed by the following equation:

$$U_r = k \cdot u_c = k \cdot ( u_{pro}^2 + u_{rep}^2 + u_{ind}^2 + u_{sta}^2 + u_{res}^2)^{1/2} \quad (1)$$

where $u_c$ is the relative combined standard uncertainty of calibration. $k$ is the coverage factor and $k$ = 2 was used in calculation of all expanded uncertainties in this paper, which correspond to the level of confidence of approximate 95 %. $u_{pro}$ is the uncertainty contribution for the reproducibility with a change in the mounting position, $u_{rep}$ is that for the repeatability without a change in the mounting position, $u_{ind}$ is uncertainty due to indication difference from the reference value, $u_{sta}$ is uncertainty due to long stability of transducer sensitivity, $u_{res}$ is that due to the resolution.

### *3.2. Reproducibility with changing mounting position*

In consideration of the influence of mounting position on the calibration results, Reproducibility with changing mounting position $u_{pro}$ was evaluated by a certain number of repeated tests in a way that after each test, the wrench and rod for loading point were reinstalled and square drive and torque transducer were rotated to a different position. The relative reproducibility with different mounting positions is calculated according to (2) as the experimental standard deviation for the measurement values of each mounting position:

$$u_{pro} = \frac{1}{\overline{S_j}} \left( \frac{\sum_{j=1}^{k} \left( S_j - \overline{S_j} \right)^2}{k-1} \right)^{1/2} \quad (2)$$

where, j and $k$ are the indexes of different mounting position and the total number of mounting positions (here $k$=10), respectively. $S_j$ is the measurement value obtained from each mounting position and $\overline{S_j}$ is the mean of $S_j$.

### *3.3. Repeatability in the same mounting position*

The repeatability with the unchanged mounting position $u_{rep}$ was evaluated by certain number of repeated tests with the wrench, rod for loading point, square drive and torque transducer in the same mounting position. The relative repeatability with the unchanged mounting position is calculated as:

$$u_{rep} = \frac{1}{\overline{S_i}} \left( \frac{\sum_{i=1}^{n} \left( S_i - \overline{S_i} \right)^2}{n-1} \right)^{1/2} \quad (3)$$

where, i and $n$ are the indexes of cycles and the total number of cycles (here $n$=10), respectively. $S_j$ is the measurement value obtained from each cycle and $\overline{S_i}$ is the mean of $S_i$.

### *3.4. Deviation due to indication*

The relative deviation due to the indication $u_{ind}$ is calculated as the difference between reference values indicated by a reference torque wrench or a weight-bar-system of higher grade and the indicating results of PTWT.

### *3.5. Stability*

The PTWT was calibrated four times by using a torque standard of higher grade over approximately one year in order to investigate its long-term stability. Relative standard uncertainty due to stability $u_{sta}$ is calculated as:

$$u_{sta} = \frac{S_{max} - S_{min}}{d_m S_{mean}} \quad (4)$$

where, $S_{max}$, $S_{min}$ and $S_{mean}$ are the maximum, minimum and mean of the 4 calibration results, respectively. Here, $d_m$ equals to 2.06 ascribed to 4 calibration results [3].

### *3.6. Resolution*

The resolution $r$ of the indication is defined as the smallest fraction of a scale division that is readable in the case of analogue scale. In the case of digital scale, $r$ is considered to be one increment of the last active number of the numerical indicator, provided that the indication does not fluctuate when the PTWT is under the non-loading condition. Relative standard uncertainty due to resolution $u_{res}$ is calculated according to (5), by determining $r$ as a half-width of the rectangular distribution:

$$u_{res} = \left(\frac{2}{3}\right)^{1/2} \frac{r}{2T_i} \quad (5)$$

where, $T_i$ means applied toque. $u_{res}$ is multiplied by $(2)^{1/2}$ considering that the measurement value is obtained as the difference between the indicated values of the loaded torque step and non-loading zero step before starting the cycle (double readings) [5].

### *3.7. Evaluation of $E_n$ number*

Intra-laboratorial comparison was conducted between PTWT and a RTS at our lab using several torque wrenches as transfer devices. The $E_n$ ratio between PTWT and the RTS for each measuring point was evaluated using the following equation:

$$E_n = \frac{S_{ptwt} - S_{ref}}{S_{ref}(U_{ptwt}^2 + U_{ref}^2)^{1/2}} \quad (6)$$

where, $S_{ptwt}$, $S_{ref}$ denote calibration results of the same point of a transfer device given by PTWT and the RTS, and $U_{ptwt}$, $U_{ref}$ are the associated relative expanded uncertainties.

## 4. Results and discussion

Uncertainty of the calibration results was evaluated at 0.5 Nm, 2 Nm, 5 Nm, 30 Nm and 60 Nm. The relative expanded uncertainty $U_r$ of every point is calculated by (1) and summarized in table 1. In all the evaluated cases, the expanded relative uncertainty of calibration results within 0.8 % could be obtained. Thus, the relative expanded uncertainty of torque realized by the PTWT was estimated to be 0.8% in the range of 0.5Nm to 60Nm, with the coverage factor $k$=2. Figure 3 shows the relationship between the relative standard uncertainties and measuring points. In general, relative standard uncertainties are larger in lower torque cases since it's harder to yield accurate calibration results when calibrating a small-rated-capacity wrench as compared to a greater one. Reproducibility due to changing mounting position have profound effect on the expanded uncertainty, while uncertainties due to resolution are sufficiently small compared with other factors which can be ignored, especially for larger rated capacity.

Figure 4 summarizes the results of the $E_n$ number evaluation between PTWT and the reference standard. The $E_n$ numbers were all less than one over the calibration range of 0.5 Nm to 60 Nm. The comparison results showed good agreement within the uncertainties in all cases. Thus, the torque realized by the PTWT was shown to be equivalent to that achieved by the reference standard.

IOP Conf. Series: Materials Science and Engineering **317** (2018) 012031 doi:10.1088/1757-899X/317/1/012031

**Table 1.** Relative expanded uncertainties.

| Torque | 0.5 Nm | 2 Nm | 5 Nm | 30 Nm | 60 Nm |
|---|---|---|---|---|---|
| $U_r$ | 0.75% | 0.64% | 0.71% | 0.66% | 0.46% |

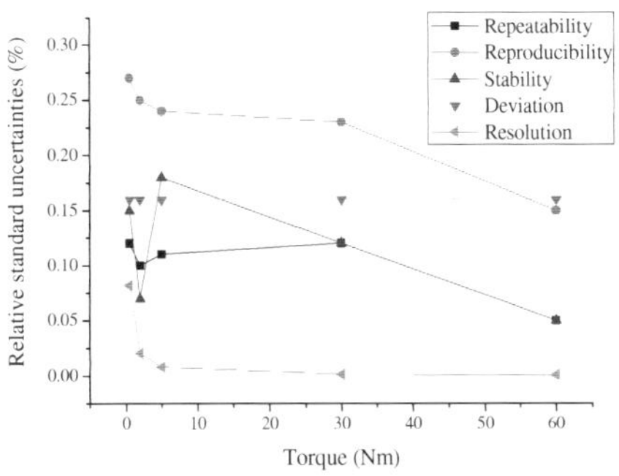


**Figure 3.** Relative standard uncertainties.

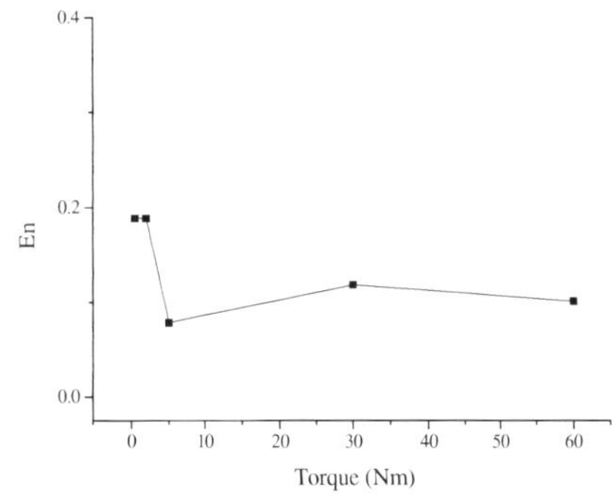


**Figure 4.** Results of $E_n$ evaluation.

## 5. Summary

In this study, a PTWT was developed and evaluated. The relative expanded uncertainty was estimated to be less than 0.8% for the calibration range from 0.5 Nm to 60 Nm, with the coverage factor $k$=2. A comparison has been performed between the PTWT and a RTS machine at our lab. The results of comparison showed quite good agreement within the uncertainties in all cases. Consequently, the consistency between the PTWT and the RTS has been verified under the claimed uncertainties.

## References

[1] Peng Y G and Wang Y 2017 *Appl. Mech. Mater.* **853** 216-20.
[2] JJG 707 2014 *Verification regulation of torque wrenches* (Chinese)
[3] GJB 2749A 2009 *General requirement of establishment and conservation of measurement standard for military metrology* (Chinese)
[4] ISO/IEC guide 98-3 2008 *Uncertainty of measurement-Part 3*
[5] Ogushi K, Nishino A, Maeda K and Ueda K 2011 *Sice Conference* (Tokyo, Japan) pp 411-416.

## Acknowledgments

Authors wishing to acknowledge financial support from the National key foundation for exploring scientific instrument of China (2014YQ350461).

IOP Conf. Series: Materials Science and Engineering **317** (2018) 012032 doi:10.1088/1757-899X/317/1/012032

# Research on the Design of Public Space Environment for Aging Society

**Gu Fang[1], Kim Chul Soo[2]**

[1]Hezhou University, Marine Convergence Design, Pukyong National University
[2]Dept. of Industrial Design, Pukyong National University

**Abstract.** This paper studies the living space environment suitable for the elderly, because the elderly and the disabled have become increasingly prominent social problems. Through the discussion of the humanistic environment design method of the elderly and the disabled, the paper puts forward a new environment design which has the traditional characteristics and adapts to the new society to care for the elderly (the disabled).By studying and analyzing the background of social aging, the theory of public space environment design and the needs of the elderly, it is pointed out that the design of public space environment in the aged society needs to be implemented in detail design. The number of elderly people in public space will increase, give full attention to the public space outdoor environment quality, for the elderly to provide a variety of environmental facilities have long-term significance.

## 1. Introduction

### *1.1. Purpose of the study*
Population aging is the inevitable trend of the development of the world today. Sociologists refer to 60-year-olds in countries with more than 10% of the total population as aging.

This paper studies the living space environment suitable for the elderly, because the elderly and the disabled have become increasingly prominent social problems.

Through the discussion of the humanistic environment design method of the elderly and the disabled, the paper puts forward a new environment design which has the traditional characteristics and adapts to the new society to care for the elderly (the disabled).

By studying and analysing the background of social aging, the theory of public space environment design and the needs of the elderly, it is pointed out that the design of public space environment in the aged society needs to be implemented in detail design. However, the detailed design comes from the physiological needs and psychological needs of the elderly Two parts. Compared with the physiological needs of the elderly, the details of the design of the care design is more focused on the psychological needs of the elderly to meet.

### *1.2. Research methods*
With the development of China's population, the aging society, the development of the elderly design, to achieve sustainable economic development, social stability, "people-oriented" concept of the formation of

an important means of pension. Learn from countries to solve the pension model, and China to the present experience of accumulation, in many pension methods only community pension approach is most suitable for China's national conditions. Providing a suitable living environment for the elderly is essentially a kind of social pension behaviour, and needs different ways of providing for the aged in different social backgrounds. Through the detailed design of the public space environment, let the elderly in the daily life of active access to self-help ability, win social respect. Care for the elderly, help the elderly, environmental design needs to be implemented in the details of the design, and the details of the design is derived from the physiological needs of the elderly and psychological needs of the two part of the analysis. Compared with the physiological needs of the elderly, the care design put forward in detail design focuses more on the satisfaction of the psychological needs of the elderly.

## 2. Theoretical background

### *2.1. Theoretical background*

According to the relevant report of the International Labor Organization, the total number of people with disabilities in the world is now more than 500 million, accounting for about 10% of the world's total population. Now the average annual increase in the number of disabled 15 million, that is, an increase of more than 40,000 people per day. In most countries, at least one out of every 10 people is disabled by physical, psychological and sensory deficiencies. At present, China has 64 million people with disabilities, accounting for 5% of the total population, is the world's most disabled countries, An average of every 10 families have a disabled or a disabled family, and their living conditions affect nearly 300 million relatives and parties concerned.

### *2.2. Domestic and overseas situation*

People with disabilities are not only a large number, but also by the physical disability and the external environment of the obstacles in the social life in a variety of unfavorable position, so that the normal role of the play by many restrictions. From the perspective of the international community, building a barrier-free environment has become one of the mainstream of urban environmental construction, urban roads, transportation and buildings in the planning and design should be reflected in the important degree of urban civilization.

In 1959, the European parliaments adopted the "Resolution on the Design and Construction of Public Buildings for the Use of Disabled Persons". Under the influence and promotion of the international community, the concept of "accessibility" began to take shape. At the same time, the development of economy also makes it possible for industrial countries to invest a large amount of manpower, material and financial resources in the popularization of barrier free environment. At that time, President Kennedy consulted on this issue and formulated the world's first "barrier free" standard in 1961".The Israeli Building Law, enacted in 1963, and the American Building Law, enacted in 1965. The United States is the first country in the world to develop "barrier-free standards", its barrier-free environment construction has a multi-level legislative protection, and has entered the field of research and education; A variety of barrier-free facilities have a full range of layout, but also with the architectural art of harmonization, first-class, and serve the United States 37 million people in the important historical and cultural buildings, such as the United Nations Building, Capitol, the White House, Libraries, the High Court, the new National Gallery, the Kennedy Center, the Aerospace Museum and so on. These buildings are accessible from outside and within the barrier-free series of facilities, in terms of location, relationship is more appropriate, functional and form is also acceptable. For the more ancient buildings of the transformation of the work is also very careful and meticulous, respect in the shape of the original style, pay attention to the overall effect. After the transformation, there is no sense of superfluous, such as the transformation of the Arlington Unnamed Military Memorial Hall. As China has a large number of people with disabilities, while the peak of aging is coming, so the rise of barrier-free construction is imperative.

### *2.3. Design questions*

*2.3.1. The physiological and psychological needs of the elderly.* Elderly physiological aging refers to the natural growth of the human body with age, degenerative changes in the body's physiological function and shape gradually appear, with the gradual decline of the ability to adapt to internal and external environment until the end of life activities.

Mainly in the following areas:

1) The body's internal system of perception system function decline

With age, the sensory capacity of the elderly will be a corresponding decline. First manifested in the auditory and visual aspects. Older people need to be stronger than the young 3 times to 7 times the light to see the object, so the intensity of light and lamp shape design must meet these characteristics of the elderly. Elderly hearing loss is mainly manifested as not sensitive to high-frequency sound. Shorten the distance with the speaker to make up for the hearing of the recession, and through touch, taste, smell to identify things, as a visual and auditory complement, affecting the elderly social space in the scale and form requirements.

Changes in the nervous system of the elderly are mainly due to the reduction of brain cells caused by slow response. Disease and aging at the same time affect the elderly to respond to external things. Elderly thinking ability is reduced, memory loss.

2) Reduced flexibility in the external body and perceived degradation of the environment

Older physical flexibility is reduced, muscle strength and control ability is also diminishing.

Bones grow with age and become brittle. Because of the slowdown in the metabolism of the elderly, reduced secretion in the body, their response to temperature, humidity and climate is less sensitive, adaptive capacity is weakened and their health is susceptible. The psychology of the elderly is mainly affected by changes in physiological function, economic income and social relations.

This effect is a combination of many factors, and in a specific period of time, with one or two factors as the leading. The elderly population is less resilient to the environment than those of other age groups. So they have a sense of security, a sense of belonging, a sense of neighborhood, privacy, comfort has a more intense demand. Elderly is expressed as hope that the group organization or others can accept themselves, and hope to be in the activities of groups and others to be respected and recognized. To create a good living environment, to provide space for the elderly activities, contribute to the enhancement of the sense of belonging to the elderly and regain. The elderly for the main place of life and the survival of the space to have a strong sense of dependence. Life in the elderly on the privacy requirements are higher, by the violation will cause mental anxiety. Old people on the comfort requirements are different, the temperature, cold, overheating, glare are very sensitive. Old people like a quiet environment, fresh air, lush vegetation.

### *2.4. Design direction recommendations*

The design of space environment is analysed for the public space of the physiological and psychological needs of the elderly. The entrance should be built wheelchair pass ramp; hall and channel should be smooth and bright; stairs to facilitate the use of disabled; staircase in the form of slope, width should be appropriate, not too steep and too narrow; To facilitate the use of wheelchair and disabled persons. The location and height of items such as garbage, public passage lighting, fire room, milk box, letter box, etc. are convenient for the use of the disabled and the elderly. Find suitable for the elderly space environment design.

*2.4.1. Design of communication environment for the elderly.* For most of the elderly, hope to meet or talk through a sense of participation, make themselves feel and have not been abandoned by society. Inconvenience of action for the elderly, sat watching the activities of others is also a way to participate. In the public space environment design, for the elderly to provide the communication space can be divided into the following two categories:

**Table 1.** Elderly social environment design table.

| | | | |
|---|---|---|---|
| Elderly communicative environment | Small - scale communication space | L-shaped seats for the elderly exchange. | In the public space, the street-facing seats are more popular with the elderly. Community entrance, building entrance, road intersection, near the activities of the square is the place of communication between the elderly. |
| | | Back to the building, walls or plant seats will make the elderly safe. | |
| | | The relative seats are suitable for the elderly. | |
| | Group communication space | The central area of public space is the largest place for the elderly in public space. Dynamic activity area with ball, boxing, dancing, boxing, jumping aerobics | The center area should be open, lively, for people to stay. The central activity area should be both time and public. It can be either an independent area or a public facility and a residential area near the green area. Consider the convenience of the elderly, while eliminating the negative impact of noise on the lives of residents. The activities of the elderly threat, interference factors to be promptly eliminated. |
| | | Static activity area can use shade, flower rack, corridor and other space for the elderly in this wait and see, sun, chat, singing and other recreational activities. | |

*2.4.2. Elderly fitness environment design.*

1) A challenging and challenging environment

The barrier-free environment design can compensate for the loss of physical function of the elderly, but also should retain a modest incentive space to maintain and train the elderly surviving independent living ability. The role of this auxiliary environment is to enable the elderly with the necessary help facilities, to complete the challenging activities independently. This requires a certain understanding of the shortcomings of the physical movement of the elderly in order to find the appropriate auxiliary facilities to complete a higher level of activities. Continuing to provide the challenge to the elderly is to keep the elderly active and self-reliant, while eliminating their worries and the timidity of the challenge.

2) Elderly fitness environment facilities

Health and exercise are many of the concerns of the elderly, but also their main reason for the use of outdoor space. A survey in the United States found that the elderly's attention to health and wellness facilities is increasing, and walking or exercising is the most popular outdoor activity. In the public space, we should provide the elderly with a variety of outdoor space and activities. Whether it is indoors or outdoors, facilities and activities should be kept diverse, giving the elderly a choice.

## 3. Elderly ornamental environment design

Elderly in the leisure time often to the outside to the sun, breathe fresh air. For many elderly people, they are peace of mind, enjoy the leisure, close to nature when you can provide a variety of ornamental content. Lawn and garden area is the place where older people prefer. Green is a symbol of life, vibrant flowers and plants can eliminate the negative emotions of the elderly, so that the elderly experience the fun of life. At the same time, if conditions permit, the elderly can participate in planting and care, while enjoying the fun of labor creation, and this is itself an opportunity for entertainment and communication.

Horticultural planting areas should ensure adequate sunshine and water supply required for watering. In the vicinity of the residential units in the small garden of the most careable care, but have a good management and guidance in order to achieve the effect of beautifying the public space environment. It should provide a raised area for the elderly with reduced mobility, preferably with several different heights, and leave a paddle area below to allow the elderly to approach.

*3.1. Ground*
The visual ability of the elderly fell faster, because the activities of the elderly cannot rely entirely on the visual. Under normal circumstances the following ways:

a) With the hand touch changes.

b) Change with the touch of the foot. Such as the intersection of the starting point, the end point, entrances and so on.

c) Help with sound.

d) With the sense of smell, air, temperature to judge.

Visual elderly people with a disabled need to have a ground mark on the ground. Linear ground marking should be used in the induction of ground material, indicating the direction of movement along the line; point-like ground mark should be used to suggest the ground material; on the way to temporarily stop or change direction, should be prompted to pay attention to the risk of height difference. In addition, the flooring of the color, material and other details of the changes can be prompted to reach the steps or traffic arteries.

*3.2. Walkway*
Most of the activities of the elderly in the day are done by walking. Elderly walking is sometimes for the purpose of necessity, and sometimes only for recreation and walking. Walking is a way more exercise for many elderly people, but pay attention to the safety of the elderly.

First of all, should ensure that the road is flat, to avoid a high degree of change, irregular bedding, the ground leaving a seam or other ground protrusions, which will threaten the safety of the elderly; followed by the floor material to ensure that the water is not slippery. Walking in the design, should strive to avoid a long and straight walking route. Meandering or changing roads can make the walking of the elderly more interesting, and curved roads can also help to reduce wind interference.

As the elderly visual and memory loss, the direction of judgment is poor, the best way to connect with each other to form a loop. In order to prevent the elderly lost or lost, in the road transition and the end should be set some markers to enhance the guiding. In the walkway, every 15 meters should be set up a seat for the poor health of the elderly to provide rest and chat places. The width of the walkway should ensure that two people pass side by side, or the width should ensure that a person and a wheelchair walk side by side, about 1.8m or so. Under normal circumstances should not be less than 2.5 meters. At the intersection of the road, the entrance of the neighborhood, and the sidewalk cut off by the edge of the stone should be set edge stone slope, the slope is generally 1: 12.

*3.3. Ramp and steps*
In order to facilitate the use of the elderly, in a highly changing place to set the steps and ramps at the same time. The slope used by the elderly should be smoother than the design criteria. The slope of the outdoor ramp should not be greater than 1/12, the width should not be less than 150cm, next to the anti-drop protection facilities.

*3.4. Outdoor seating*
Good recreational space design is the prerequisite for many activities of the elderly. The old man is not just outdoors in the outdoors, more in the outdoors, chatting the sun, watching and so on. So the old man sitting time is still more, therefore, to provide good rest space and location is very important. The seat position in the space is usually selected under the tree, under the corridor of the public building, near the entrance of the building, the intersection of the traffic flow lines in the district. Seat position should be

well ventilated, plenty of sunshine, but not in the tuyere. Outdoor sitting space to have continuity, for the elderly to provide a convenient rest, watch the Department. In the rest space, the elderly can listen to others, such as children playing, pedestrians, crowds, beautiful scenery, but also leisurely watch vivid and smooth visual effects. But the use of wooden seats in the outdoor poor durability, poor maintenance and easy to damage, so the actual operation of concrete and other hard materials made of. Seat size should be fully taken into account the characteristics of the elderly, the appropriate height between 30 ~ 45cm. The seat is too low for the elderly to take inconvenience, too high and uncomfortable. The width of the seat should be guaranteed between 40 ~ 60cm. For the elderly, the seat of the comfort and practicality is very important, the seat is easy to sit, but also comfortable to sit for a long time. At the same time, the seat should have a better match with the table to meet the needs of the elderly playing cards, playing chess and other activities.

### *3.5. Lighting*
In the public space entrance and building entrance, and so on the flow distribution center should use high brightness lighting. Lighting should be set up near the area where people are often used to define the floor line of the floor. Lighting as far as possible using light down the lighting facilities, rather than light up or out of the lighting facilities to avoid glare.

### *3.6. Outdoor logo*
Elderly people always feel at ease, always want to repeatedly confirm, therefore, and at each intersection must be set clear signs. For older wheelchair users, in the wheelchair cannot pass the road, to set the notice at the intersection. For the wheelchair can be used in the toilet, telephone, etc. should also be marked with a special sign out. For the elderly with visual disability, the need to voice, foot feel, feel and so on to guide. Amblyopia of the elderly is marked the text to be large and contrasting, easy to identify. Logo color should be more yellow, red and other warm colors.

## 4. Conclusions
In the next few decades, due to the improvement of quality of life, the requirements for the living environment has gradually increased. The number of elderly people in public space will increase, give full attention to the public space outdoor environment quality, for the elderly to provide a variety of environmental facilities have long-term significance.

## Reference
[1] Gu F 2017 *Design and Layout of Environmental Space* Beijing China Construction Industry Press.
[2] Hu R L 1995 *Elderly living environment design* Nanjing Central South University Press, pp. 12.
[3] Nomura H 1990 *For the disabled and the elderly building safety design* Beijing China Construction Industry Press, 10.
[4] Claire M, Caroline F, 2001 *Humanity - urban open space design guidelines* Beijing: China Construction Industry Press.
[5] Wu C P 1999 *Social Elderly* Beijing Renmin University of China Press, 10.
[6] Yang G, 1992 *Communication and space* Beijing China Construction Industry Press.

IOP Conf. Series: Materials Science and Engineering **317** (2018) 012033 doi:10.1088/1757-899X/317/1/012033

# Architecture for Business Intelligence in the Healthcare Sector

**Sang Young Lee**[1]

[1]Department of Health Administration Namseoul University, Korea

**Abstract.** Healthcare environment is growing to include not only the traditional information systems, but also a business intelligence platform. For executive leaders, consultants, and analysts, there is no longer a need to spend hours in design and develop of typical reports or charts, the entire solution can be completed through using Business Intelligence software. The current paper highlights the advantages of big data analytics and business intelligence in the healthcare industry. In this paper, In this paper we focus our discussion around intelligent techniques and methodologies which are recently used for business intelligence in healthcare.

## 1. Introduction

Business intelligence (BI) and healthcare analytics are emerging technologies that provide analytical capability to help healthcare industry improve service quality, reduce cost, and manage risks [1]. However, such component on analytical healthcare data processing is largely missed from current healthcare information technology (HIT) or health informatics (HI) curricula [2, 3]. Recently the term BI means two main components, namely; data-centric and process-centric [4, 5]. The data-centric uses BI systems to combine operational data with analytical tools to present complex and competitive information to planners and decision makers. Therefore BI is mainly used to understand the capabilities available in the organization. The process-centric component notes a major shortcoming in this inherent data-centricity. BI therefore should be used to integrate the information world with the process world in order to facilitate decision making with an all-embracing information basis In this paper we focus our discussion around intelligent techniques and methodologies which are recently used for business intelligence in healthcare.

## 2. BI applications for healthcare sector

BI applications in healthcare can be categorized in two major sets of solutions. Technology Solutions are as follows [6, 7].

- Decision Support Systems (DSS): Support managerial decision making, usually day-to-day tactical.
- Executive Information Systems: Support decision making at the senior management level which provide and consolidate metrics-based performance information.
- Online Analytical Processing (OLAP): Support analysts with the capability of perform multi-dimensional analysis of data.
- Query and Reporting Services: Provide quick and easy access to the data with predefined report design capabilities.
- Data Mining (Predictive Model): Examines data to discover hidden facts in databases using different techniques
- Operational Data Services: Collect data from end users, organizing data, establishing solid data structures and store them in different databases, retrieve data from multiple databases.

- Integration Services: Design and implement of process flow of data extracting, transforming, and loading to the data warehouse.

Business Solutions: Business focused analytical applications, as follows:

- Patient Analysis: Focuses on analysis of patients' demographic and satisfaction processes.
- Electronic Health Record Analysis: Focuses on analysis of the quality of clinical data
- Performance Analysis: Streamline and optimize the way that a business uses its resources.
- Fund Channel Analysis: Devise, implement, and evaluate fund strategies, then use the corporate metrics to continuously monitor and enhance the fund process
- Productivity Analysis: Focuses on building business metrics for activities such as quality improvement, risk mitigation, asset management, capacity planning, etc.
- Behavioural Analysis: Understanding and predicting trends and patterns that provides business advantage.
- Supply Chain Analysis: Monitor, benchmark, and improve supply chain activities from materials ordering through service delivery.
- Wait Time Analysis: Focuses on the factors that are associated with longer waiting times and the effects of delays in scheduling and operation.

## 3. Business intelligence architecture

Figure 1 represents the essence of BI with the proper workflow of the interdependent components as follows:

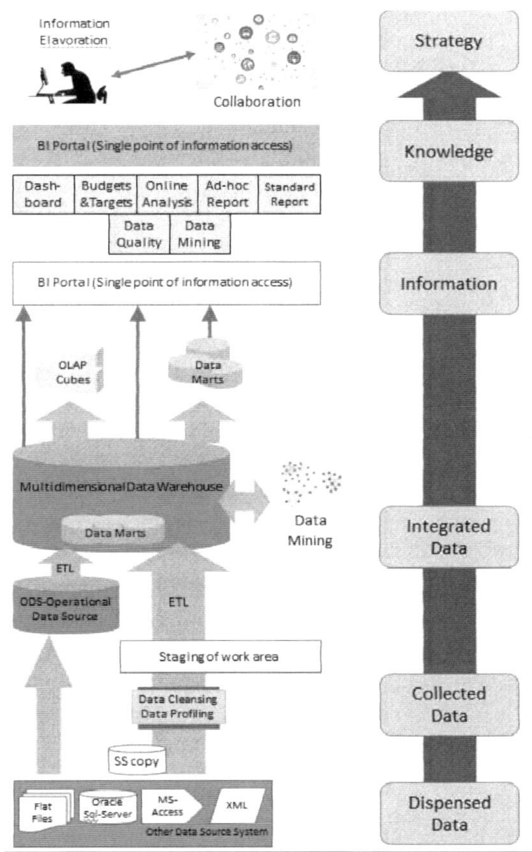


**Figure 1.** BI architecture.

The initial proposed coverage of BI within Healthcare IT is based on the conceptual analysis of general BI systems and processes. The application of BI in healthcare sector is expected to correspond to the full spectrum of business intelligence system (data management, analysis, presentation, and delivery) from both technical and healthcare perspectives. The technical focus emphasizes healthcare BI system development, deployment, and administration. The healthcare focus emphasizes healthcare service provider performance management, healthcare business analysis, and usage of user oriented reporting applications. Figure 2 illustrates the traditional solution architecture and workflow.

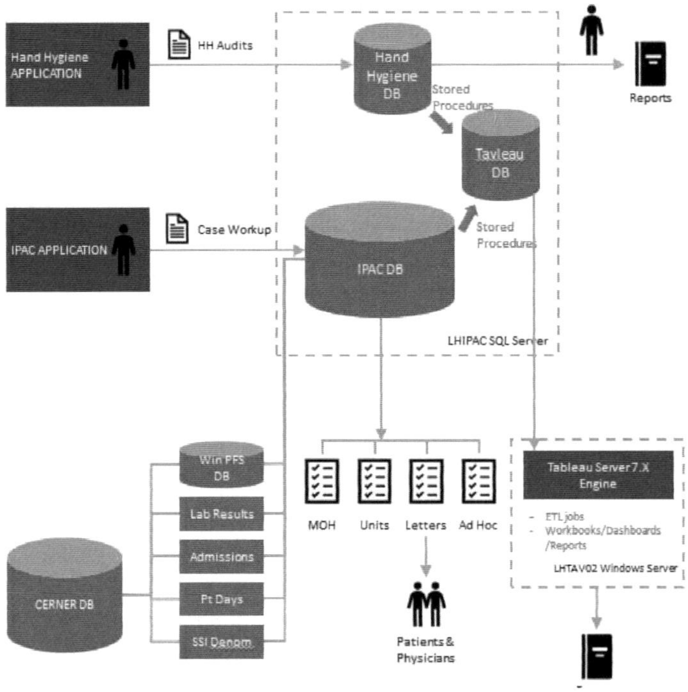


**Figure 2.** Diagram of data systems and infrastructure of the healthcare sector.

Healthcare processes. For example, the advanced analytics may help suggest treatment options to doctors, based on medical records, patient history, research databases and other sources, and predict whether patients with chronic conditions like diabetes are following doctors' orders based on patient records and claims histories. Systems like these need a complete solution from data collection to results delivery. Many care providers can also use BI system to identify and monitor who is at risk of a hospital admission. Effective prediction of who was going to be hospitalized could reallocate resources to prevent unnecessary hospitalization and put the resources to use for cure rather than care.

## 4. Conclusion

Many healthcare organizations struggle with the lack of access and ability to utilize data collected through nonintegrated traditional systems for data mining and decision making. For successful healthcare organizations, it is important to empower staff and management for strategic decision making through data warehousing based on critical thinking. In this paper, the general architectural approaches for BI solution have been outlined and its major development components have been introduced in order to give the reader a high level picture with some important details regarding healthcare informatics.

## References

[1] Mettler T and Vimarlund V 2009 *Health Inform. J.* **15** 254-64.
[2] Sanders D 2002 *J. Healthcare Inform. Manag.* **16** 17-21.
[3] McKinney C, Hess R, and Whitecar M 2012 *Implementing business intelligence in your healthcare organization* Chicago, IL: Healthcare Information and Management Systems Society (HIMSS).

[4] Wixom B and Watson H 2010 Int. *J. Bus. Intell. Res.* **1** 13-28.
[5] Nassif A B, Ho D and Capretz L F 2013 J. Syst. Softw. **86** 144-60.
[6] Nassif A B, Capretz L F and Ho D 2011, *Estimating software effort based on use case point model using sugeno fuzzy inference system*, 23rd IEEE International Conference on Tools with Artificial Intelligence, Florida, USA, pp. 393-8.
[7] A project charter, *Data Management System for Infection Prevention and Control*, IPAC Project Team, London Health Sciences Centre, 2011.

**Acknowledgments**
This work was supported by the National Research Foundation of Korea (NRF) grant funded by the Korea government (2017R1D1A1B03027873).

IOP Conf. Series: Materials Science and Engineering **317** (2018) 012034 doi:10.1088/1757-899X/317/1/012034

# Engineering of Impulse Mechanism for Mechanical Hander Power Tools

**Anatoliy Nikolaevich Drozdov**[1]

[1]PhD in Technical Sciences, assistant professor in Moscow State University of Civil Engineering

**Abstract.** The solution to the problem of human security in cities should be considered on the basis of an integrated and multidisciplinary approach, including issues of security and ecology in the application of technical means used to ensure the viability and development of technocracy. In this regard, an important task is the creation of a safe technique with improved environmental properties with high technological characteristics. This primarily relates to mechanised tool — the division of technological machines with built in engines is that their weight is fully or partially perceived by the operator's hands, making the flow and control of the car. For this subclass of machines is characterized by certain features: a built-in motor, perception of at least part of their weight by the operator during the work, the implementation of feeding and management at the expense of the muscular power of the operator. Therefore, among the commonly accepted technical and economic characteristics, machines in this case, important ergonomic (ergonomics), regulation of levels which ensures the safety of the operator. To ergonomics include vibration, noise characteristics, mass, and force feeding machine operator. Vibration is a consequence of the dynamism of the system operator - machine - processed object (environment) in which the engine energy is redistributed among all the structures, causing their instability. In the machine vibration caused by technological and constructive (transformative mechanisms) unbalance of individual parts of the drive, the presence of technological and design (impact mechanisms) clearances and other reasons. This article describes a new design of impulse mechanism for hander power tools (wrenches, screwdrivers) with enhanced torque. The article substantiates a simulation model of dynamic compression process in an operating chamber during impact, provides simulation results and outlines further lines of research.

## 1. Introduction

Both, conventional percussive mechanisms and various types of impulse sockets are used for generating torque pulses in hander power tools (such as wrenches and screwdrivers). The latter have enhanced reliability and better ergonomic properties. The main disadvantage of many impulse mechanisms is their low volume energy which limits their field of application to threaded connections with small bolt diameter. [1-4].

The article introduces a new impulse wrench mechanism design providing enhanced torque although having a small size and low weight. The mechanism combines both, the advantages of a conventional bumping block mechanism, and a pulse socket which implies the absence of a spring, damping impact energy with oil, and reduced axial vibration due to low weight of the hammer piston. [5, 6].

## 2. Design and operating principle

Figures 1 and 2 show a 3D model of the mechanism with parts of the impact pair: a shaft 1 and a hammer 2, and auxiliary return ring 7, casing 11, and a cover 12. When assembled, the interior of the casing is filled with oil and separated with the cam into two sealed chambers: for High pressure and for Low pressure. Return ring 7 has a square hole and is fastened to the shaft with square 6. The cam and the casing interact via ballscrew (parts 9, 10, and 13). The shape of the groove in cam 9 provides repeating reciprocating action of the cam when the casing moves against it. During operation, the casing is connected directly to the gear, while the shaft is connected to the bolt (shaft, etc.) via square 3 and an operating head.

A crossover valve (not in the pictures) is installed inside of the hammer. The valve provides free flow of fluid from the low pressure chamber into the high pressure chamber but prevents back flow.

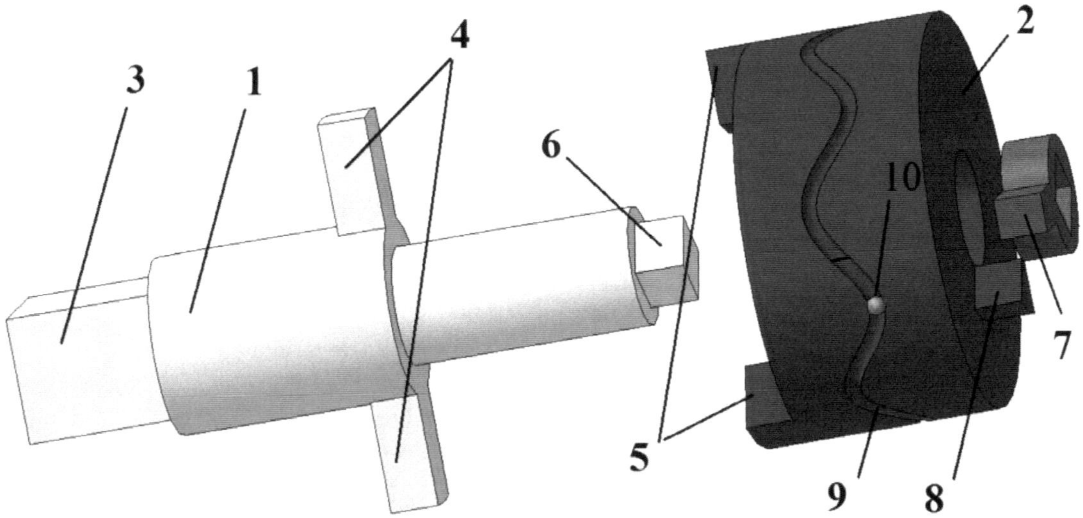


**Figure 1.** Parts of the impact pair of the impulse mechanism.

1 – shaft; 2 – hammer; 3 – square for attaching operating heads; 4 – shaft cams;
5 – hammer operating hams; 6 – square for attaching return ring; 7 – return ring with a cam;
8 – hammer return cam; 9 – ballscrew groove; 10 – ballscrew ball

The mechanism operates as follows: during the first stage when the threaded connection is tightened with low strength, hammer 2 is in the lower position, cams of shaft 4 and cams hammer 5 are interlocked, the crossover valve is closed, and the torque is transmitted to the hammer via ballscrew connection. This is the static operation mode where all parts of the mechanism rotate as a whole.

As the resistance increases, the shaft and the hammer are trigged, and the hammer start moving to the upper position gradually disengaging the shaft. During progressive motion, the hammer is subjected to resistance due to the change of volume in the HP chamber, the displaced fluid flows into the LP chamber solely via the capillary channel formed by the surface of the hammer and the casing inner wall. (During progressive motion, the hammer is subjected to resistance due to the change of volume in the HP chamber, the displaced fluid flows into the LP chamber solely via the bypass channel of (not shown in the pictures).)

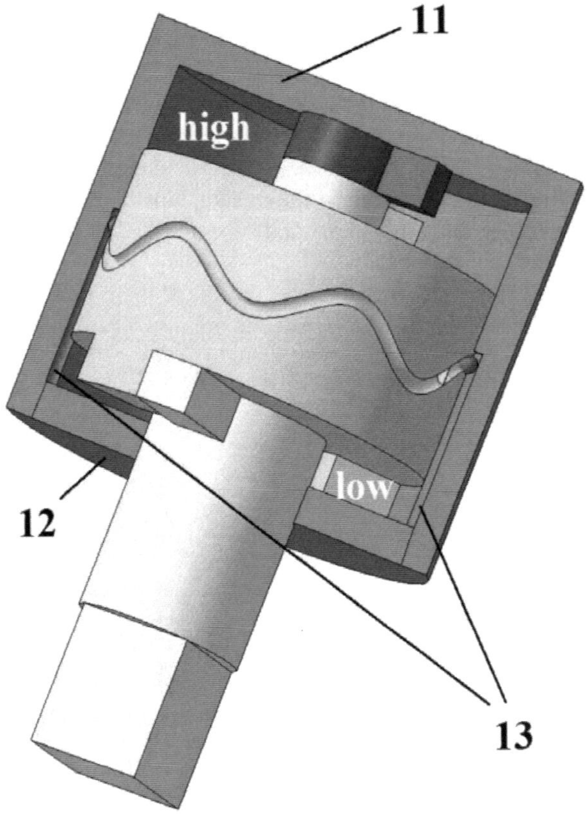


**Figure 2.** Complete impulse mechanism

11 – casing; 12 – balls of the casing ball screw gear; 13 – casing cover;
high – HP chamber; low – LP chamber

When the hammer reaches its hight point, it disconnects from the shaft. The casing and the hammer start to move freely to the engagement point of the cam of return ring 7 and the return cam of casing 8. The hammer is trigged and pressed into the lower position; during this, the crossover valve is open due to which the LP chamber furnishes practically no resistance. In the lower position, the hammer and the casing, move freely again, till the point where the shaft is engaged. After this, the process repeats. This is the dynamic operating mode of the mechanism.

## 3. Substantiation of the simulation model

The general simulation model of dynamic compression process for the mechanism constitutes a continuity equation for HP chamber expressing equality between the rate of chamber volume change and the oil flow from the capillary channel considering the fluid compressibility: [7-10].

$$\frac{d\Delta V}{dt} = Q + \frac{d\Delta V_{сж}}{dt} \tag{1}$$

where ΔV is the change in the HP chamber volume; Q is fluid flow rate; $\Delta V_{сж}$ is change of fluid volume due to compression.

Assuming that casing rotation rate is constant and the law of axial motion for the cam is $Z = R_п \omega t$, let's expand members of (1):

$$\Delta V = 2(\pi R_n^2 - \pi R_\kappa^2)h - (\pi R_n^2 - \pi R_\kappa^2)R_n \omega t,$$

$$\frac{d\Delta V}{dt} = -(\pi R_n^2 - \pi R_\kappa^2)R_n \omega,$$

where $R_п$ is hammer radius; $R_к$ is radius of the return ring; h is height of cams; ω is angular speed of the casing.

$$\frac{d\Delta V_{сж}}{dt} = \frac{\Delta V}{E}\frac{d\Delta P}{dt},$$

where E is fluid elastic modulus liquid; ΔP is chamber pressure.

$$Q = \left[\frac{\Delta P}{12\mu e^{b\Delta P} m}\delta^3 - \frac{R_n \omega \delta}{2}\right] 2\pi R_n,$$

where μ is oil dynamic viscosity; b is dependency ratio between viscosity and pressure; m is height of the hammer (length of capillary gap); δ is the size of opening between the casing wall and the cylindrical surface of the cam (depth of capillary gap).

In order to obtain the theoretical form of the impact pulse, (1) is integrated over the interval:

$$t = \left[0 \ldots \frac{h}{R_n \omega}\right].$$

## 4. Simulation results

Figure 3 shows the design form of the pressure pulse in the chamber for the following parameters of the mechanism:

$R_п$ = 0.02 m; $R_к$ = 0.02 m; h = 0.02 m; ω = 104 rad/sec; E = 1350 $10^6$ Pa; μ = 0.0388 N m/sec; ρ = 850,2 kg/m$^3$ – density of oil; b = 1.5 $10^{-7}$; δ = 0.0005 m; m = 0.04 m.

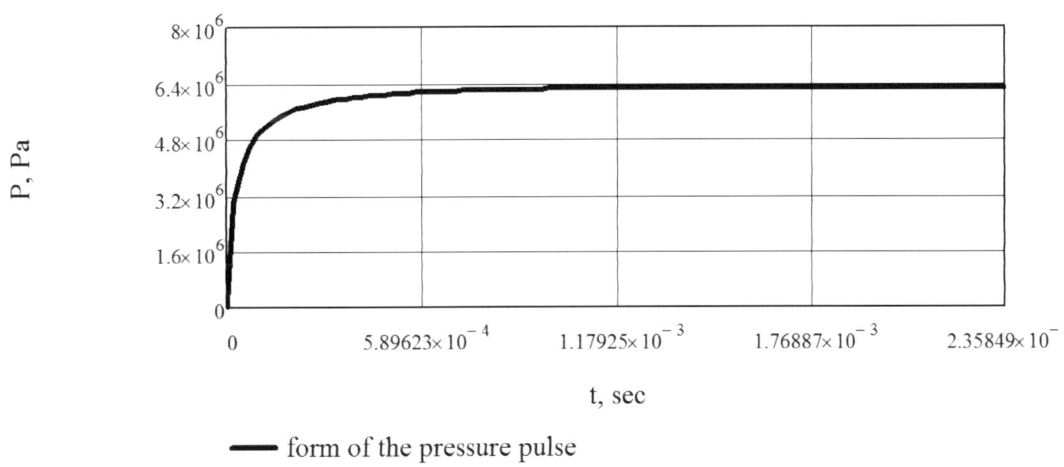


**Figure 3.** The calculated dependency of the impulse mechanism chamber pressure.

## 5. Conclusions and further line of research

- The article describes a new design of the impulse mechanism characterized by enhanced torque due to great surface area of the hammer piston;
- The new mechanism has small size and a relatively simple design, and the pressure in the chamber depends on a few parameters;

- Further research of the new mechanism implies substantiation of optimization techniques for its design parameters at a given torque, and manufacturing a prototype model for bench testing.

## References

[1] Vodolazsky N In. Vodolazsky E G, Iskritsky V M 2012 *Vestnik SevNTU* **128** 22–6.
[2] Vodolazskaya N In. Vodolazsky E G, Iskritsky V M 2011 *Vestnik SevNTU* **117** 32–7.
[3] Drozdov A N, Stepanov V V 2013 *Electric impact wrench. Dynamics: The Monograph.* Moscow: MGSU, 118.
[4] Drozdov A N 1999 *Manual machines for construction works (the device and the basis of calculation).* Moscow: MGSU, 252.
[5] Ushakov H P 2013 *Hydraulic impact mechanisms. World experience of analysis and design.* Germany. Ed. House. Palmarium Academic Publishing 280.
[6] Nikitin Y F 2010 *Hydraulics and hydropneumatic.* Moscow: MGTU named after Bauman, 414.
[7] European Power Tool Association 2009 *Procedure 05/2009 – Measurment of the single impact energy of rotary hammers and breakers* Frankfurt: European Power Tool Association.
[8] Doyle S 2012 *Impact Energy Testing – Trails and Triumphs,* Brookfield: Milwaukee Electric Tool Corporation, Internal Document.
[9] Jatsun S F, Lupehina I V, Panovko G J, Volkova L Yu, 2011 *Dynamics of the vibration driven tool at its interaction with the processing material* Proc. 13th World Congress in Mechanism and Machine Science, Guanajuato, Mexico, 19-25.
[10] Zhang S L, Tang J 2016 *J. Manuf. Sci. Eng.* **138(11)**.

# Efficiency Improvement of Some Agricultural Residue Modified Materials for Textile Dyes Absorption

**P Boonsong[1] and J Paksamut[1]**

[1]Department of Social and Applied Science, College of Industrial Technology, King Mongkut's University of Technology North Bangkok, Bangkok, 10800, Thailand

**Abstract.** In this work, the adsorption efficiency was investigated of some agricultural residue modified materials as natural bio-adsorbents which were rice straw (*Oryza sativa* L.) and pineapple leaves (*Ananas comosus* (L.) Merr.) for the removal of textile dyes. Reactive dyes were used in this research. The improvement procedure of agricultural residue materials properties were alkali-acid modification with sodium hydroxide solution and hydrochloric acid solution. Adsorption performance has been investigated using batch experiments. Investigated adsorption factors consisted of adsorbent dose, contact time, adsorbent materials and pH of solution. The results were found that rice straw had higher adsorption capacity than pineapple leaves. The increasing of adsorption capacity depends on adsorbent dose and contact time. Moreover, the optimum pH for dye adsorption was acidic range because lowering pH increased the positively charges on the adsorbent surface which could be attacked by negatively charge of acid dyes. The agricultural residue modified materials had significant dye removal efficiency which these adsorbents could be used for the treatment of textile effluent in industries.

## 1. Introduction

Among the consequences of this rapid of mankind, society, science and technology growth is environmental disorder with a big pollution problem. Toxic organic and inorganic pollutants are compounds of great concern due to toxicity, persistence, bioaccumulation in animals and persist in living organism, which cause several environmental problems to environment [1]. Industrial developments in recent years have left their impression on the environmental society. Many industries, such as textile, paper, plastics and printing industry are using more than 100,000 commercially available dyes to color their products and thus produce wastewater containing organics with a strong color, where in the dyeing processes the amount of dye lost wastewater is up to 50% of the dye [2]. One of the important classes of the pollutants is dyes, and once they enter the water it is no longer good and sometimes difficult to treat as the dyes have a synthetic origin and a complex molecular structure which makes them more stable and difficult to be bio-degraded [3]. Wastewater remediation is considered to be one of the main approaches for water recycling that provides the deficiency of water in different regions in daily life. Of these dyestuffs, 5–10% is lost in industrial effluents, which is released into water resource such as river, seas and groundwater. Global and local agencies have therefore established certain limits on the quantities of dyestuffs being discharged into environment when the concentration is as low as 1 ppm [4].

The removal of dyes from waste effluents becomes environmentally important because even a small quantity of dye in water can be toxic and highly visible. Therefore, efficient techniques for the removal of dye from wastewater, such as biological processes, photo catalytic degradation, coagulation, chemical oxidation processes and adsorption, which depend on the types of dyes to be removed, their composition, concentrations and production flow into the wastewater. Adsorption is a physical–chemical treatment, which is used as top quality treatment procedures for the removal of dissolved organic pollutants like dyes from industrial waste water. Adsorption is defined as concentration of materials on the surface of solid bodies (adsorbents). Adsorbents are mainly derived from natural and unnatural sources such as zeolites, charcoal, clays and other waste resources. Natural bio-adsorbents prepared from waste resources used include coconut shell, rice husk, petroleum wastes, tannin-rich materials, sawdust, fertilizer wastes, chitosan, algae, fruit wastes, rice straw and pineapple leaves etc. [1, 5].

Natural bio-adsorbents are comprised of lignocellulosic matters, could be an excellent adsorbent for wastewater treatment due to low cost, biodegradability, eco-friendliness and high stable nature to most organic solvents. Furthermore, these types of lignocellulosic matter have high density of hydroxyl groups and easy to modify with specific functional groups, e.g., carboxyl, amino, sulfo and cyclodextrin groups for removal of specific pollutants [6]. Nevertheless, natural bio-adsorbents have not always a very high adsorption capacity, which is one of the main restrictions for its use in water remediation. However, some research report that the use of simple chemical modifications was increase the adsorptive capacity without increasing the cost of final product and, thereby, its widespread use may not be impeded. Consequently, natural bio-adsorbents are chemically treated with chemical reagents in low cost solutions such as acids or bases, causing an increase in the pollutant-removing capacity. Several studies report that chemical modifications of bio-adsorbents may introduce functional groups within the structure of the adsorbents or increase their porosity, with an increase in their adsorption capacity [7]. The agricultural residues materials for removing effluent waste are rice straw and pineapple leaves as adsorbent. Rice straw has 69.2% cellulose, 10.2% hemicellulose and 4.9% lignin [8]. Pineapple leaves have 43.5% cellulose, 21.9% hemicellulose and 13.9% lignin [9]. Rice straw and pineapple leaves have the advantage of high lignocellulosic material, large-scale production, low cost, and easy access.

Thus, the aim of this study was to investigate the adsorption efficiency of some agricultural residue modified materials as natural bio-adsorbents which were rice straw and pineapple leaves for the removal of textile dyes from aqueous solution. The influence of process parameters including: adsorbents dose, contact time, bio-adsorbent types and pH were studied.

## 2. Materials and methods

### *2.1. Preparation of modified materials*

Bio-adsorbent materials used in this study as rice straw and pineapple leaves were collected from a farm in Kanjanaburi, Thailand. Firstly, the materials were washed with distilled water to remove dust and soluble impurities and grinded using a laboratory mill. Secondly, the ground bio-adsorbent materials were soaked in distilled water for several times until the water was colorless. Following, the ground bio-adsorbent materials were dried at 85 ± 2°C for 12 hours and sieved into size of about 0.5 mm.

The bio-adsorbent materials are modified with hydrochloric acid (HCl) and sodium hydroxide (NaOH) at difference concentrations (0.1%, 0.3% and 0.5%). The ground bio-adsorbent materials were soaked in the solution at room temperature for 2 hours and filtered off. The modified bio-adsorbent materials were washed with distilled water for several times until the pH of the filtrate reached between 6 – 7. Finally, the modified bio-adsorbent materials were dried at 85 ± 2°C for 12 hours and stored in plastic bags at room temperature and dry place for further use.

### *2.2. Adsorption experiment*

Adsorption experiments were carried out using rice straw and pineapple leaves as bio-adsorbents. The influences of several parameters (adsorbent/dye solution ratio, contact time and pH) were studied in batch experiment. In order to establish the effect of bio-adsorbent (g) : dye solution (mL) ratios as 1:5, 1:10, 1:15 and 1:20 g/mL were used, with 0.1% reactive blue dye solution. The effect of contact time on the adsorption capacity of rice straw and pineapple leaves bio-adsorbent was studied for 30, 60, 90, 120, 150, 180 and 240 min at an initial concentration of 0.1% reactive blue dye solution. The optimal conditions of adsorbent/dye solution ratio and contact time were performed for dye adsorption of bio-adsorbents (rice straw and pineapple leaves) and modified bio-adsorbents (modified rice straw and modified pineapple leaves). The effect of pH on the adsorption of dye solutions was studied in a range of 4–8 with resolution of ± 0.05 pH. The solutions were pH adjusted with 0.1 M HCl or 0.1 M NaOH using a pH meter to perform the measurements. In all experiments, different bio-adsorbents were added to dye solution as wastewater sample. The mixtures were stirred at 500 rpm with magnetic stirrer at room temperature. The bio-adsorbents were separated by filtration through, Whatman qualitative No. 4, filter paper. The concentration of residual dye concentration was determined using UV-VIS spectrometer at suitable wavelength (λ=597 nm for dye solution). All measurements were carried out in triplicate and the results were averaged. The removal efficiency percentage was calculated from the following equation:

$$\text{Removal dye (\%)} = (C_i - C_t) / C_i \times 100$$

where $C_i$ and $C_t$ are the reactive blue dye concentrations at the beginning of an experiment and that at time t, respectively.

## 3. Results and discussions

### *3.1. Effect of bio-adsorbent:dye solution ratios and contact times*

The effect of bio-adsorbent:dye solution ratios on removal percentage of reactive blue dye was shown in figure 1. The percentage of removal gradually increased with enhancing bio-adsorbent dosage. Reactive blue dye removal efficiency for all types of bio-adsorbents was found to increase with increasing amount of rice straw and pineapple leaves. The increase of removal efficiency was possibly because of increasing availability of reactive blue dye capturing sites, adsorption sites of bio-adsorbent and surface area of bio-adsorbent with increased amount of bio-adsorbents [6, 10]. Moreover, the chemical structure of lignocellulosic materials as bio-adsorbents consists of many reactive chemical functional groups for removal of specific pollutants such as alcoholic and phenolic hydroxyl, carboxyl, carbonyl, ether, ester groups, etc. [11]. The maximum percentage removal of reactive blue dye from aqueous solution by rice straw and pineapple leaves were 1:10 of bio-adsorbent:dye solution ratio.

Contact time represents the time of interaction between the bio-adsorbent and reactive blue dye in the solution (figure 1). Generally, increasing the contact time between the bio-adsorbent and the reactive blue dye increases the amount of reactive blue dye uptake from their solutions. The reactive blue dye uptake was rapidly within the first period and reached an equilibrium state at 120 min due to the large number of the free active adsorption sites on the bio-adsorbents and the relatively high concentrations of the reactive blue dye in the solution [12].

The optimum bio-adsorbent: dye solution ratios and contact times were selected as 1:10 and 120 min for other experiments.

### *3.2. Effect of different bio-adsorbent types*

The chemical modifications play an important role to improve the adsorption capacity of the biomass. The comparison of all pretreatments is displayed in figure 2. The results showed that the enhancement in the uptake capacity of rice straw and pineapple leaves was achieved when NaOH and HCl were used for pretreatment. The chemical modifications of bio-adsorbents may introduce functional groups within the structure of the adsorbents or increase their porosity, with an increase in their adsorption capacity [7]. The rice straw modified with 0.5% HCl and pineapple leaves modified with 0.5% HCl showed much improvement in the uptake capacity of bio-adsorbents. The possible explanation for significant improvement in the uptake capacity of bio-adsorbents for reactive blue dye is that upon modification with inorganic acids might rupture the lignocellulosic fibers and reduce of impurities from the lignocellulosic fiber, which could release polysaccharides, natural fat and wax from lignocellulosic fiber surface, thus revealing chemically reactive functional groups and offer increase porosity and surface area for adsorption and also display high affinity for certain reactive blue dye [13,

14]. The rice straw modified with 0.5% HCl and pineapple leaves modified with 0.5% HCl were selected for the effect of solution pH experiments.

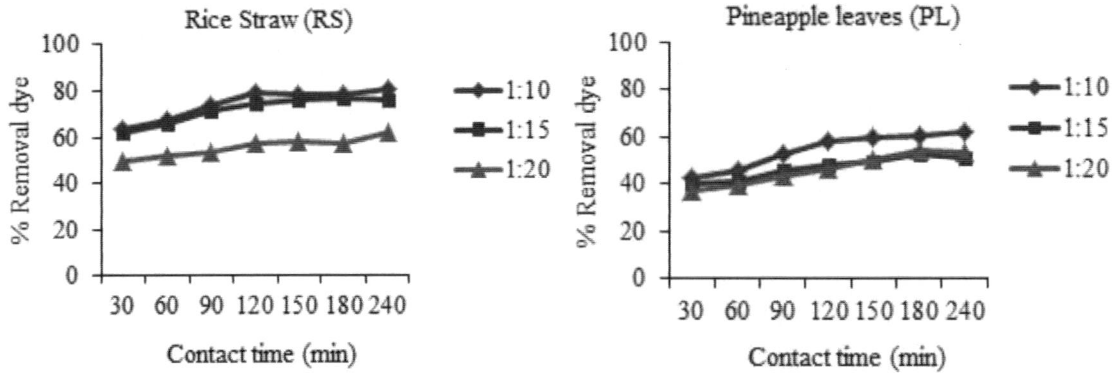


**Figure 1.** Effect of bio-adsorbent: dye solution ratios and contact times onto rice straw and pineapple leaves as bio-adsorbents for reactive blue dye adsorption.

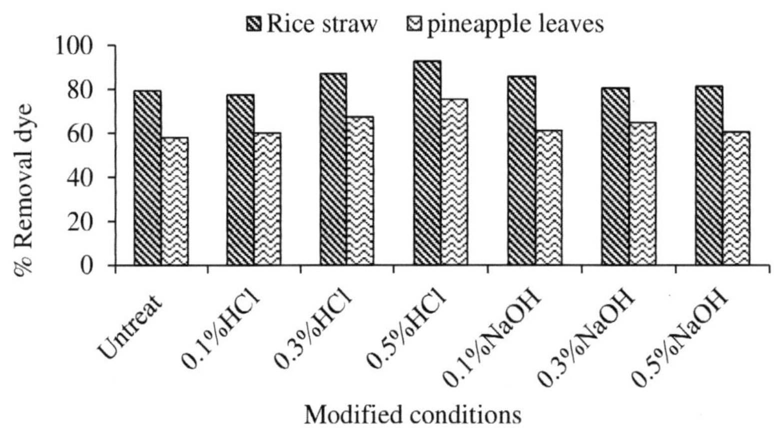


**Figure 2.** Effect of different bio-adsorbent types for reactive blue dye adsorption.

*3.3. Effect of pH*

The effect of initial solution pH on the adsorption of reactive blue dye onto rice straw modified with 0.5% HCl and pineapple leaves modified with 0.5% HCl was examined and the results are presented in Figure 3. The pH value determines the surface charge of the adsorbent and the ionic form of the dyes molecules, both influencing the amount of reactive blue dye adsorbed by bio-adsorbent. Removal efficiencies were the greater at acidic pH because the surface of the bio-adsorbent become positively charged from protonation of hydroxyl and carbonyl groups on the surface which easily attract the negatively charged of reactive blue dye (anionic dye) ions and remove them from water [6, 15].

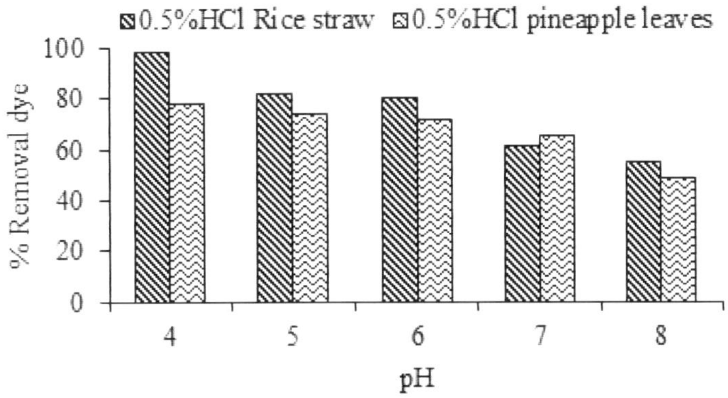


**Figure 3.** Effect of the solution pH on the adsorption of reactive blue dye onto bio-adsorbent.

## 4. Conclusions

The adsorption efficiency of rice straw and pineapple leaves on chemically modified materials as natural bio-adsorbents which were for the removal of reactive blue dye from aqueous solution. The influence of process parameters including: adsorbents dose, contact time, bio-adsorbent types and pH were studied. The optimum of bio-adsorbent : dye solution ratios and contact times were 1:10 and 120 min. The rice straw modified with 0.5% HCl was the outstanding bio-adsorbent for the removal of reactive blue dye from aqueous solution. The acidic pH solution indicated the greater removal efficiencies.

## References

[1] Rashed M N 2013 Adsorption Technique for the Removal of Organic Pollutants from Water and Wastewater *Organic Pollutants–Monitoring, Risk and Treatment* (Croatia: InTech Publisher) chapter 7 pp 167–194.
[2] Adegoke A K et al. 2015 *Water Res. Ind.* **12** 8–24.
[3] Gupta V K et al. 2009 *J. Environ. Manage.* **90** 2313–42.
[4] Zhang S et al. 2016 *Proc.. Environ. Sci.* **31** 3–11.
[5] Vital R K et al. 2016 *J. Bioremediat. Biodegrad.* **7** 371–4.
[6] Manna S et al. 2017 *Process Saf. Environ. Prot.* **107** 346–356.
[7] Schwantes D et al. 2016 *J. Chem.* **2016** 1–15.
[8] Sriariyanun M et al. 2015 *Kasetsart J. (Nat. Sci.)* **49** 146–154.
[9] Pardo M E S et al. 2014 *J. Agric. Chem. Environ.* **3** 53-56.
[10] Wu Y et al. 2016 *Biochem. Eng. J.* **105** 27–35.
[11] Manciulea I et al. 2016 *Bull. Transilv. Univ. Brasov Ser. I* **9** 49–54.
[12] Badawi M A et al. 2017 *Int. J. Biol. Macromolec.* **99** 465–476.
[13] Thakur V K 2014 *Lignocellulosic Polymer Composites: Processing, Characterization, and Properties* (New Jersey : John Wiley and Sons).
[14] Zafar M N et al. 2015 *J. Taiwan Inst. Chem. Eng.* **46** 82–88.
[15] Suteu D et al. 2015 *Environ. Eng. Manag. J.* **14** 525–532.

# Reliability of Hull Girder Ultimate Strength of Steel Ships

**Gao Da-wei[1] and Shi Gui-jie[2*]**

[1]School of Mechanical Engineering, University of Shanghai for Science & Technology Shanghai 200093, China
[2]Collaborative Innovation Center for Advanced Ship and Deep-Sea Exploration, State Key Laboratory of Ocean Engineering, China Strategy Institute of Ocean Engineering, School of Naval Architecture, Ocean & Civil Engineering, Shanghai Jiaotong University, Shanghai, 200135, China

**Abstract.** Hull girder ultimate strength is an evaluation index reflecting the true safety margin or structural redundancy about container ships. Especially, after the hull girder fracture accident of the MOL COMFORT, the 8,000TEU class large container ship, on June 17 2013, larger container ship safety has been paid on much more attention. In this paper, different methods of calculating hull girder ultimate strength are firstly discussed and compared with. The bending ultimate strength can be analyzed by nonlinear finite element method (NFEM) and increment-iterative method, and also the shear ultimate strength can be analyzed by NFEM and simple equations. Then, the probability distribution of hull girder wave loads and still water loads of container ship are summarized. At last, the reliability of hull girder ultimate strength under bending moment and shear forces for three container ships is analyzed by using a first order method. The conclusions can be applied to give guidance for ship design and safety evaluation.

## 1. Introduction

In the whole life time, container ships are mainly suffered to two kinds of loads, such as wave induced loads and still water loads. Wave loads are always changing and inconstant for different ocean areas. Even for the same ship, the draught, sea-going angle or stern-stem trim can also have effect on the wave induced loads on hull girder. Relatively speaking, the still water loads on hull girder should not have lots of fluctuation owing to fixed hull weight and cargo weight. However, each container weight is not same as the declared weight at the port and even the gap between actual weight and declared weight is more than 7.5%, as pointed out in Class NK report [1]. So, the accurate and reasonable evaluation of wave loads and still water loads on hull girder is very important for the safety of container ships.

The ultimate strength of hull girder may, in principle, be calculated according to any selected computational method. Different analysis methods should be firstly compared with and the potential bias in the calculation should be properly covered in evaluation criteria. In addition, the uncertainties of yielding strength of hull girder members have much influence on ultimate strength. Geometrical dimensions can be modeled as deterministic values, as pointed out in CSR OT background document [2]. So, ultimate strength of hull girder should evaluate the probability distribution and analyze the reliability of actual ships under rules design.

Fang and Das [3] studied the relationship between the risk evaluation and structural reliability, and then reviewed the evolution of structural reliability applied to ship structures. The failure probabilities of damaged ship were obtained based on Monte Carlo simulation technique.

Hussein and GuedesSoares [4] studied the reliability of damaged tankers considering the increase in the still-water bending moment due to damage and the loss in ultimate strength. The sensitivity analysis showed that the ultimate strength and the corresponding uncertainty have the highest importance. The still-water bending moment and the wave-induced bending moment have almost the same importance.

Zhang and Tang [5] studied time-variant reliability analysis of FPSO hull girder considering corrosion based on statistics. The results not only provide the reliability under different corrosion conditions, but also do well for further inspection and maintenance research. The results provide necessary foundation for deciding inspection intervals and maintenance measures, which has practical sense to improve the general safety level of ocean engineering.

Gao [6] summarized common crack types of container ships and analyzed the crack effect on the residual ultimate strength. Based on the numerical results, simple equations are fitted to predict residual ultimate strength of cracked container ship. And also assuming the corrosion additions ruled by classification societies ABS, CCS and DNV as the most serious corrosion, the corrosion effect on residual ultimate strength of container ship was analyzed.

Firstly, this paper discusses different methods of calculating hull girder ultimate strength. The traditional increment-iterative method (hereafter abr. as Smith), firstly proposed by Smith [7], is used to calculate the hull girder ultimate strength of oil tankers. However, container ship has special characteristics, such as large opening in cargo hold and extreme thick plating of higher strength in upper deck and coaming. The correctness of Smith method applied on container ships ultimate strength should be verified by nonlinear finite element method (NFEM).

Secondly, container ships always suffer to large shear forces due to non-uniform weight distribution and therefore it is needed to check the hull girder ultimate strength under shear. A simple equation of hull girder shear ultimate strength, which has considered shear buckling strength of each plating in hull cross section proposed in GL rules [8], is compared with the results by NFEM.

Then, this paper summarizes the probability distribution of hull girder wave loads and still water loads of container ship based on available data. At last, the reliability of hull girder ultimate strength under bending moment and shear forces for three container ships is analyzed by using a first order method.

## 2. Analysis of bending ultimate strength

### *2.1. Nonlinear Finite Element Method (NFEM)*

NFEM is a powerful tool to solve the problems of complex engineering structures. Hull ultimate strength analysis involves material nonlinearity and large deformation nonlinearity, hull stiffness decreases as external loads increase, so the NFEM in ABAQUS software is chosen to analyze the ultimate strength of container ship and also track the hull collapse modes after the ultimate strength.

The steel in container ships is assumed as a kind of perfect elastic and small plastic harden material, so the lower limit of ultimate strength from material view can be obtained. The elastic modulus of steel is 206GPa and the Poisson's ratio is 0.3. The middle parallel body of container ship is analyzed and one frame length of hull girder is fabricated in FE model.

In the analysis, the hull structures of container ship can be seemed as thin-walled structures, which are suitable for using the plane stress element, such as S4R element, to build a finite element model. S4R is a kind of 4 nodes shell element, has 6 freedoms at each node, adopts reduced integration to avoid volume self-locking, and also considers finite membrane strain and flexible rotation. The mesh size in the middle frame is about 100mm.

Simply supported conditions are applied on the both ends of FE model. In order to apply boundary conditions, point A of one end and point B of another end are separately created at intersection point of neutral axis. The cross section nodes of one end are constrained by point A, and another end is constrained by point B.

Initial imperfection of first buckling mode of each panel is applied in FE model to trigger the most possible collapse model of hull girder. However, residual welding stress is not considered in this analysis.

Subjected to hogging or sagging moment, part of hull structures are in compression and other part are in tension. As the external loading increases, local buckling may occur in compression panels and then the strain energy in this region will diffuse immediately to neighboring structures. The hull girder ultimate strength under hogging moment is $1.87\times10^7$kN·m, and the hull girder ultimate strength under sagging moment $2.17\times10^7$kN·m.

### *2.2. Increment-Iterative Method (Smith)*

Using NFEM analyse hull ultimate strength will need plenty of time to fabricate models and do calculations. The most effective method to avoid this problem is to reduce the number of freedom during calculation process, that is to say, decrease the matrix order of finite element stiffness. One kind of method is using large structure element, such as stiffener element, stiffened plating element and hard corner element. The cross section of ship will be divided into a series of separate structure elements, which are considered to act independently and fail in their own damage modes. In order to assure the effectiveness of this method, the structure element should be reasonably divided and their damage mode should be accurately defined.

The moment–curvature curve is obtained by means of an incremental–iterative approach based on the principles of Smith's method [7]. The bending moment that acts on the hull girder cross section increases due to the imposed curvature ki for each step of the incremental procedure. This imposed curvature corresponds to an angle of rotation of the hull girder transverse section about its effective horizontal neutral axis, which induces an axial strain e in each hull structural element. In the sagging condition, the structural elements below the neutral axis are lengthened, whilst elements above the neutral axis are shortened. As the external moment applied at two ends of the model increases, the curvature of hull cross section and neutral axis location will be calculated based on given damage mode of each structure element. This procedure will be repeated until the moment-curvature curve reaches the peak, which is corresponding to the hull ultimate strength value.

The software of Mars2000 provided by BV Classification Society adopts the Smith method to analyze the ultimate strength of hull girder between two adjacent frames [6]. The cross section will be divided into two kinds of structure element: stiffener attaching plating element and hard corner element. The former will present buckling or yielding damage mode when in compression; however, it will present only elastic-plastic damage mode when in tension. The latter, constituting plating crossing, collapses mainly according to elastic-plastic failure mode.

The model of target ship is fabricated in Mars2000. Based on the Smith method, the ultimate strength under hogging moment is $1.87\times10^7$kN·m and the ultimate strength under sagging moment is $1.89\times10^7$kN·m.

### *2.3. Verification of analysis method of bending ultimate strength*

The results of hull ultimate strength of the target ship under bending moment, obtained by NFEA and Smith method, are summarized in table 1. The hogging ultimate strength by the two methods shows a good consistence. However, the scatter of sagging ultimate strength is bigger than the hogging. This may be because of different inelastic buckling strength for upper thick plating obtained by various analysis methods. The thickness of upper deck in container ship is 68mm, which is more than 2~3 times common hull plating. The deck has a much larger elastic buckling stress than the yielding stress and will present an inelastic buckling when the hull reaches the sagging ultimate state. Generally, the smith method is conservative in analyzing bending ultimate strength of container ship.

**Table 1.** Bending ultimate strength of target container ship. [kN·m]

| Load case | NFEA | Smith | NFEA/ Smith |
|---|---|---|---|
| Hogging | $1.87\times10^7$ | $1.87\times10^7$ | 1.00 |
| Sagging | $-2.17\times10^7$ | $-1.89\times10^7$ | 1.15 |

## 3. Analysis of shear ultimate strength

### *3.1. Nonlinear Finite Element Method (NFEM)*

Based on the FE model described in Sec. 2.1, the shear ultimate strength of the hull girder can be analyzed by NFEM. However, the boundary conditions should be changed to facilitate the application of shear loads and shear deformation on hull girder. One end of the hull girder is constraint all three translation degrees and one ration degree, UX, UY, UZ and URX. At the other end, point A is created at intersection point of neutral axis, which is used to link the cross section nodes of the end.

As described in Sec. 2.1, initial imperfection has some influence on reducing hull girder ultimate strength. The shear buckling mode along the diagonal line in stiffened panel may bring about the greatest reduction on shear ultimate strength. However, the compression buckling mode, that is a wave pattern along the length direction, may also reduce the shear ultimate strength. Considering the initial imperfection only triggering the most probable failure mode of hull girder, the lateral deformation has a bigger influence. So the compression buckling mode as Sec. 2.1 is used to calculate the shear ultimate strength.

As the external shear load increases, the structures in side shell and inner longitudinal bulkhead of hull structures present shear buckling along diagonal line in each panel. Local buckling only occurs in compression direction and then the strain energy in this region will diffuse immediately to neighboring structures. However, the upper deck and double bottom show low stress level. The hull girder ultimate strength under shear is $5.41\times10^5$kN.

### *3.2. Simple equation*

Compared with bending ultimate strength of hull girder, the shear ultimate strength is a much simple issue. Only the vertical plating in hull girder is seemed as effective under shear load, that is, the side shell, inner longitudinal bulkhead, bottom girder and bilge plating. The other structures including stiffener and deck/bottom plating can be excluded in calculating shear ultimate strength.

For each stiffened panel, the shear ultimate strength is equal to the inelastic buckling strength and only the vertical plating is effective for shear load, but the stiffener can always be kept straight and non-deformed. So the boundary conditions of each panel can be kept unchanged during the increase of external hull girder shear load. From this point, each panel can seemed to be independent on resisting shear force. The summation of shear inelastic buckling strength of each panel should be equal to the hull girder shear ultimate strength. The rules requirements provided by GL Classification Society use the same idea to verify hull girder shear ultimate strength of Bulk Carriers, Oil Tankers and Container Ships [8].

It should be noted that the bilge plating in connection of side shell and outer bottom belongs to strengthen longitudinal structures, which should be included in calculating shear ultimate strength by using projected area to the vertical plane. The shear ultimate strength of hull girder can be calculated by (1).

$$F_U = \frac{1}{\sqrt{3}} \cdot 10^{-3} \cdot \sum_{i=1}^{n} C_{\tau i} \cdot b_i \cdot t_i \cdot R_{eHi} \text{, kN} \tag{1}$$

where,

$n$ ——plating number for resisting shear load;

$C_{\tau i}$——$i^{th}$ plating reduction factor for shear buckling, referred to CSR rules;

$b_i$ ——$i^{th}$ plating breadth, mm;

$t_i$ ——i[th] plating thickness, mm;

$R_{eHi}$ ——i[th] plating yielding strength, MPa.

*3.3. Verification of analysis method of shear ultimate strength*

The results of hull ultimate strength of the target ship under shear load, obtained by NFEA and simple equation, are summarized in table 2. The shear ultimate strength by the two methods shows a good consistence. It proves that the simple equation described in Sec. 3.2 can be used to analyze shear ultimate strength of container ship.

**Table 2.** Shear ultimate strength of target container ship. [kN]

| NFEA | Simple equation | NFEA/ Simple equation |
|---|---|---|
| $5.36\times10^5$ | $5.41\times10^5$ | 0.991 |

## 4. Reliability analysis of ultimate strength

The limit states are conditions that will cause a hull girder to experience performance failure and ultimate strength is corresponding to the limit state of hull girder. The limit state designs consider various conditions/factors under which the hull girder may fail to function, and account for the uncertainties associated with determining the safety margins. Reliability analysis of ultimate strength is able to evaluate the uncertainties, including still water loads, wave loads and ultimate strength for container ships.

The reliability of hull girder bending ultimate strength of container ship can be calculated by inducing the following limit sate equation [4].

$$g(x) = X_R M_u - [X_{st} X_{nl} M_w + X_{sw} M_{sw}] \tag{2}$$

where,

$M_u$ is the bending ultimate strength with a model uncertainty factor $X_R$;

$M_w$is the wave bending moment with model uncertainty factors;

$X_{st}$ is the linear response calculation and $X_{nl}$for non-linear effects;

$M_{sw}$ is the still-water bending moment with a model uncertainty factor $X_{sw}$.

Based on technical background of Bulk Carriers and Oil Tankers Common Structure Rules [9], the values of the above mentioned uncertainties are taken as presented in table 3.

**Table 3.** Distribution of uncertainty factors.

| Uncertainty factor | Distribution | Mean, μ | Standard deviation, σ |
|---|---|---|---|
| $X_R$ | Normal | 1.05 | 0.1 |
| $X_{st}$ | Normal | 1 | 0.1 |
| $X_{nl}$ | Normal | 1 | 0.1 |
| $X_{sw}$ | Normal | 1 | 0.1 |

For hull girder of container ships, the bending moment is an integral of shear forces in each cross section along ship length and the ship length can be seemed as a fixed value. Therefore, the model uncertainty factors for shear forces can be assumed the same as bending moment. The limit sate equation of hull girder shear ultimate strength of container ship can be described as the following,

$$g(x) = X_R F_u - [X_{st} X_{nl} F_w + X_{sw} F_{sw}] \tag{3}$$

where,

$F_u$ is the shear ultimate strength with a model uncertainty factor $X_R$;

$F_w$is the wave shear force with model uncertainty factors;

$X_{st}$ is the linear response calculation and $X_{nl}$for non-linear effects;
$F_{sw}$ is the still-water shear force with a model uncertainty factor $X_{sw}$.

*4.1. Stochastic model of ultimate strength*

In general, the following are considered as uncertainties on hull girder ultimate strength, including: Scantlings of structural members, Structural arrangements, Structural details, Yield stresses of steel, Effect of welding residual stress, Effect of initial imperfection, Lateral loads, such as sea pressure and container loads.

Among these factors, yield stress of steel has the greatest influence on the ultimate strength. Therefore, only yield stress is seemed as variables, and the other factors are assumed as definite factors, which are taken in consideration on calculation model to estimate the hull girder ultimate strength of container ships. Based on studies in CSR OT [2], the distribution of yield stress for different grade steel can be approximated as lognormal and their probability characteristics are summarized in table 4.

**Table 4.** Probability characteristics of yield stress.

| Yield stress | Mean, μ | Standard deviation, σ | Coefficient of variation, COV |
|---|---|---|---|
| ReH=235 | 269 | 21.52 | 0.08 |
| ReH=315 | 348 | 20.88 | 0.06 |
| Reh=355 | 391 | 21.66 | 0.06 |

In the process of calculating ultimate strength, only the factor of yield stress is seemed as variable, so the ultimate strength can also be roughly assumed as lognormal distribution. The mean value of bending ultimate strength of container ship hull girder is calculated by Smith method as validated in Sec. 2 and the mean value of shear ultimate strength is calculated by (1) as validated in Sec. 3. The COV of ultimate strength are conservatively taken as 0.08 [10], the maximum value of yield stress COV. The stochastic model of all the ultimate strength for the three container ships is summarized in table 5.

**Table 5.** Stochastic model of ultimate strength.

| | Hogging [kNm] | | Sagging [kNm] | | Shear [kN] | |
|---|---|---|---|---|---|---|
| | μ | σ | μ | σ | μ | σ |
| Container ship 1 | 1.18E+07 | 9.44E+05 | 1.19E+07 | 9.52E+05 | 3.97E+05 | 3.18E+04 |
| Container ship 2 | 1.57E+07 | 1.26E+06 | 1.56E+07 | 1.25E+06 | 4.21E+05 | 3.37E+04 |
| Container ship 3 | 1.85E+07 | 1.48E+06 | 1.72E+07 | 1.38E+06 | 5.41E+05 | 4.33E+04 |

*4.2. Stochastic model of wave loads*

The wave loads can be based on the use of RAOs for the container ship, and follows the procedure described in the IACS background document [9]. There is another method, which is based on the rule formulae for wave bending moment and wave shear force in IACS unified requirements. In this paper, the latter is chosen. The distribution of the extreme values of the annual wave loads is obtained as a Gumbel law [10].

The mean value and standard deviation of wave loads can be estimated by the following equation [11].

$$\mu_w = \delta(\ln N + 0.577) \tag{4}$$

$$\sigma_w = 1.283\delta \tag{5}$$

where,
$N$ is the number of wave cycle during ship operation, $10^8$;

$$\delta = \begin{cases} M_{wdesign}/19 \text{ for bending moment} \\ F_{wdesign}/19 \text{ for shear force} \end{cases}$$

$M_{wdesign}$ and $F_{wdesign}$ is the rule formulae value in IACS unified requirements.
The stochastic model of all the wave loads for the three container ships is summarized in table 6.

**Table 6.** Stochastic model of wave loads.

| | Hogging [kNm] | | Sagging [kNm] | | Shear [kN] | |
|---|---|---|---|---|---|---|
| | μ | σ | μ | σ | μ | σ |
| Container ship 1 | 3.44E+06 | 2.32E+05 | 4.26E+06 | 2.88E+05 | 3.09E+04 | 2.09E+03 |
| Container ship 2 | 5.79E+06 | 3.91E+05 | 6.88E+06 | 4.65E+05 | 4.16E+04 | 2.81E+03 |
| Container ship 3 | 6.68E+06 | 4.51E+05 | 7.85E+06 | 5.30E+05 | 4.76E+04 | 3.21E+03 |

### *4.3. Stochastic Model of Still Water Loads*

There are mainly two methods of statistics for still water loads. The first one is assuming that the still water loads can be seemed as normal distribution, and the mean value and standard deviation have some relationship with the allowable still water loads in IACS rule requirements for the target ships. For example, the mean value is equal to 60% of the allowable still water loads and the standard deviation is equal to 40% of the allowable. The second one is assuming that the mean value and standard deviation have some relationship with the maximum still water loads in loading manual of the target ships. For one year period of seagoing, the still water load of each loading condition can be fitted by normal distribution.

In reliability analysis of this paper, the still water moment and shear force is set as normal distribution, the mean value is assumed as the 70% of the maximum value in loading manuals, and the standard deviation is assumed as 20% of the maximum value in loading manuals. The stochastic model of all the still water loads for the three container ships is summarized in table 7.

**Table 7.** Stochastic model of still water loads.

| | Hogging [kNm] | | Sagging [kNm] | | Shear [kN] | |
|---|---|---|---|---|---|---|
| | μ | σ | μ | σ | μ | σ |
| Container ship 1 | 2.67E+06 | 5.35E+05 | 2.67E+06 | 5.35E+05 | 8.61E+04 | 1.72E+04 |
| Container ship 2 | 3.98E+06 | 7.95E+05 | 7.00E+05 | 1.40E+05 | 4.06E+04 | 8.12E+03 |
| Container ship 3 | 4.68E+06 | 9.37E+05 | 7.00E+04 | 1.40E+04 | 6.30E+04 | 1.26E+04 |

### *4.4. Reliability results*

According to the stochastic models presented in tables 5–7 the reliability index of container ships under bending moment and shear force can be calculated by a first order method and the results are presented in table 8. The consistent reliability level for each load condition can be attributed to using the IACS rule value which is already calibrated to achieve certain level of reliability.

Generally, hogging condition is the most dangerous for container ship ultimate strength and shear condition is relatively safer. Compared with the results derived from other researches [12-14], the range of reliability index coincide with the general situation of ship reliability, which indicates the comparability of this study.

**Table 8.** Reliability results of container ships.

| | Hogging | | Sagging | | Shear | |
|---|---|---|---|---|---|---|
| | β | P(f) | β | P(f) | β | P(f) |
| Container ship 1 | 3.90 | 4.82E-05 | 3.29 | 4.99E-04 | 6.20 | 2.74E-10 |
| Container ship 2 | 2.93 | 1.69E-03 | 3.99 | 3.34E-05 | 8.04 | 4.62E-16 |
| Container ship 3 | 3.00 | 1.36E-03 | 4.10 | 2.02E-05 | 7.84 | 2.18E-15 |

## 5. Conclusions

The results of hull ultimate strength of the target ship under bending moment, obtained by NFEA and Smith method. The hogging ultimate strength by the two methods shows a good consistence. Generally, the smith method is conservative in analyzing bending ultimate strength of container ship.

The results of hull ultimate strength of the target ship under shear load, obtained by NFEA and simple equation. It proves that the simple equation described in Sec. 3.2 can be used to analyze shear ultimate strength of container ship.

The stochastic models of ultimate strength, wave bending moment and shear force, and still water bending moment and shear is analyzed in this paper. Then, the reliability index of container ships ultimate strength under bending moment and shear force is calculated by a first order method. The reliability results show that hogging condition is the most dangerous for container ship and shear condition is relatively safer.

## References

[1] Class N K 2014 *Investigate report on structural safety of large container ship.*
[2] IACS-CSR-OT 2006 *TB-Design verification hull girder ultimate strength.*
[3] Fang C L, Das P K 2005 *Ocean Eng.* **32(3-4)** 293-307.
[4] Hussein A W, Guedes Soares C 2009 *Ocean Eng.* **36(17-18)** 1446-59.
[5] Zhang D, Tang W, Zhang S 2007 *China Ocean Eng.* **21** 197-205.
[6] Gao D, *Eng. Fail. Anal.* **25** 316-28.
[7] Smith C 1977 *Influence of local compressive failure on ultimate longitudinal strength of a ship's hull*, Proceedings of the International Symposium on Practical Design in Shipbuilding, pp. 73-9.
[8] GL-Rules 2013 *Structural Rules for Container Ships.*
[9] IACS-CSR. 2014 *TB-Hull Girder Residual Strength.*
[10] Guedes Soares C 1984 *Probabilistic Models for Load Effects in Ship Structures. Division of Marine Structures*, The Norwegian Institute of Technology, Report UR-84-38.
[11] Wirsching P H, Ferensic J, Thayamballi A 1997 *Mar. Struct.* **10(7)** 501-18.
[12] Sun H H, Guedes Soares C 2006 *J. Ship Res.* **50(2)** 171-80.
[13] Liu X 2014 *Qual. Reliab. Eng. Int.* **30(8)** 1293-1300.
[14] Liu X, 2014 *Qual. Reliab. Eng. Int.* Online.

## Acknowledgments

This project is supported by National Natural Science Foundation of China (Grant No.51305269). However, the views expressed in this paper are those of the authors and do not necessarily reflect the official views of affiliations.

# Research on Application of FMECA in Missile Equipment Maintenance Decision

**Wang Kun**[1]

[1]China Ship Development and Design Center, Wuhan, China

**Abstract.** Fault mode effects and criticality analysis (FMECA) is a method widely used in engineering. Studying the application of FMEA technology in military equipment maintenance decision-making, can help us build a better equipment maintenance support system, and increase the using efficiency of weapons and equipment. Through Failure Modes, Effects and Criticality Analysis (FMECA) of equipment, known and potential failure modes and their causes are found out, and the influence on the equipment performance, operation success, personnel security are determined. Furthermore, according to the synthetical effects of the severity of effects and the failure probability, possible measures for prevention and correction are put forward. Through replacing or adjusting the corresponding parts, corresponding maintenance strategy is decided for preventive maintenance of equipment, which helps improve the equipment reliability.

## 1. Introduction

With the development of science and technology and the demand of modern military struggle, the speed of equipment innovation continues to increase. Meanwhile, the fighting style changes, which brings new challenges to the quality of equipment maintenance personnel, the equipment maintenance management mode and the traditional maintenance methods. Make full use of the product information from FMECA, and apply to the maintenance decision-making of the products, which is the main content of this article. On the basis of the equipment failure mode and effect analysis, by technical analysis instead of experience, use the results of technical analysis as the basis of maintenance decision.

## 2. FMECA

Failure mode effects and criticality analysis (FMECA) is an inductive-type analysis method which analyzes all possible failure modes and effects of each unit in the system, and according to the probability of each failure mode and its severity classifies the failure modes [1-3]. The functional FMECA is one branch that can analyze potential failure modes and possible effects and criticality of failures in the conceptual design phase of products. It can find out the potential weaknesses of the design process to facilitate the improvement of design before the determination of product structure, which is an effective method to guarantee the quality of reliability design.

The functional FMECA is mainly used in the demonstration and project phase, as well as the early stage of the engineering development. The purpose is to analyze the defects and weak links of the product's functional design, and provide the basis for the improvements of the product's functional design. According to GJB/Z1391-2006, the functional FMECA process is shown in figure 1.

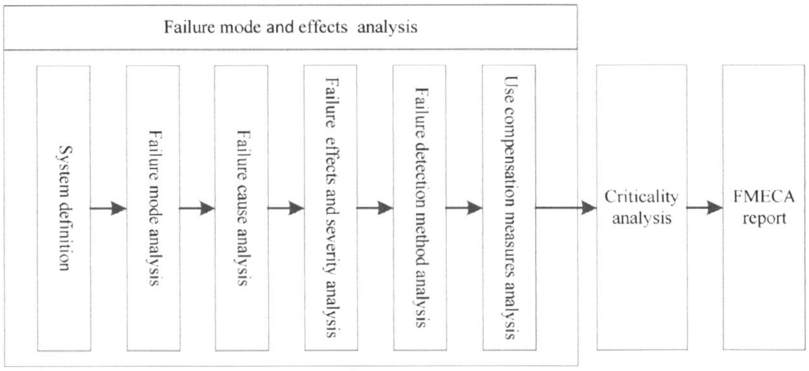


**Figure 1.** The FMECA process.

In engineering practice, the functional FMECA is often combined with the hardware FMECA. They are generally carried out by filling the FMECA form [4], and the commonly used FMECA table is shown in table 1.

**Table 1.** Function and hardware failure mode and effects analysis table.

The initial agreement level: Mission: Audit: Page:
The agreement level: Creater: Ratify: Creat date:

| Number | Function | Potential Failure Mode | Potential Cause(s) of Failure | Mission Phase | Potential Effect(s) of Failure | SEV | Current Controls Detection | Current Controls Prevention | Actions Taken |
|---|---|---|---|---|---|---|---|---|---|
| | | | | | | | | | |

## 3. Characteristics of missile equipment maintenance support

Missile equipment generally adopts three-level maintenance system, corresponding to three-level support. The first level maintenance is the grassroots level maintenance, and at this stage the field security personnel use outfield equipment to complete the routine testing and maintenance of missiles. The second level maintenance is the intermediate level maintenance, and at this stage the security personnel use support equipment to complete the missile's storage, packaging, transportation and preparation to be transmitted. The third level maintenance is the depot level maintenance, and at this stage the security personnel can complete a full range of missile maintenance in the repair center.

Missiles are long-term storage and single-use weapons. At storage or standby stage missiles belong to repairable system, but at launch or flight stage missiles belong to repair-free system. Therefore in the three-level maintenance system, the preventive maintenance work is completed during the first level maintenance and the second level maintenance; the corrective maintenance work is completed during the third level maintenance.

## 4. Missile equipment maintenance decision based on the FMECA technology

Maintenance decision-making process based on the technology of FMECA is as follows: the first step is to partition the system level; Then Failure Mode and Effect analysis is taken to determine the risk priority number; And products with or without important functions are divided according to their risk priority numbers; Then reliability-centered maintenance decision analysis is taken for products with important functions; Finally according to the severity of the failure consequences, maintenance task types and maintenance interval are determined to make up the maintenance work outline[5-7]. The flow chart is shown as figure 2.

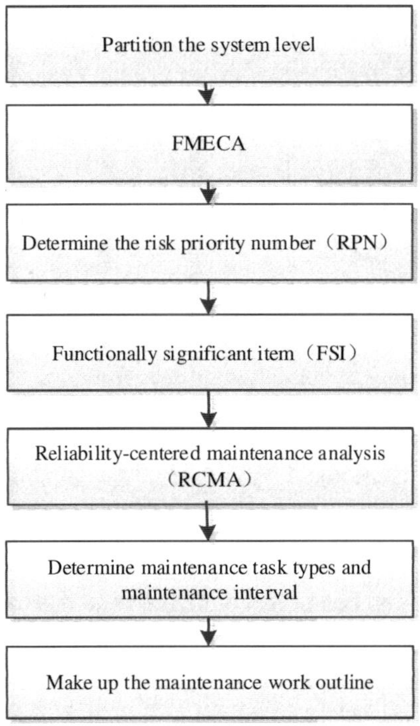


**Figure 2.** The maintenance decision-making process based on the technology of FMECA.

In this paper, with a certain kind of missile as research object, we choose the terrain matching function branch of the flight control stage as a case study. First conduct the functional analysis; then use functional tree to decompose the missile function forward and get the complex branch of each function. Figure 3 takes the main function of the missile as an example mapping functional tree, focusing on the research object "terrain matching function" branch with other parts omitted [8, 9].

The main task stage of the missile includes ground test stage, launch control stage and flight control stage. Terrain matching's main function is playing in the flight control stage, so below we select the flight control phase to analyze. The power-on self-test, data binding, and data loading in the figure 3 belong to the launch control stage, so they are left out in the following analysis. Analyzing the terrain matching branch we get the functional failure mode analysis table as is shown in table 2.

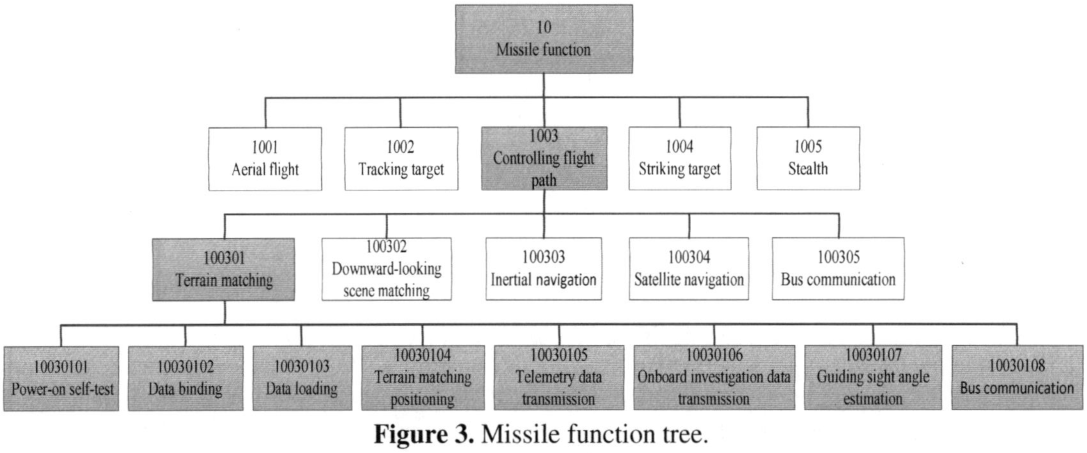


**Figure 3.** Missile function tree.

**Table 2.** The functional FMEA table of terrain matching(part)

The initial agreement level: Missile function Mission: ** Audit: ** Page: **
The agreement level: Terrain matching Creater: ** Ratify: ** Creat date: **

| Number | Function | Potential Failure Mode | Potential Effect(s) of Failure | SEV | Key Function | Actions Taken | Current Controls Detection |
|---|---|---|---|---|---|---|---|
| 10030104 | Terrain matching positioning | Terrain matching positioning failure | Terrain matching accuracy decrease | 8 | Yes | System software fault tolerance design | BIT |
| | | Terrain matching positioning accuracy decrease | Terrain matching accuracy decrease | 8 | Yes | System software fault tolerance design | BIT |
| | | Terrain matching positioning synchronous signal failure | Terrain matching positioning pulse output error | 8 | Yes | The optimization design of the hardware interface circuit | BIT |

Then we conduct the functional criticality analysis of the key function in the terrain matching function branch and get the terrain matching functional criticality analysis table as is shown in Table 3.

**Table 3.** The functional criticality analysis table of terrain matching (part).

The initial agreement level: Missile function Mission: ** Audit: ** Page: **
The agreement level: Terrain matching Creater: ** Ratify: ** Creat date: **

| Number | Function | Potential Failure Mode | SEV | Failure Mode Probability ( $*10^{-6}$ ) | Occur | RPN | Risk Level | Achieve fault detection requirements |
|---|---|---|---|---|---|---|---|---|
| 10030104 | Terrain matching positioning | Terrain matching positioning failure | 8 | 28.04 | 4 | 32 | R3 | Yes |
| | | Terrain matching positioning accuracy decrease | 8 | 15.10 | 4 | 32 | R3 | Yes |
| | | Terrain matching positioning synchronous signal failure | 8 | 8.71 | 3 | 24 | R3 | Yes |

FMECA analysis of missiles can help obtain functions with larger risk priority numbers, and the corresponding Functionally Significant item (FSI) can be determined, followed what the reliability centered maintenance theory is analyzed in detail. Based on 7 types of preventive maintenance policies (including maintenance, operation personnel monitoring, inspection, functional detection, timing overhaul, timing scrap, integrated work), and reasonable maintenance interval determined due to previous maintenance support data, a complete outline of maintenance work is made up finally.

## 5. Conclusion

In this paper we conduct the FMECA analysis for a certain type of missile system, and then determine the maintenance decision based on the FMECA analysis. We give the maintenance decision processes based on the technology of FMECA, which can not only improve the reliability of the missile system, also can enhance the scientific, effectiveness of maintenance work. This method improves the work efficiency and relevance of maintenance, avoiding excessive maintenance, which reduces the maintenance cost, with certain military, economic and applied value [10, 11].

## References

[1] Zeng S K, Zhao T D et al. 2001 *System Reliability Design and Analysis Tutorial.* Beijing: Beijing University of Aeronautics and Astronautics Press.
[2] Lu X Y, Zheng P Z 1995 *Reliability Analysis and Design* Beijing: National Defense Industry Press.
[3] Stamatis D H, Chen X T, Yao S H 2005 *Failure mode and effects analysis(FMEA) from theory to practice* National Defence Industry Press.
[4] Kang R, Shi R D 2006 *FMECA technology and its application* National Defence Industry Press, 129-130.
[5] Barkai J 1999 *Automatic generation of a diagnostic expert system from failure mode and effects analysis(FMEA) information* SAE Technical Paper Series.
[6] Meng J, Zhang W J 2013 *Research on Missile PHM Design based on FMECA* Prognostics and System Health Management Conference, Italy.
[7] Hawkins P G, Woollons D J 1998 *Artif. Intell. Eng.* **12(4)** 375-97.
[8] Garcia P A A, Schirru R A 2005 *Prog. Nucl. Energ.* **46(3)** 359-73.
[9] Bowles J B 1998 *The new SAE FMECA standard* Reliability and Maintainability Symposium, Proceedings. Annual. IEEE, 48-53.
[10] Zhang P, He J, Wang Y D et al. *Chinese J. Ship Res.* **7(4)** 6-10
[11] Zhang D, Zhang P, Fang Q et al. 2013 *Chinese J. Ship Res.* **8(1)** 102-06.

# Methodology and Results of Mathematical Modelling of Complex Technological Processes

**Nataliya V. Mokrova**[1]

[1]Moscow State University of Civil Engineering, Russia, Moscow, 129337, 26, Yaroslavskoye Shosse

**Abstract.** The methodology of system analysis allows us to draw a mathematical model of the complex technological process. The mathematical description of the plasma-chemical process was proposed. The importance the quenching rate and initial temperature decrease time was confirmed for producing the maximum amount of the target product. The results of numerical integration of the system of differential equations can be used to describe reagent concentrations, plasma jet rate and temperature in order to achieve optimal mode of hardening. Such models are applicable both for solving control problems and predicting future states of sophisticated technological systems.

## 1. Introduction

Today, any production process in any sector, be it engineering or construction or economic or social, could be considered as a complex process packed full of variables and connections. A scientific approach involves studying a wide class of related interconnected phenomena that obey certain laws, rather than a particular elementary process. By modelling, we mean a scientific cognition method aiming to analyze an object or a system of objects by building models and studying their essential properties identified in the mathematical description modelling process [1]. The modelling process, much less the practical application of mathematical and simulation models of complex systems, must be iterated numerous times, and at each stage its attributes are refined, modelling priorities are set and the ways to build newly built objects are rationalized. The mathematical description scheme obtained in the course of modelling together with the parameters identified as a result of interpolation, reflect the most essential features of the object or process and allow us to understand the mechanism of the phenomenon and make it possible to forecast changes in the phenomenon and its development trend. Different models, connections and relationships between individual components or subsystems, which are considered most relevant for this research, can correspond to the same process. They are written down using mathematical relations, which are checked for adequacy at the final stage of modelling, and, finally, the developed mathematical model is validated for accuracy and usability.

## 2. Methodology of complex technological processes modelling

Production processes in various industries have a number of features: they are mostly continuous and stochastic; chemical transformations and heat and mass transfer in technological processes depend both on the internal state of an object and external environment. The interaction of the object of research with the external environment is characterized by the loss of mass and energy and, therefore, by the process quality. To increase the efficiency of production, it is necessary to ensure optimal modes for individual processes and stability of external conditions.

Owing to the complexity of a specific technological process, for which a mathematical description is developed; the need for complex processing of raw materials; the need for rational use of energy and materials and recovery of side-line material and energy flows a systematic consideration of complex multi-stage processes characterized by homogeneous and heterogeneous stages in the chemical and related hi-tech industries will be necessary.

The system analysis is based on decomposition of a complex system into subsystems and establishment of quantitative relationships between such subsystems [2]. Subsystems (levels) are allocated not only depending on the complexity of the object under consideration, but also on the information about the object, previous study of process physical and chemical regularities, availability or possibility of obtaining a mathematical description of each stage [3]. If a subsystem with its I/O flows is considered independently, it is possible to estimate the potential of flows and subsystems and identify loss sources and ways to minimize such losses, and determine the reserve for increasing the efficiency of individual devices and the entire system.

The system analysis methodology applied in the development of a mathematical model of the generalized technological process is shown in Figure. 1. The research proposes a mathematical description of the plasma-chemical process and fine purification of cobalt solutions from iron and copper.

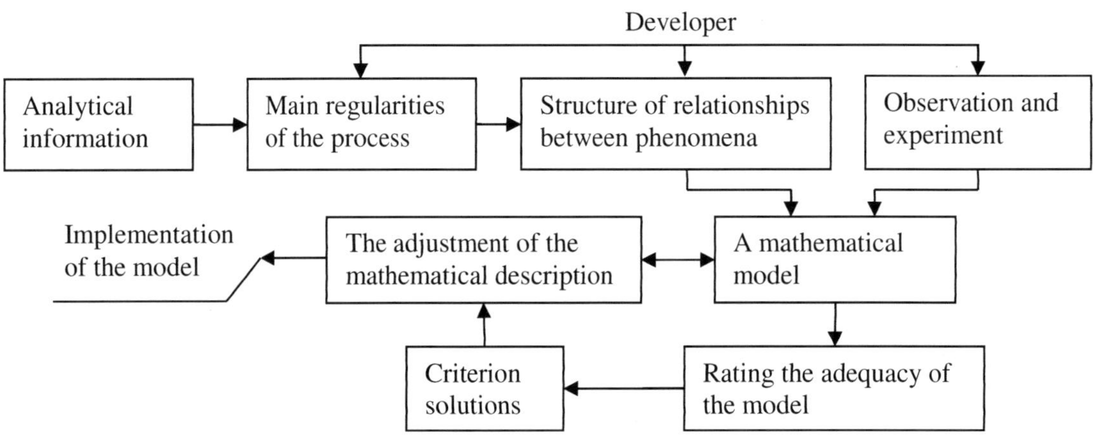


**Figure 1.** System analysis methodology applied in the development of a mathematical model.

The use of low temperature plasma in chemical engineering makes it possible to run processes at high temperatures and direct exposure to electric and magnetic fields and accomplish transformations that do not occur at all or are extremely time consuming (hundreds or thousands of times) under normal conditions [4-7].

## 3. Results of mathematical modelling of plasma-chemical processes

The analysis allows us to draw a conclusion about the prospects of complex processing of natural gas and other hydrocarbons in the plasma to produce coherent nitrogen, acetylene and hydrogen. The process is carried out at $2000^0$C and atmospheric pressure; methane pyrolysis is carried out in nitrogen plasma to produce simultaneously up to 10,5 % of hydrogen cyanide and up to 13,5 % of acetylene. It should be noted that methane pyrolysis carried out in a jet of hydrogen plasma at $4000^0$C can produce an acetylene concentration in the reaction products of up to 15,5 % by volume, with a total degree of methane conversion reaching 85 % and the 'methane-to-acetylene' conversion reaching 71.5 % at a total energy consumption of 9,5-10 kWh per 1 $m^3$ of acetylene.

A plasma-chemical plant is a sequence of a plasmatron, plasma-chemical reactor and a quenching device. A complete mathematical description of such plants is very sophisticated and requires knowledge of all the elementary processes occurring in the system at all levels. The plasmatron mathematical model can be represented in the form of a system of partial differential equations comprising the following equations: material balance by components, mass continuity, pulse and quantity of motion conservation, plasmatron arc dynamics at minor arc deviations from its equilibrium state, and equation of state [8].

In practical modelling, a system of assumptions is applied, taking into account the most essential processes occurring in the system. The equations describing electromagnetic phenomena are not considered when describing the reaction node. Plasma-chemical process kinetics equations should be written down for all reacting components, and, at the same time, a simplified scheme of the mechanism of thermal decomposition of methane with the formation of acetylene at high temperatures should be developed and explained [8]. Write equations of reactions:

$$\begin{aligned} &2 \cdot CH_4 \rightarrow C_2H_4 + 2 \cdot H_2 \\ &C_2H_4 \rightarrow C_2H_2 + H_2 \\ &C_2H_2 \rightarrow 2 \cdot C + H_2 \end{aligned} \tag{1}$$

Hydrodynamic phenomena play an important role in the description of acetylene formation processes, because the processes run in a flow reactor at high flow rate. It should be noted that the laws of an ideal gas are applicable to thermal plasma even at pressures on the order of 1000 atmospheres, because the density of plasma-forming particles is very small due to high temperatures.

The system of differential equations describing plasma-chemical processes is stiff. High temperatures, low reaction rates, exponential dependencies for the calculation of reaction rate constants (for example, the methane residence time in the reaction zone corresponding to the maximum yield of acetylene which equals to $7{,}21 \cdot 10^{-4}$ seconds) complicate the search for a system solution by known numerical methods, and impose high standards on setting system parameters, refining ranges of coordinate variation; additionally, any additional experiment with the object of modelling is associated with certain technical and production difficulties.

It is important to note that the production of acetylene from hydrocarbons is a process where the final product is intermediate in the chain of reactions, therefore it is necessary to interrupt the reaction at the desired stage by quenching. It has been established experimentally that in order to produce the maximum amount of the target product we should focus on the quenching rate and initial temperature decrease time; it has been established that a $2 \cdot 10^{-3}$-sec quench delay leads to a 15,5 .. 10 % decrease in the acetylene concentration.

The importance of maintaining the optimum temperature profile throughout the entire technological process can be explained by the following: acetylene polymerizes vigorously at temperatures below $600^0$C to form solid and liquid substances; both decomposition and polymerization occur at $600\text{-}1000^0$C; a slow decrease in the gas mixture temperature leads to the decomposition of acetylene into hydrogen and soot, especially at $1200\text{-}1300^0$C. So, to quench plasma-chemical reaction products and shift the system to the temperature range on the order of $2000^0$C to make it stable for acetylene, a higher rate of mixture cooling will be necessary. The experimental studies and modelling results show that both intensive mixing of flows and dramatic uniform decrease in the temperature of reaction products can significantly increase the yield of the desired product.

The results of numerical integration of the system of differential equations obtained in the course of mathematical modelling of the plasma-chemical process can be used to describe what reagent concentrations, plasma jet rate and temperature should be opted for to achieve optimal quenching of reaction products. The analysis of changes in the concentration of reaction mixture components has revealed that the concentration of the initial substance - methane - decreases in time monotonously in accordance with the law of kinetics of simple reactions. Ethylene and acetylene concentration curves are described by an extremum dependence. The position of points of the maximum value of product concentration corresponds to the sequence of their formation. The temperature curve has an extremal form with a selected minimum point, which is reached in the interval of $10^{-5}\text{-}10^{-3}$ seconds and almost coincides with the instant of formation of the maximum acetylene concentration; and reaching a minimum, the temperature begins to increase dramatically to make acetylene decompose at a growing rate. Forced quenching of the reaction mixture at the optimum time allows to prevent decomposition of the target product and ensure a sufficiently high yield of acetylene. Modelling of plasma-chemical plants provides an opportunity for building systems that are able to help control the behavior of similar objects in an optimal way.

## 4. Mathematical modelling of purification processes

The system analysis methodology has allowed to solve the problem of modelling a multistage chemical and technological process of cobalt production [9, 10]. Cobalt solutions are purified in a cascade of 3 continuous reactors; primary products pass through the devices to react with caustic ash (main agent) with subsequent precipitation of iron and copper impurities and cobalt decline. The purification mathematical model is developed using the building-block concept.

The problem of optimal control over the purification process is formulated with a strict restriction with regard to the concentration of impurities. According the research chosen, methods for identifying a mathematical model of a cascade of reactors at the fine purification stage were proposed and a successive approximation method was adopted to identify the characteristics of the subsystems. When building a mathematical model of dynamics of the fine purification process, we used experimental data and coefficients obtained in the analysis of the process statics. The system of differential equations that characterize the dynamics of fine purification (decontamination) of cobalt solutions is described by the quartic Runge-Kutta method at a constant step. The analysis of dynamic processes in the cascade of reactors allows us to conclude that it is important to solve control tasks by means of the last reactor in the cascade as this helps control the set acidity value. Such models are applicable for solving control problems and predicting future states of systems.

## 5. Conclusion

Thus, in accordance with the system analysis methodology, the levels of the hierarchical structure of the system under consideration are distinguished for each of the technological processes considered in the research; we examined their interrelations, determined the aspects of their mutual influence, elucidated physical and chemical phenomena in each device and took into account characteristics of main state variables and controlling and disturbing influences. We put forward a mathematical description of the stages, which helped us find more rational ways of running the process and solve optimization problems. Such models are applicable both for solving control problems and predicting future states of sophisticated technological systems.

As already noted, the system analysis is based on decomposition of a complex system into subsystems and establishment of quantitative relationships between such subsystems. Subsystems (levels) are determined not only depending on the complexity of the object under consideration, but also on the information about the object, previous study of process physical and chemical regularities, availability or possibility of obtaining a mathematical description of each stage. If a subsystem with its I/O flows is considered independently, it is possible to estimate the potential of flows and subsystems and identify loss sources and ways to minimize such losses, and determine the potential for increasing the efficiency of individual devices and the entire system. A quantitative system analysis of complex chemical processes employing state-of-the-art computing facilities and mathematical modelling is necessary both for the development of new processes and intensification of existing ones.

## References

[1] Lasdon L S 2011 *Optimization Theory for Large Systems* (New York: Dover Publications).
[2] Peregudov F I and Tarasenko F P 1989 *Introduction to System Analysis* (Moscow: High School).
[3] Mesarovic M D and Takahara Y 1975 *General Systems Theory: Mathematical Foundations* (Amsterdam: Elsevier Science).
[4] Polak L S 1975 *Theoretical and Applied Plasma Chemistry* (Moscow: Nauka).
[5] Polak L S and Lebedev Yu A 1991 *Low-temperature plasma (Chemistry of Plasma* vol 3*)* (Novosibirsk: Nauka).

[6] Polak L S and Lebedev Yu A 1998 *Plasma Chemistry* (London: Cambridge International Science Publishing).
[7] Hanz S N, Melnik A P and Parhomenko V D 1969 *Plasma in Chemical Technology* (Kiev: Tehnika).
[8] Artamonov A G, Volodin V M and Avdeev V G 1989 *Mathematical Modelling and Optimization of Plasma-chemical Processes* (Moscow: Himia).
[9] Mokrova N V and Volodin V M 2007 *Chem. Oil Gas Eng.* **2** 17–9.
[10] Mokrova N V and Volodin V M 2006 *Bull. TSTU* **12** 22–8.

IOP Conf. Series: Materials Science and Engineering **317** (2018) 012039 doi:10.1088/1757-899X/317/1/012039

# Features and perspectives of automatized construction crane-manipulators

**Mikhail A. Stepanov[1], Peter A. Ilukhin[2]**

[1]PhD. of Tech. Sci., Assistant Professor, Moscow State University of Civil Engineering
[2]Postgraduate student, Moscow State University of Civil Engineering

**Abstract.** Modern construction industry still has a high percentage of manual labor, and the greatest prospects of improving the construction process are lying in the field of automatization. In this article automatized construction manipulator-cranes are being studied in order to achieve the most rational design scheme. This is done through formulating a list of general conditions necessary for such cranes and a set of specialized kinematical conditions. A variety of kinematical schemes is evaluated via these conditions, and some are taken for further dynamical analisys. The comparative dynamical analisys of taken schemes was made and the most rational scheme was defined. Therefore a basis for a more complex and practical research of manipulator-cranes design is given and ways to implement them on practical level can now be calculated properly. Also, the perspectives of implementation of automated control systems and informational networks on construction sites in order to boost the quality of construction works, safety of labour and ecological safety are shown.

## 1. Introduction

Modern construction industry is highly mechanized, however, it is still has a high percentage of manual labor, due to the limited possibilities for the further mechanization in many kinds of construction works.

The greatest prospects of increasing the speed of construction and raising the quality of work are lying in the field of automatization and robotization of construction. Such trends can be seen for a while now [1]. Significant progress has been made in recent years in computing, electric drives, control systems, batteries and other, important for robotics, fields of science and technology. Reducing of the costs and expansion of the possibilities of robotic systems allows to consider the possibility of their practical application and implementation in the construction process. In addition to improving the speed and quality of work, which is a direct challenge for robots, other positive effects of their implementation are even more important, related primarily to the increasing of safety of human labor.

Automatized construction manipulator-cranes, or ACM-cranes, are seem to be the most perspective way to improve construction processes [2, 3]. Design of modern cranes haven't changed for decades, and this design has a number of problems – high percentage of human labour, insufficient positioning quality, poor observing conditions for operator and a number of human safety issues, linked with these problems. But ACM-cranes can solve this problems, reducing the amount of human labour, increasing the quality of work and it's safety, including ecological safety through better optimization of work and lower emission levels [4].

Also, ACM-cranes are fitting well with automated control systems (ACS), which, in turn, can be united into single network, improving safety and quality of construction processes even further.

At the same time it is clear that the use of ACM-cranes has some features associated with specific requirements for robots [5, 6]. And the most important problem, that has to be solved before their implementation is their rational design scheme. The objective of the current study was to find out and validate some specific construction scheme for ACM-cranes.

## 2. Materials and methods

In order to formulate an ACM-crane design features, several steps of research were made:

First of all, there are some general features, essential for ACM-cranes. These features are clearly seen from analysis of working processes and construction site conditions [7], but they had to be formulated in order to be used in further work.

Secondly, the most rational kinematic scheme had to be found. With the set of already formulated general conditions, with an addition of some specialized conditions for kinematic analysis, certain general kinematic schemes were taken for a deeper analysis.

The analysis consisted of two parts. In the first part, a set of different kinematical schemes was evaluated with Lee-Young theorem [8]. This theorem states [9]:

$$K = \frac{V}{\left(\sum_{i=1}^{m} L_i\right)^3} \tag{1}$$

where K – is the kinematic quality of construction, bigger number means that the construction is better optimized; V – is the volume of the working area of the robot; $L_i$ – is the length of i-th link of the robot; m – is the amount of links.

The second part of the analysis used the kinematical schemes with the most kinematic quality for relative dynamical analysis. All studied schemes were remade as a simplified 3d models, with certain conditions:

- Equality of working zone volumes
- Equality of overall model mass
- Equality of payload capacities, speeds and velocities

The characteristics were taken equal to those of tower crane KB-403B. Loading conditions for models with horizontal boom were simultaneous lifting of payload and turning of the crane around its vertical axis. For models with jointed boom – simultaneous extension of the boom and turning of the crane around its vertical axis. The time length of modeling was 60 seconds.

The analysis itself was performed in Dassault Systemes Simulia Abaqus CAE 2016. Four types of parameters were taken:

- Stress level, measured by Mises
- Maximum vibration amplitudes
- Maximum linear velocity amplitudes
- Maximum angular velocity amplitudes

This parameter was used to create six sets of data for crucial points of construction. The gathered data was then used to build an approximation graphs in MS Excel. The approximation functions were found via MS Excel, with the following limitations:

- Equation degree should be less or equal to 3
- Approximation coefficient $R^2$ should be bigger than 0,9

This functions were used to achieve average comparative values $C^{av}$ for each analyzed type of construction. These values were an arithmetic mean (2) of two values – the maximum of approximation function $F^{max}$ and the definite integral for t=60 seconds of approximation function $I^t$.

$$C^{av} = \frac{F^{max} + I^t}{2} \tag{2}$$

These two values were calculated (3,4) as arithmetic mean of six values for maximum and integral calculations respectively, turned in comparative scale, where the biggest result is taken as equal to 1, and all other results are taken proportionally.

$$F^{max} = \frac{\sum_{k=1}^{6} F_k^{max}}{F_{max}^{max} \times 6} \quad (3)$$

$$I^{t} = \frac{\sum_{k=1}^{6} I_k^{t}}{I_{max}^{t} \times 6} \quad (4)$$

And through these average comparative values the most rational scheme was found.

## 3. Results

General features, which are essential for ACM-cranes:

- High load capacity (more than 3 tons).
- Large volume of the working area (more than 10 meters in each dimension).
- Higher speed then of modern crane equipment, through the use of a rigid hand-capture instead of the cable suspension.
- The relatively high positioning accuracy, providing a very high quality of the geometry of structure being built, bringing the basic geometry errors to the errors of the materials used.
- Increased demands on the reliability and safety, because robotic manipulators, while reducing manual labor within the construction process, does not exclude it entirely.

With these general conditions formulated, several kinematic conditions were formulated:

- Kinematically rigid construction of crane
- The kinematic scheme should have the least possible amount of the simplest kinematical pairs.
- The kinematic scheme should have the least possible kinematical links.
- The kinematical scheme should allow the crane to work on all heights of the mounting horizon
- The output link should have the least possible mass

This set of conditions allows us to figure out the construction of ACM-crane, which would be rationally simple, but able to perform all the necessary tasks without unwelcomed limitations.

As a result of applying these conditions to a whole variety of different kinematical schemes, and their evaluation through the Lee-Young theorem, four different schemes were taken for further analysis, as shown on figure 1:

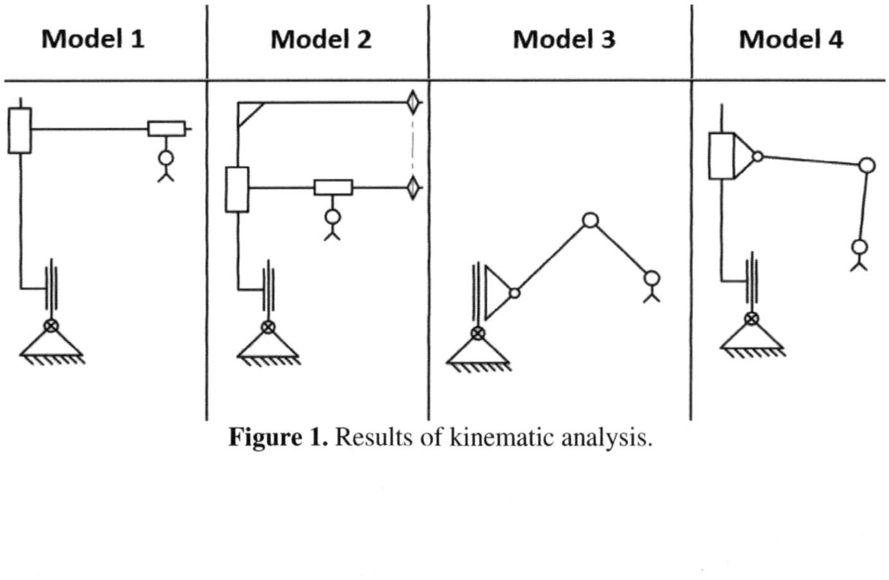


**Figure 1.** Results of kinematic analysis.

The results of dynamical analysis data processing via the methods shown above can be seen on figure 2:

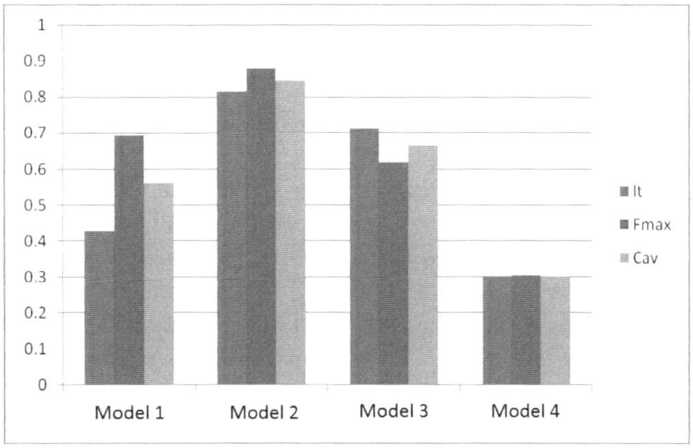


**Figure 2.** Results of dynamic analysis.

Smaller numbers mean smaller stresses and loads, and therefore it is clearly seen, that the model №4 (shown on figure 3) has the best results, and is the most rational for ACM-cranes.

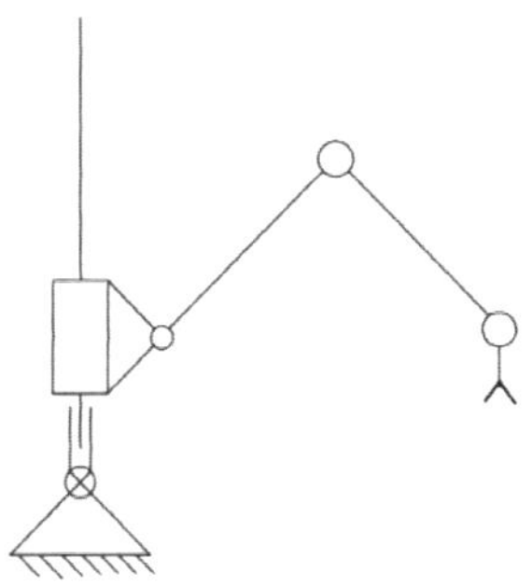

**Figure 3.** The principle scheme of ACM-crane.

## 4. Discussion

As the most rational design scheme for ACM-cranes was found, it is now possible to calculate more particular features of these machines in order to see their technical, economical, ecological and other properties. With such complete analysis, it would be possible to create and design such machines in practice and implement them on the construction site properly.

But also, these machines can become a perfect platform for first wide spread ACS, which would allow further improvement of construction process and conditions of work.

ACS of construction machines has a number of unique features. Firstly, it has rather low requirements for independence because of the large number of characteristic objects that are suitable for use as markers to help orient the robot. Secondly, the ACS should provide highly accurate and fast control of the robot drive. Third, because of the large dimensions of the building elements, ACS must be able to quickly and accurately calculate the overall dimensions, mounting dimensions and location of the mounting fixture on the building structure itself.

In addition, widespread usage of ACS opens the possibility of creating a full-fledged information network within the construction site. This network will allow improving the interaction between machines and workers, and the level of control over the safety measures and quality of work done.

However, ACS - is only the first step towards a full robotization of construction process. In perspective, the creation of a highly specialized form of artificial intelligence for the control of machines on a construction site seems most likely.

The main difficulty of creating classic automated control systems for different machines and mechanisms at the site lies in the fact that, unlike the laboratory and even the factory floor, construction site - is the environment with unpredictably changing conditions. Here is the human factor, and weather conditions and constantly changing work places and diversity of types of work. Standard ACS can hardly bear such a variety of tasks.

But robots equipped with Artificial Intelligence are capable of more. Picture and sound recognizers, the ability to orient in an ever-changing space will allow the robots to work without danger for themselves and others working at the construction site. AI can operate with non-specialized instruments with universal grips, it can calculate how to properly and securely use them in the absence of hard-coded application program. AI can safely and efficiently interact with humans directly, which is very important in the conditions of the construction site. Creating of self-learning program module for such an agent would be relatively easier than writing a rigidly programmed ACS [10].

Over time, the AI self-learning module, in theory, will allow robots to perform their work more and more efficiently with the accumulation of information and education. Creation of the network between agents will allow them to co-educate, speeding up the process of self-improving. This self-learning can lead to the appearance of heuristic properties of the AI, which will allow him to look for new, non-trivial solutions of tasks [11, 12].

## 5. Conclusions

The results of this research are giving a basis for a full scale design of ACM-cranes. This will require more specific technical analysis, especially in its drive systems, and an overall economical observation of project, but now these studies can be made on a rationalized basis. Also, more research on the creating of the ACS for such cranes is needed, for broadening the possibilities and boosting the characteristics of such machines – and construction machines in general.

Practical implementation of such machines can greatly improve quality and especially safety of construction process, as well as it's ecological aspect through better optimization of different construction works and lowering of emission levels. And now the conditions of this implementation can be calculated properly.

In wider perspective, progress in the field of AI research will lead to a further improvement of construction process via self-educating and creation of organized machine network on the construction site.

## References

[1] Romashko A M 2014 *Lift Transport. Eng.* **3-4** 24-9.
[2] Vil'man U A and Stepanov M A 2006 *Mech. Constr.* **1** 3-8.
[3] Stepanov M A and Hartenstein K 2012 *Lift Transport. Eng.* **4** 2-5.
[4] Vil'man U A 1989 *Basics of robotization in construction* High School.
[5] Bulgakov A G et al. 2006 *Automatization of construction* Novochercassk: URGTU
[6] Bulgakov A G, Vorobiev A V, Evtushenko S I and Parshin D Ya 2013 *Automatization and robotization of construction industry: Textbook – 2-d ed.* Moscow: RIOR: INFRA-M.
[7] Stepanov M A and Ilukhin P A 2015 *Mech. Constr.* **10** 9-12.
[8] Stepanov M A and Ilukhin P A 2016 *Mech. Constr.* **11** 54-7.
[9] Shahinpoor M 1987 *A Robot Engineering Textbook* Harper&Row, Publishers, Inc.
[10] Stepanov M A and Ilukhin P A 2016 *Mech. Constr.* **11** 58-60.
[11] Bostrom N and Yudkowsky E 2011 *The Ethics of Artificial Intelligence* Cambridge University Press.
[12] Russell S and Norvig P 2009 *Artificial Intelligence: A Modern Approach* Pearson.

IOP Conf. Series: Materials Science and Engineering **317** (2018) 012040 doi:10.1088/1757-899X/317/1/012040

# Issues of using Longer Heavier Vehicles on Roads

**R Matuszkova[1], M Heczko[1], J Cepil[1] and M Radimsky[1]**

[1]Brno University of Technology, Faculty of Civil Engineering, Veveri 331/95, 602 00 Brno, Czech Republic

**Abstract.** Many logistics companies aim to save on freight costs. Recently, not only on Czech roads and on motorways, longer and heavier vehicles that exceed dimensions' limits appeared. For these vehicles, it is necessary to apply for a special permit, which is, however, much more liberal than the permit for oversized and overweight load transport. This paper informs about checking routes of these vehicles by swept path analysis and finding locations on roads that can generate both safety risks and traffic fluency problems.

## 1. Introduction

Companies are trying to save their costs when transporting freight, which is the reason why we have more and more frequent encounters with vehicles exceeding weight and dimension maximum allowed limits on the motorway and road network. Thanks to these vehicles, road and motorway pavements are being damaged faster and the necessary repairs of the roads are required more frequently. "The crossing of one axle of 10 t has approximately the same effect as 125 axle crossings of 3 tons" [1]. The damages done by overloaded trucks are mainly ruts or longitudinal depressions. According to the Act 13/1997 Coll. on Roads as amended is a fine for exceeding the permitted values during the weighing control from CZK 15,000 to CZK 500,000 [2].

The possibility to increase the statutory limits is an oversized and overweight load transport vehicle in accordance with Act No. 341/2002. However, obtaining a one-off permit is a relatively demanding process and has a number of conditions (such as transport escorts) that have to be met to pass through. That is why companies have recently begun to use an intermediate level between standard freight transport and oversized load transport. Using vehicles exceeding the permitted dimensions, but all other parameters are in accordance with the legal conditions. There is no uniform name for heavy long cargo vehicles; they are called Road Train, LCV, LHV (Longer Heavier Vehicle), Gigaliner, EuroCombi or Megatruck, but there are many others. In the Czech Republic the maximum allowed length of the road vehicle is 22 m, LHV usually has 25.25 m [3]

For these vehicles, the weight of the load is distributed between more axles, and therefore the axle pressure on the road is much smaller. Using LHV will due to that save not only the costs of the number of used vehicles and drivers for transport, but also the costs of repairing damaged roads. With the LHV, every third ride is saved comparing to a standard heavy truck, but it is worth mentioning that it will increase the investment in the trailer and suspension equipment and will increase the fuel consumption by 15 to 20%, but the transport costs are still lower than using ordinary heavy truck [3].

When using LHVs, however, there is a problem with their dimensions, as most junctions on the motorway and road network are not designed to enable passing such long vehicles. This paper deals with the issue of passing LHVs through selected junctions using swept path analysis.

## 2. Formatting the title, authors and affiliations

The use of LHVs is common in the US, Canada, or Australia, where lengths of such LHVs are up to 50 m and weight up to 100 tons. In most European countries, however, they are forbidden or allowed only as exception with necessary permit. LHVs are permitted in Scandinavia and the Netherlands, their permitting is considered in Belgium [3].

The Ministry of Transport grants permissions for such vehicles longer than allows the Decree on Technical Approval and on Road Vehicle Technical Conditions No. 341/2014 Coll. Operators of these vehicles must submit application. Permitting conditions include the static test, i.e. the inspection of individual vehicle parts with a focus on the coupling. The maximum weight must not exceed 48 tonnes. The LHV may only use directionally divided roads and the arrival at the point of loading or unloading must be within 10 km of the motorway and the route must not intersect with the railroad at-grade.

After fulfilment of all requirement and payment of administrative fee of 6,000 CZK, the license to operate LHV is granted for a maximum of 3 months. [4] By September 2017, the 46 LHVs were allowed to operate.

Many initiatives are against the inclusion of extra-long trucks into normal traffic, including railway organizations. The most common reason is that the LHVs are non-ecological because their permission would bring an even greater freight transport from trains and boats to roads. Traffic safety may also be a problem, for example overtaking LHvs may be a problem or driving in urban areas. The question is also the unsatisfactory infrastructure and the high cost of its rebuilding, like the construction of parking lots with different parameters or the reinforcement of the bridges [3].

## 3. Analysis of critical location

During the task, critical locations were identified on the routes of these vehicles. Swept path analysis was done by using the AutoTURN software created by Transoft Solution. Swept paths are enveloped by curves that result from the vehicle's external contours and axles' position. Their particular shape, depends on the position and layout of the axles, on the position of the coupling points, on the type of trailer and, to a great extent, on the assumption of behaviour of the drivers.

### *3.1. Vehicle used for simulation*

One of the companies producing such vehicles is Krone. Company Hesti, the importer of Krone vehicles to the Czech Republic, provided detailed drawings of their LHV for this paper.

The vehicle assembly consists of a 10.17 meter truck with a 13.65 meter long trailer coupled by a drawbar, a clear gap between cargo spaces is 1.10 meters. The width of the vehicle without mirrors is 2.55 meters with a total height of 4 meters. The vehicle is 24.92 meters long when connected. The internal radius of rotation is 5.3 m and the outer radius of rotation is 12.5 m.

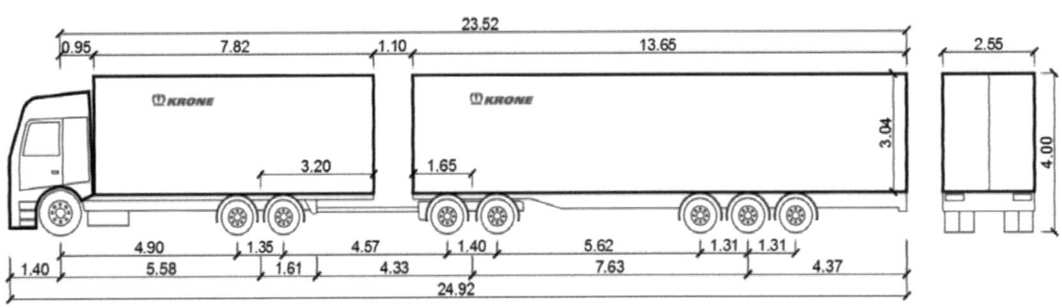


**Figure 1.** KRONE LHV dimensions.

### *3.2. Critical locations*

Approximately 40 critical locations were selected, located on our well-known LHVs' routes, and located also on the important nodes of the Czech Republic's road network, such as arrivals to production halls or airports.

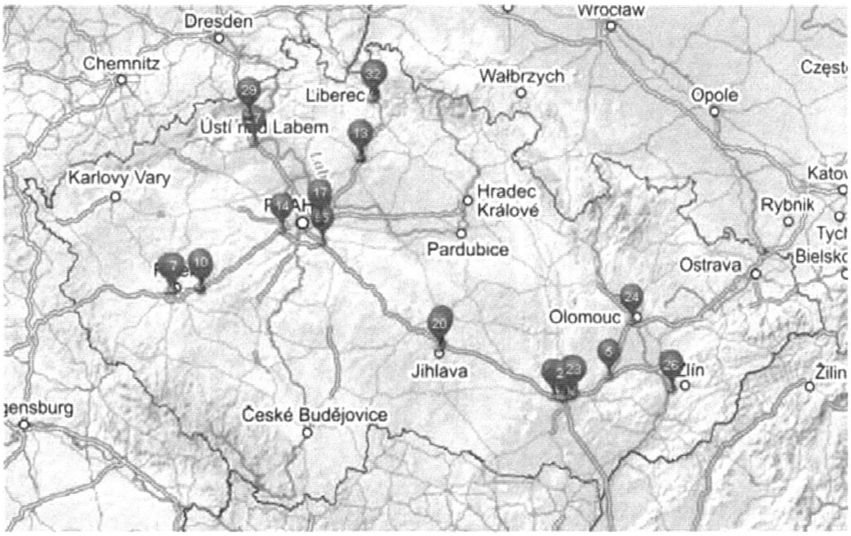


**Figure 2.** Selected interchanges and intersections.

### *3.3. Inputs of simulation*

At junctions, the emphasis was on the requirements of the Police of the Czech Republic, which in most cases do not allow the overhanging of the swept paths to the opposite traffic lane or the contour of the body outside the pavement. Overlaps are only allowed in the case of oversized or excessive loads under a special permit. However, these types of vehicles are not considered.

In addition, the rounding of the pavement at intersections, the widths of traffic lanes in the curves, the width of the pavement at the point of directional islands were examined. Width of the lanes, ring's width, ramps' width and roundabout junctions.

Vehicle characteristics in the software were set according to the production drawings provided by the vehicle manufacturer, and a speed of 15 $kmh^{-1}$ was set as the driving speed for the LHV, without the possibility of turning the wheels at zero speed. 20 different simulations were performed at each junction for each movement, and the individual differences in the problem spots were then averaged.

### *3.4. Analysis evaluation*

Of all the examined junctions, the following examples appeared to be interesting, demonstrating problematic locations at junctions. The first example is the intersection of the Svedske valy street and Vlastimila Pecha street in Brno city, which is located in a technology park that grows every year with new companies. If the LHV's effort is to stay in its traffic lane and not to interfere with other traffic lanes, there is a problem with the right turning, where the vehicle interferes the pedestrian lane, exactly on the pedestrian crossing. The overhang of the pavement is up to 1.5 m. If the LHV tries to avoid walkways, which is more likely solution, it will have to use the space of other traffic lanes, which will increase the risk of traffic accidents.

A similar example is another roundabout junction at the intersection of road II/183 and B. Nemcova street in Rokycany city, which is located near the industrial area and transfer the traffic from the D5 motorway to the area. There is a similar problem as in the previous case, the LHV is forced to drive over the sidewalk to miss the dividing island at the pedestrian crossing.

**Figure 3.** Intersection of Svedske valy street and Vlastimila Pechy street in Brno city.

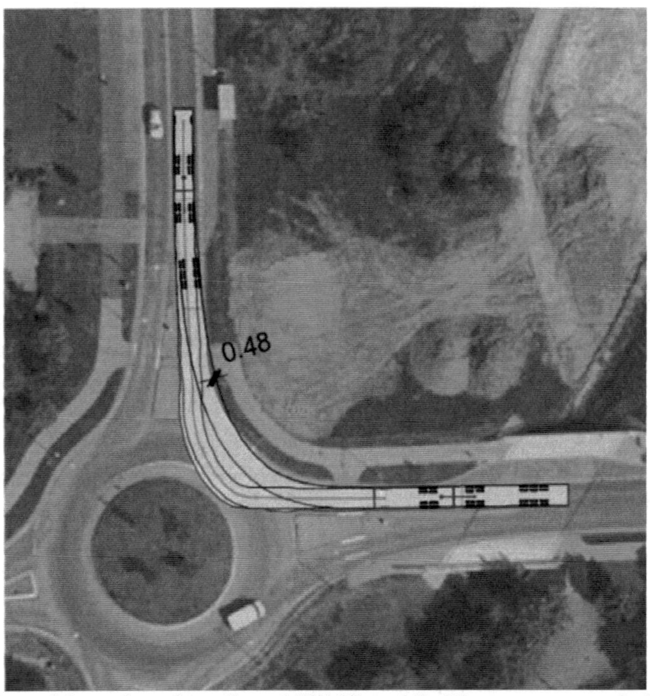


**Figure 4.** Roundabout junction of road II/183 and B. Nemcova street in Rokycany city.

Another example is a set of intersections near the Skoda Auto complex in Mlada Boleslav city. There is a roundabout junction and a turbo roundabout junction. At an interchange junction, designed according to CSN standards, there was no problem with the passing of LHV. At the roundabout junction, which has two traffic lanes, there was a significant interfering of LHV from one traffic lane to another, up to 3 meters. However, this type of roundabout junction with two traffic lanes is not suggested to use in the Czech Republic, as drivers do not use the potential of the second lane. Therefore, interference with the minimally used second lane does not mean a very high safety risk due to the minimal traffic intensity on the second traffic lane.

**Figure 5.** Roundabout junction of road I/38 and Škoda Auto's grounds road in Mlada Boleslav city.

However, the turbo roundabout intersections can be a major safety risk, where the inner lane is used to a by much higher percentage of vehicles. Such a junction is in Figure 6, where the LHV overlaps to the adjacent traffic lane by width of up to 3 m.

**Figure 6.** Roundabout junction of road II/430 and road I/50 entry ramp.

## 4. Conclusion

In the case of interchange junction, it is possible for LHV to pass through both directional and non-directional ramps without any problems, but the problems arise at intersections. At these places, when turning right the rear part of the LHV gets over the edge of the pavement, even when comfortable radius of corner rounding is used and vehicle is not entering opposite traffic lane. Overlap of the vehicle at selected intersections was up to 1 meter. A slight improvement occurred where compound curve is used to round the edge of pavement, but even in that cases the outline of the vehicle sometimes exceeded the pavement. The possibility is to allow LHVs to enter the opposite traffic lane, which, however, generates safety risk and worsen the traffic flow of the junction during not-restricted traffic.

The LHVs passing is also not being easy due to dividing and directional islands. Which, for normal traffic, regulate traffic and improve the junction clarity. Islands also prevent oversized vehicles to adjust the trajectory, so that the rear part of the vehicle can stay on the pavement or in case of left turn do not hit the traffic island itself.

Another problem occurred at roundabouts especially on the spiral (turbo) roundabout arrangement where the LHV overlap to the next traffic lane, which again causes a safety risk and slows the traffic flow of the junction.

The solution to these problems is to change the design elements of the junctions, like changing the approach to directional islands design, roundabout intersections and lane widths. However, this change would subsequently affect the driving habits of other drivers, who could then pass the junctions at greater speeds while rising traffic safety risks.

## References

[1] Zajicek J 2014 *Road Construction Technology* (Prague: CKAIT) ISBN 978-80-87438-59-6.
[2] Law n. 13/1997 Sb., *The Road Act*. In: Collection of Laws Czech Republic. Breclav: Moraviapress. ISSN 1211-1244

IOP Conf. Series: Materials Science and Engineering **317** (2018) 012040 doi:10.1088/1757-899X/317/1/012040

[3] *Advantages and risks of kits Road Train* (Prague: Logistics) 2013. https://logistika.ihned.cz/c1-61410210-vyhody-i-rizika-souprav-road-train.
[4] Law n. 634/2004 Sb., *Law on Administrative Fees*. In: Collection of Laws Czech Republic. Breclav: Moraviapress. ISSN 1211-1244.

**Acknowledgements**
This paper has been worked out under the project No. FAST-J-17-4767 Examination of the possibilities of moving oversized vehicles on Czech routes and project No. LO1408 AdMaS UP - Advanced Materials, Structures and Technologies, supported by Ministry of Education, Youth and Sports under the National Sustainability Programme I.

IOP Conf. Series: Materials Science and Engineering **317** (2018) 012041 doi:10.1088/1757-899X/317/1/012041

# A Doubly-Curved Piezoelectric Composite with 1–3 Connectivity for Underwater Transducer Applications

**Yanjun Zhang[1, 3], Likun Wang[1*], Lei Qin[1, 2, 3], Qingwei Liao[1], Chao Zhong[4]**

[1]Beijing Key Laboratory for Sensor, Beijing Information Science & Technology University, Beijing, China
[2]Key Laboratory of Modern Measurement & Control Technology, Ministry of Education, Beijing Information Science and Technology University, Beijing 100192, China
[3]Beijing Key Laboratory for Optoelectronic Measurement Technology, Beijing Information Science & Technology University, Beijing, 100192, China
[4]School of Electronic Engineering, Beijing University of Posts and Telecommunications, Beijing, China

**Abstract.** Aim to increase the horizontal and vertical beam width of the high frequency transducer simultaneously, we present a doubly-curved 1-3 piezoelectric composite element. It consists of 54% piezoelectric ceramic volume fraction and two phases polymer matrix. The finite element analysis (FEA) is used to evaluate the dynamic response of composite. Electroacoustic response in water was measured for the doubly-curved composite being considered as underwater transducer. An underwater transducer was fabricated using the doubly-curved 1-3 piezoelectric composite element. The -3 dB full angle beam width of transducer is approximately 106° and 36° in the horizontal and vertical plane respectively. Both the FEA simulations and experimental results show the potential of a broad covered area of the composite transducer in underwater environment.

## 1. Introduction

Piezoelectric materials have been extensively applied in transducers [1, 2] and smart structure [3, 4]. Due to its excellent piezoelectric properties and easier to fabricate than single crystals, piezoelectric ceramics are currently commonly used in underwater acoustic transducers [5-7]. However, a drawback of piezoelectric ceramics is the higher acoustic impedance than water [8]. Also, being brittle and stiff, piezoelectric ceramics are difficult to conform to doubly-curved surface. They also have many hybrid resonance modes for monolithic piezoelectric ceramic. Which affect the electromechanical properties of transducers. In order to alleviate those problems, by combining a piezoelectric ceramic and a polymer matrix with the concept of "connectivity" developed by Newnham [9], an array of piezocomposites has been designed and fabricated. The piezoelectric ceramics exhibited excellent piezoelectric properties, the softer polymer phase can reduce the acoustic impedance and allows the composites to more flexible, so it can conform to doubly-curved surface [8]. Piezoelectric composite with 1-3 connectivity consists of piezoelectric ceramics connected in one dimension embedded in a polymer matrix. The softer polymer makes the piezoelectric ceramic pillars in a strong $k_{33}$ mode, such a composite can yield a high $k_t$ approach the $k_{33}$ of the piezoelectric ceramic.

Recently, high frequency transducers have been widely used in underwater detection and imaging with high resolution [5]. In an underwater environment, a broad beam width can enhance the detection range of transducer. The existing cylindrical transducer presents an open emission pattern in the horizontal plane [1], so it can achieve an omnidirectional coverage in horizontal plane, but the vertical beam width is relatively narrow. It is required to increase vertical beam width of the transducer to achieve a large coverage in underwater environment. Being flexible, piezoelectric ceramic/polymer composites can be easily conformed to doubly-curved shape for increasing the beam width of high-frequency transducer. Simultaneously, low acoustic impedance and high coupling factor were obtained for transducer.

Our research designed a doubly-curved 1-3 piezoelectric composite to increase the horizontal and vertical beam width of the transducer simultaneously. In next section, the design of the doubly-curved composite is detailed. The harmonic analysis using finite element method is performed to evaluate the resonant mode of composite and acoustics fluid-structure interaction (FSI) analysis is applied to predict the beam width of composite transducer. Finally, a doubly-curved composite transducer is presented, and the experimental results of transmitting voltage response and beam directivity index of composite transducer in water are shown in the Section III. These experimental results are shown that the doubly-curved 1-3 piezoelectric composite is a promising material for fabricate broad beam transducer.

## 2. Design and Development of Doubly-Curved 1-3 Composite Transducer

### *2.1. Design of doubly-curved 1-3 piezoelectric composite*

Considerable data have been reported on the electromechanical properties of 1-3 composite. Earlier work by Smith et al. [10] demonstrated composite's properties vary with the ceramic volume fraction, samples with volume fraction ranging from 50% to 80% could reach a high electromechanical coupling factor, but the acoustic impedance could increase linearly as ceramic volume fraction. The high acoustic impedance of ceramic cause acoustic impedance mismatch with that of water and body tissues (1.5 Mrayls). Therefore, the doubly-curved 1-3 composite with volume fraction of 54% was selected to investigate. For the cylindrical transducer, it can present an open emission pattern in horizontal plane [1]. To increase the vertical beam width, the proposed composite design as a doubly-curved structure as is shown in figure 1, $\theta_a$ and $\varphi_a$ are the bending angle in the colatitudinal and circumferential direction in the spherical coordinate system [11]. The outer radius of composite is $R$. The $t$ and $t_e$ refer to the radial thickness of composite and epoxy resin. Hence, resulting thickness of silicone rubber equals $t$-$t_e$. The PZT-5H is chosen as piezoelectric phase, and the epoxy resin and silicone rubber are chosen to prepare the polymer matrix. In the composite, piezoelectric ceramic columns are poled along the radial direction of the composite.

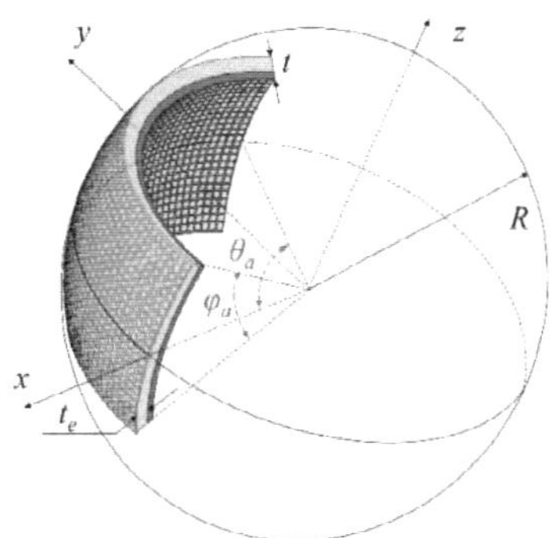


**Figure 1.** Schematic of a doubly-curved 1-3 composite (light blue: piezoelectric ceramic columns, green: epoxy resin, red: silicone rubber).

### *2.2. Numerical study*

In this subsection, we devoted to the finite element analysis (FEA) using the commercial FEM package ANSYS (ANSYS, Inc., Canonsburg, PA). This numerical calculation is performed to study the beam width varies with the magnitudes of the bending angle of the curved composite. To simplify the modeling, the FE model was established using the 2-dimensional element. The proposed curved composite is bending by monolithic piezoelectric composite plate with the same dimensions, so the arc length of the different composite is set as a fixed value (46 *mm*) in the simulation. For the thickness of polymer, $t$ and $t_e$ are considered to be constants as designed ($t$=5 *mm*, $t_e$=3.5 *mm*). Aronov et al. studied the spherical shell transducers for underwater acoustics. The analytical and experimental results showed the increase of the bending angle were effective in terms of improving the beam width of transducer [12]. For comparison purposes, eight curved composites are defined with different bending angle (i.e., the central angle), as shown in figure 2. Table 1 summarizes the eight configurations.

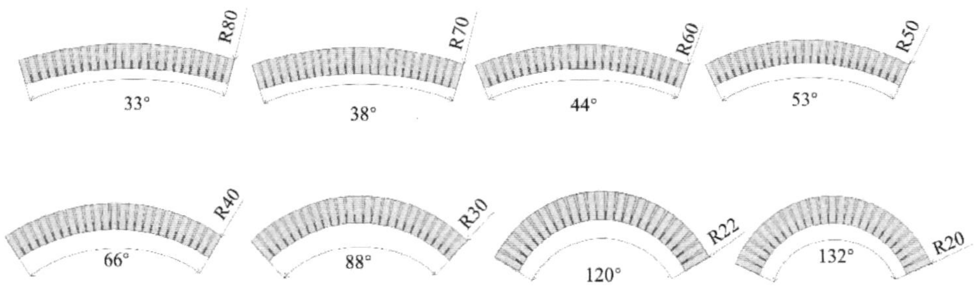


**Figure 2.** Increase of bending angle of 1-3 composites.

Acoustics fluid-structure interaction analysis was carried out to study the electroacoustic performance of composite in underwater environment. An FE model of composite transducer in the fluid was constructed, as shown in figure 3. To simplify the calculation, 1/2 fluid domain is modeled using the symmetry conditions [13]. The external parts such as acoustic backing layers were simplified [14]. The infinite boundary condition was applied to the spherical surface of fluid domain [14, 15]. Figure 3 shows that the eight numerical transmitting voltage response (TVR) of the eight composite in the frequency range 250 KHz-400 KHz. All the maximum transmitting voltage response, bandwidth values and beam width (at -3 dB) are summarized in table 1. As expected, the beam width is influenced by the bending angle of the composite. The greatest beam width is 120° which is obtained from sample 1. In general, the increase of the bending angle of composite can improve the beam width. However, it will also decrease the TVR of sample. For the curved piezoelectric composite, a trade-off then must be made between the TVR and the beam width, as illustrated in figure 4.

**Table 1.** Geometrical specifications and numerical results of the eight doubly-curved composites.

| No. | R (*mm*) | bending angle (°) | maximum transmitting voltage response (dB) | $BW_{-3\ dB}$ | beam width$_{-3\ dB}$ (°) |
|---|---|---|---|---|---|
| 1 | 20 | 132 | 153 | 296-351 | 120 |
| 2 | 22 | 120 | 153.8 | 299-353 | 108 |
| 3 | 30 | 88 | 155.5 | 277-362 | 74 |
| 4 | 40 | 66 | 158.4 | 289-357 | 58 |
| 5 | 50 | 53 | 160.2 | 285-350 | 42 |
| 6 | 60 | 44 | 162.5 | 289-350 | 38 |
| 7 | 70 | 38 | 164.2 | 284-344 | 30 |
| 8 | 80 | 33 | 165.6 | 294-357 | 28 |

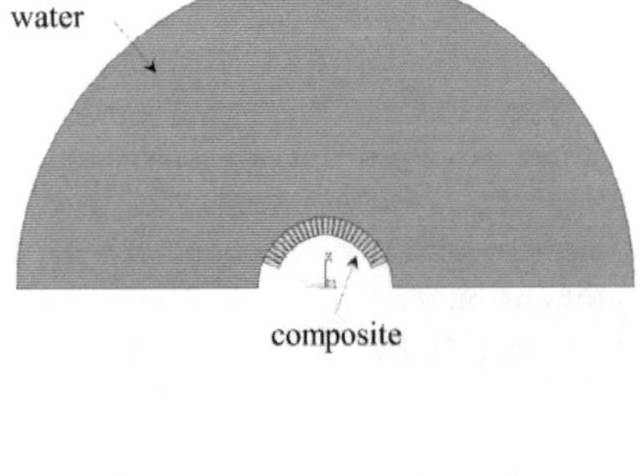


**Figure 3** FE model of the composite transducer in the fluid.

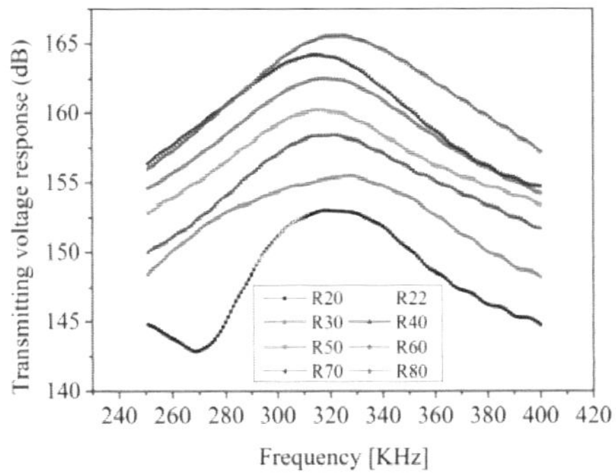


**Figure 4.** Transmitting voltage response of the curved 1-3 composite transducer.

These numerical results clearly show the possibility of a broad beam width of the doubly-curved 1-3 composite transducer in underwater environment.

## 3. Performance test of doubly-curved 1-3 composite transducer

Figure 5 shows the fabrication procedures of the doubly-curved 1-3 composite. The "dice and fill" method was used to fabricate the doubly-curved 1-3 composite. Frist, a PZT ceramic is cut along the length and width directions of the ceramic block, and a common base is retained. The thickness of PZT ceramic is $t$, and the thickness of common base is $t_e$. The rubber is then cast into the PZT ceramic pillars. After casting, the PZT ceramic sample is inverted, and the ceramic base is cut precisely along the grooves between the PZT ceramic pillars. The cutting depth is $t$-$t_e$. Then, fix the sample in a doubly-curved-shaped mould. After fixed the sample, pour the epoxy resin into the mould. After solidifying, polished the surfaces of the sample and cleaned by acetone. Finally, a doubly-curved 1-3 piezoelectric composite element will be complete, as shown in figure 6.

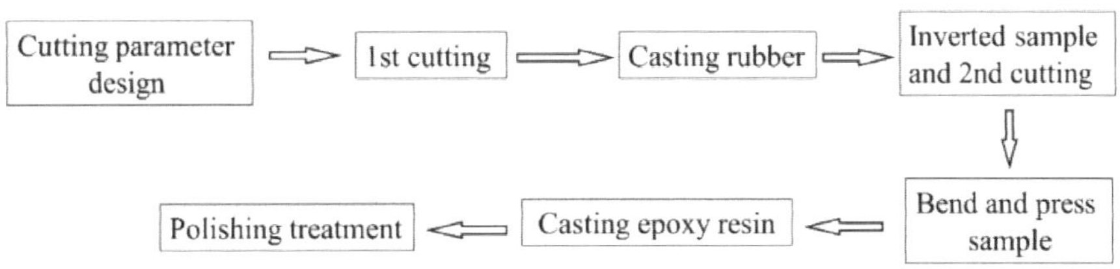


**Figure 5.** Fabrication procedures for a doubly-curved 1-3 composite

The material properties of the three phases were the same as those used in the previous numerical calculation. The outer radius of composite is 50 *mm*. The radial thickness of composite and epoxy resin are 5 *mm* and 3.5 *mm*. The bending angle in the colatitudinal and circumferential direction are 120° and 53°.

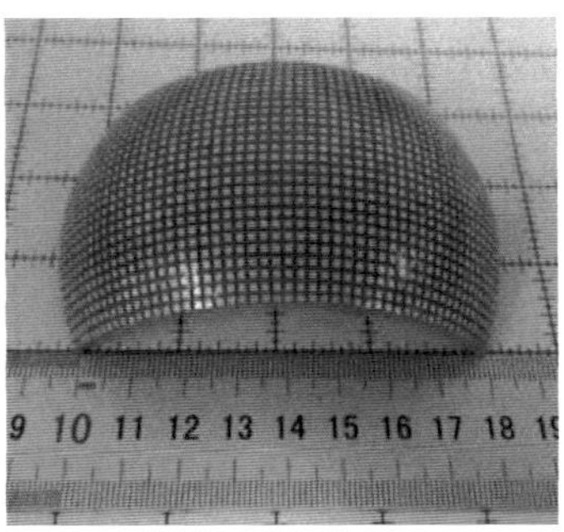


**Figure 6.** Photographs of the fabricated doubly-curved 1-3 composite.

An underwater transducer was fabricated with the doubly-curved 1-3 piezoelectric composite as the driving element. Figure 7 shows the structure of the doubly-curved 1-3 composite transducer. The performance such as the transmitting voltage response and beam directivity index were measured at room temperature. The transducer was connected to an underwater acoustic measurement system (The Institute of Applied Acoustics, Hangzhou, China). The measured transmitting voltage response

(250KHz-350KHz) of the composite transducer is shown in the figure 8. The maximum transmitting voltage response is 163.5 dB. The -3dB frequency bandwidth of the transducer is 274kHz-307kHz.

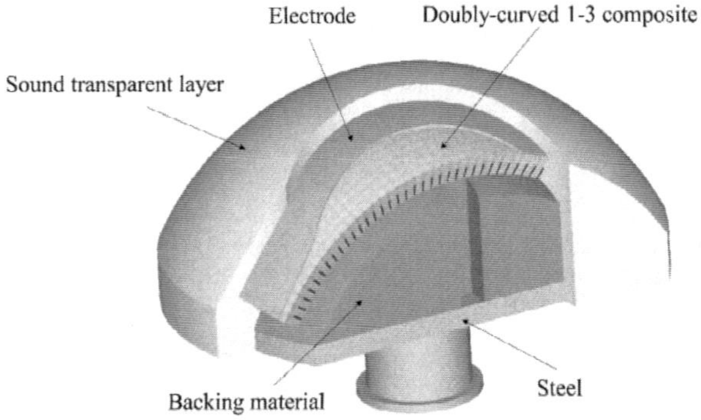


**Figure 7.** Structure of the doubly-curved 1-3 composite transducer.

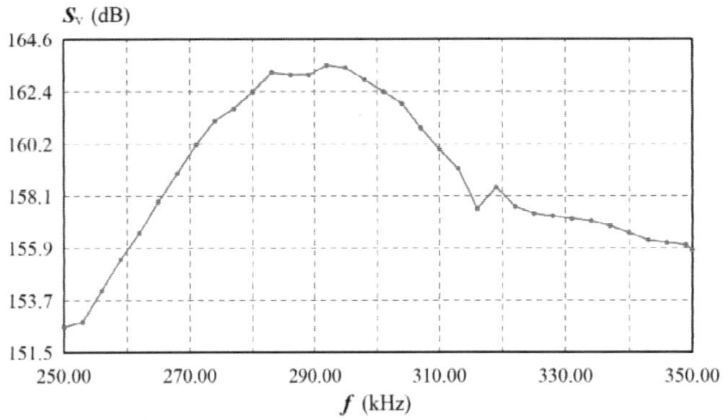


**Figure 8.** The transmitting voltage response of the doubly-curved 1-3 composite transducer.

The measured beam directivity index of the transducer is presented in figure 9. It can be seen that the -3 dB full angle beam width is approximately 106° in the horizontal plane. Moreover, the -3 dB full angle beam width is approximately 36° in the vertical plane. The results show that the doubly-curved composite transducer can achieve a broad beam width in the horizontal and vertical plane simultaneously in a relatively high operating frequency (300KHz).

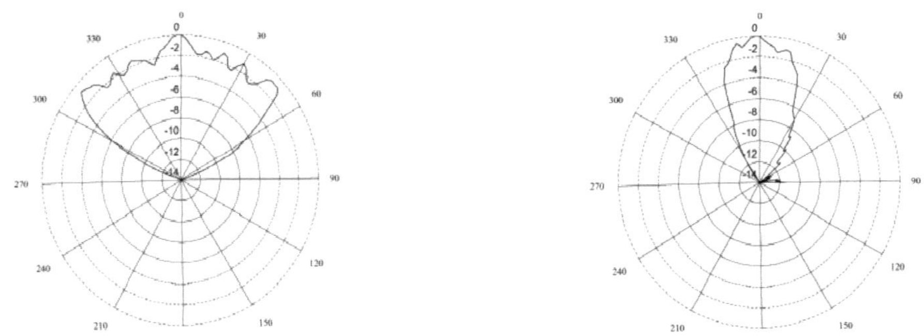

(a) Horizontal beam directivity index

(b) Vertical beam directivity index

**Figure 9.** Experimental beam directivity index of the doubly-curved composite transducer.

## 4. Conclusions

A doubly-curved piezoelectric composite with 1–3 connectivity has been successfully designed and fabricated. Finite element analysis is performed to evaluate the its electroacoustic performance in underwater environment. According to the results of FEA simulation, an underwater transducer using doubly-curved 1-3 composite with volume fraction of 54% is fabricated. The outer radius of composite is 50 *mm*. The radial thickness of composite and epoxy resin are 5 *mm* and 3.5 *mm*. The bending angle of composite in the colatitudinal and circumferential direction are 120° and 53°. In the operating frequency (300KHz), the -3 dB full angle beam width of transducer is approximately 106° and 36° in the horizontal and vertical plane. Consequently, this composite is a promising geometry for making broad beam underwater transducer.

## References

[1] Villadangos J M, Urena J, Mazo M, Hernandez A, Marziani C De, Jimenez A and Alvarez F 2007 *Improvement of Cover Area in Ultrasonic Local Positioning System Using Cylindrical PVDF Transducer* IEEE International Symposium on Industrial Electronics, 1473–7.
[2] Li S, Fu Z, Zhang X, Wang Y and Hu J 2013 *Smart. Mater. Struct.* **22(1)** 015005.
[3] Ray M C 2010 *Smart. Mater. Struct.* **19(3)** 035008.
[4] Ray M C 2006 *IEEE Trans. Ultrasonic. Ferroelectr. Freq. Contr.* **53(11)** 2152-65.
[5] Yun C H, Ko N Y, Moon Y S and Choi H T 2014 Oceans, **2014** 1-5.
[6] Nishamol P A and Ebenezer D D 2015 *Underwater Tech.* **2015** 1-6.
[7] Butler S C 2017 *Appl. Ferroelectr.* **2017** 19-23.
[8] Li H, Deng Z D, Carlson T J, Deng Z D and Carlson T J, 2012 *Sens. Lett.* **10(10)** 679-97.
[9] Newnham R E, Skinner D P and Cross L E 1978 *Mater. Res. Bull.* **13(5)** 525-36.
[10] Smith W A and Auld B A 1991 *IEEE Trans. Ultrasonic. Ferroelectr. Freq. Contr.* **38(1)** 40-7.
[11] Lee J 2017 *Int. J. Mech. Sci.* **122** 53-62.
[12] Aronov B, Brown D A, Bachand C L and Yan X 2012 J. Acoust. Soc. Am. **131(3)** 2079-90.
[13] Rouffaud R, Hladky-Hennion A C, Pham-Thi M, Bantignies C and Levassort F 2012 *Influence of 1-3 piezocomposite fabrications on lateral modes.* IEEE International Ultrasonics Symposium, 1-4.
[14] Kim J and Roh Y 2014 *Sens. Actuat. A Phys.* **206(3)** 97-106.
[15] Zhao W, Ming A and Shimojo M 2014 *Dynamic analysis and optimization of soft robotic fish using fluid-structural coupling method.* IEEE International Conference on Robotics and Automation IEEE, 1474-9.

## Acknowledgments

This work was supported by the National Natural Science Foundation of China (No.61471047, 61671068, 11604363), and the Beijing College Innovation Capability Promotion Plan of Beijing Municipal Institutions (No. TJSHG201510772015). The project is also supported by The Importation and Development of High-Caliber Talents Project of Beijing Municipal Institutions (No. CIT&TCD201504053).

IOP Conf. Series: Materials Science and Engineering **317** (2018) 012042 doi:10.1088/1757-899X/317/1/012042

# Principles of control automation of soil compacting machine operating mechanism

**Tikhonov Anatoly Fedorovich, Anatoly Drozdov**

National Research Moscow State University of Civil Engineering, 129337, 26, Yaroslavskoye Shosse, Moscow, Russia

**Abstract.**The relevance of the qualitative compaction of soil bases in the erection of embankment and foundations in building and structure construction is given.The quality of the compactible gravel and sandy soils provides the bearing capability and, accordingly, the strength and durability of constructed buildings.It has been established that the compaction quality depends on many external actions, such as surface roughness and soil moisture; granulometry, chemical composition and degree of elasticity of originalfilled soil for compaction.The analysis of technological processes of soil bases compaction of foreign and domestic information sources showed that the solution of such important problem as a continuous monitoring of soil compaction actual degree in the process of machine operation carry out only with the use of modern means of automation. An effective vibrodynamic method of gravel and sand material sealing for the building structure foundations for various applications was justified and suggested.The method of continuous monitoring the soil compaction by measurement of the amplitudes and frequencies of harmonic oscillations on the compactible surface was determined, which allowed to determine the basic elements of facilities of soil compacting machine monitoring system of operating, etc. mechanisms: an accelerometer, a bandpass filter, a vibro-harmonics, an on-board microcontroller. Adjustable parameters have been established to improve the soil compaction degree and the soil compacting machine performance, and the adjustable parameter dependences on the overall indexhave been experimentally determined, which is the soil compaction degree.A structural scheme of automatic control of the soil compacting machine control mechanism and theoperation algorithm has been developed.

## 1. Introduction

In modern construction, the filled soil compaction is an important part of engineering construction technology: building and structure foundation bases for various applications; hydraulic dams and dikesembankment; roads and railways; main pipeline trenches, etc.

This task is particularly relevant in the context of the rapid development of oil and gas production and transportation in recent years, as the pipeline rupture due to subsidence through the filled soil poor compaction can lead to an environmental disaster [1].

The bearing capability, strength and durability of constructed building structures depend on the quality of compactable gravel and sand base. It is established that up to 60% of damage occurs due to poor compaction of bulk construction materials [2].

Hence, there are prerequisites for control automation of filled soil compaction technological process to compensate for FB with maximum possible continuous output to the engineman's control panel of the objective assessments of the compaction degree quality [3].

The lack of domestic and foreign automatic control and control of the soil compaction degree systems is that they are information (light display) and have no ability to control the working mechanism (WM) of the soil compacting machine depending on the optimum degree of soil compaction and taking into account external disturbing actionsFB [4, 5].

To solve the problem on implementation of the automatic control system (ACS) of soil compacting machine WM, it is necessary to determinethe compaction method and facilities of objective assessment of the compaction and to develop an algorithm for automatic optimization of the compaction technological process, which provides the signal transformation of the vibroacoustic sensors in the automatic equipment with application of modern computer engineering micro processing means [2].

## 2. Technique

Studies have shown that the basic and effective way is compaction with vibrodynamic effect of machine WM on the compactible sand and gravelbase surface, as the dynamic methods of soil compaction using vibration machines make up 80% of the total number of the compactible means [3].

A dynamic method of compaction is provided by a combination of static pressure and vibration effect on the soil [6].

When soil compaction, the displacementand convergenceof sand and gravel rigid particles occurs, as well as destruction, deformation and adhesion of brickle and cloggy components, which increases the density and strength of the soil foundation, the displacement of softening and weakening liquids and gases. According to experimental data, the bearing capability of soils due to compaction increases by 40% [1].

In the process of soil compaction under the influence of vibrodynamic loads from (WM),the harmonic oscillations of the soil base surface with a certain amplitude and frequency depending on the compaction degree are formed [7, 8].

Depending on the adjustable parameters, the WM changes in magnitude of soil ultimate deformation $R_o$, which leads to different values of the harmonic amplitudes $A_1$ and $A_2$ and, respectively, of the frequencies $S_1$ and $S_2$.

Experimentally obtained oscillograms of harmonic oscillations shows that the harmonics $A_0$ and $A_1$practically does not change depending on the machine passnumber, but they are much subject to change theharmonic of amplitude $A_2$, associated with the soil compaction degree. If to determine the amplitude level ofharmonics $A_1$ and $A_2$, then with high confidence we can get the main indicator, which is the current value of the soil compaction degree–$\rho_{ф}$.

Uncompactable soil has the smallest deformation modulus, which characterizes the soil elasticity degree and therefore the harmonics of vibrational oscillations have a very weak signal. With increase the compaction degree, soil becomes more "hard" and the harmonic of oscillations $A_1$ and $A_2$ appears.

Automatic analyzer selects the actual frequency from the input signal that has a harmonic frequency spectrum– the first harmonic and the second harmonic of the same frequency.After detection of these signals, the ratio of second harmonic to the first harmonic is calculated; this characterizes the compaction degree depending on the machine pass number.

Sinusoidal input signal is determined using the following formula:

$$A(\text{K}) = \frac{1}{N}\sum_{n=0}^{N-1} A(\text{n})(\text{Cos}\frac{2\pi}{N}mK - \text{J}_m\text{Sin}\frac{2\pi}{N}mK) \quad (1),$$

where A – sinusoidal input signal amplitude; K= 1, 2 – harmonicnumber; N– number of reports during the experiment; m–number of intermediate current report; $\text{J}_m$ – imaginary part of complex plane.

From the formula (1) will receive the frequency of the 1st harmonic:

$$S_1(\text{к}) = \sqrt{[R_eA_1(K)]^2 - [\text{J}_mA_1(K)]^2} \quad (2)$$

where $R_e$– real part of complex plane.

The frequency of the 2nd harmonic is determined by a similar expression:

$$S_2(\text{к}) = \sqrt{[R_eA_2(K)]^2 - [\text{J}_mA_2(K)]^2} \quad (3)$$

The result of the actual soil compaction degree is determined using the following formula:

$$\rho_{ф} = \frac{S_2(к)}{S_1(к)} < 1, \left[\frac{г}{см^3}\right] \quad (4)$$

The solution to the problem on the optimization of the actual compaction degree is possible only on the basis of the WM automated control system with the use of modern computer technology.

In addition to operational information on the engineman's control panel on the soil compaction degree, an important technological aspect is the automatic control of the vibrating machine WM on the sensor signals, providing the optimum compaction degree.

Adjustable oscillation frequency of the WM eccentric creates perturbing force: $F_{\omega}=m_o*v_m*\omega^2$(figure 1,characteristics of $F_{\omega}$), wherem$_o$– eccentric vibrating mass, $v_m$– machine traverse speed, $\omega=2*П*f$– eccentric rotation speed, f=15-50 Hz – eccentric oscillation frequency.

Operatingvibrating machine WM forms the other adjustable parameters: $F_y$ – compaction device impact force of the soil surface (figure 1, characteristics of $F_y$), $\tau_y$ – impact number on a single bay (timer) (figure 1, characteristics of$\tau_y$).

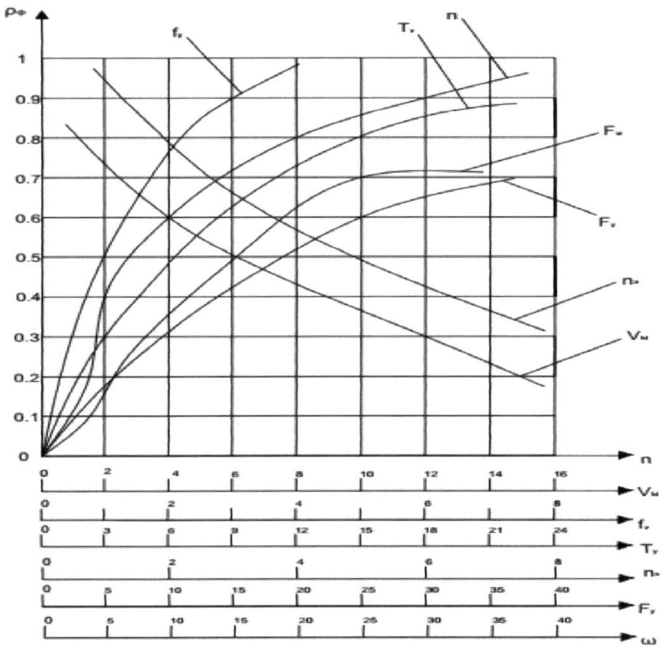


**Figure 1.**Characteristic of degree of soil compaction depending on the controlled parameters.

The criterion for soil quality determination on the vibrodynamic method of control is universal, because it does not measure absolute values of harmonic oscillation amplitudes, but their averaged values within the compactablearea of a given size, defined by the machine pass number.

Since, apart from $S_1$ and $S_2$, there are other harmonics that are characterized by their amplitudes Si=f(Ai), then during the pulse width modulation, low harmonics, which does not reflect the compaction degree, are smoothed and we get the sine waves of the required frequencies $S_1$ and $S_2$, which with increasing of $\rho_{ф}$ transferred to nonlinear characteristics (sections 1-2 and 3-4), as the value of the $S_1$significantly decreases, but $S_2$increases with $\rho_{ф}$ growth.

## 3. Discussion

Adjustable (compensating) impacts Fρduringthe automated control in the continuous process of soil compaction include: n–number of machine passes with one bay, $V_M$– machine speed, $\tau_y$- time (impact number) of compactionon a local area, $F_y$– compactable impact force, $f_y$– operating device impact frequency, $F_{\omega}$–perturbing force from the eccentric rotation rate. The determination ofvalues and time of action of the regulatory parameters, which are included in the Fρ group and impact assessment on $\rho_ф$ of the unregulated external actions $F_B$, has allowed obtaining reliable information about the increase degree of the actual soil density $\rho_ф$.

The issues for determination of the adjustable parameters actions Fρ and uncontrolled impacts $F_B$ in such a dynamic system that can automate the compaction process, is possible only with the development and implementation of a microprocessor control system with flexible software, for implementation of which on the basis of active experiment the values of the adjustable parameters affecting the main parameter of $\rho_ф$ were obtained (figure 2).

Operating performance and high compaction degree largely determine the technical and economic indicators when performing construction works. Criteria for optimal operation of the compacting machine are maximum productivity with minimum number of passes and increasing of machine movement speed by ensuring optimal soil density.

Operating performance is determined by the following formula:

$$П_э = \frac{1000(B-b)V_m{*}Fw{*}fy{*}\tau y}{n} * K_{B,}\left[\frac{м^3}{ч}\right] \quad (5)$$

where B – compaction b and width, [m]; b=0,1 m – related b and overlap width; $V_m$ – machine movement speed during compaction, [m/hour]; n – number of machine passes on one bay; $K_B$=0,8…0,85 – machine utilization (fuelling, maintenance).

In the formula (5) the coefficients B, b, $K_B$, $H_0$ does not depend on the operation mode of the machine, but the coefficients n and $V_m$ are functionally associated with the compaction degreeρ $_ф$ (figure 1).

Therefore, $П_Э$ should be determined depending on the main indicator of $\rho_ф$, because there is no technological need for the maximum provision of $П_Э$ in case of poor compaction $\rho_ф$.

Increasing the value of "n" leads to decreasing of $П_Э$, but increasing $\rho_ф$, and the speed increasing $V_m$ improving $П_Э$, but dramatically reduces $\rho_ф$ (5). If you do not adjust the $F_P$ parameter, for reaching a given value of $\rho_{ф,}$ it is necessary to significantly reduce the performance of $П_Э$ (5), by reducing the machine speed and increasing the number of its passages, which leads to a decrease of techno-economic indicators. The ACS allows resolving this contradiction by adjusting the parameters of $F_y$, $F_w$, $f_y$, $\tau_y$, the changes of which allow increasing the value of $\rho_ф$ (5), without reducing the operational performance of the machine.

The formation of the control signals from the harmonic oscillation amplitudes, arising from the effects of the adjustable parameters $F_y$, $F_w$, $f_y$, $\tau_y$ is as follows (figure 2). During the filled soil compactionwith a vibrating operating device 1 at the output of the accelerometer 2 a harmonic-type signal occurs, which is close tosinewave. This signal, passing the amplifier 3, is sent to a band-pass filter 4, containing information on the process of soil compaction 25. This signal is being tuning outfrom both high-frequency and low-frequency noise.

With the output of the band-pass filter 4, a valid signal through the amplifier 6 is sent to the frequency converter in the analog signal 6 and the algebraic adder 7, on the input of which the signal from the adjustment 8 of the vibration amplitudes is sent, which determine the specified compaction degree. With the output of the adjustment 8, this signal comes on the analyzer 14.

The fixed signal from the adjustment 8, the algebraic adder 7 and the frequency converter in the analog signal 6 is sent to an analog-to-digital converter 9, which at its output generates a code and sends it to the memory block 13 and from its output to the input of the comparator 10, with "Output 1" of which the encoded signal is fixed at "Input 2" of the analyzer 14.On the "Output 2" of the comparator 10 the indicator 5 is connected. The signal from band-pass filter 4 in the form of pulses is sent to the complementing flip-flop 11, and from it is sent tothe pulse shaper 12, which converts the commands so that on the comparator output rectangular pulses occur, arising from various harmonics of the compactable soil 25.

In a controlled technological process several signals from the harmonic frequencies are formed, which arise from various adjustable parameters and are functionally associated with the actual soil compaction degree: $\rho_ф = f(V_M$, n. $F_y$, $F_w$, $f_y$, $\tau_y)$. The choice of optimal parameter values from the $F_p$ group produces the controller software 22.

IOP Conf. Series: Materials Science and Engineering **317** (2018) 012042 doi:10.1088/1757-899X/317/1/012042

At the input of the comparator 10, the steady-state value of a monitored signal is fixed, for example harmonics amplitude signal, related to $F_y$ that is sent to the "Input 1"of comparison analyzer 14, and at the "Input 2" the signal comes from the adjustment 8, where the actual value of the soil compaction degree is being compared specified: $\rho_{\Phi}$ with $\rho_0$.

In case of $\rho_{\Phi} \neq \rho_0$unbalanced signal generates $\Delta\rho$, which is sent at the "Input 2" of the microprocessor controller 22, and with the "Output" a control signal generates through the normalizer 23 to change the operation mode of regulating actuating mechanic 24. Simultaneously, the multichannel normalizer 21 for connecting multiple converters generates a signal, corresponding to the impact 17, and in the case of the unbalance signal of the microprocessor controller 22 occurs, the following and the most effective optimal operation mode of regulating actuating mechanism 24 is installed on the adaptive program, for example, by changing the $F_w$ parameter. Thus the monitoring signals, received from transmitters 15-20 with the use of multi-channel normalizer 21 happens. When the equality is achieved $\rho_{\Phi} = \rho_0$ ($\Delta\rho$=0), the automatic control system switches to the steady-state mode, showing that the compaction is completed in a given area.

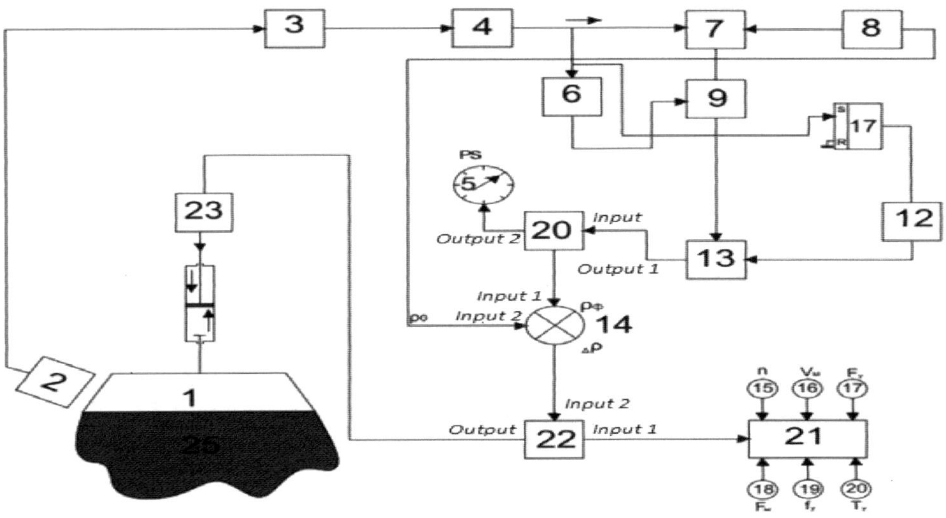


**Figure 2.** Structural arrangement of hardware components of automatic control system with the control mechanism of the soil compacting machine.

## 4. Conclusions

Preliminary analysis shows that the automation of the control process will increase the performance of work on soils compaction by at least 25% only due to the reduction in the number of passes on one grip and the increaseof the speed of the machine movementunderinsignificant environmental impacts owing to the created vibrational load.

In addition, the automatic control system allows improving the quality of soil compacting operations, reducing the consumption of combustibles and lubricants, and in some cases completely abandoning the traditional approaches to the control of the degree of compaction by taking soil samples and relieving the workers performing these operations.

## References

[1] Vaskovsky A M, Vorobiev A A 1998 *Automation of compaction of soil by vibration rollers* Moscow, Izvestiya VUZ - Institute of Higher Education, 36.

[2] Kascheev A A 2005 *Technological substantiations of automation of the process of compaction of bulk soils by vibrodynamic method* Sat scientific Proceedings of "Interstroymech-2005", Moscow.

[3] Shilkov V A 1992 *Automation of vibro impact impact machines for soil compaction and piling* Moscow, Stroiizdat.
[4] Anashkin Z T, Balkireyeva V Ya 1994 *A method for continuously monitoring the density of the material to be laid when the vibrating roller is operating* Moscow, Construction and road machines, **6** 95.
[5] Barkan D D, Shekhter O Ya 2002 *Theory of surface compaction of soil* Moscow, Gosstroizdat, **51** 89.
[6] Akishkin A I, Voloshchin A P 1996 *The influence of the impact velocity on the compaction of a bound soil with impulses is less than optimal* Moscow Soyuzdornii, **26** 42.
[7] Vladimirov V N 1996 *Investigation of the propagation of stresses and deformations in densified soil under the action of a vibrator* Moscow, Soyuzdornii, **24** 41.
[8] Vladimirov V N 1998 *Studies of the propagation of vibrations, stresses and deformations in a compacted soil under the influence of a vibrator* Moscow, Stroyizdat.

IOP Conf. Series: Materials Science and Engineering **317** (2018) 012043 doi:10.1088/1757-899X/317/1/012043

# Automatic Control of the Concrete Mixture Homogeneity in Cycling Mixers

**Tikhonov Anatoly Fedorovich[1], Anatoly Drozdov[1]**

[1]National Research Moscow State University of Civil Engineering, 129337, 26, Yaroslavskoye Shosse, Moscow, Russia

**Abstract.** The article describes the factors affecting the concrete mixture quality related to the moisture content of aggregates, since the effectiveness of the concrete mixture production is largely determined by the availability of quality management tools at all stages of the technological process. It is established that the unaccounted moisture of aggregates adversely affects the concrete mixture homogeneity and, accordingly, the strength of building structures. A new control method and the automatic control system of the concrete mixture homogeneity in the technological process of mixing components have been proposed, since the tasks of providing a concrete mixture are performed by the automatic control system of processing kneading-and-mixing machinery with operational automatic control of homogeneity. Theoretical underpinnings of the control of the mixture homogeneity are presented, which are related to a change in the frequency of vibrodynamic vibrations of the mixer body. The structure of the technical means of the automatic control system for regulating the supply of water is determined depending on the change in the concrete mixture homogeneity during the continuous mixing of components. The following technical means for establishing automatic control have been chosen: vibro-acoustic sensors, remote terminal units, electropneumatic control actuators, etc. To identify the quality indicator of automatic control, the system offers a structure flowchart with transfer functions that determine the ACS operation in transient dynamic mode.

## 1. Introduction

Technological requirements to the quality of dosing in the production of concrete mixtures are determined by the need to ensure the consistency of the specified properties of mixtures, such as homogeneity. The concrete mixture homogeneity is a percentage according to the given composition of components. The mix is considered homogeneous if the samples taken from different places in the mixer contain the individual components of the mixture in equal percentages. The concrete mixture homogeneity is associated with the strength of concrete and, accordingly, building structures and directly depends on the correction of water supply to the cyclical mixer while mixing components in it [1, 2].

It has been established that the correlation between the mixing components and the dosed quantity of water on which the water-cement ratio ($^{W}/_{C}$), and therefore the mixture homogeneity depends, is not maintained due to the variable moisture content of aggregates, and especially sand [3, 4].

Practice has shown that the introduction of the known devices for automatic control of the concrete mixture homogeneity during the mixing of components (by changing the current of the mixer motor, monitoring the electrical conductivity of the concrete mixture depending on the amount of water contained in it, monitoring the moisture content of aggregates with neutron moisture meters) does not ensure the desirable homogeneity of the concrete mixture (the error is up to 30%), because of the

IOP Conf. Series: Materials Science and Engineering **317** (2018) 012043 doi:10.1088/1757-899X/317/1/012043

imperfection of the devices for controlling the mixture homogeneity and the associated mixing time of the components [5]. Therefore, the production requires efficient and reliable automatic devices that allow to determine the amount of unaccounted moisture in inert aggregates based on modern computer equipment and, accordingly, promptly adjust the consumption of the water batcher, which ensures the desired homogeneity of the mixture and the optimal mixing time of components, which is related to the productivity of the cyclic concrete mixer [6].

## 2. Study methodology

Concrete mixture is a compound mass containing fine cement particles, coarse aggregate grains, water and air involved in the process of preparing the mixture [7]. During mixing in the mixer, component particles are given a whirling motion, thus causing friction and collision forces between the particles, the body and the blades of the mixer, which creates elastic modes, leading to the body vibration and the generation of vibroacoustic signals. With the introduction of a controlled dose of water into the mixer, the cement paste gradually converts from the elastic medium into a viscoelastic one with the passage of time of the process of mixing components, which affects the characteristics of vibrational and, accordingly, electric acoustic signals that are functionally related to a change in the concrete mixture homogeneity [8].

The theoretical studies have constituted a ground for the development of an adaptive (with fuzzy logic) automatic control system for the concrete mixture homogeneity, taking into account the previously unknown environmental perturbations $F_{ENV}$: variable humidity of inert aggregates, mixing time of components and their grain size composition, a set of variable mix formulation depending on the type of construction, wear of rotating blades of the mixer, bearing friction of the mixer drive, etc. [9, 10].

The mixing time of the components in the mixer will be a variable unknown value along the mixture homogeneity, which depends on environmental influencing factors $F_{ENV}$. The main control device $CD_M$, together with the control object CO, is a conventional non-adaptive automatic system (connection 1 – opening of the mixture discharge gate $Y_2$ after the specified mixing time and switching on the mixer drive M). The adaptation control device $CD_A$ constantly adjusts the operation of the$CD_M$ by regulating the water supply of the valve $Y_1$, depending on the concrete mixture homogeneity, the parameters of which are controlled by vibro-acoustic sensors VAS1 ... VAS4. Signals from these sensors arrive at the $CD_A$ through the $RTU_{IN}$ and in this case the correction circuit 2 is connected. When the optimum mixture homogeneity is reached, a command is made to open the gate of the concrete mixer $Y_2$, which indicates the readiness of the dose (figure 1).

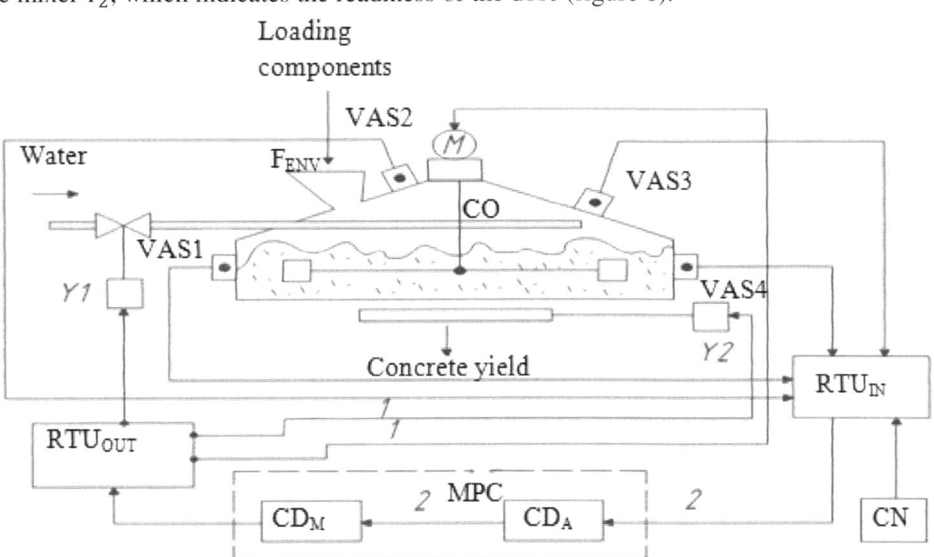


**Figure 1.** Structure of the elements of the adaptive system for automatic control of the concrete mixture homogeneity with the operational correction of the water supply in a concrete mixer.

VAS1...VAS4 – vibro-acoustic sensors; $CD_M$ – main control device; $CD_A$– adaptation control device; CN – controller of concrete trademark and minimum mixing time; $RTU_{IN}$, $RTU_{OUT}$ remote terminal unit – input and output; $Y_1$– actuator for correcting the water supply to the concrete mixer; $Y_2$– actuator for discharging the mixture when homogeneity is achieved; M – mixer motor; MPC – microprocessor controller; $F_{ENV}$ – environmental perturbations; CO – control object.

To obtain the dependence of the root-mean-square value of vibration accelerations as a function of the mixing time when the homogeneity changes, it is necessary to simulate this process based on the following perturbing factors: change of humidity variable from 0 to 18%, different grades of concrete mixture (100, 200, 300, 400, 500), taking into account the wear of the rotating blades to 65% and bearing friction of the mixer drive with $C_{FR}$=0,86. To this end, a structure flowchart of the automatic control system ACS of the concrete mixture homogeneity during its mixing in a concrete mixer has been created, which shows the transfer functions of all system elements in accordance with the structure in (figure 2).

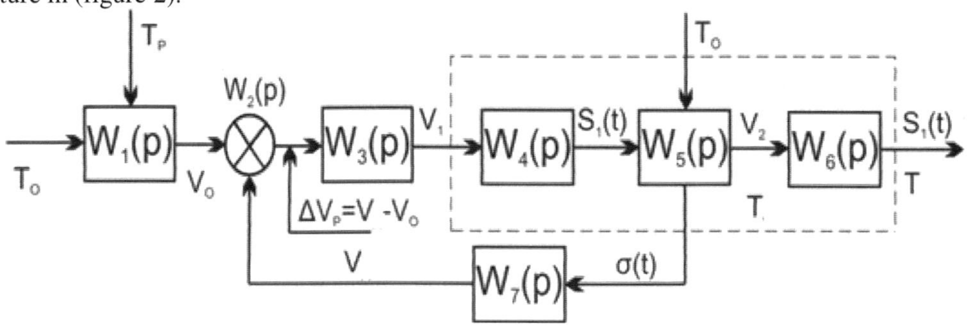


**Figure 2.** Structure flowchart of the ACS of the concrete mixture homogeneity.

$W_1(p) = F_1$– controller; $W_2(p) = \frac{U_D}{U_0} = F_2$ – reference element;

$W_3(p) = \frac{C_{MP}\cdot(T_{MI}+1)}{T_M}$ – microprocessor controller; $W_4(p) = \frac{F_{AC_1}}{T_{AC_1}p+1}$ – water control valve; $W_5(p) = \frac{l^{-\tau p}}{T_C p+1}$ – mixer; $W_6(p) = \frac{F_{AC_2}}{T_{AC_2}p+1}$ – discharge gate of the ready mixture;

$W_7(p) = \frac{C_{VAS}}{T_{VAS}p+1}$ – vibroacoustic sensor;

$W_O(p) = \frac{W_1(p)\cdot W_2(p)\cdot W_3(p)\cdot W_4(p)\cdot W_5(p)\cdot W_6(p)}{1+W_1(p)\cdot W_2(p)\cdot W_3(p)\cdot W_4(p)\cdot W_5(p)\cdot W_6(p)}$ – general transfer function of ACS elements;

$T_O$ – mixing time of the mixture components specified by the operator; $T_{OPT}$ – optimal mixing time, determined by homogeneity; $F_{AC_1}, F_{AC_2}$ – air cylinder gain factors; τ – delay time; $C_{VAS}$ – conversion coefficient of vibroacoustic signals into electrical;σ – mixer body vibration frequency; $T_{MP}, T_{AC_1}, T_{AC_2}, T_c, T_{VAS}$ – element time constants; $F_P$, $F_{ENV}$– operator-controlled external influence.

## 3. Discussion

By setting different values of the transfer functions of individual elements $W_i(p)$, the necessary quality parameters of the ACS regulation are achieved in the transient dynamic mode by mathematical modeling of the technological process of mixing concrete mixture components in the expression $W_O(p)$.

The operator provides information on the trademark of the ready concrete mixture, its required volume and the minimum mixing time through the input device (controller). The microprocessor controller takes the required mix formulation from the data bank and calculates the mass of each component without taking into account the actual moisture content of the aggregates. In this case, the supply of water by the main batcher to the mixer is 15-20% less than the preset one according to the formulation, which allows to take into account the moisture content of the aggregates. As the mixture reaches the optimum value in accordance with the homogeneity controlled by the vibro-acoustic sensors, water is added (by the valve $\gamma_1$) directly to the mixer during the mixing until the dose is ready.

The vibro-acoustic sensors with the static characteristic convert the vibrations of the mixer body into an electrical signal that will be a variable depending on the readiness of the dose determined by the vibration frequency of the mixer body. When the concrete mixer is idling (not loaded with components), the voltage from the vibration sensors during the time T will be a constant value. Experimental studies have shown that when the mixer is filled with components, the direct dependence of the points 1-2 during the mixing time turns into a broken curve (p. III – p. IV, figure 3). In this case, the nature of the change in the curve III-IV, and hence the output voltage $U_{OUT}$, depends on the time of arrival of each component in the concrete mixer.

The nature of the dependence $U = f(T)$ (figure 3) reflects the technological process of preparing the mixture from the loading of components to the delivery of the ready mixture, taking into account the changes in vibroacoustic signals determining the degree of optimization of homogeneity, and hence the mixture readiness.

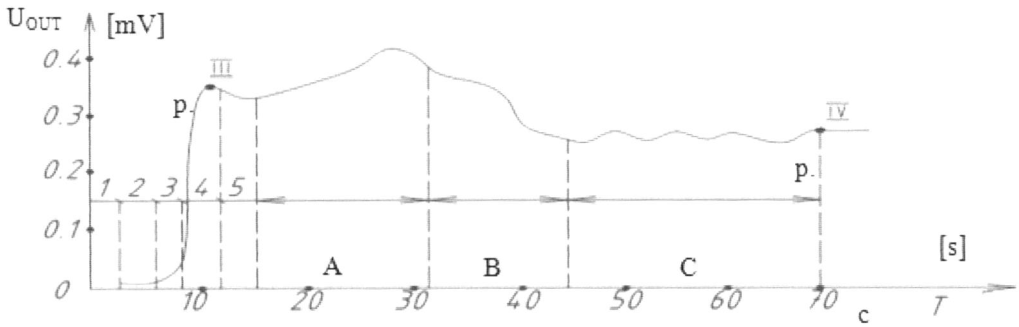


**Figure 3.** Dependence of the output voltage on the vibration acceleration of the mixer body during the mixing time of the concrete mixture components.

1, 2 – idling mode of mixer motor operation;
3...5 – mode of loading and simultaneous mixing of the mixture components;
A – regulation of the mixture homogeneity by supplying water to the mixer;
B – mixing of components with homogeneity control;
C – readiness of the mixture and the command to open the mixer gate $Y_2$ to discharge the concrete mixture into vehicles.

In the idle mode of the mixer (not loaded), $U_{OUT}$ varies in the range of points 1 ... 4 (figure 3), and when mixing sand (it pours first) and crushed stone (it pours second), the vibration intensity rises sharply (section 4-5) and reaches the value of p. III. When loading cement (it pours third) and mixing with sand and crushed stone, $U_{OUT}$ decreases (section 4-5), which can be attributed to the damping of elastic vibrations by the cement. When water is supplied, $U_{OUT}$ again increases (section A), and then begins to decrease sharply (section B), which characterizes the process of cement paste formation and gradual conversion of the medium from elastic to viscoelastic. After a while (about 45 seconds after loading all components, including water), the decrease in the value of the $U_{OUT}$ stops and a steady-state vibration mode is established, which corresponds to obtaining the mixture homogeneity and ending the mixing process. At the same time, a command is made to open the gate $\gamma_2$ to discharge the ready concrete mixture into vehicles.

## 4. Conclusion

The results confirm the possibility of determining the concrete mixture homogeneity and the optimal duration of mixing by measuring the characteristics of the oscillatory field of the mechanical system: mixer – a concrete mixture and can be used for the development of corresponding measuring and controlling devices in establishing the microprocessor control system of concrete mixture and solution technologies in mobile installations and central plants for the production of building structures.

## References

[1] Korolev K M 1996 *Intensification of the preparation of concrete mix* Moscow: Stroiizdat 144.

[2] Kaiser L A, Levin L I 1992 *Technological fundamentals of automatic correction and optimization of the concrete mix composition* Moscow: Stroyizdat, 182.

[3] Korolev K M, Malinina L A 1998 *Actual questions of automation of processes of preparation of concrete mixes* Moscow: NIIZhB, 130.

[4] Berkut A I, Zakharov Ya V, Goryunov I I 2002 *Experimental study of the process of preparation of building mixtures for optimization purposes* Moscow: MGSU, **1** 54.

[5] Tikhonov A F 1997 *Device for continuous control of uniformity of a freshly prepared concrete mix Report on research work* Moscow: TsNIIOMTP, 78.

[6] Gayle J B, Lacey P M, Gary J H 1997 *Ind. Eng. Chem. so.* **1279** 1997.

[7] RulnovA A, Makarova E V 2000 Automation of the continuous process of mixture formation on the basis of metering-flow integrators 2000 Moscow: Higher educational institutions herald "Construction", **7** 31.

[8] Khayutin Yu G, Levye E B, Sovalkov I G 2003 *Statistical analysis of the uniformity of a concrete mixture* Moscow: Stroiizdat, 192.

[9] Tikhonov A F 1996 *Automation of quality control of products made of concrete and reinforced concrete* Moscow: Stroiizdat, 300.

[10] Weidenвант S S 1998 *Advances in chemical Engineering*, **II** Academic Press, New York.

IOP Conf. Series: Materials Science and Engineering **317** (2018) 012044 doi:10.1088/1757-899X/317/1/012044

# Machine Learning Methods for Production Cases Analysis

**Nataliya V. Mokrova[1], Alexander M. Mokrov[2], Alexandra V. Safonova[2] and Igor V. Vishnyakov[3]**

[1]Professor, Moscow State University of Civil Engineering, Russia, Moscow, 129337, 26, Yaroslavskoye Shosse
[2]Software Engineer, CJSC NORSI-TRANS, Russia, Moscow, 127015, 12/15, Bolshaya Novodmitrovskaya Ulitsa
[3]Senior Lecturer, Bauman Moscow State Technical University, Russia, Moscow, 105005, 5, 2-Ya Baumanskaya Ulitsa

**Abstract.** Approach to analysis of events occurring during the production process were proposed. Described machine learning system is able to solve classification tasks related to production control and hazard identification at an early stage. Descriptors of the internal production network data were used for training and testing of applied models. k-Nearest Neighbors and Random forest methods were used to illustrate and analyze proposed solution. The quality of the developed classifiers was estimated using standard statistical metrics, such as precision, recall and accuracy.

## 1. Introduction

Nowadays machine learning methods are widely used in the analysis of a wide variety of data sets. This technology is based on mathematical statistics, probability theory, optimization techniques, and other classical mathematical disciplines [1].

The modern production process uses a lot of equipment to be monitored, as well as many automated systems exchanging data among themselves. This fact allows us to collect these data and to carry out statistical research.

Machine learning methods help us to analyze the collected data and use the results to build an online monitoring system. Such systems allow us to reform many tasks related to the quality of the production process. The problem of analyzing data consisting of many parameters is solved.

The described approach allows us to build a warning system for emergencies in the production process basing on the problems arose before, and to identify new risks at an early stage. It also helps to identify the possible consequences of accidents at hazardous facilities and to prevent them.

The introduction of methods of machine learning in the system of quality control of products will reduce the amount of manufacturing defect. This is the way to identify low-quality products in the production process. This will reduce the time to eliminate problems and the number of damaged products.

## 2. Methodology

Automated systems help us to reduce the cost of production. However, these systems are subject of hacker attacks while the important data is transferred to manage the system over a network. The task is to protect the system. Modern equipment uses data encryption, but this is often not enough to provide a high degree of protection [2]. Invasion of an attacker into the system and changes in the production process can lead to an accident or large financial losses. There are many similar precedents in all areas

IOP Conf. Series: Materials Science and Engineering **317** (2018) 012044 doi:10.1088/1757-899X/317/1/012044

of industry, ranging from banal electricity supply shutdown, ending with the provocation of contamination with potent toxic impurities. Let's consider the application of machine learning methods in this example. The main task is to classify the transmitted data to detect atypical behavior to prevent possible intrusions.

The creation of a system begins with collecting and analyzing existing data. We represent the obtained data in the form of blocks. In our case, we consider the data block as a network packet. To create a universal classification system based on the type of traffic of network data, the parameters of the transmitted data will be used but not their content. This approach allows us not to bind to a specific area and bypass the problems associated with the encryption of the transmitted data [3].

The problem is to select the classes to which the data blocks will be assigned, to mark the available data according to the selected classes, to choose the statistical class description method, to implement the classifier, to test the resulting system.

Any data transmission over a network is associated with the connection of the client to the host and the transmission of some type of data. One session of such a transmission is several network packets having different directions, but having common parameters that allow them to be merged into a data transfer session. Each session can be described as a piece of data sent from the client host and a piece of data which is the host's response to the client. The main most effective characteristics of the session used for statistical description are the parameters of the packet sizes in both directions, the duration of the connection and the ratio of these parameters [4].

The following classes are selected: typical automatic actions, typical actions of the system operator and atypical actions. Atypical actions represent the possible actions of the attacker.

After receiving the data and presenting it in the form of sessions the results are marked out. Each object is assigned to one of the classes and the numerical indicators of the selected data descriptors are obtained. The selected classes are the most distinguishable from each other. This fact allows us to qualitatively perform the preparation of a set for machine learning. If necessary in the future, it is possible to break the data into subclasses to create a binding to a domain.

To solve this problem, the methods of machine learning k-Nearest Neighbors [5] and Random forest [6] are used. These methods show satisfactory results for this type of problem [7].

The most natural way to assess the quality of classification is accuracy (A):

$$A = \frac{N_t}{N} \quad (1)$$

where $N_t$ – the number of correct classifications, $N$ – the total size of the sample to be classified.

To obtain more accurate information about the quality of the classifier's work, let's consider the ratio of correct system responses to all positive responses, precision (P):

$$P = \frac{T_p}{T_a} \quad (2)$$

where $T_p$ – the number of correct classifications for one class, $T_a$ – the total number of cases when the classifier made a choice in favor of this class.

Also, we consider a value showing the relationship between the correct system trips and all the sample elements that belong to the selected class, recall (R):

$$R = \frac{T_p}{N_c} \quad (3)$$

where $T_p$ – the number of correct classifications for one class, $N_c$ – the true size of this class according to expert assessment.

For the sake of convenience, consider the F-score. This value shows the averaged value of completeness and accuracy and calculated as the average harmonic value. Using these estimates, it is possible to draw conclusions about improving or deteriorating the quality of the classifier, depending on the changes that are made to its implementation.

## 3. Experimental results

Now we will form the training and test sample in the ratio of 30% and 70% respectively. The results obtained using the classification by the methods of k-Nearest Neighbors and Random forest are described in Table 1 and Table 2 respectively.

**Table 1.** Precision, Recall, F-score, Accuracy using k-Nearest Neighbors on initial sample.

| | Precision, % | Recall, % | F-score, % |
|---|---|---|---|
| Automatic actions | 86,4 | 77,6 | 81,8 |
| Actions of the system operator | 34,3 | 32,4 | 33,3 |
| Atypical actions | 47,5 | 57,3 | 52,0 |
| Accuracy, % | | | 62,7 |

**Table 2.** Precision, Recall, F-score, Accuracy using k-Nearest Neighbors on initial sample.

| | Precision, % | Recall, % | F-score, % |
|---|---|---|---|
| Automatic actions | 82,3 | 82,6 | 82,4 |
| Actions of the system operator | 40,8 | 32,1 | 35,9 |
| Atypical actions | 52,1 | 58,8 | 55,2 |
| Accuracy, % | | | 65,3 |

The overall accuracy was not very high. Satisfactory results are obtained for the predominant class of automatic actions. The change in the descriptors did not lead to better results. The actions of the system operator are clearly regulated and often like the actions of the automatic system. Therefore, we can manually perform the removal from the automatic actions class such operations to improve overall accuracy. The data obtained after such manipulation are presented in Table 3 and Table 4.

**Table 3.** Precision, Recall, F-score, Accuracy using k-Nearest Neighbors on filtered sample.

| | Precision, % | Recall, % | F-score, % |
|---|---|---|---|
| Automatic actions | 90,1 | 93,2 | 91,6 |
| Actions of the system operator | 92,8 | 85,8 | 89,2 |
| Atypical actions | 75,9 | 81,5 | 78,6 |
| Accuracy, % | | | 86,9 |

**Table 4.** Precision, Recall, F-score, Accuracy using k-Nearest Neighbors on filtered sample.

| | Precision, % | Recall, % | F-score, % |
|---|---|---|---|
| Automatic actions | 92,9 | 95,2 | 94,0 |
| Actions of the system operator | 91,3 | 87,5 | 89,4 |
| Atypical actions | 78,7 | 81,4 | 80,0 |
| Accuracy, % | | | 87,9 |

Such data sample changes allowed to improve the accuracy of the classification.

The system was tested using cross-validation [8]. The initial sample was divided into 20 random variants of the training and test samples, the test sample was 70%, and the training sample – 30%. As a result of training and testing, the average values of the classifier ratings yielded the results presented in Table 5 and Table 6.

**Table 5.** Precision, Recall, F-score, Accuracy using k-Nearest Neighbors for cross-validation.

| | Precision, % | Recall, % | F-score, % |
|---|---|---|---|
| Automatic actions | 89,7 | 94,5 | 92,0 |
| Actions of the system operator | 93,1 | 86,7 | 89,7 |
| Atypical actions | 79,0 | 81,9 | 80,2 |
| Accuracy, % | | | 87,6 |

**Table 6.** Precision, Recall, F-score, Accuracy using Random forest for cross-validation.

| | Precision, % | Recall, % | F-score, % |
|---|---|---|---|
| Automatic actions | 92,5 | 95,8 | 94,1 |
| Actions of the system operator | 90,9 | 87,1 | 88,9 |
| Atypical actions | 79,4 | 80,8 | 79,9 |
| Accuracy, % | | | 87,8 |

## 4. Conclusion

The result of this work showed the possibility of successful application of the statistical approach for data analysis in production. The results demonstrate the effectiveness of the implementation of such systems to optimize production and protect systems from outside interference. The resulting accuracy does not guarantee maximum safety and does not allow to completely replace the human resources, but it makes possible to detect problems at an early stage and simplify the production process.

## References

[1] Mitchell T 1997 *Machine Learning* (New York: McGraw-Hill Education).

[2] Krawczyk H 2001 *The Order of Encryption and Authentication for Protecting Communications (or: How Secure Is SSL?)* Advances in Cryptology – CRYPTO 2001 (Santa Barbara, California) 310–31.

[3] Erman J, Arlitt M and Mahanti A 2006 *Traffic Classification Using Clustering Algorithms* Proc. of the 2006 SIGCOMM Workshop on Mining Network Data (Philadelphia) 281-6.

[4] Tiwari D, Mallick B 2016 *Int. J. Comput. Appl.* **147(3)** 1–5.

[5] Altman N 1992 *Am. Statist.* **46(3)** 175–85.

[6] Ho T K 1998 *IEEE Trans. Patt. Anal. Mach. Intell.* **20(8)** 832–44.

[7] Lim Y, Kim H, Jeong J, Kim C, Kwon T and Choi Y 2010 *Internet Traffic Classification Demystified: On the Sources of the Discriminative Power* Proc. of the 6th Int. Conf. on Emerging Networking Experiments and Technologies (Philadelphia) 9–20.

[8] Kohavi R 1995 *A Study of Cross-validation and Bootstrap for Accuracy Estimation and Model Selection* Proc. of the 14th Int. Joint Conf. on Artificial Intelligence (San Mateo, California) **2(12)** 1137-43.

# Design and Implementation study of Remote Home Rehabilitation Training Operating System based on Internet

**Jin Zhuo[1], Jang Chung Gun[2]**

[1]Bozhou University, 236800, Marine Convergence Design, Pukyong National University

[2]Dept.of Visual Design, Pukyong National University

**Abstract.** The proportion of rehabilitation doctors and patients mismatch is very grim in the context of social aging. The Family Rehabilitation System captures the profound information of the trainer's movements through the kinect bone tracing technique, allowing the doctor to remotely master the patient's training progress. With the help of computers and the Internet, the patient can consult a physician, while the physician can remotely guide and launch the training "prescription" through the Internet according to the training effect. Patients can have rehabilitated training at home. The results of the test showed that the system has a positive effect on the rehabilitation of the patient.

## 1. Introduction

The number of patients with motor dysfunction caused by hemiplegia and stroke increased. In order to promote better recovery of their body muscles, patients are still required to perform rehabilitation exercises in the community or family after the treatment of discharge. However, there are still some difficulties in community rehabilitation for patients with motor dysfunction: (1) The number of therapists on-site services is scarce and expensive; (2) In the absence of standard and systematic action guidance, the patients ' own training is not only the science is not high and the effect is limited. (3) Patients need to be trained in special environments such as rehabilitation centers, and wearing complex training equipment is inconvenient for them. The family rehabilitation system collects the depth information of the trainer's movements through the Kinect skeletal tracking technique; With the help of computers and the Internet, patients can consult physicians, and the doctor through the Internet remote guidance and open training action "prescription" according to the training effect so that patients at home can be rehabilitation training. This liberated the physician's labour force and formed a network community that was closely linked to the hospital and regularly received "training prescriptions" to improve patient rehabilitation.

## 2. The foundation of rehabilitation theory

Exercise therapy is the use of physiological stimulation to regulate the function of multiple systems and organs, promote the body remodelling and regulate the organization function to adapt to the needs of the treatment of movement. The basic mechanism is to improve body metabolism, nervous system, heart and respiratory capacity, so as to maintain or restore the patient's original motor function, thereby helping the body to form a new compensation mechanism. Exercise therapy is the foundation and core of physiotherapy and it is also the obvious method of effect. It combines the disciplines of

IOP Conf. Series: Materials Science and Engineering **317** (2018) 012045 doi:10.1088/1757-899X/317/1/012045

biomechanics, kinematics and neuroscience. Active motion refers to the active use of their own strength to complete rehabilitation training rather than using external forces for training. It is in the recovery period of the body, the muscles have a certain strength based on the use of human active movement to improve muscle strength and control of muscle movement. Training program from simple to complex, power from small to large, gradually improve the difficulty.

## 3. Design and implementation of the system

The system hardware consists of a driving vehicle, a Kinect camera and a computer. Display of the rehabilitation action to guide the disabled system to complete rehabilitation training. The Kinect sensor is installed in the range of 0.4m~8 m, which is used to capture limb and body tilt angles in patients with limb rehabilitation movements. The computer receives the automatic capture data obtained by the Kinect sensor and contrasts with the standard rehabilitative action, giving the correct voice reward and incorrect voice prompts. The computer is transmitted to the external computer through the network, and the doctor can judge the training completion condition according to the data, then through the computer remote instruction (figure 1). When the patient is standing in training, the hydraulic spring will lift the human body to relieve the lower extremity endurance. When the patient stands up, the elastic reduction band and the folding reduction structure will lift the seat vertically, increasing the space of the patient's walking training. When the patient is tired in training, they can turn the seat 90 degrees horizontally and rest on it. At the end of the training, the car will be folded to reduce the space occupied. This system is especially suitable for the training of patients with lower extremity strength.

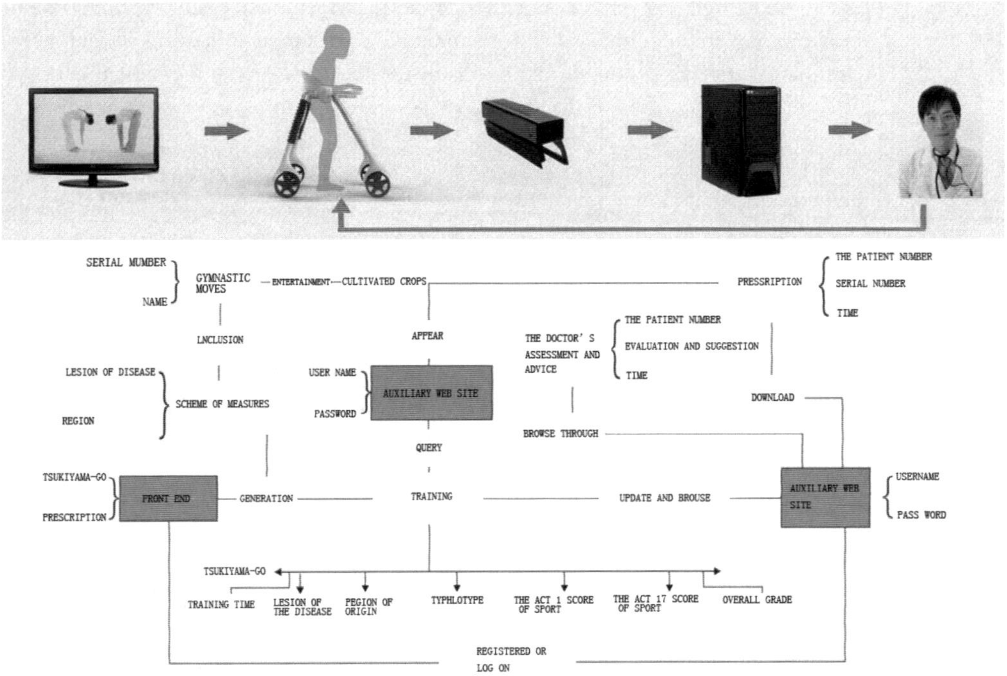


**Figure 1.** Design and implementation of the system.

The system solves the following problems: 1, through the extraction of Kinect depth information guidance training. The equipment and environment without any high requirements. So the patient was freed from the limitations of training equipment and environment. 2. It is more scientific to design training action according to medical principle. 3, unlike other game training or only a few methods of training action, the system has multiple action, multiple mode of training. 4, this system has the rehabilitation training front-end module, the Center website module, the Network Exchange module, forms a network of community which patients and hospitals are closely linked.

## 4. The system structure and workflow

### *4.1. Front-end system*

The front-end system consists mainly of Kinect sensors and monitors. Kinect is a highly-performing sensor launched by Microsoft. Its core components include four parts of infrared projector, color camera, infrared depth camera and chip, and capture range of 1 to 8 meters. The data information that Kinect sensors can provide is: depth information, video, and audio three kinds. The functions of Kinect in the system are as follows: the input and storage of patient data, the setting up of rehabilitation plan and the voice reminder during training according to the side of sports obstacle, rehabilitation part and training mode. The monitor mainly displays the action template and the real-time display patient training picture and result appraisal after the training completes through the Chat window and the PDF document feedback training.

### *4.2. Background real-time data processing module*

After the data acquisition and initial processing, the system carries on the real-time score to the patient's movement data according to the angle value and the position range value of the corresponding joint connection. In the Data acquisition module, the angle between the 3 joint dots is calculated by using formula, such as formula (1):

$$\theta = \arccos\left(\frac{D(M,N)^2 + D(N,P)^2 - D(P,M)^2}{2 \cdot D(M,N) \cdot D(N,P)}\right) \tag{1}$$

Through the formula (2), when the Kinect captures the joint motion, the system can obtain the relative angle between the nodes at any time, thus calculating the angle change $\Delta\theta$ of the joint point and the center joint point of the motion in the range of $\Delta t$, and suppose that the motion joint point corresponds to the angle range of the variable K, K determined through the formula (2):

$$K = \begin{cases} 1, & |\Delta\theta| \geq \theta_t \\ 0, & |\Delta\theta| \geq \theta_t \end{cases} \tag{2}$$

In each data acquisition process, the system calculates the angle change $\Delta\theta$ between current time and the front $\Delta t$ moment, then obtains the data of K to calculates. When the k= false, continue to capture the position of the joint point of movement information, when the k= really display the corresponding audio information and text message reminder, end and start the next action of information collection.

### *4.3. Design of system action template library*

Action Template Library consists of action template and scoring algorithm, which is not only an important part of the whole system, but also the core of the rehabilitation training front. The action template and scoring algorithm have the function of providing training guidance for patients and provide the basis to evaluate the user training. According to the characteristics of each action to select the points of grading, such as the angle of the key position, the range value of the joint position and the relationship between different joint points and so on. The action template gallery as shown in table 1

**Table 1.** The action template gallery as shown.

| Type of action | Action Name | Action explanation |
|---|---|---|
| First joint action | Flexion of the shoulder joint | Arm straight forward slowly up, Motion range 0°~180° |
| | Shoulder joint stretching | Arm straight back slowly up, Motion range 0°~50° |
| | Shoulder Joint Outreach | Arm unbend along the midline side of the body slowly up, motion range 0°~180° |
| | Flexion of elbow joint | A small arm bends along the elbow, and the range of motion 0°~180° |
| | Palmar flexion of wrist joint | The forearm is horizontally straightened and the palms rotate downward along the wrist joint, ranging 0°~90° |
| | Dorsal flexion of wrist joint | The forearm is horizontally straightened, and the palms rotate upwards along the wrist joint, ranging 0°~70° |
| Compound action | Flexion of the shoulder joint+Palmar flexion of wrist joint | flexion of shoulder joint 90°,rotate palms downward |
| | Translation+Palmar flexion of wrist joint | The palms move horizontally and then rotate downward. |
| | Translation+Flexion of the shoulder joint | flexion of shoulder joint 90°, then do the horizontal move |
| | Shoulder Joint Outreach+Flexion of elbow joint | Shoulder joint outreach 90°, then do elbow flexion movement |

*4.4. Patient dynamic real-time match person display module*
The system set the Kinect collection information for 30 frames/sec, and in the monitor dynamic real-time playback contains 20 of the human body joint point of the match person's screen, patients can refer to the match person's screen to know their training action is standard and timely adjustments. A total of 20 joint points were collected by the Kinect bone tracking technique.

*4.5. Center website and Network communication module*
The purpose of the development center website is to facilitate patients ' access to personal training, physician advice, download new "rehabilitation prescriptions" and doctors can check patient information and distance guidance training. The site consists of three parts: (1) Public section, including home page, landing pages and registration page. (2) The patient user part, including the patient after landing the total page and each function page set on the page, as shown in figure 4. (3) The Doctor User section, including the total page after the doctor landed and the collection on the page's various functional pages.

*4.6. Assisted driving*
It is mainly used to assist the patients to rise, walk and rest in situ. When the patient is standing in training, the hydraulic spring will lift the human body to relieve the lower extremity endurance. When the patient stands up, the elastic reduction band and the folding reduction structure will lift the seat vertically, increasing the space of the patient's walking training. When the patient is tired in training, they can turn the seat 90 degrees horizontally and rest on it. At the end of the training, the car will be folded to reduce the space occupied.

## 5. System testing

In order to test the stability and reliability of the system, the elbow flexion, the combination of two hands, the flexion kapyong shift of elbow joint, the 360-degree rotation of the head were selected and 4 actions were tested 100 times. Then compare the correctness of the actual action and the system recognition. The results show that the system can fully recognize the flexion of elbow joint and the movement of both hands, and the flexion kapyong shift of the elbow and the rotation of the head 360 degrees 5 times are not recognized. The reason is that the tester moves faster and causes the system recognition error. Overall, the stability and reliability of the system can meet the needs of patients.

## 6. Concluding

In the background of the aging of the Society, the increase of the patients with motor dysfunction and the shortage of the medical staff, the family rehabilitation training system can liberate the doctor's labor force and improve the patients' rehabilitation effect. It has a positive effect on the rehabilitation of patients on the basis of good interaction between human and computer and doctor and patient. The test results show that the stability and reliability of the system conform to the design requirements and have certain application value.

## Reference

[1] Yao Q, Wang L, Hou X T 2015 *The World Electr. Prod.* **8** 32.
[2] Yang Y, Xu Y, Zhang J J et al. 2014 *Micro Comput. Appl.* **3** 56.
[3] Jiang Z X 2015 *Research and design of human rehabilitation system based on Kinect*. Ocean University of China, pp. 21.

IOP Conf. Series: Materials Science and Engineering **317** (2018) 012046 doi:10.1088/1757-899X/317/1/012046

# Multimedia Image Technology and Computer Aided Manufacturing Engineering Analysis

**Song Nan**[1]

[1]College of Arts and Media in Shenyang Institute of Technology 113122

**Abstract.** Since the reform and opening up, with the continuous development of science and technology in China, more and more advanced science and technology have emerged under the trend of diversification. Multimedia imaging technology, for example, has a significant and positive impact on computer aided manufacturing engineering in China. From the perspective of scientific and technological advancement and development, the multimedia image technology has a very positive influence on the application and development of computer-aided manufacturing engineering, whether in function or function play. Therefore, this paper mainly starts from the concept of multimedia image technology to analyze the application of multimedia image technology in computer aided manufacturing engineering.

## 1. Introduction

In the new period of increasingly advanced science and technology, multimedia video technology generally appears in the public perspective, it can be said that in the new era of socialist modernization, multimedia image technology becomes more and more important. So what kind of high-tech technology is the multimedia system? In short, the normal operation and development of the multimedia system is done with the help of the computer, it has comprehensive and integrated treatment of text, language, music and picture, etc. by the medium of computer, which greatly improves the quality of life and the level of people's lives [1]. Therefore, the better research on multimedia imaging technology has the positive significance of improving the quality and quality of the computer aided manufacturing engineering in China.

### *1.1. Conceptual analysis of multimedia image technology and computer-aided manufacturing engineering*

There is a great correlation between multimedia image technology and computer aided manufacturing engineering. To explore and analyse the application of multimedia image technology in computer aided manufacturing engineering, the first thing is to have a careful grasp of the concept of multimedia image technology and computer aided manufacturing engineering.

*1.1.1. The concept of multimedia image technology.* Multimedia imaging technology, in a nutshell, is included in the multimedia system. The basic principle of its operation and working process is that the comprehensive processing of information such as text, graphics, image, sound, animation, video, etc by the medium of computer, and establishing a logical relationship based on its inherent relationship and characteristics, in order to maximize the human-machine interaction. For example, in most schools today, every classroom basically has a multimedia system, teachers use the basic systems for lessons and use the courseware PPT slides, video and other forms for display, which is the simple application of the typical multimedia image technology in real life. It is noteworthy that the real multimedia technology involves the creation of computer technology, and other simple things, such as movies, TV and acoustics, do not belong to the category of multimedia technology [2]. Therefore, it is important to take a detailed grasp of multimedia image technology to make better use of it, in order to promote the progress and development of computer aided engineering field and multimedia imaging technology in China.

*1.1.2. Function introduction of multimedia image technology.* Only understanding the concept of multimedia image technology and its function, can give full play to its inherent advantages and promote the development of productivity. Specifically, multimedia image technology has many functions under the guidance of multimedia system. First, from the angle of audio reception, it can receive and further store the audio files of audio microphones or other systems. Second, from the perspective of image transmission and preservation, it can receive, process and further store images of scanners or other systems. Third, from the perspective of the movement TV signal, it can search, receive and further store the movement TV signal to a maximum extent. Fourth, from the perspective of integrated information processing, it can comprehensively process and integrate the moving video, the audio information and the video information to make their information more logical and relevant. Fifth, from the perspective of text, graphics, sound, etc., the new media imaging technology can integrate the information of the three and realize full synchronization playing, for example, when we watch a movie in a movie theater, there is the combination of subtitles, sound and video. The function of the sixth aspect is mainly reflected in the animation aspect, which can make and play the animation, such as the cartoons we have saw in the market. The function of the seventh aspect mainly focuses on the input and output of external devices such as compact disc and video tape, which can transmit information about video, audio, image and so on by using them. For example, now many teachers copy the courseware to the U disk, then copies the courseware to specific playing equipment through its connection with multimedia equipment, which is a good show of making full use of its function.

*1.1.3. The characteristics of the multimedia image technology*

The multimedia image technology that based on the high and new science and technology has many characteristics and advantages in the process of concrete use and display. First, for the multimedia image technology, its storage of information is very large, which has great breakthrough compared with the traditional information storage technology. Second, it can be fully realized the functions of real-time push and display under the technical support, that is to say, the real-time sharing of information is very good. For instance, enterprises in the process of the meeting can use multimedia technology to display the content of the meeting, and some of the latest information related to the meeting can also realize real-time transfer to each other, which greatly improves the efficiency and save the cost of the meeting. Third, the multimedia image technology in the digital era can be directly supplemented in the process of information replenishment. Fourth, from the point of view of space, it takes small space, which is advantageous to the conference, exhibition, etc. Fifth, the multimedia image technology has a strong interactivity, in other words, the user and the machine is a relationship of mutual choice, users can choose their own favorite video, which gives full play to the subjective initiative of people.

*1.1.4. Conceptual introduction of computer-aided manufacturing engineering.* Computer-aided manufacturing engineering is a broad concept of engineering. Specifically, it needs to be on the basis of the computer aided software to conduct a comprehensive design and development of its content, and in the process of design and development, depending on the degree of demand and the application of specific software products to adjust their work. Therefore, in general, computer-aided manufacturing engineering is a software design and development project relying on computer technology, which plays a huge role in real life.

## 2. Application research of multimedia image technology in computer aided manufacturing engineering

### *2.1. Application research of multimedia image technology based on CAD/CAM engineering education*

With the continuous development of our socialist modernization construction, the development of education and training industry is more and more quickly, the application of multimedia technology is becoming more and more popular, which is developing towards the direction of more and more mature. In the process of specific training education, the project models that generated by multimedia image technology are often used to present course information. Multimedia technology under the image system is particularly suitable for engineering education of CAD / CAM, the reason is that in the process of specific training education, the course information and content are complex, there are not only some the pictures, video need to show, but also some complex operation method, etc. Therefore, in the process of CAD/CAM engineering education, we should make full use of multimedia education system to complete the education of theoretical knowledge and practical to promote the further development of CAD/CAM engineering education in China.

### *2.2. Application research of multimedia image technology based on product sales*

For product sales, multimedia image technology is an important link in engineering manufacture, in the process of specific product sales, multimedia image technology can make a system analysis for the market demand, product after-sales of manufacturing engineering, which has a positive significance for promote the production quality and level. In the process of its concrete operation, depending on the specific multimedia technology to select product sales system, under the system, the factors such as user demand, market competition can be comprehensively counted, on this basis, the enterprise can be targeted to do product sales plan, which greatly reduce the risk of market competition, improve the degree of targeted product sales.

### *2.3. Application research of multimedia image technology based on touch technology*

Under the development trend of diversification, multimedia technology not only has wide application in the CAD/CAM engineering education, product sales, but also has very big development space in touch technology. In particular, touch technology is actually an important part of multimedia technology, its core content is to use the touch screen to input and analyze their specific information [3]. The screen of touch technology, users simply click on the screen, and then choose the services they need, they can complete the basic operation. For example, at the information age today, almost everyone has a smart phone, which is the result of make full use of multimedia image technology. Therefore, the combination of multimedia technology and touch technology has improved the level of modernization of our country, and it also plays a big role of promoting economic development and social progress.

### *2.4. Application research of multimedia image technology based on engineering document management*

For specific computer engineering, the basic file management is very important, in the process of enterprise production and management, there will be large amounts of information and data, for example, in the electric power enterprise's file, it will contain basic employee information, financial information, and the power equipment information, etc. So the information integration work of data and documents is very important. In this case, some computer engineers use of multimedia video technology development and design with a special file management system, under the system, it can not only classify the information, but also can systematically save the images, video and so on. When users viewing the file, they can search according to the specific requirements, which greatly improve the efficiency of file management.

### *2.5. Application research of multimedia image technology based on dynamic engineering*

In addition to the operations of saving, transferring pictures, videos, and so on, the application of multimedia technology in the field of computer aided engineering, dynamic engineering has great advantages. Dynamic engineering is to rely on a certain system, such as CAD graphics, data information automatically generates dynamic images and the cartoon that is currently on the market is a good example. Advanced multimedia animation technology not only can be in the form of animation display images, can also reflect quality and fluency of animation video to the greatest extent [4].

### *2.6 Application research of multimedia image technology based on CAD/CAE engineering*

Under the multimedia system of engineering products, with the increasing of user demand and market competition factors, CAD/CAE engineering gradually appear in the public's view, and, in the actual operation process, computer engineers have developed different system for different users. CDI company, for example, has developed a special computer system according to the demand of itself, under this system, it can realize the setup of graphics and entity model, then on this basis, comprehensive data analysis is carried out on the information, data, etc. Therefore, the application of multimedia technology system based on CAD/CAE engineering has a great positive significance of the development of new products and the adaption of the new demand.

## 3. Conclusion

In the fierce market competition, science and technology more and more become a powerful competitive factors. In recent years, the multimedia image technology is widely used in computer aided manufacturing engineering, which provides a great convenience for engineering design and development. Therefore, we should attach great importance to multimedia video technology and give full application of it.

## References

[1] Lv Y M 2010 *Mod. Propert. (Middle periodical),* **9(04)** 43-44 + 71.
[2] He F 2013 *Application research of multimedia image technology in engineering machinery display design.* Hunan University.
[3] Sun X Y 2008 *The invasion of multimedia image technology*. Shanghai Theater Academy.
[4] Dai Y, Li T, Hu D M 2000 *J. Anhui Radio & TV Univ.* **1** 86-91.

IOP Conf. Series: Materials Science and Engineering **317** (2018) 012047 doi:10.1088/1757-899X/317/1/012047

# Application of multimedia image technology in engineering report demonstration system

**Jiang Lili**[1]

[1]College of Arts and Media in Shenyang Institute of Technology 113122

**Abstract.** With the rapid development of global economic integration, people's strong desire for a wide range of global exchanges and interactions has been promoted, and there are more unprecedented convenient means for people to know the world and even to transform the world. At this stage, we realize that the traditional mode of work has become difficult to adapt to the changing trends of the world and informatization, multimedia, science and technology have become the mainstream of the times. Therefore, this paper will mainly analyze the present situation of the project report demonstration system and the key points of the work and put forward with pertinence specific application strategy of the integration with multimedia image technology.

## 1. Introduction

Application of multimedia image technology in engineering report demonstration system can solve the disadvantages of traditional demonstration process to a great extent (such as destruction of environment, time consuming, demonstration rigidity and lack of attraction, etc.), adding to the attraction of rigid engineering presentations filled with data, mapping, etc. The use of PPT, short video, and even holographic projection technology greatly restores the engineering process and engineering results and is more intuitive; at the same time, a large number of pictures, images take the place of traditional paper-based documents and digital data appears in report demonstrations, increasing the interest and attraction of the demonstrations and making them fresh and impressive.

## 2. Present situation and key points of project report presentation

### *2.1. Status of engineering report presentation*

Project report presentation is a complicated system engineering that requires the consumption of a large amount of manpower, financial resources and material resources, in addition, there are the characteristics of engineering construction itself (such as long cycle, large investment, strong comprehensiveness of technology, and the likeliness to be influenced by topography, geology, hydrology, meteorology, transportation, social economy, etc.), therefore, it is very important to improve and perfect the project presentation system scientifically and reasonably [1].

At present, in the process of project report presentation, the main methods of collecting data and materials are estimation statistics, manual reports, document delivery and photo recording. This method is with heavy workload and low efficiency, and it is difficult to ensure the timeliness and efficacy, nor can it macroscopically manage the construction progress of the whole project.

### *2.2. Keynotes and difficulties of project report presentation*

Large engineering building area, complex structure and difficulty in accurate measurement. In the process of large-scale engineering construction, there is often involvement of large construction area and very complicated structure; athe time of traditional data collection and mapping, manual fixed point mapping and manual drawing are the main measures; not only does this impose a high requirement on the quality, ability and knowledge system of professional surveying and mapping personnel, but it is time and energy consuming with huge workload. And because it's manual calculation, mistakes are very likely to happen, leading to the late mapping and data compilation waste and in the final engineering report presentation system, mature results are hard to be actualized [2].

### *2.3. Plenty of working high above the ground that is hard to observe for blind spots*
In the whole process of data acquisition, aerial photography is often needed, which requires the specialization and precision of equipment; meanwhile, manual aerial photography is not perfect, often there will be blind spot phenomenon; as rarely can the perspective switching be perfect, there can be passive impact on the final results.

### *2.4. The report lacks visualization*
Traditional engineering presentations are mainly based on paper as the main medium, and report demonstration are performed on paper files of a large amount of printed documents, materials, even mappings. But in the actual engineering, there will be some special circumstances that are difficult to be intuitively reflected on paper documents. For example, the special nature of the terrain cannot be directly reflected in the paper, but report demonstrations are without the possibility of field observation; this contradiction is one of the most important and difficult points in the project presentation system at this stage.

### *2.5. It is difficult to realize paperless office*
In the preparation phase of the project reporting demonstration system, we need to collect a large amount of data (mostly paper-based), and we will be surveying and mapping the project site, producing draft 1, 2 and even 3; in the early preparation process, if there is a simple mistake (for example, data entry mistake in the early stage), all the documents need to be printed again, which adds to the difficulty of preparation. Even if final documentation is formed, more papers can be needed in the process of batch printing and compiling into books due to multiple numbers of leaders in the report demonstration. At this stage, with the continuous reduction of limited resources and the needs of building a "resource-saving and environmental protection" society, the implementation of paperless office is inevitable choice, but as mentioned in the previous paper, traditional engineering demonstration system is very demanding for the quantity of paper, so it is extremely unfavorable for the implementation of paperless office.

## 3. Multimedia image technology
With increasing requirements on engineering report demonstration system and multimedia production of electronic images, as in the concept generation of multimedia "image multimedia" in narrow sense; as in presentation of design and engineering results via text dubbing, graphic representation, audio and video productions. From the perspective of the system, there are many types of multimedia presentations in architectural or engineering presentations, including multimedia bidding demonstration, PPT bidding demonstration, video bidding, 3D demo bidding, demonstration of indoor and outdoor design scheme, building decoration scheme demonstration, landscape design bidding demonstration, planning scheme demonstration, project bidding demonstration, interior decoration project demonstration, planning and design report presentation, post competition report demonstration, project achievement report presentation, landscape design scheme bidding, government procurement bidding, exhibition hall design bidding, demonstration of the achievement award of scientific research project, scheme optimization report, multimedia demonstration of urban planning scheme and multimedia demonstration, etc., the production requirements and development of project report demonstration system change rapidly.

## 4. Application of multimedia image technology in engineering report demonstration system

### *4.1. Application of 3D scene model in engineering process management*

3D dynamic visualization digital model of construction schedule is established basing on the 3D visual digital modeling of construction site topography and various kinds of ground objects, solving key issues concerned by engineering management decision makers and urgent to be solved - visualization of engineering progress. 3D modeling based on oblique photography put all the management objects in a real 3D world, realizing the visualization of data and the true meaning of "WYSIWYG" in the sense of management. There is no need for staff to go to the scene personally and the construction site, construction schedule and construction problems are clear; for areas inconvenient for staff to come (such as the bottom of the pier and near-water steep slope, etc.), the environment and project information can be easily checked. In addition, the 3D scene model can provide abundant and detailed information about the water temperature, geomorphology, environment, humanity and traffic around the construction site, which is conducive for the comprehensive evaluation of construction conditions by the construction site managers and the formulation of targeted resource scheduling and configuration scheme.

If oblique photography 3D model is adopted, then there is no need for the installation of any software and the images can be uploaded to Wish3D Platform, and the 3D data publishing, display and sharing function can be utilized; there is also integration of flight browsing, tagging, measurement and other characteristics of the application. Large errors can occur in traditional surveying and mapping when the project site is large, time-consuming and requires massive labor, the results can have difficulty in terms of statistical data, and it is difficult to make timely analysis in case of emergency. Real time measurement of ground objects can be carried out by UAV oblique photography modelling, 3D spatial analysis can also be conducted, especially in complex terrain engineering. The powerful oblique photography 3D model enables a panoramic view of the situation without the need to present at the site, bringing new construction management mode and operation decision mode and solving the problem of visualization of construction schedule, it can not only help with the presentation and report of the executants but provide decision basis for the managers.

### *4.2. Improve the professional quality of the reporters and actualize bidirectional promotion of the multimedia imaging technology and reporters.*

To improve the professional quality of the reporters, a full understanding of the presentation plan is needed with skilled and magical craftsmanship, the use of multimedia imaging technology should also be mastered to ensure a smooth progress of the reporting process [3].

Reporting program should be with "transposition thinking" and understanding of the thoughts of reporters, this can make communication more efficient. And the so-called thinking, of which a large part is - data. It's better for each of our decisions to be supported by correspondent data. Pictures produced based on this data logic are able to withstand test outs. Another content of thinking is logic. For large enterprises, brief introduction of the entrance position and flow line organization are not enough, or simply several color flow charts. The causal relations between the location of the entrance and exit and the surrounding land and the urban road should be clearly identified and expressed. Drawing based on deduction logic is a graph that can stand up to scrutiny. Fully understand what the value of design is: finding problems - analysing problems - solving problems at different stages of project advancement. This is the core value of the design, and our report should follow the same train of thought. In addition, the focus should be placed on improvement of professional quality, now the professional quality of the architects is more than column grid arrangement and space partition, on the basis of mastering basic skills, knowledge of relevant building types such as big business, hotel, etc. should be dabbled at; this will greatly improve the efficiency of communication and realize the bidirectional promotion of multimedia imaging technology and reporters.

### *4.3. System design should follow the logical process sequence*

Logicality is an important factor in the reporting demonstration system. Many people don't know where to start when they design a demo process or elaborate on a project, then the following design order is recommended, and it can be deleted or subtracted according to specific requirements and actual situation.

First, a brief account of engineering background to be presented with data and pictures. For example, where is the project, the size of it, what it is and why the project, is there something influencing the project and the problems to pay attention to; the second is the explanation of the project starting point, what is the starting point of the project? What problems need to be solved (or your idea); Third, start the planning layout of the project with multimedia image technology, axis, entrance, node, road, house layout, style, skyline, city facade, etc. need to be exhibited; fourth, introduce the general principles and specific space and the consideration on orientations; fifth, expound the engineering advantages and "scientific and technological natures" such as energy conservation, environmental protection, greenness, barrier free and materials, etc. meanwhile, in the process of system design, it is necessary to increase or reduce some parts according to the response of the reporters, if the owner is familiar with the terrain, no elaboration is needed; if the owner is particular about multi-building monomer modelling, this should be indicated in the design in detail.

### *4.4. Integration of holographic technology into engineering report presentation system*

Holographic technique is the technique of recording and reproducing real 3D images of objects by means of interference and diffraction, it can not only produce 3D aerial illusion but make the illusion interact with the performer, completing the performance and produce stunning effect. Different from the holographic glasses of Microsoft hololens, holographic cultural experience is no need for wearing any accessory equipment, and the position is "unique immersive holographic cultural experience", it is more three-dimensional, realistic and the visual effect is more shocking, and the effect of expression is more immersive. Adoption of holographic technology for engineering report demonstration and recording holograms by means of electronic components omit the later chemical processing and plenty of time, realizing real time processing of images. Meanwhile, quantitative analysis can be down on digital images on computer, intensity and phase distribution of image can be obtained via calculation and simulation of superposition of multiple holograms and other operations can be conducted, greatly restoring the actual situation of the project and construction process and providing visual impact of authenticity and intuition for people who watch presentations.

## 5. Conclusion

Along with the rapid development of computer multimedia technology, multimedia is with more attention and application by means of its own advantages, and it has been influencing many aspects of our lives imperceptibly. In the application process of multimedia image technology in project report demonstration system, it has developed rather mature in terms of both hardware and software. However, there are still some shortcomings in the targeted development and more in-depth technical research. Only by realizing the more standardized and mature development of multimedia imaging technology, can we better serve the engineering report demonstration system.

## References

[1] Wang E Q 2017 *World Nonf. Met.* **16** 284-5.
[2] Dong X J 2015 *Comprehensive application of 3D spatial imaging technology in Geological Engineering*. Chengdu University of Technology.
[3] Wang Y, Wang H X, Wu C S 2003 Northeast Water Conserv. Hydropow. **9** 53-4.

IOP Conf. Series: Materials Science and Engineering **317** (2018) 012048 doi:10.1088/1757-899X/317/1/012048

# The Analysis on the Integration and Embodiment of Traditional Cultural Element in Environmental Artistic Design

**Bao Ling**[1]

[1]He University, 110163

**Abstract.** For this phase of environmental artistic design, the traditional culture element is one very precious design element, but it has difficulty in breaking out of its shell, and that looks too outdated, however, the traditional culture element would be more peculiar if ponderously adding some elements. This paper will further analyse the integration and manifestation of traditional culture element which from the environmental artistic design, it aims to integrate the tradition and modernity perfectly and give the spectators a refreshing and unconventional sense of design.

## 1. The overview of environmental artistic design

The environmental artistic design has been an emerging subject in recent years in our country, it is widespread in landscape design, the house decoration design and street design, etc. The environmental art design utilizes some means of organizations and enclosure, brings the space interface (the indoor-outdoor cylinder wall, floor, ceiling and windows) to an art form (form, color and texture), applies natural light, artificial illumination, furniture, decoration and modeling etch design language, and the configuration of plants, flowers, waters, gardens, sculptures, to make the indoor and outdoor space environment of buildings can embody a style and specialized atmosphere, to meet people's functional and visual aesthetics needs.

## 2. The characteristics of traditional culture element

In the field of environmental art design, the traditional culture mainly characterized by its a variety of classical design philosophy, and those application examples based on different traditional culture elements and environmental art which are extended by that. The designers must keep the traditional culture elements integrating with modern environment when they excavate out it, especially in the process of inheriting and carrying forward traditional culture elements, the modern environmental artistic connotation should be excavated and presented, and the designers should expound the cultural features of modern environmental artistic design based on the application of traditional culture elements.

### *2.1. Most of materials are woods.*

For Chinese, they have a special feeling for the wood. In the five elements, the wood is the only organic matter and vital one among earth, water, metal, fire and wood. It means sunlight, warm and harmony, so it is mostly used in the Chinese furniture. This is probably related to people's thinking habit, Chinese thoughts lay particular stress on the moderation, harmony and sensibility; the westerners are usually biased towards the practicability and rationality, so eastern and western countries have their own preference of residential architecture. The ancient revers heath heaven and earth than modern, the five elements are the soul of life. The architectures take the meaning of wooden to keep its life, integrate the living environment and nature [1]. Chinese architectures has few decorative patterns, the woods to be chosen are based on the interaction between space and time of shape, color and taste. When choosing those woods with hundreds of years or thousands of years, its longterm fate has been considered. Later, more and more decorative patterns on the wooden architecture has based on it.

*2.2. Strong color*

In Chinese traditional cultural elements, the relationships between the color and craftsmanship, color and art, color and poem, color and customs are closely. The urban construction, mural and painting has widely applied the colors. For instance, after the Ming dynasty, in former capital ----Beijing, only the influential officials and King children can live in the building with red walls and yellow tiles, the ordinary people's building is built by black bricks and grey tiles. But on the painting pillars and carving beams, the folk buildings are mostly black roof and white walls. In the Dunhuang Caves with over 1500 years of history, ten thousands of precious murals are preserved in it, the mural colors are different in different periods. For instance, in the Norther Wei dynasty, the reddish brown with blue and black are commonly used; in Tang Dynasty, yellow is added, the application of colors are multifarious, dazzling and flamboyant; in Song Dynasty, the blue-green is used in common. Chinese paintings shows the charm by the shades of ink, for "there are five colors in Chinese ink" "being dazzling without painting", it can expresses the transcendent artistic state among the shade.The five colors of ink are charred, thick, grey, light and clear.

*2.3. Most styles are filled with nationality and regionalism*

In the traditional civilization of agriculture and feudal society, many territories have formed their own group. The Mongol located in the northwest, the Mongolians have moved to the prairie and lived on nomadic production, so their cultural characteristics have presented an bold and unrestrained attitude; but the Han Chinese that live in the central China always pay more attention to study cultural knowledge, they are determined to work on the official career, so they are more gentle and cultivated. If the style of strong regional characteristics and national characteristics applied in the design, that would be same as that. The most of nomadic people live in the Mongolian yurts, the patterns are mostly the eagles and wranglers; but the people in central plains pay more attention to paint the ornamental plants, such as flowers, birds, fish or bugs, it is relatively common that carve a bird on the roof.

## 3. The issues of traditional cultural elements in the modern environmental artistic design

*3.1. The inelasticity and stiffness*

The design itself has owned its fundamental rules, to a large extent, the tradition and modernity are "labels". So I think the starting point of designing is the rules of design, we do not have to worry about so-called tradition or modernity, this question is really a positioning error. What we need to discuss are those regular or irregular elements. These designs for so-called "modernity" are results gotten by the design rules, the design and its rule are modern products, so many ancient elements are inconsistent with modern design rules. China in particular, if you apply these elements into design mechanically, it must be difficult to be self-consistent. So it is not difficult to draw the conclusion that the biggest difficulties of integration between traditional cultural elements and environmental artistic design is in the change of traditional cultural elements, if change too much, you cannot experience the heaviness and historical feeling of tradition. On the contrary, the whole design would look like gruff. So what we should to make efforts to realize is to look for an equinoctial point between traditional and modern design, not only make the traditional elements to be felt but also adjust the rules of modern design. Hence, we should integrate and embody the traditional elements in the realistic environmental artistic design through some obscure, natural and harmonious methods.

*3.2. The color itself has no stationary*

The ancient color name are not widely used in modern, they are strange for people, and those strange ones may give people a sense of high-class, this seems like some designers like to use English.

Why we rarely use these colors? That was not because we didn't protect our traditional culture, but is they could not embody a fixed color. For the color of light yellow, it represents the color of gosling's villus, but the colors of two goslings are not the same, the colors of every villa of a gosling are not exactly alike; the mineral yellow is an ancient mineral pigment it is divided into realgar and orpiment, they are with great chromatic aberration, but the pigment produced by different sulfide oil are differ from each other, so these colors only have a rough scope, these values of RGB is marked to solve the chromatic aberration in different displayers and prints.

The differences between traditional color chip made by different websites are great, for example, the "fish belly white", some websites show the light blue, some are light green, some are light orange, It depends on how you look at it, indeed different color names have different color scope in different dynasties, for the color of cyan, it is represented as green, blue-green or dark blue and light blue.

## 4. The technique of expression of traditional culture in modern environmental artistic design

*4.1. The symbolic method*

There are only a few examples to apply the symbolic method, in most cases, people attempt to reserve those features of "tradition", but to make compromises on the issue of design rules.

**Figure 1.** The work of Ikko Tanaka.

Take the work from Ikko Tanaka, a famous Japanese environmental designer as the example, this work presents a design style of minimalism, the angular graph and strong color symbolism are his typical features. Only from its color and composition we cannot analyse the existence of traditional elements. (The only Japanese traditional element is the head at the top right corner, the gold thread on his head is a typical symbol of Japanese Lin School pattern), but the structure of the picture can give spectators space for imagination and association, people can fell the traditional elements unknowingly.

This is a handling method of traditional elements, refers to the symbolic method, integrated in modern design concept can make people associate with traditional elements, rather than applying traditional elements into design mechanically.

### *4.2. The integration method*

From the daily design case diagram, we can find the reason that we have directly applied some traditional elements in one modern poster, but it didn't do something wrong. Because these traditional works have some basic rules and characteristics of modern design, even it directly participation in the construction of modern design theory (If the Japanese do not blow their own horn. This is embodied in the application of slash in the composition), that is modality (the space cut by river); the emphasis of decoration of pattern (The lines and ripples of trunk); the purity and transformation of modeling, to pursue the purity of modeling and beyond the realism (this is embodied in the river modeling); disregard the distance method, to pursue its complication; the technique of collage patterns (Collage, the patterns of river and trunk). So you can see the application of traditional element in the modern works which is established by removing "non-modern" part in the tradition. However, its final presentation is still being modern design-oriented, that is fundamental principle. The position of traditional elements often plays a role of embellishment and symbolization. In a word, you cannot directly move them into your design without reservation. If we summarize it from the view of visual "tradition" "modernity", I think the modernity principle about design is definite, but the modernity in traditional elements is ambiguous, Only be needed to be explored and remold can it be used in the design. Therefore, to integrate well, the traditional arts need to be equipped with strong modern quality, so the wide gap between traditional elements and modern design would be not difficult to cross.

## 5. The analysis on integration and embodiment of traditional culture in modern environmental artist design.

### *5.1. The application of traditional thinking modes in the design*

Chinese culture has a long history, Chinese nation has approximately five thousands years' organic cultural histories, under this circumstance, comparatively speaking, we have suffered the more cultural impacts and shocks. From the "Letting all flowers blossom and all schools of thought contented" in Spring and Autumn period to the later "rejecting all kinds of theoretical schools but Confucius", it is not hard for us to see the Confucius has a great of impact on our thinking mode [2]. The goals of Confucianism want to achieve is: everyone is a real man and the people of whole society can live in harmony with each other. Therefore, in the process of integrating traditional thinking mode into modern environmental artist design, we should try to follow the sense of harmony between man and nature, to pay attention to the combination of colors, the spirits conveyed by it are also consistent. When we design the interior environments, we do not only select some popular fashion products, what's more important is to determine an environmental overall tone, to choose the style of simple, luxury, European or Chinese, next, choose an environmental design and layout based on its core tone. If a sculpture of "Venus de Mil" being put in an antique environment, the whole artistic design would be uncoordinated, even its aesthetic would be destroyed.

### *5.2. The application of traditional coloring in the design*

Almost all books are the same in theory, the color bathing is a broader concept, it belongs to the color composition, one of the artistic theory. If started from color wavelength and optical knowledge is too prolix, so I will expound the fundamental knowledge as follows.

There are three attributes in color, the hue (the most obvious feature of color, refers to the appearance of color, we commonly use hue circle to represent it.), the lightness (refers to the level of lightness of color, we commonly use the lightness axis to represent it), the purity (refers to the degree of purity of color, we can use the purity stage to represent it) [3]. The harmony and contrast of above three attributes have formed the rhythm of color, just like the note of music. For any design objects, the color does not exist in isolation, hence, if you want to present a better color, you take account of various comprehensive factors. The proper color matching is conducive to the design, it can form unique color style and culture.

### *5.3. The application of characters in the design*

Chinese characters are evolved from painting, so it is highly valuable in aesthetic art. Ancient China has “Six Scripts Theory” on the coinage, the ways of form of Chinese characters include self-explanatory, pictographic, echoism, knowing, mutually explanatory characters and borrowing characters. The development of Chinese characters form experience these stages: oracle, bronze inscriptions, big seal characters, small seal characters, official script and regular script, the boundary between ancient and modern characters are the stage from the small seal character to official script, its tendency is the characters are being stereotyped and symbolic, it is changed from the line to stroke, the round to the square. The seal script mainly uses the round brush, its pictographic degree is high and character pattern is archaic and abstruse. By the end of the Warring state till to the Qin dynasty, because the continual wars and over scheduled of the secretaries, they needed a kind of character for use which can be quickly written. On the basis of seal character in Warring state, through a large number of blending, to change the round of it to the square. But the marks of seal characters hadn’t yet been removed, we also can find out a great deal of marks of official script. We should integrate the Chinese characters into the practical environmental artist design, not only can feel more cultural deposits, but also can add glamour of history to the design.

## 6. Conclusion

The traditional decoration elements cannot exist with no change forever, but it is tough to change its image. The issues need to consider are much more than those pure innovative points, such as: what original element should be remained? What is new about designers? How to combine both of elements naturally? If the newborn design exist in both traditional and modern space, how to make it doesn’t look incompatible? These issues need us to do more research on the integration and embodiment manner of environmental art design and traditional cultural element, to realize a more and harmonious integration, to provide a better driving force to develop our traditional culture and arts.

## Reference

[1] Wu Y 2017 *The Design* **20** 130-1.
[2] Hu S H 2017 *Art Sci. Tech.* (2017-10-31).
[3] Chen C 2014 *The Pack. World Mag.* **04** 104-5.

# Research on Visualization Design Method in the Field of New Media Software Engineering

**Hu Deqiang**[1]

[1]Associate Professor, School of Arts and Media of Shenyang Institute of Technology 113122

**Abstract.** In the new period of increasingly developed science and technology, with the increasingly fierce competition in the market and the increasing demand of the masses, new design and application methods have emerged in the field of new media software engineering, that is, the visualization design method. Applying the visualization design method to the field of new media software engineering can not only improve the actual operation efficiency of new media software engineering but more importantly the quality of software development can be enhanced by means of certain media of communication and transformation; on this basis, the progress and development of new media software engineering in China are also continuously promoted. Therefore, the application of visualization design method in the field of new media software engineering is analysed concretely in this article from the perspective of the overview of visualization design methods and on the basis of systematic analysis of the basic technology.

## 1. Introduction

Under the guidance of scientific outlook (thinking) on development, the development of our country is increasingly moving towards the bond of direction of science and technology; we often say that science and technology are the primary productive forces, and productive forces determine the relation of production. That is to say, in the context of socialist modernization, great importance ought to be attached to science and technology, and only in this way can the effect of "half the work with double results" be achieved in the right direction. In recent years, the visualization design methods are gradually applied to various trades and professions with great promoting efficacy, especially in the field of new media software engineering. Therefore, in the development process, the positive role of visualization technology should be paid full attention to and it should be applied and practiced scientifically and reasonably.

## 2. An overview of visualization design methods

To better explore the relation between the field of new media software engineering and visualization design methods, the first thing to do is have a clear grasp of the related concepts of visualization design methods. Knowing what a special method that visualization design method is can impose a huge influence on the software development of new media, etc.

### *2.1. The meaning of visualization design method*

Visualization is a technical method, in simple terms, understanding of it can start from the literal internal basic concepts. That is, present the data, information and materials analyzed by computer to users in need in the form of images, helping the professional technical personnel complete some other observations and operations. Therefore, in certain sense, visualization is not only a technical method that help people complete their work, but also one that promote their orderly work, and it plays an irreplaceable positive role in the field of computer information.

*2.2. Conceptual analysis of visualization design methods*

Visualization method refers to the fact that after the initial data is converted to visualization elements, complex resource content is indicated in vivid and intuitionistic manifestation so that the user understanding can be deepened. Normally there are scatter diagram, histogram, time axis and tree graph, etc [1].

Visualization design method is a technical assistance involving various fields; to be more specific, it concerns various fields including computer graphics, image processing, computer vision, CAD, etc.; in actual work, it is a comprehensive technical means which can not only provide accurate basis for the analysis of data representation and data processing but it can influence the analysis of a series of important decisions to the largest extent. Visualization technology was first used in computational science, and formed an important branch of visualization technology -- visualization of scientific computing. Visualization in scientific computing can transform scientific data, including measurements, images, or digital information involved in calculations into physical phenomenon or physical quantity that is represented by graphic information and changes with time and space in front of the researchers and enable them to observe, simulate and calculate. Visualization in scientific computing was discovered and formally put forward in 1987, since its publication, it has been widely used and developed in various fields of engineering and computation.

*2.3. Classification of visualization design methods*

There are many ways to implement visualization, which need to be chosen according to different types of tasks. Firstly, the visualization methods can be preliminarily screened according to whether the original data is consecutive or not. For example, scatter diagrams are suitable for displaying discrete data such as resource topics, and time axises are suitable for displaying continuous data like the time of resource release. In the next place, as the key of visualization is mapping operation, controllable or uncontrollable factors in mapping often fail to achieve complete conversion from data to visualization elements. Therefore, information loss is bound to occur. The loss of information can be classified into unintentional and intentional loss. Unintentional information loss occurs due to the technical limitations of the visualization system and the cognitive limitations of the users and intentional information loss is the information selection conducted for the overall structure of resources rather than specific details. There are two typical categories in distinguishing visualization methods: direct conversion method and abstract conversion method.

*2.3.1. Direct conversion method in visualization design methods.* Direct conversion methods only involve simple data processing, and the visualization results presented to users are closer to the original features of data. In the case of unintentional loss of information, linear mapping of discrete data is the most direct method of conversion. For instance, in the design and development of new media software, collection and arrangement of information on the consumption types and consumption habits of different groups before developing ordering software for users; the information is relatively clear and easy to organize, and direct conversion method can be adopted to process the visualization computer image so as to complete the follow-up design and development work.

*2.3.2. Abstract conversion method in visualization design method.* Data obtained after abstract conversion tend to be abstract compared to the initial data after multi-link and multi-level processing. This method emphasizes the overall readability of data, usually requiring sacrificing individual data in exchange for an understanding of the overall data set. That is to say, in this case of information collection, factors for information to require technology integration increase; in the case of intentional information loss, implementation of nonlinear mapping on continuous data is the most abstract conversion method. Abstract conversion reduces the trivial of data compared with direct conversion and improves the readability. For example, in the development of everyday software such as the development and utilization of online supermarket APP, in the process of its design, it is necessary to conduct detailed distinguished statistics between offline groups and online groups; in this case, if a simple direct conversion method is adopted, systematic analysis on specific situation cannot be conducted, therefore, the existence and development of abstract conversion method is very necessary.

### *2.4. Demand analysis of new media engineering field on visualization design methods*

With the development of digital technology, digital resources grow exponentially; faced with large-scale, rich-content, various and multilingual resources, how to effectively obtain the required resources has become a difficult problem. At present, most of the information service institutions tend to conduct resource navigation and display via text-based user interface, but this traditional resource organization and presentation method cannot meet the needs of users for rapid positioning. Users need an intuitive and vivid interface with certain opinions, huge digital resources can be fully displayed effectively navigated in limited interface space. The application of information visualization is a worthwhile solution, and it has made some progress.

## 3. Significance analysis of visualization design method in the field of new media software engineering

In recent years, with the continuous development and progress of the Internet, new media gradually appear in people's vision [2]. What is the new media? What is the relationship between new media and the traditional old media? In fact, in simple terms, new media means the new media form under the support of advanced science and technology, and it can be said that new media for disseminating information contributes to the emergence of new media, for example Digital TV and computers, etc. in our homes, there is a big difference between new media and the traditional media in terms of propagation medium. In this background, in the software development, it is very necessary to have advanced science and technology as support, as a technical means with great practicability and compatibility, visualization design method plays an important role in the development and design of specific new media software. It not only plays a role in technical assistance in the software development by professional staff but provides an intuitive and understandable data analysis result, promoting the development schedule and design efficiency of the software so as to further promote the progress and development of new media software engineering in China.

## 4. Capacity utilization assessment of visualization design method in the field of new media software engineering

In an era of APP popularity, various application software come out in an unending flow, in specific new media software development, scientific and rational application of advanced visualization design methods is positive to improvement of the practicability and development rationality of the software. From this point of view, there is a clear understanding of the systematic evaluation of visual design methods in the application of new media software engineering capabilities, and this plays an important role in grasping the advantages and significance of the visualization design methods.

Specifically, the design method of visualization is accompanied with the emergence of graphical user interface, in this process, the proportion of its use in specific software system is increasingly larger, some can be more than 70 percent; the reason for this problem is that the generation of graphical interface elements is very inconvenient and time consuming; in this case, visualization design methods have been applied to the software development of new media. Its specific ability application is mainly embodied in the actual software design and development process: use visualization development tools and some operating interface elements, such as menu bar, dialog box, list box, etc. on the graphical user interface provided and automatically generate application software with visualization development tool system. In fact, the automatically generated APP here is actually driven by a specific event, it will use the basic functions to conduct data analysis by computer and forms of images or videos converted from function analysis so that engineering designers can intuitively grasp the designed software and improve the quality and level of the software.

## 5. Application analysis of visualization design method in the field of new media software engineering

In the previous section, the design method of visualization is studied and discussed systematically, in this process, not only do we have a more intuitive and clear understanding of visualization design methods, but we also have a new understanding of software design and development in the new media environment [3]. To apply theoretical knowledge to real life, an overall grasp of the application of visualization design methods in the field of new media software engineering is needed, and only by this can promoting function of advanced technology means be given full play in the new period of continuous development of science and technology; the visualization design method is a good manifestation.

### *5.1. Application research based on visualization design methods of information interaction*

Under the support of visualization design methods, software users and software designers can communicate before and after design, the communication here is not a real face-to-face communication, but an exchange of information usage feedback. Before the software development, software designers should make a systematic investigation and research on the practicability and marketability of their software, in this process, the "first communication" between the two is realized, that is, the user's requirements survey results are presented to the software designers in a visualization form, and after the completion of software development, users can give usage feedback to the designers in APP internal program with visualization tools in real use, and this is "the second communication" between the two.

### *5.2. Application research of visualization design methods based on image processing*

As we all know, with the technical support of visualization and in the process of software development, software engineers can use visualization development software to view the needed data in visualization forms, such as images. In this case, the software engineers can grasp the development information accurately, which is helpful to improve the quality and level of the development.

### *5.3. Application research of visualization design methods based on software data information management*

In the process of specific software design and development, data information management in software is also very important; in the actual development process, effective management of the data can be actualized with visualization design methods and management of visualization can be realized and presented in images with its functional advantages given full play, which is conducive to further monitoring on software.

## 6. Conclusion

In the new era of increasingly developed science and technology, especially in the new media era, the extensive application of advanced visualization technology has positive significance for promoting the development and utilization of software in our country. Therefore, we should attach great importance to the visualization technical means and apply them scientifically and rationally to promote the progress and development of software design and development in the new media era in China.

## References

[1] Zhu M Y 2017 *Art and Tech.* (2017-10-27).
[2] Li M, Cai M J, Jiang K W, Wang Y L 2013 *E-Edu. Res.* **34(03)** 16-22.
[3] Chen J L, Liu H X 2017 *Libr. Inform. Work* **1-7** (2017-09-26).

IOP Conf. Series: Materials Science and Engineering **317** (2018) 012050 doi:10.1088/1757-899X/317/1/012050

# The Application of Visual Illusion in the Visual Communication Design

**Tao Xin[1], Han You Ye[1]**

[1]He University, 110163

**Abstract.** With the development of our national reform, opening up and modernization, the science and technology has also been well developed and it has been applied in every wall of life, the development of visual illusion industry is represented in the widespread use of advanced technology in it. Ultimately, the visual illusion is a phenomenon, it should be analyzed from the angles of physics and philosophy. The widespread application of visual illusion not only can improve the picture quality, but also could maximize peoples' sense degree through the visual communication design works, expand people's horizons and promote the diversity of visual communication design works.

## 1. Introduction

In the new era of technology which being more and more developed, now increasingly, the visual communication design has appeared in our life, to be specific, the visual communication design is a design to send some messages to spectators through visual media. It is "a design for all to appreciate and it is formative one". But for explaining why would the visual illusion phenomenon exist, after all, it might all come down to this principle, in short, that is human eyes can consciously balance what they see , and this kind of balance would lead to the visual illusion. For example, the background around the black one would be brighter, and that around the white would be darker. Therefore, applying this visual illusion in the visual communication design works.

### *1.1. The concept analysis of visual illusion*

The visual illusion, it is not hard to understand from the literal, the visual communication designers utilize some techniques of expression to show figures and pictures through abnormal forms in their specific works. The process of formation of illusion in vision is actually a process of "the figure itself produces ambiguity", that also is a misconception of people to those external objects.

## 2. The manifestation of visual illusion in the visual communication design.

The visual illusion occupies a significant position in the visual communication design, in view of the novelty and innovation, its advantages are obvious [1]. The widespread application of visual illusion is conducive not only to improve the vitality and figurativeness in visual communication design, but also to meet the basic needs of spectators to the works' artistry and novelty. The visual illusion has abundant manifestation forms in visual communication design, in the process of studying and analyzing it, we should study it from different aspects and viewpoints, and that it would deepen our understanding to the visual illusion in the visual communication design, and we can give the spectators the works with a mighty sense of vision impact and novelty, then to accelerate the development of our country's visual communication industry.

*2.1. The manifestation of visual illusion in figure creativity*

In the process of designing some specific works of visual communication. The proper utilization of creativity of figure is conducive to the application of visual illusion phenomenon. As we know, the work that visual communication design, its figures normally are shown by a pattern of planar space, for this situation, if applying the visual illusion in its plane figure, some unexpected results and experience would be presented skillfully. In particular, there are different aspects of manifestation in the figure creativity, the next sections would expound it systemically.

*2.1.1. The isomorphic visual illusion of graphics.* For visual illusion, isomorphic image is a creative manifestation, analyzing it from the angle of concept, generally speaking, it is an operation to make isomorphic of images for those elements with internal characteristics and common features, integrate them systematically to become a new image to appear in the eyes of spectators. For example, in the display of some concrete works, the metal display object can be replaced by opposite material, for wooden one. Putting some leaves and fruits on the wood, this could enhance their visual enjoyment and experience provided that they would create an illusion. Therefore, from the view of isomorphic, the visual illusion plays an important role in visual communication works.

*2.1.2. The visual illusion graphics with its background reversion.* Apart from the above the isomorphic in visual illusion, the manifestation of graph background reversion are common, in the process of practice, it has strong practicality and operability. About the graph background reversion, the graph is that we knew, background means something against the figure, as we all know, in the concrete visual communication design, the graph and background are mostly not separated. In a sense, the graphic background reversion is exchanging each position and sequence, making them become one manifestation with common features, that is easy for spectators to form an impressive effects and experience when they appreciate the works [2]. So the graphic background reversion is irreplaceable in the presentation of visual communication design works, we need to take highly of its application.

*2.1.3. Multidimensional visionary visual illusion graphics.* The multidimensional visionary visual illusion graphics is related to the dimensionality and space in the concrete manifestation forms. Generally speaking, the graphics and matters that we see are all two- dimensional, however, applying the multidimensional visionary patterns in works reasonably and efficiently can realize the three - dimensional even four - dimensional visual enjoyment and experience. For two- dimensional space, using the basic principle visual illusion and its manifestation form to break through its inherent setting and forms, then fully combine with active visual language, integrate those disorder matters together, and to present something with both internal thinking logic and a certain content to spectators, thus to produce illusion. Hence, multidimensional visionary manifestation are vital to the graphic creativity, it not only make spectators feel something different, but also promote spectators receive the designers' emotion when they appreciate the works, and enhance their experience and feelings on works, so to promote our visual communication industry's progress and development.

*2.2. The manifestation of visual illusion in color application.*

In our daily life, there exists a rich of colors, in the visual communication works, the selection of hue is also an art, and the requests to colors are varied for different visual communication works. This also requires designers must have a basic understanding mastery of color and hue in practice. For bright colors would give people a sense of free and easy, but dark color would make people fell heavy. Similarly, the application of visual illusion is the same as it. In general, it can be divided into the illusion application of color visual communication which caused by color itself and one of contrasts between colors.

*2.2.1. The illusion application of color in visual communication which caused by color itself.* In terms of classification of colors, there exits many colors in our life, what's more, thinking on the hue, the color can be divided into "cold-warm" and "bright - dark". On this level, the manifestation content of its hue and the design of visual communication are combined together is benefit for designers to apply visual illusion into the works [3]. The concrete principle of application is to match those different hues to hit the vision, and make it create an illusion. Such as, in our daily life, we all know one in black looks thinner than in white, actually, that takes advantage of the visual illusion phenomenon between colors see below.

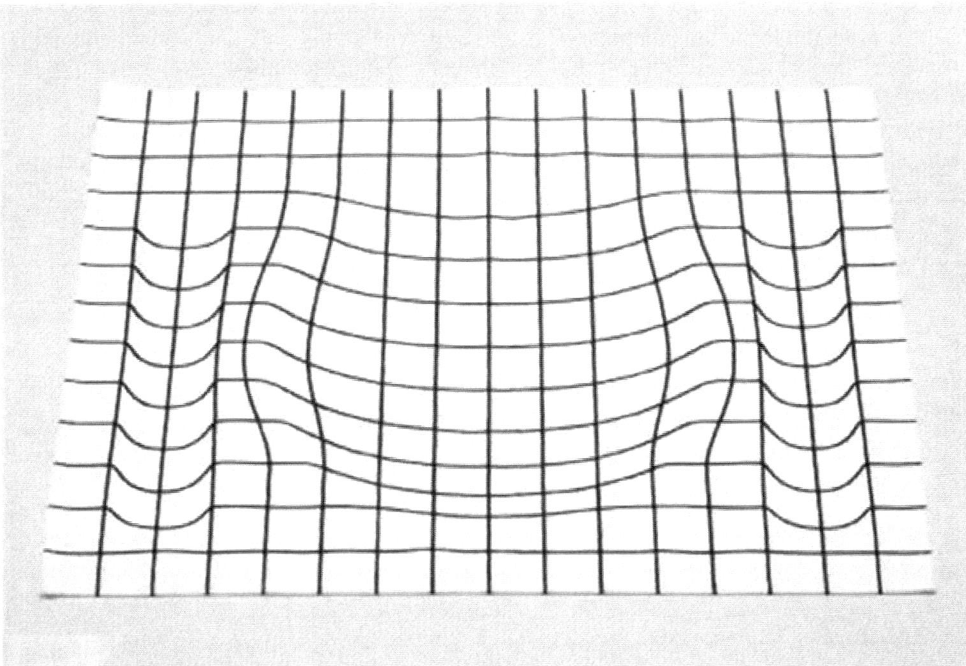

**Figure 1.** Visual Illusion Pictures.

*2.2.2. The contrasts between colors.* Apart from the above visual illusion phenomenon which caused by color itself, there is also one caused by the contrasts between colors. The contrast technique is not only commonly used in literature, but also is widely used and of core value in aesthetics. When presenting a work, there are some essential differences between using color contrast or single color on its representation effects. For instance, matching the black with red, yellow and golden, not only can create a visual impact, but also maximize its function to against the dominant hue. This shows different color contrast can bring different visual experience and feelings to people.

### *2.3. The manifestation of visual illusion in the graphic design*

The basic manifestation of visual illusion can be transformed to a concrete manifestation by graphic design. In the new era of China's socialist modernization, under the tendency of economic globalization, the whole society develop in the trend of multiplication, hence, people put forward higher requests for enjoying concrete visual communication works, the designers would combine with some arbitrary and spontaneity elements in the process of adding elements, integrate them into a visual sign to present to spectators, this process is a process completing graphic design. For example, in modern works, adding some traditional elements or things would take people to another state, and through making creative combinations on the graphic to achieve the perfect performance.

## 3. The meaning of the widespread use of visual illusion in visual communication design

The widespread application of visual is vital to the charm showing and expression of emotion of visual communication design. For it can not only improve the levels and aesthetic perception of works, what's more, it can realize more visual impact aesthetics and mental or physical experience by some simple operation. Therefore, designers should become better acquainted with the visual illusion, widely apply them to improve the visual communication works' quality when they design concrete works.

IOP Conf. Series: Materials Science and Engineering **317** (2018) 012050 doi:10.1088/1757-899X/317/1/012050

### *3.1. The analysis and research based on the meaning of visual impact.*
In the socialism market competition, the works of visual communication are very common, and all levels of works exist in the market. Most visual communication designers always apply some traditional techniques and skills of expression when they design works, that is easy to cause visual fatigue to spectators. In this case, skillfully applying the visual illusion can relieve spectators' long term visual fatigue [4]. Making full use of visual illusion not only can cause giant impact on their vision, but also can give a shock to them, thus to enhance appreciation effects.

### *3.2. The analysis and research based on the meaning of graphical simplification function*
For graphical simplification, it is, in essence, not a simple operation to simplify the graphs, but to show the nature of the concrete object by the manner of simplifying implementation according to it. To present the profound truth on the basis of simplification. Hence, the widespread use of visual illusion has a great influence in the graphical simplification.

### *3.3. The analysis and research based on the meaning of improving graph's space sense*
In the above discussion, we have actively discussed the two-dimensional space of gray, with the continuous improvement of people's aesthetic requirements, so the two-dimensional space cannot fully satisfy the basic needs of spectators. Therefore, to apply visual illusion skillfully, you can make fully use of some methods and skills to excavate three-dimensional space even the four-dimensional space, so as to present more comfortable works to spectators on the basis of increasing its sense of layer.

## 4. The analysis and research on application strategy of visual illusion in visual communication design.

### *4.1. The application analysis of visual illusion based on the angle of communal shape*
From the point of communal shape, the basic application of visual illusion is to integrate those graphs with different properties, so as to achieve the final result. That is to say, adopting the communal partial shape or edge to create a new implementation model, and this new graphs may show something different by visual illusion.

### *4.2. The application analysis on visual illusion from the angle of graphics transformation.*
For a graph, if it is analysed from different angles and different effects would be gotten. In other words, to transfer the background of graph and the implementation itself, and on this basis, to apply hyperspace to effectively present the object of works.

## 5. Conclusion
In the new era, while the society and economy is being developed constantly, the technology is also being improved continually, in this circumstance, the manifestation forms of visual communication design appear in public become more widespread. In the process of its development and progress, the appreciation needs of spectators become higher and higher, and designers' design level are being improved continually. Combining visual illusion with visual communication design works is conducive not only to improve the levels of designers, but also to maximize the progress and development of visual communication industry, thus improve our country's construction level of socialist modernization.

## Reference
[1] Luo Y 2017 *Art Sci. Tech.* (2017-10-25).
[2] Li X Y 2012 *The Applied Research of Visual Illusion in the Visual Communication Design* Henan University.
[3] Wang J 2013 *Study Arts Edu.* **22** 47.
[4] Lu Q M 2016 *Popular Culture & Arts* **07** 120.

IOP Conf. Series: Materials Science and Engineering **317** (2018) 012051 doi:10.1088/1757-899X/317/1/012051

# The Integration Method of Ceramic Arts in the Product Design

**Wang Shuxin**[1]

[1]He University, 110163

**Abstract.** As one of the four ancient civilization countries, the firing technology of ceramic invented by China has made a great contribution to the progress and development of human society. In modern life, even the development of technology still needs the ceramics, there are large number of artists who take the ceramics as carrier active in the field of contemporary art. The ceramics can be seen everywhere in our daily life, this paper mainly discusses the integration means of ceramic art in the product design.

## 1. Introduction

The ceramic art is one typical artistic feature in our country, it has a long history. It is the combination of technology, materials and art, many elements of it is of great significance for modern product design, especially the integration of ceramic material characteristics, decorative patterns and modern product design, which can effectively improve the quality and cultural connotation of it. As ceramic art has its own charm and value, artistic quality and cultural connotation, so this requires modern designers to inherit and develop, avoid copying the elements and features of ceramic art entirely, but to combine them skillfully.

There is a deeper connection between ceramic art and product design, the household porcelain of ceramic belongs to the industrial design, the excavation and application of ceramic culture which can reflect out traditional culture would make modern product design filled with ethnic characteristics and Chinese characteristics, promote Chinese design better.

## 2. The ceramic art

Just like Chinese calligraphy and painting, the ceramics are one of Chinese favorite arts. It has brilliant achievements with thousands of years' history in China, its unique linguistic form and abundant expressiveness has developed its particular aesthetic value. From ancient times to now, the ceramic arts has been peoples' collection items [1]. We can see the aesthetic value of ceramic from those poems left by ancient literati, the Tang Dynasty poet Meng jiao had wrote "after drinking Mengshan Tea, the lotus leaves in Yue kiln are empty "Song dynasty poet Li Qingzhao had also wrote "in the bedstead, the cool saturate the pure pillow", it is praise for bluish white porcelain of Jingdezhen,; Su Dongpo has wrote" the tea after decoction poured into the white gazed tea set, which carved decorative pattern ", the poems have praised the caramel glazed porcelain and black glaze of Ding kiln and Jian kiln. It is observed that the ceramic has always been the aesthetic objects among Chinese.

The ceramic as an art, it is not common. Ordinary people would take it as a household tool, maybe a decoration. In the art world, it doesn't have high value. For ones with good appearance, practical and durable are meet their needs. They never appreciate it purposely. Therefore, the problem is on what kind of people does it want to attract. I am not saying that appreciating the ancient items is bad, after all the highbrow will end our further ways. As a quintessence, I think the ceramic art should be entered into the market and introduced to our life, the art root in the life, but also need to serve for the life. The art that can be appreciated by most people would be filled with vitality in the modern life, which with the development of market economy.

## 3. The state of ceramic art in the product design

### *3.1. The high plagiarize rate*
Because of the imperfection of patent protection system at the present stage, as people's increasing material and cultural needs and economic condition have a great demand for ceramic products, this unequal relation had led to the emergence of plagiarism in producing ceramic products [2]. It's common that when a designer of a brand or manufacture complete a product, at the beginning of it is put on the market, the product must be plagiarized by other manufactures, over time, it is hard for the original designers to find motivation and confidence to keep on creating, this disorder phenomenon had the negative effects for the long- term development of market.

### *3.2. The crudeness of design*
In the process of combining the ceramic art with product design, we often ignore artistic attribute of ceramics, we just think that the blue and white stand for the whole ceramic art, and then design a product by slavishly copied. Hence, the later ornamental value and aesthetic of products must be greatly reduced. So we must pay attention to the integration of two elements in the products design.

### *3.3. A lack of connotation*
The connotation is the sum of essential attribute that reflects an object's concept, the connotation of product design is the embodiment of designers' sour reaction and thinking consciousness, it is hard to convey the connotation in the assembly production, the consumers can not feel the artistic aesthetic that through the ancient and modern, we have to admit this is an inevitable products in the industrial era.

### *3.4. The rough craftsmanship*
According to the field lists and investigation, it is not difficult to find that there are lots of ceramics in the supermarkets and markets, although the widespread application of ceramics is gratifying, some disadvantages are brought by it. With the essence of economy is for profit, in the economical production, very few people would think by its artistry, so the most of ceramics in the market are with uneven glazing, scratch and rough surface. This roughness highly kill our affection to ceramics, this has obstructed the development road of ceramics in product design.

In conclusion, I think there are five points should be noticed in designing household ceramics: at present, domestic household ceramics often put artistry as the first consideration rather than practicability; whether household ceramics or artistic ceramics, its artistry should be first considered; consider its scientific and material attribute; both of them should be particular about its sales; both of them should pay attention to economic benefits.

## 4. The application of ceramics art properties in the product design

### *4.1. The mental feelings and product design of ceramic materials*
Due to the difference of culture, quality and history environment of materials, adopting different materials would bring various feelings, for ceramics, when its firing temperature differs from matching, rematching other materials would produce different mental feelings, this kind of mental feelings can be effectively applied in the modern product design.

*4.1.1. The application of rustic mental feelings in the modern product design.* Because the surface color of ceramic is white and slippy, so it can make people associated with the era of the ancestors, afterwards a sense of warm would arise spontaneously. Therefore, it is mostly used in households for presenting a kindness.

*4.1.2. The application of technologically sophisticated mental feelings in the modern product design.* The high-tech ceramics often give the spectators a feeling of science and technology, the difference of production mode would make the shape of products differ from each other. Modern production modes often improves the accuracy of ceramics, in the application of it in the modern products design, the elaborate design, size, structure and appearance will become the symbolization of its science and technology.

*4.1.3 The application of mental feelings with white in modern product design.* A lot of ceramics with white and clean surface would bring people a sense of holiness, one reason is that its texture is clear, the other reason is its bright surface represents clean and holy, and its chemical properties are relatively stable. So lots of high-class restaurants would select ceramics as tableware.

### *4.2. The physical property of ceramic material and modern product design*

It is well known that ceramics are featured as high temperature resistance, high hardness and easier to clean, these properties are consistent with modern products on the pursuit, hence it is widely applied in modern product design. For example, the dishes, spoons and vases in our life are all the important embodiment of ceramic art.

*4.2.1. The application of high hardness in modern product design.* Compared with traditional handicraft era, after precise matching, the hardness of ceramic has improved significantly, just because of this, the ceramics can be effectively and widely used in modern product design, at the end of the last century, Japan has developed ceramic tools, for its hardness is higher than half of that of steel, so the tools remain popular among consumers. And the hardness of Chanel ceramic watch are ten times than traditional watch, but its weight is lighter, so it is favoured by users and designers.

*4.2.2. The application of easy-clean in modern product design.* With the boringness of surface, high hardness of ceramic, it is easy for cleaning, this characteristic has been widely used in the field of kitchen utensils and bath appliances, for the hand washing sink, teapot. About small appliances, some microwave oven or induction cooker which are easy to be dirty are all use the ceramic as the panel, the reason to this is because of the characters of easy-clean of the ceramic. What's more, it can be a plus for the ceramic products, such as the ceramic watch, it is not susceptible to rust than traditional watch, even though the watch is in dust, it is easy to clean.

*4.2.3. The application of high temperature resistance in the modern products design.* The ceramic materials not only being hard, but also endure the high temperature. Because the ceramic is produced in the temperatures of thousands of degree, it has high temperature resistance. In modern products design, many designers have modified the technology to make the ceramics can be used in the modern kitchen, we put food into the ceramic products to cook, that is good for people to digest the food nutrition. Hence, the characteristic of the ceramic itself owned has been adopted by many modern designers, many modern kitchen appliances also have applied the ceramic elements.

## 5. The application of ceramics artistic decorative patterns in the product design

For ceramic decorative patterns, it can effectively embody its artistry, the level decorative pattern may well determine the level of ceramic art, and a number of ceramics artists have spent a large amount of time and efforts in the decorative patterns [3]. There are mainly plants patterns and geometric patterns in the traditional ceramic decorative pattern, these patterns can be used in modern product.

### *5.1. The application of plant patterns in modern product design*

In plant patterns, the curve of Tang pattern is seductive and of high degree of freedom, its composition mainly derived from the trees and branches, according to the plant's crimp shape and its transformation, or some wavy pattern based on it, to create the regular and creative Tang pattern. For a complete set of tableware with Tang pattern, it often has a light and elegant characteristics, if people have dinner by ceramic tableware, that add a special flavor to food. What's more, some plants have

IOP Conf. Series: Materials Science and Engineering **317** (2018) 012051 doi:10.1088/1757-899X/317/1/012051

been widely used because of its special symbolic meanings, for peony pattern represents rich, the lotus pattern represents holiness, the wintersweet represents haughty etc, precisely because of this connotation, the plant patterns get richer meanings. In modern product design process, the plant pattern is always used in household decoration, it is also used for decorating dressers, because plant patterns are more popular for women, so it is more appropriate to use and adorn the plant patterns.

### *5.2. The application of geometric pattern i modern product design*

The geometric patterns of ceramic decoration mainly include the "palindromic pattern" and "eight treasures pattern", for the latter, it mainly points to the pattern surrounded by eight auspicious treasures or utensils, the true eight treasures patterns are complex and cockamamie, many artists or designers focus on the use of "shape" in the process of designing modern products, at the same time to give consideration to its allegoric and spirits extension. So-called shape is to recreate on the basis of origin, and integrate other related materials to make related elements perfect. For the former, a geometric pattern "Hui" formed by surrounding backwards and forwards, the palindromic the ceramic is being used as the assistant pattern in the process of designing modern products, for example, all the expression of luxury would use the palindromic.

## 6. Conclusion

Above all, for ceramic art, it has a greater variety of contents, it represent the glorious development of our national craftsmanship to some extent, although the ceramic art itself has a certain limitation on the design concept and craftsmanship, however, there also has its own charm, value, artistic aesthetic and cultural connotation, this requires designers to inherit and develop its advantages, cannot blindly copy the elements and features of ceramics, we should subtly combine both of them, take the essence and discard the dregs, utilize its advantages, and combine the aesthetic characteristics and life style which are peculiar in the modern society, make the ceramics be widely used in our life and production, promote the modern products design towards an innovative road with Chinese characteristic.

## Reference

[1] Cai K Z 2016 *Pack. Eng.* **37(18)** 118-20.
[2] Sun T S 2014 *The Applied Research Based on the Ceramic Design in the Product Extension.* Shenyang Ligong University.
[3] Tu Y X, Yu Y J 2015 *Beaut. Time (First)* **12** 110-2.

IOP Conf. Series: Materials Science and Engineering **317** (2018) 012052 doi:10.1088/1757-899X/317/1/012052

# Design of Remote Monitoring System of Irrigation based on GSM and ZigBee Technology

**Zheng Xiao xi[1], Zhao Fang[1] and Shao Shuaifei[2]**

[1]Zhengzhou Technical College, Zhengzhou, Zheng Road 81, CHINA
[2]Sippr engineering group co., LTD, Zhengzhou, Central Plains Road No. 191, CHINA

**Abstract.** To solve the problems of low level of irrigation and waste of water resources, a remote monitoring system for farmland irrigation based on GSM communication technology and ZigBee technology was designed. The system is composed of sensors, GSM communication module, ZigBee module, host computer, valve and so on. The system detects and closes the pump and the electromagnetic valve according to the need of the system, and transmits the monitoring information to the host computer or the user's Mobile phone through the GSM communication network. Experiments show that the system has low power consumption, friendly man-machine interface, convenient and simple. It can monitor agricultural environment remotely and control related irrigation equipment at any time and place, and can better meet the needs of remote monitoring of farmland irrigation.

## 1. Introduction

In the process of agricultural production, the growth, development and yield of crops are closely related to the soil moisture temperature. Suitable soil temperature and humidity can not only improve crop yield, but also improve the quality and taste of crops. At present, most agricultural irrigation in our country uses flood irrigation or artificial watering, not only wasting water resources, but also may result in the phenomenon of not timely irrigation, insufficient irrigation or over irrigation, thus affecting the growth of crops. Traditional manual measurement of soil temperature and humidity for irrigation mode, not only wasting manpower and time, it is difficult to do real-time monitoring, do not take advantage of the production of farm crops. The traditional method of measuring soil temperature and humidity for irrigation is not only wasteful of manpower and time, but also difficult to monitor in real time and not use farmland crop production.

ZigBee is a two-way wireless communication technology, with low power consumption, low cost and strong networking of devices, which can be used for a near distance wireless connection [1, 2]. GSM global mobile communication system has the characteristics of good communication quality and strong anti-interference ability. In this paper, a remote monitoring system for farmland irrigation based on GSM and ZigBee is designed. Through the real-time monitoring of soil temperature and humidity, the system can automatically obtain real-time data without going to the site, and timely and accurate understanding of soil environmental parameters, and according to the demand of crops automatically adjust the valve switch, on-demand irrigation of crops, not only greatly avoided the waste of agricultural water, reduce the cost of agricultural products, but also increased the yield of crops [3-5]. The system uses the wireless sensor network to collect soil temperature and humidity information in real time, and uses GSM network to realize remote transmission of soil data information, the staff and the host computer can analyse the data remotely to complete the real-time control of the amount of water applied to the drip irrigation or sprinkler terminal, which not only saves water but also greatly saves labor, improves work efficiency and increases farmers' income.

## 2. Overall plan of remote monitoring system

This irrigation remote monitoring system mainly includes data collection function, data processing function, electromagnetic valve control function, power management function, remote monitoring function. It mainly consists of sensor nodes (including soil temperature, humidity sensor, light sensor, etc.), solenoid valve node, coordinator node and monitoring center. Its structure is shown in figure 1. The monitoring part consists of temperature, humidity sensor, light sensor, etc., and it is arranged on the surrounding farmland, mainly responsible for periodically gathering information about farmland environment. The coordinator node has the instructions to send and receive the soil information and to receive a PC or cell phone. PC and mobile phones are used to display field data and to analyse and determine whether irrigation is needed. From time to time, the system collects the information of soil temperature and humidity, illumination, state of the electromagnetic valve and other information in the farmland through the sensor and collects the collected information to the coordinator node through the ZigBee network. The coordinator node collects Information uploaded to the designated phone and host computer.

When the mobile phone or the upper computer gets the definite soil parameter information, the irrigation plan can be made automatically by the program, the control instruction is sent to the coordinator node of the corresponding farmland via the GSM network. The coordinator node in the farmland transmits the control instruction through the ZigBee network to the control node, the actuator to achieve the appropriate control of the valve, for precision farmland irrigation. The manager receives the soil parameters information through the GSM network and remotely controls the irrigation equipment, so that the information of the temperature and humidity of the farmland soil can be monitored conveniently on the host computer or the administrator's mobile phone. When the soil temperature and humidity parameters exceed the set value, the system has alarm function. The upper machine USES configuration software to realize data management and human-computer interaction to achieve the purpose of user monitoring. The system is characterized by low power consumption, high performance, simple maintenance, low transmission cost and high reliability.

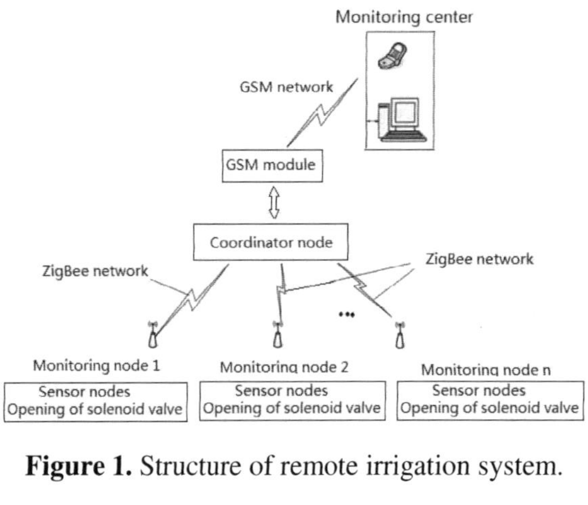


**Figure 1.** Structure of remote irrigation system.

## 3. System detection circuit design

### *3.1. The choice of ZigBee module*

The ZigBee module of the system uses TI chip CC2530 as the core module, the module supports the IEEE802.15.4/ZigBee protocol, the working frequency range is 2.4GHz. This chip integrates a high performance direct sequence 2.4GHz RF transceiver core, the enhanced 8051 controller with low power consumption, temperature sensing function and battery monitor function; There are 2 USART, 8 timer and 16 timer each. It supports 2 ~ 3.6V power supply; Working temperature is minus 40 ~ 125 °C; It supports the working mode of hibernation, sending, receiving and so on, with short conversion time and low current power consumption, which can meet the requirements of low power consumption of the system. The CC2530 minimum system circuit connection diagram is shown in figure 2.

The minimum system of CC2530 can be realized only by a few components. The crystal oscillator 1 is 32.768KHz, the crystal oscillator 2 is 32MHz, and the RF transceiver needs 32KHz crystal oscillator when transmitting and receiving functions; The digital power interface 40 pipe joints with a 1uF decoupling capacitor to improve the stability of the power supply; RF-N and RF-P adopt inductance and discrete capacitance realize the connection with RF matching network part (including matching capacitor, antenna, etc.); The purpose of crystal vibration-load capacitance C2, C3, C4 and C5 is to ensure that the crystal vibration has precise and stable oscillation frequency.

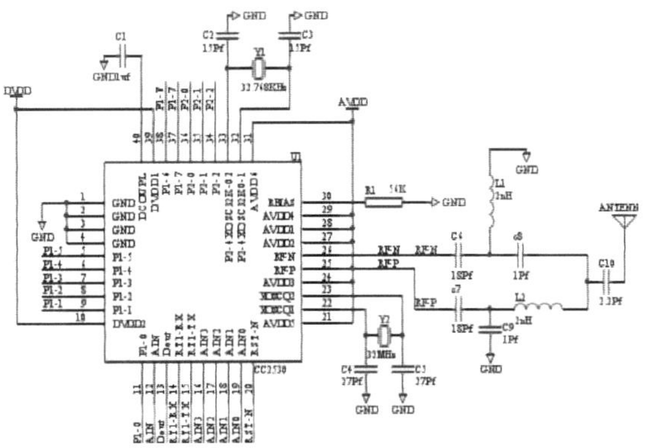

**Figure 2.** CC2530 peripheral hardware circuit diagram.

### *3.2. GSM module design*

The function of the GSM module is to send the collected data to a remote host or mobile phone [6]. Because ZigBee network can only realize short distance data transmission, and GSM network has signal stability, can realize global roaming and networking, high coverage, so the system uses GSM network to send data to the remote. The GSM module of the system adopts the TC35i module of Siemens, which is a GSM debug demodulator, and the supply voltage is 3.3 ~ 5.5V, which supports the 900MHz/1800MHz dual frequency [7]. Providing RS232 serial port in PC communication, using AT format instructions.

GSM module circuit includes TC35i module, power module, SIM interface circuit and RS232 serial communication circuit with CC2530 to realize GSM communication function. The interface circuit of the TC35i module is shown in figure 3.

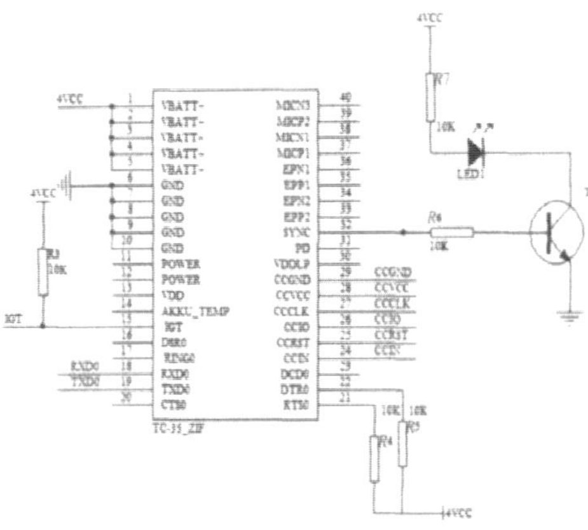

**Figure 3.** TC35i module interface circuit.

1 ~ 5 pin is the power pin in parallel. The 6-10 pin is a parallel grounding pin; In order to ensure the stability of TC35, the parallel capacitance between the earth and the power supply is decoupled.

The 15 pin is the boot foot, then the work state of TC35i is controlled by the P2-0 of CC2530. 16 ~ 23 pin is the data input/output function pin, which conforms to itu-trs232 interface standard, and is connected with the 16/17 foot of CC2530. The 32 pin controls LED lights, and the TC35i work state can be determined by displaying the light.

Access to the GSM network must have a SIM card, The SIM card is connected to the TC35 module from 24 ~ 29.The circuit diagram is shown in figure 4.

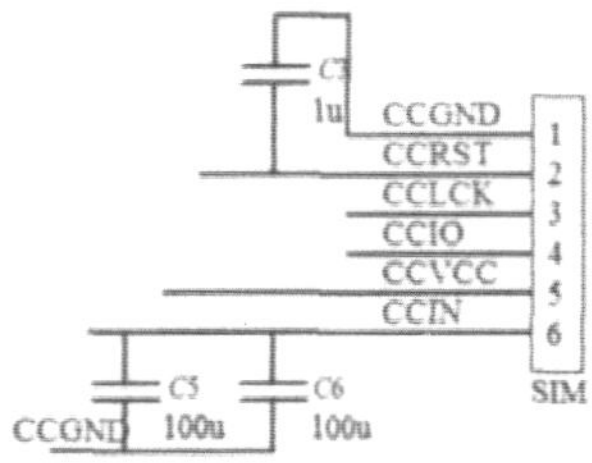


**Figure 4.** Interface circuit between TC35i and SIM card.

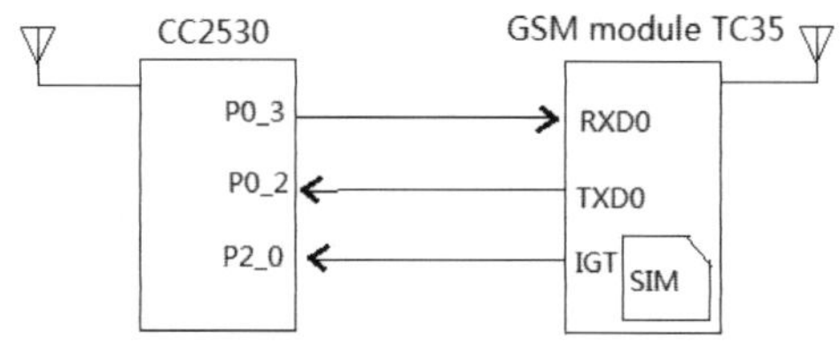


**Figure 5.** Connection between coordinator and GSM module.

CC2530 and GSM module use the RS232 protocol for serial communication. The connection between CC2530 and GSM module is shown in figure 5.

*3.3. CC2530 charging circuit*

Because the ZigBee coordinator requires uninterrupted work and consumes a lot of power, the ZigBee coordinator recharges the battery using a solar panel. The solar charging circuit is shown in figure 6.

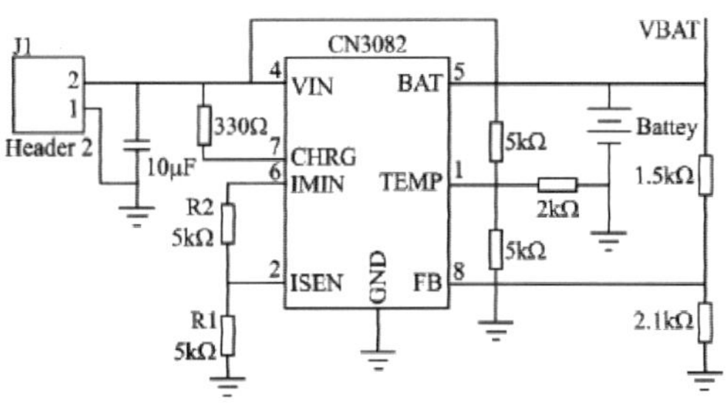


**Figure 6.** Solar charging circuit.

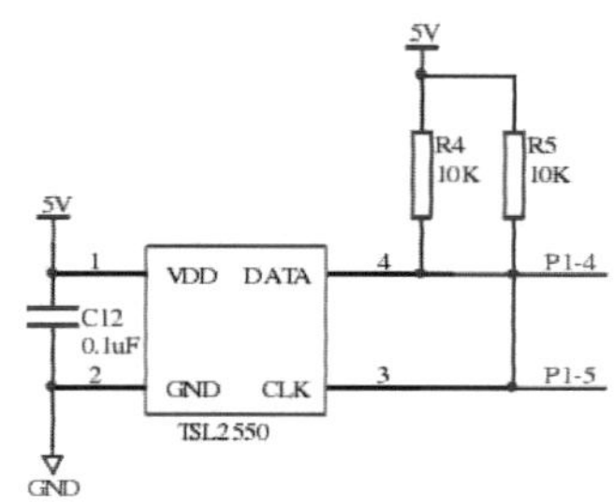


**Figure 7.** Interface circuit of illumination sensor TSL2550.

*3.4. Sensor selection*

The selection of sensors in this system should be determined according to the actual farmland.

*3.4.1 Selection of soil moisture sensor*

Soil moisture is the main factor determining whether crops are irrigated or not. Soil moisture is too high, affecting soil aeration, prone to lodging and other phenomena; soil moisture is too low, so that crop water shortage, drought affect crop growth, serious water shortage may lead to crop wilt or even death. The FDS-100 sensor is used to collect the soil moisture of the system, the FDS-100 sensor outputs analogue signals, the working current is 25mA, the working voltage is 5 ~ 12V, the measuring accuracy range is about 3%, and the probe length is 5.3cm.

*3.4.2 Selection of soil temperature sensor*

DSl8B20 temperature sensor is used in soil temperature measurement. The sensor has the characteristics of fast transmission speed, small volume and low power consumption. It can directly convert the measured temperature into serial digital signal and transmit it to the host computer, which has unique wire interface mode. The range of temperature measurement is -55 ~ +125°C, and the precision is 0.5 C at -10 ~ +85°C. It can realize high precision temperature measurement, and is suitable for field temperature measurement in harsh environment.

### *3.4.3. Selection of light sensor*

The lighting of farmland also affects the amount of irrigation for cropland crops. This system adopts the TSL2550 module as a light sensor, and the light intensity is detected by the potentiometer on the module, which converts the light signal collected to the digital signal and sends it to the connected node. The working power of TSL2550 is 2.7 ~ 5.5 V, and the working current is 0.35 mA, small volume and low power consumption. Light sensor TSL2550 1 pin to connect power; The number 2 pin is grounded, the 4 pin is the output of the signal, and the signal is transmitted to the I/O port of the CC2530 module. The specific interface circuit is shown in figure 7.

## 4. System software design

According to the system monitoring requirements, the system software design is mainly to solve the collection and processing of soil environmental information in farmland, and the shutdown and opening of the field solenoid valve; Control monitoring nodes to transmit real-time data through intermediate nodes to GSM communication module, and sent to the control center through GSM network administrator mobile phone, When the soil information of the collected farmland exceeds the set value, the control center's upper computer and the administrator's mobile phone send control instructions to the terminal equipment to complete the control operation. The software of monitoring center mainly has functions of remote setting soil temperature and humidity, displaying soil temperature and humidity information, inquiring temperature and humidity function, temperature abnormal prompt, alarm record and historical data inquiry, printing, etc. The design of the system software enables the staff to realize the automatic or manual control without needing to go to the farm site. It not only saves human resources, but also improves the reliability of operation. The system software design flow is shown in figure 8.

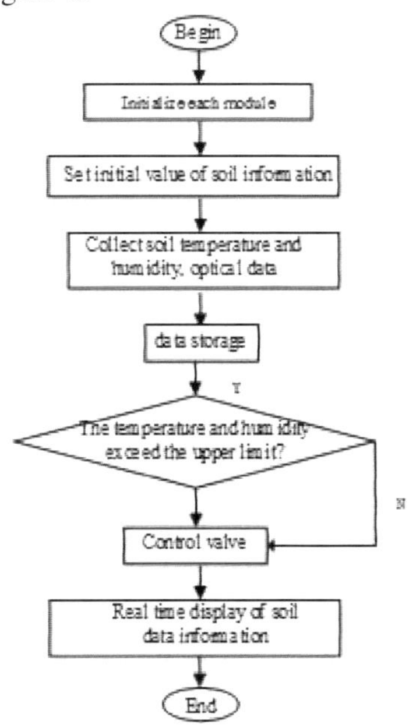


**Figure 8.** Flow chart of system software design.

## 5. Conclusion

This system uses GSM technology and ZigBee technology to realize accurate irrigation and remote monitoring of farmland irrigation system, and low power consumption. The monitoring nodes in the farmland regularly collect the status information of temperature and humidity and light intensity in farmland, and transmitted to the intermediate node, intermediate nodes to transmit data through the GSM network to the remote control center or mobile phone users, After data analysis, the control center sends instructions through GSM network to realize irrigation of farmland. Through testing, the system can realize precise irrigation, save resources, and have certain practicability and applicability.

## References

[1] Li Y, Dong S T, Huang D D 2015 *Agric. Mech. Res.* **2** 226-9.
[2] Zhang G S, Gao D S, Hou J L 2012 J. Shandong Agric. Univ. (Nat. Sci. Ed.) **43(3)** 377-80.
[3] Bao Q M, Chen Y M, Su B L 2014 *Comput. Measur. Contr.* **10** 3171-3.
[4] Dai J, Shi Z C, Wu F, Wang C Z, Jiang F 2015 *Autom. Instr.* **30(9)** 38-42.
[5] Fan Y, Yu Y, Li Y Y, Ding Y X 2016 Labor. Res. Explor. **01** 80-4.
[6] Zhang M, Fang J L, Han Y 2013 *J. Agric. Eng.* **29(25)** 171-6.
[7] Li W 2009 *Low-Volt. Electr. Appl.* **12** 37-9.

IOP Conf. Series: Materials Science and Engineering **317** (2018) 012053 doi:10.1088/1757-899X/317/1/012053

# Advanced Twisted Pair Cables for Distributed Local Area Networks in Intelligent Structure Systems

**Andrey Semenov**[1]

[1]Moscow State University of Civil Engineering (MGSU) National Research University, 129337 Yaroslavskoye Shosse, 26, Moscow, Russia

**Abstract.** The possibility of a significant increase in the length of cable communication channels of local area networks of automation and engineering support systems of buildings in the case of their implementation on balanced twisted pair cables is shown. Assuming a direct connection scheme and an effective speed of 100 Mbit/s, analytical relationships are obtained for the calculation of the maximum communication distance. The necessity of using in the linear part of such systems of twisted pair cables with U/UTP structure and interference parameters at the level of category 5e is grounded.

## 1. The necessity of changing the requirements to the twisted pair cables of contemporary information systems

Contemporary real estate, originally designed or adapted for permanent or prolonged stay of people inside them is compulsorily equipped with an information and telecommunication system (ITS). The latter one is organizationally divided into a number of separate components, which, for reasons of serviceability, simplicity of expansion, increased operational flexibility, and other similar criteria, are built on common or ideologically very close to each other principles.

Traditional subsystems of the ITS are a local area network and a telephone network. The physical layer of these subsystems based on ISO/IEC 11801 standards is implemented on the basis of a structured cabling system (SCS). The horizontal subsystem of the SCS is built on the twisted pair cables, whereas on the backbone levels, fiber-optic technology is primarily used. The SCS, which serves the LAN and the telephone network of the enterprise, has a hierarchical structure.

Recently, the list of the ITS subsystems has expanded significantly. They were added by the subsystem of video surveillance, access control and several others. All of them are built on the basis of ip-technology [1], however, unlike their classical counterparts they gravitate toward a centralized construction scheme. In such a situation, the possibilities of typical balanced cable channels, taking into account the known 100-meter specification length of communication, do not meet the requirements of practice.

Typical for a LAN method to increase the limiting length of communication by switching to fiber-optic technology is unprofitable for the following reasons:

- higher cost of the line as a whole due to the additional conversion of the electric signal to optical one and vice versa;
- the complexity of ongoing maintenance and development of the system by the forces of operational staff;
- difficulties in implementing multiplexing due to low density of terminal devices;
- the impossibility of remote power supply of terminal devices with appeal for this purpose to the mass-market technology of PoE and PoE+.

In addition, we will point out that the transport level of most new ITS subsystems does not require high performance and it can be implemented on the basis of 100Base-TX network interfaces.

## 2. The prerequisites for extending the length of the line and the design model

The following considerations are underlying the technical possibility of extending the length of a symmetrical cable channel in excess of the 100 m value specified in the IEEE 802.3u standard:

- Fast Ethernet network interface specifications were initially developed taking into account the capabilities of the Category 5 cable engineering, whereas a typical contemporary engineering provides significantly better parameters corresponding to the Category 5e;
- it is possible to use simple channel structures of the direct connection type;
- due to the unilateral nature of the transmission over individual twisted pairs and the use of a two-pair scheme of the transmission path, the noise level at the input of the receiver is significantly reduced.

Simultaneously, the transmission channel of the ITS subsystems under consideration is not the part of the SCS. It is not subject to the limitations of standards in the length of not more than 100 m.
In the future we proceed from the fact that

- the network interface complies with the IEEE 802.3u specification;
- the information transfer rate is 100 Mbit/s (application in the line of block code type 4B5B increases the speed of the line signal up to 125 Mbit/s);
- the transmission channel is built according to the direct connection scheme;
- the parameters of the interference of the line and patch-cord cables are the same, and the attenuation coefficient of the line cable is 1.5 times smaller compared to the patch-cord one;
- as part of the channel, the ISO/IEC 11801 standard connection hardware was used.

Linear signal of the LAN-interface has uniform spectrum [2]. This makes it possible to use the theory of K. Shannon and seek the limiting length of the channel as a solution to the following equation [3].

$$W(l, f_b) = \int_0^{\infty} log_2[1 + ACR(f, l)]\, df, \tag{1}$$

where ACR = NEXT - αl – signal protection; NEXT - the near end cross talk loss; α is the attenuation coefficient; l is the "electrical" length of the channel.

## 3. The determination the communications capacity of the channel

For convenience of calculation, it is convenient to represent expression (1) in a three-term, figure 1:

$$\mathrm{W} = \mathrm{I_1} + \mathrm{I_2} + \mathrm{I_3} = \int_0^{f_b} \log_2 \mathrm{ACR(f)df} + \int_0^{f_b} \log_2\left[1 + \frac{1}{\mathrm{ACR(f)}}\right]\mathrm{df} + \int_{f_b}^{\infty} \log_2[1 + \mathrm{ACR(f)}]\mathrm{df}. \tag{2}$$

The upper frequency limit fb is the root of equation

$$\mathrm{NEXT} - 15\,\mathrm{lg(f)} - \alpha\mathrm{l}\sqrt{\mathrm{f}} = 0. \tag{3}$$

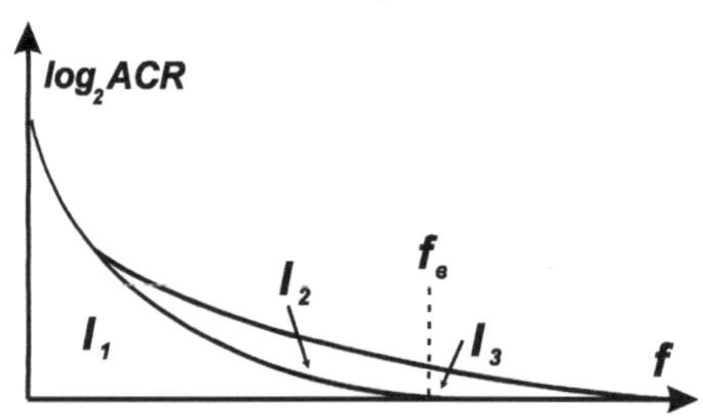


**Figure 1.** To calculation of the Shannon's capacity of a symmetric channel.

The solution of (3) has the form

$$\mathrm{f_b} = (\frac{\mathrm{NEXT}}{\alpha\mathrm{l}} - \frac{13}{\alpha\mathrm{l}} \cdot \ln\frac{\mathrm{NEXT}}{\alpha\mathrm{l}} / (1 - \frac{13}{\alpha\mathrm{l}}))^2. \tag{4}$$

IOP Conf. Series: Materials Science and Engineering **317** (2018) 012053 doi:10.1088/1757-899X/317/1/012053

Expression (4) provides approximately 10 percent accuracy of calculation. If necessary, the iterative process is repeated.

In expression (4), the total NEXT of the near end of the balanced channel appears. The cross-talk interference in the balanced channel on which NEXT depends is created by the line and patch-cord cables, as well as by the near end connecting hardware. Omitting the cumbersome intermediate calculations, taking into account the equality of the parameters of the interference of the line and patch-cord cables [4], and also the construction of the patch-panel by the interconnect scheme, we write the value NEXT in the form

$$NEXT = NEXT_0 - 15lgf - 20\lg\left[1 + 0.13f_b^{0.25}\right],$$

where $NEXT_0$ is NEXT loss coefficient of the cable.

The calculated relationships, which can be used to determine the individual terms (2), are given in table 1.

**Table 1.** Calculated relationships for computing *W*.

| Term | Calculated relationships | Note |
|---|---|---|
| $I_1$ | $0.166 \cdot NEXT \cdot f_b - 1.08 \cdot f_b \cdot (lnf_b - 1) - 0.11 \cdot f_b^{1.5} \cdot \alpha l$ | Lower bound |
| $I_2$ | $\frac{2.88}{0.115\alpha l} \cdot 10^{-\frac{NEXT}{20}} \cdot e^{0.115\alpha l\sqrt{f_b}} \cdot [f_b - \frac{2\sqrt{f_b}}{0.115\alpha l} + \frac{2}{(0.115\alpha l)^2}]$ | Upper bound |
| $I_3$ | $\frac{2.88 \cdot 10^{NEXT/20}}{0.144 \cdot \alpha l \cdot f_b^{1.75}}$ | Upper bound |

## 4. Analysis of results

The results of calculations are shown in figure 2 and indicate the possibility of a significant, multiplied increase in the length of the channel. Two criteria were used for the calculation:

- congruence of Shannon's capacity *W* to 185 Mbit/s, which corresponds to the 33 percent service stock recommended by the IEEE Institute for operational inventory;
- reaching the upper limit frequency of 70 MHz.

For typical for the equipment of categories 5e NEXT values and α for individual *W*, we have $\frac{I_1}{I_1 + I_2 + I_3} \approx 0.6$. Hence we can state that $I_1$ is the main part of *W* and can be used as an effective estimate of the real communications capacity of the balanced cable channel over which the Fast Ethernet network interface operates.

From the expression for determining $I_1$ for the band $[0;f_b]$ we have

$$\frac{\partial^2 W}{\partial f^2} = -\frac{1}{f}(1.08 + 0.15\alpha l) < 0. \tag{5}$$

It follows from (5) that the communications capacity *W* of the channel is determined by its parameters in the low-frequency area and the increase in the accuracy of determining $f_b$ with respect to the value (4) makes no sense when performing engineering calculations.

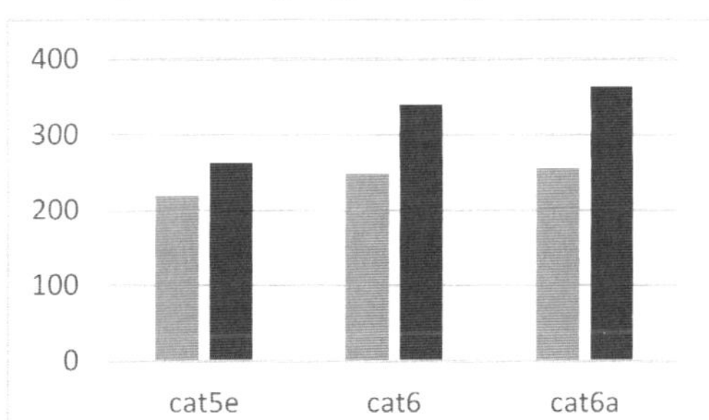


**Figure 2.** Limiting calculated path lengths: left column - 70 MHz criterion, right - 185 Mbit/s.

From the calculated relationship for (3) it follows that for typical values for contemporary equipment NEXT and α it is valid that

$$\sqrt{f_b} \cong \frac{NEXT}{2\alpha l} \tag{6}$$

Hence it follows that the main directions for increasing $l$ at $f_b$ = const are a decrease in α and an increase in NEXT. It is easy to obtain from (6) that

$$\left|\frac{\partial l/\partial \alpha}{\partial l/\partial NEXT}\right| = \frac{NEXT}{\alpha}. \tag{7}$$

Conclusion (7) does not take into account the dependence of $f_b$ on NEXT and α. It can be shown that its presence does not have a significant effect on the estimate (7).

Equipment of categories 6 and 6a has the typical increase in NEXT is usually around 15%. Given that the "electrical" length of the channel is determined primarily by a line cable, it follows that

- the main means of increasing the extent of the balanced channel is to reduce the attenuation coefficient α, since the efficiency of this method based on (7) is approximately NEXT/α ≈ 30 times higher;
- to achieve the desired effect, the attenuation coefficient α of horizontal cable must be reduced by at least 15%, which corresponds to units dB/100 m at a frequency of 100 MHz.

This situation is confirmed qualitatively by figure 2. From the data presented there it follows that when moving to the next category of equipment (typical increase in NEXT by 10 dB), the insignificant 15% increase in the limiting length of the channel is observed.

## 5. The attenuation of a twisted pair and the way of its reduction

The attenuation in the twisted pair is determined by the losses in: the conductors metal, insulation, the screen, separator, shell material and other similar cable components. The relationship between these components is shown at figure 3.

The main contribution to the final value of α is made by losses in the metal $\alpha_1$, the fraction of the remaining components does not exceed 12-15%/

Effective reduction of the value of α can be achieved by a rejection from the screens and a reduction in losses in the conductors metal. The attenuation coefficient of a twisted pair with a U/UTP structure is

$$\alpha = \frac{4.34R}{Z} + \alpha_2, \tag{8}$$

where $Z$ is the impedance, a $R$ is the active resistance, $\alpha_2$ – loss in the insulation ($\alpha_2 = 0{,}1\ \alpha$).

An estimate of the expected value of α for different initial data is given in table 2. When performing calculations, the parameters of the standard [5] were used.

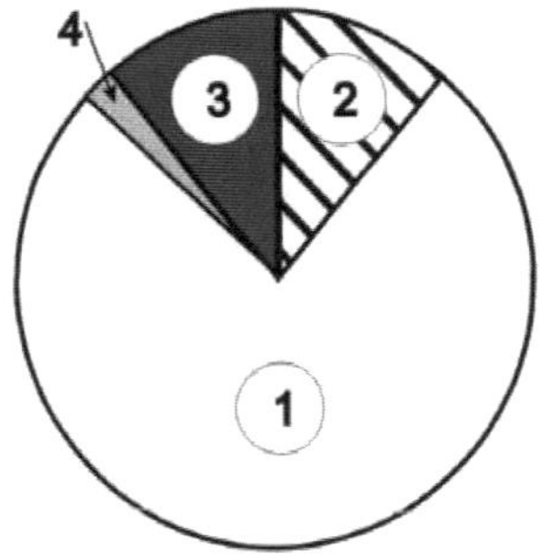


**Figure 3.** Ratio of individual components of the loss of a twisted pair:
1 – metal; 2 – the insulation of conductors; 3 – the screen (if any); 4 – other polymer cable components.

**Table 2.** The attenuation coefficient and channels length of Fast Ethernet network interfaces

| Conductor diameter, mm | 0.5 | 0.52 | 0.55 | 0.6 | 0.64 |
|---|---|---|---|---|---|
| Z = 100 Ohm | 21.5 | 20.0 | 18.2 | 15.6 | 14.0 |
| Length, m | 185 | 202 | 226 | 258 | 288 |
| Z = 120 Ohm | 18.3 | 16.3 | 14.8 | 12.8 | 11.5 |
| Length, m | 221 | 251 | 268 | 311 | 345 |
| Z = 150 Ohm | 15.1 | 13.8 | 12.6 | 10.9 | 9.9 |
| Length, m | 263 | 293 | 315 | 366 | 402 |

From the data of Table 2 we can see that the decrease in extinction due to the increase in the diameter of the conductor leads to a noticeably greater effect compared with the increase in impedance.

## 6. Conclusion

The obtained results make it possible to state the following

1. The length of the balanced cable channel of subsystems of contemporary ITS can be increased at least in two or three times.
2. Unlike classical office LAN, the function of the main means of achieving specified quality indicators goes from NEXT to the attenuation coefficient α.
3. It is expedient to achieve a reduction in α by increasing the diameter of the conductor.
4. Line cables must have a U/UTP structure and characteristics according to NEXT, RL and other similar parameters at the category level 5e.
5. In the case of a simultaneous increase in the diameter of the conductor and the impedance of a twisted pair with a U/UTP structure at a frequency of 100 MHz, attenuation coefficient of less than 10 dB / 100 m at a frequency of 100 MHz are attained, which makes it possible to increase the estimated channel length up to 400 m.
6. For the reduction of the cost of implementing the ITS, it is possible to use 2-pair versions of U/UTP cables.

## References

[1] Semenov A B, Strizhakov S K, Suncheley I R 2002 *Structured cabling systems* Moscow: DMK-Press, 640.
[2] Gerber M, Klasse E A and Klasse F A 2012 *Cabling for 10GBase-T and 40GBase-T* Reichle & De-Massari AG. – White paper. 11.
[3] Semenov A B 2012 *The prospects for a structured cabling system, class FA* Cable-news **6** 56–9.
[4] Semenov A B 2016 *Classic structured cable systems* Moscow: Goryachaya liniya Telecom, 462.
[5] ISO/IEC 11801:2011, *Information technology – Generic cabling for customer premises.* International standard. 194.

# Possibility of Coal Combustion Product Conditioning

**Tomasz Z. Błaszczyński[1], Maciej R. Król[2]**

[1]Technical University of Poznań, Piotrowo 5, 60-965 Poznań, Poland
[2]Technical University of Koszalin, Śniadecki 2, 75-453 Koszalin, Poland

**Abstract.** This paper is focused on properties of materials known as green binders. They can be used to produce aluminium-siliceous concrete and binders known also as geopolymers. Comparing new ecological binders to ordinary cements we can see huge possibility of reducing amount of main greenhouse gas which is emitted to atmosphere by 3 to even 10 times depending of substrate type used to new green material production. Main ecological source of new materials obtaining possibility is to use already available products which are created in coal combustion and steel smelting process. Most of them are already used in many branches of industry. They are mostly civil engineering, chemistry or agriculture. Conducted research was based on less popular in civil engineering fly ash based on lignite combustion. Materials were examine in order to verify possibility of obtaining hardened mortars based of different factors connected with process of geopolymerization, which are temperature, amount of reaction reagent and time of heat treatment. After systematizing the matrices for the basic parameters affecting the strength of the hardened mortars, the influence of the fly ash treatment for increasing the strength was tested.

## 1. Introduction

Need of new binders searching is strictly connected with problem of climate changes. Process of temperature rising which effect on climate change, and state of environment is without doubt undesirable process. Derivative of temperature rise is amount of carbon dioxide in the atmosphere [1]. Cement industry is being responsible on up to 9 % of global anthropogenic $CO_2$ emission (data for year 2016) [2, 3]. Lately this amount had risen from 3 in year 2000 [4]. Future perspectives for reduction are not looking acceptable. In some ways, cement industry has been doing more to the green agenda than other great producers of greenhouse gasses like fossil fuel energy companies or large utilities. Cement companies owners have signed up to voluntary emissions reductions, reassured they will do more than in last years to produce the materials needed for a low carbon environment. It is strictly connected with building hydropower dams or fixing wind turbines. In less than 15 years from now even the fact of stopping acceleration of cement production and introducing technologies of $CO_2$ emission like carbon capture and storage (CCS) [5], speculation about amount of cement production due to worlds carbon dioxide production does not look satisfactorily (figure 1).

Possibilities of making the cement industry more economical and at the same time ecological are plural. Unfortunately, not all treatments offer significant reductions in greenhouse gas emissions. Relative reduction of energy consumption, transport and its use in the process and maintenance of the industry in the form of electricity, even by 20 or 30% will reduce the total reduction by only 2-3%. In addition, the calcination-related production process, which cannot be changed, does not provide the opportunity to reduce almost half of the $CO_2$ value attributed to clinker cement.

Future possibilities of reducing carbon dioxide production are limited. Three different forecast describing future possibilities of reducing emission have been introduced (figure 2). Carbon share projections have been developed on the basis of predicted cement production data for years 2017-2030. The carbon dioxide emissions per 1t of cement produced were set at 2016. Global production has assumed a constant emission from fossil fuels. Forecast I does not include significant introduction of geopolymer binders. Their share is limited to the production of specialized materials and production by small companies already using the technology of geopolymer cements.

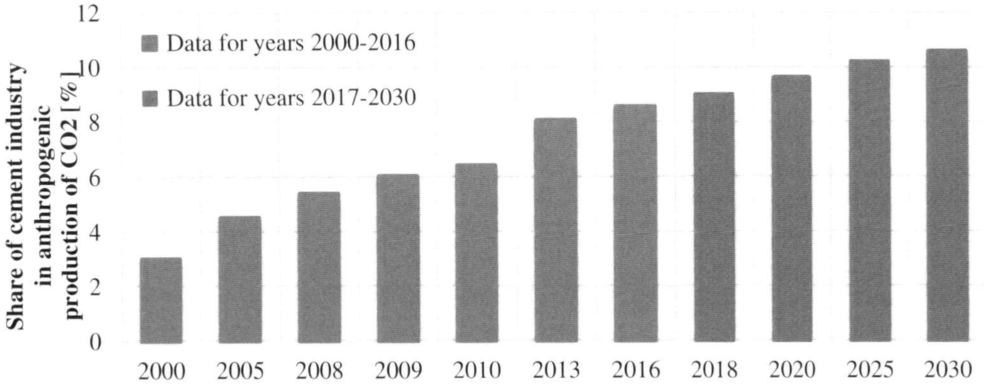


**Figure 1.** Actual and perspective of future share of cement industry in anthropogenic production of $CO_2$ [2- 4, 6].

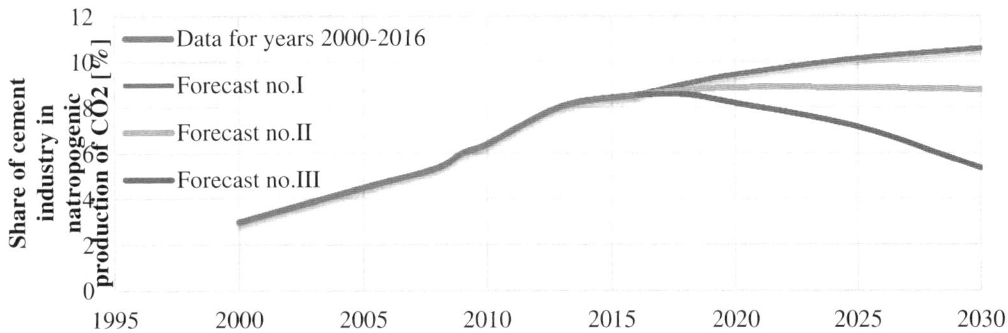


**Figure 2.** Actual and perspective of future share of cement industry in anthropogenic production of $CO_2$ with forecasts assuming replacing Portland cement by green binders [2-6].

Forecast concept II takes into account the replacement of 25% of clinker cement produced with green binder produced from waste substances such as fly ashes and blast furnace slags from current production. Concept III assumes that at least half of cement production will be replaced by geopolymer binder produced both from waste materials and dedicated material, i.e. produced specifically for the production of geopolymers. As can be seen from forecast II, the growth of carbon dioxide production is halted, while in forecast III, the value of greenhouse gas emitted into the atmosphere, despite huge quantities of cement produced, is significantly declining.

## 2. Research base

The fly ash derived from the combustion of lignite with significant calcium content was used for the study. In the samples identified, the CaO content ranged from 8 to 15%. Samples were not cleaned of unburned carbon, which was in the composition of about 3 to 5% (table 1). The only treatment for ash was the reduction of grain size by ash grinding. During the grinding process, a noticeable change in the screening curve was observed (figure 3).

**Table 1.** Chemical content of subjected for research calcium fly ash.

| $SiO_2$ | $Al_2O_3$ | $Fe_2O_3$ | $TiO_2$ | MgO | CaO | $Na_2O$ | $K_20$ | Ignition loss |
|---|---|---|---|---|---|---|---|---|
| **44,17** | 21,79 | 4,58 | 1,85 | 1,49 | 21,06 | 0,23 | 0,19 | 4,64 |

IOP Conf. Series: Materials Science and Engineering **317** (2018) 012054 doi:10.1088/1757-899X/317/1/012054

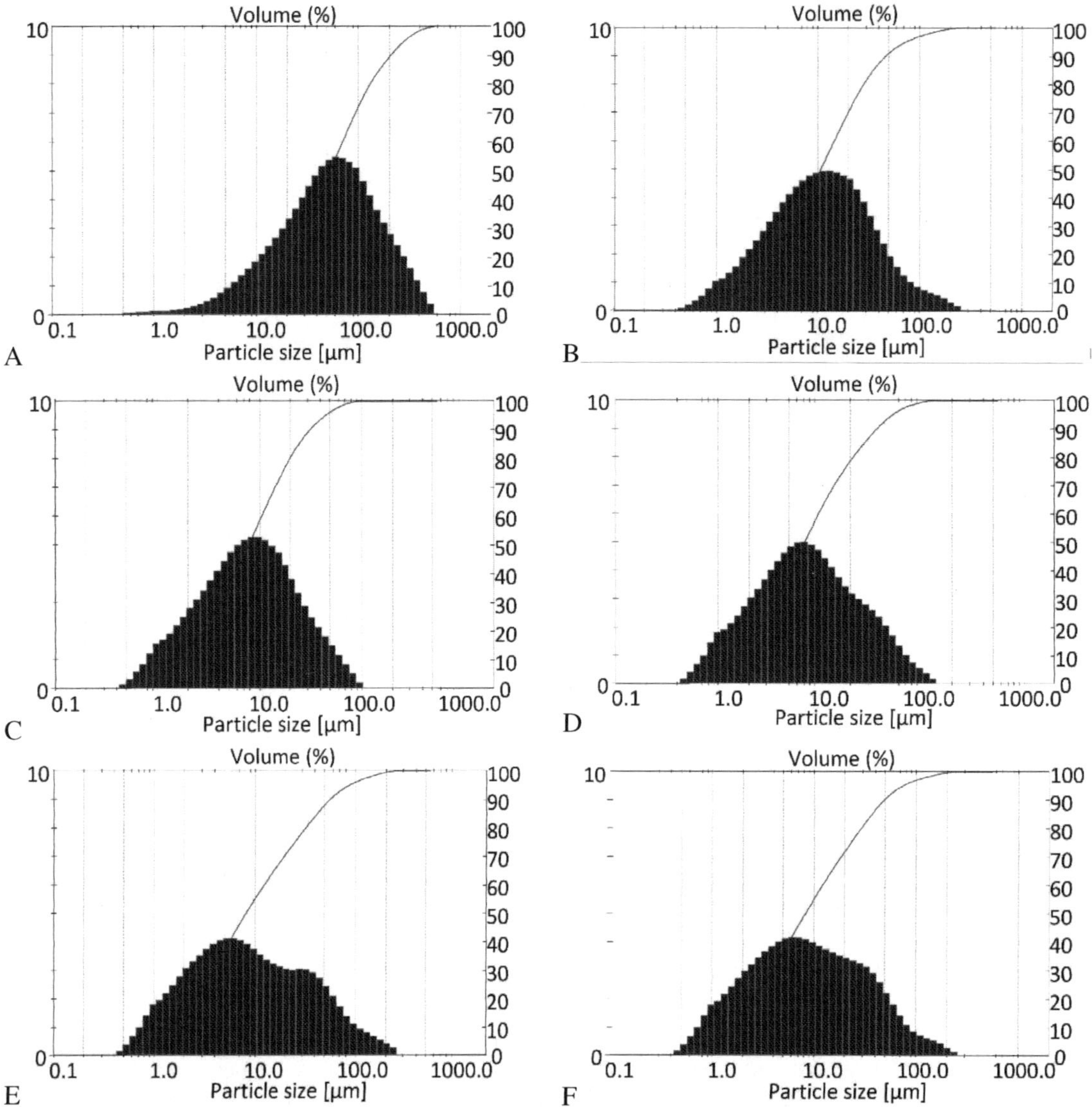


**Figure 3.** Graphs of grinded and subjected to ultrasound wave fly ash used for research A) original, not subjected to grinding sample., B) sample grinded for 30 s., C) sample grinded for 60 s., D) sample grinded for 120 s., E) sample grinded for 180 s., E) sample grinded for 240 s.

Most preferred in terms of maximum fragmentation was milling in time of 60 seconds. In the course of fractions it gives the highest ash content of less than 45 microns. A further increase in grinding time resulted in the opposite effect, in the material began to form agglomerates. The next stage of treatment was to subject the material to an ultrasound wave that broke up the formed agglomerates, making the sample the most crushed to be the sample milled by 120 seconds. The degree of fragmentation described by the specific surface in this case corresponded to the obtained compressive strength of the test samples.

## 3. Results

Grounded fly ash was used to create geopolymer binders. Activation took place with aqueous NaOH solution. The binder was used to form beams 40x40x160 mm. The most optimal water-binder ratio was W/B = 0.50 and the molar concentration of NaOH in water was 8M. For the first 12h of maturation, they were placed in a low temperature furnace at 85 °C, then after 24 hours they were disassembled and subjected to testing. Rest of samples assigned for subsequent tests, were placed in air-dry conditions with a relative humidity of 50% and a temperature of 20 °C. According to conducted earlier tests (figure 4)

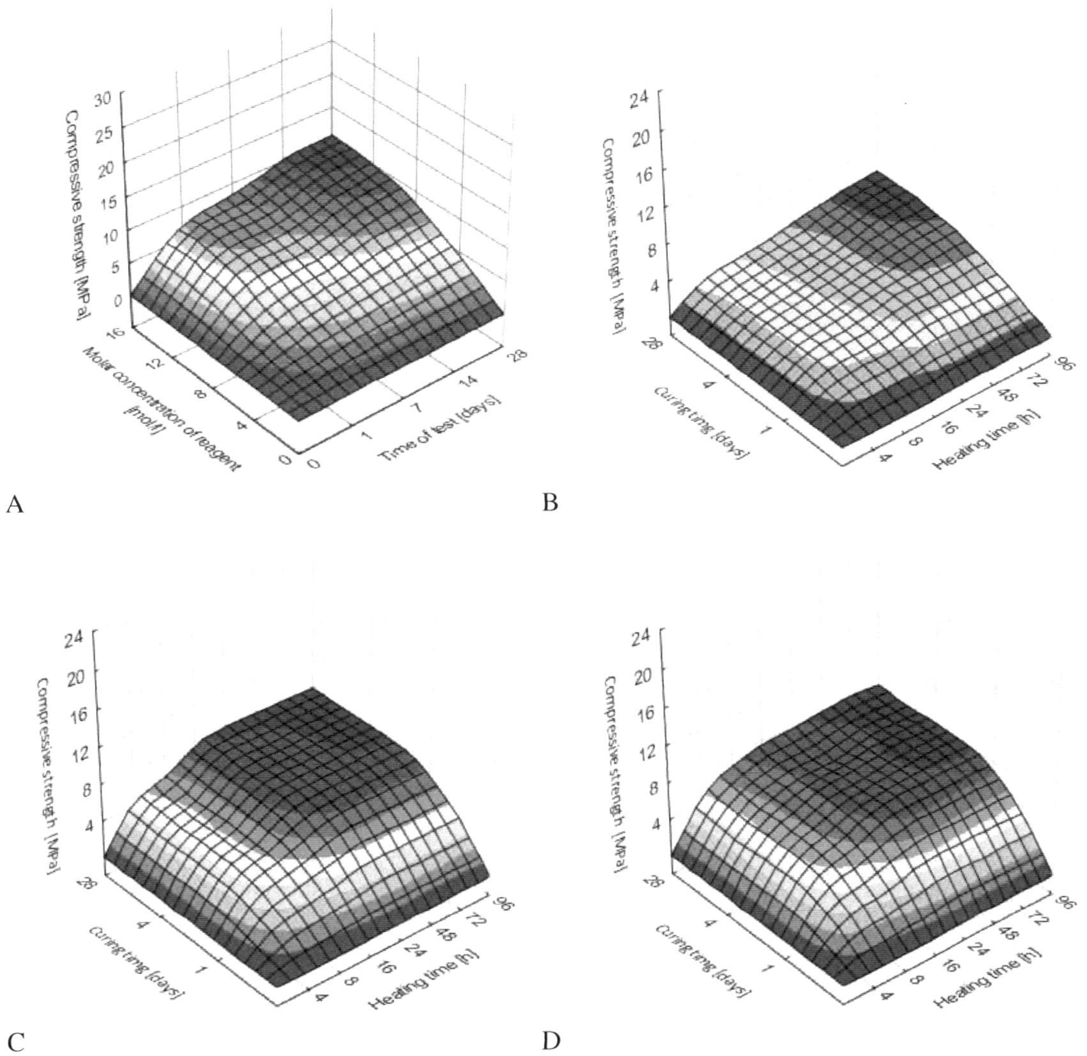


**Figure 4.** Influence of A) molar concentration, B) temperature 65$^0$C, C) temperature 80$^0$C, D) temperature 95$^0$C, at the compressive strength of tested calcium fly ash.

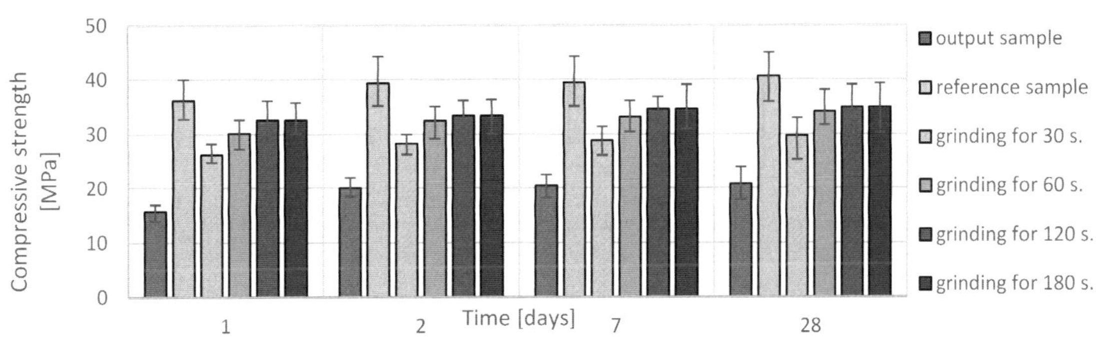


**Figure 5.** Compressive strength of grinded fly ash.

Together with the reduction of grain size, both compression strength and bending tensile strength were simultaneously increased. Significant increase in the tensile strength of the reference sample, which was the mixture of 50% of the milled material for 120 seconds and 25% of the milled material for 30 and 240 seconds respectively. The material obtained after mixing these components was characterized by relatively the smallest particle size, with a particle size less than 10 [μm] in relation to the amount of material less than 45 [μm], and the most even distribution of grain sizes in the range of 10-45 μm. This allowed the polymerization reaction to be more uniform, depending on the grain size. This has a positive effect on the compressive strength to flexural tensile strength ratio (figure 5, 6).

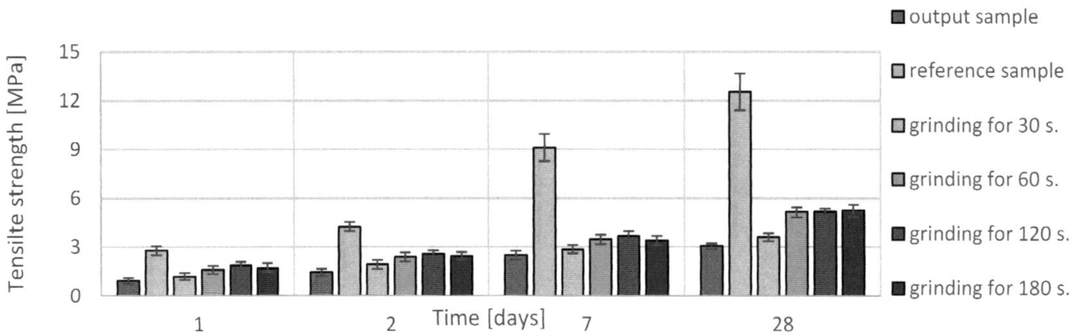


**Figure 6.** Compressive strength of grinded fly ash.

In comparison to ordinary cements highly specialized materials based on alumino-silicate binders can easily reach equal strength and much better physical properties than Portland cements [7,].

## 4. Conclusions

At the Paris climate summit which was held in 2015, one of the biggest cement company and 16 other cement manufacturers went further and promised to decarbonise their production by 2030. This exertion will lead to an even 10 percent worldwide reduction in carbon dioxide emissions connected with cement production. As we can see (figure 1, 2) this already taken into consideration amount of $CO_2$ reduction won't be enough to bring level of greenhouse gasses emission to proper level. Additional resources introduced on global, national and even local fields would really accelerate research that is needed to perform new green binders implementation. Till then we will need to believe that a lots of work that has been spent on environment awareness will spill in social effect [9].

## References

[1] GRID Arendal, *Ahead of the Curve*, Annual Report, Norway 2016,

[2] Cembureau, Activity Report 2016 *The Cement Sector a Strategic Contributor to Europe's Future Brussels.*

[3] Denman K L, Brasseur G 2007 *Couplings Between Changes in the Climate System and Biogeochemistry United Kingdom and New York*, NY, USA, Cambridge University Press. ISBN 978-0-521-88009-1.

[4] UNEP/GRID-Arendal-Publications, Vital Climate Change Graphics. Norway, 2005.

[5] Walker B 2016 *Can cement clean up its act*, Chinadialogue.

[6] Humphreys K, Mahasenan M 2002 *Toward a Sustainable Cement Industry: Climate Change*, World Business Council for Sustainable Development.

[7] Blaszczynski T, Krol M 2017 *Alkaline Activator Impact on the Geopolymer Binders*, IOP Conf. Series: Materials Science and Engineering, **245**, 2017. 022036 doi:10.1088/1757-899X/245/2/022036.

[8] Blaszczynski T, Krol M 2017 *Durability of cement and geopolymer*, IOP Conf. Series: Materials Science and Engineering **251** 2017. 012005 doi:10.1088/1757-899X/251/1/012005.

[9] Blaszczynski T, Krol M 2015 *Proc. Eng.* **122** 296-301.

IOP Conf. Series: Materials Science and Engineering **317** (2018) 012055 doi:10.1088/1757-899X/317/1/012055

# Property Specification Patterns for intelligence building software

Seungsu Chun[1]

[1]Research Fellow, Business Planning & Budget Division, Korea Institute of S&T Evaluation and Planning

**Abstract.** In this paper, through the property specification pattern research for Modal MU(μ) logical aspects present a single framework based on the pattern of intelligence building software. In this study, broken down by state property specification pattern classification of Dwyer (S) and action (A) and was subdivided into it again strong (A) and weaknesses (E). Through these means based on a hierarchical pattern classification of the property specification pattern analysis of logical aspects Mu(μ) was applied to the pattern classification of the examples used in the actual model checker. As a result, not only can a more accurate classification than the existing classification systems were easy to create and understand the attributes specified.

## 1. Introduction

Model checking is a method for logically verified whether the intelligence building software meets the desired properties [1]. To test the model, the user must express the property to be examined. This property is called the specification. There were tense used in the specification means a logical, automata, various expressions, including visual language, it is the most widely used among these temporal logic [2]. Because, temporal logic is because it is excellent in expressive power can specify a variety of attributes. However, the property has a temporal logic specification created by the questions that could be easily understood by users. Conversely, there is the question of exactly what we can attribute to specify a given temporal logic. For example consider the following attribute specification written in a logical aspects Mu(modal μ-calculus) [2].

$$\nu Z.\ [emergency](\ \mu X.\ [alarm](\ \mu Y.\ [alarm](\ \nu Z.\ [alarm]ff \wedge [-]Z)\ \langle - \rangle tt \wedge [-alarm]Y)\ \langle - \rangle tt \wedge [-alarm]X) \wedge [-]Z$$

In order to interpret the above specifications, it is required to understand the meaning and symbols used in logic, and background knowledge. For a particularly aspect mu logic it is essential to understand the maximum and minimum fixed point fixed point. The meaning of the above specification is "continuously when an emergency situation occurs two times and the alarm must be sounded."

If you called a temporal logic CTL (Compu-tation Tree Logic) and LTL (Linear Temporal Logic) advanced Specification Language, MU logical aspect is the same as assembly language. Therefore, it is not easy to write and interpretation of the specification. This study was research to help the creation of specifications and logical aspects Mu understand.

In the past, study or the like dedicated language specification [3], using the pattern [4] was to solve the problem related to the attribute specification. Survey of existing research results specified in the

Dwyer pattern property was found to be the best in our research direction. However, the pattern of Dwyer applies only CTL, LTL, GIL (Graphical Interval Logic). In this study, the mapping to the aspect targeted mu logic did not address. In addition, CTL and LTL covers only path to each state and GIL, on the other hand because it is a modal logic to accommodate both the concept of the state and behavior not accept the logic that his classification system deals only with aspects Mu($\mu$) events.

So, this study was subdivided according to the classification system of modal logic Dwyer, was applied to the classification of the sample which was used in the model checker. The results were classified more accurately than existing classification systems. This suggests that it is helpful to understand the specification creation and use of the property specification pattern.

## 2. Pattern based property specification

In this study, it chose the following aspects MU logic for the same reason. First, it is the most highly expressive. Second, it is possible to express both the state and behavior. Third, CTL, LTL, CTL * can be converted into a logical aspects Mu. That aspect mu logic is that to provide a single framework for the temporal logic.

Classification of property specification pattern for pattern MU logic is shown in figure 1. A large category are the Occurrence and Order, there are eight patterns of classification, as shown below.

Absence, Universality,
Bonded Exsitence, Existence, Response,
Precedence, Chain Response, Chain Precedence.

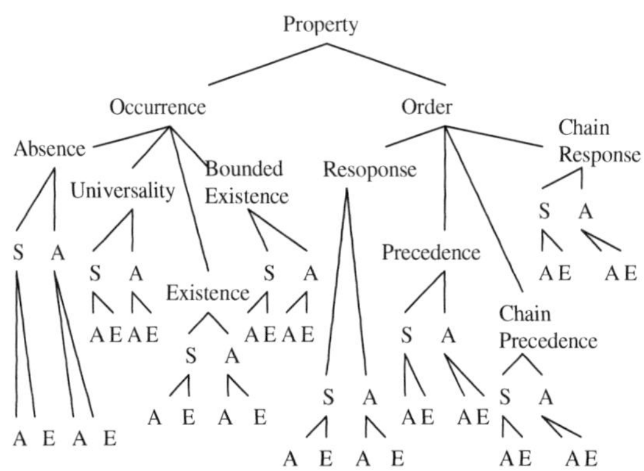


**Figure 1.** Pattern's classification.

Modal MU logic is broken down into these state (S) and the action (A), because all of the possible representations of the state and behavior. If the property is satisfied in all the paths is referred to as a 'strong', and as, if there is a path that satisfies the attribute 'weak'. Thus, the state and action expression is subdivided into strong (A) and weak (E).

For example, let's look at the pattern indicating the safety Universality property. The meaning of this pattern is an "attribute is always satisfied." It is broken down according to the state and the behavior as follows:

The pattern is further divided as follows depending on whether satisfactory in all the properties of the path state or behavior-oriented in the satisfaction of one or more paths:

- State oriented(S): $\nu Z.\Phi\wedge[-]Z$
- Behavior oriented (A): $\nu Z.[K]ff\wedge[-]Z$

- State oriented strong(SA): $\nu Z.\Phi\wedge[-]Z$
- State oriented weakness(SE): $\nu Z.\Phi\wedge\langle-\rangle Z$

- Behavior oriented Strong(AA): $\nu Z.[K]ff \wedge [-]Z$
- Behavior oriented Weakness(AE): $\nu Z.[K]tt \wedge \langle - \rangle Z$

The detail of such means based pattern classification enables the expression of the correct meaning. On the other hand, Universality pattern is expressed by a fixed point of maximum = f Z (Z). Its meaning is "always complete set has a value of true (True) S 'is unfolded from the top down (Top-down) method from the first state. In releasing strong (AA) in the center of the path, repeat calculation as follows:

It means the following expression is 'on all paths, an infinite series of movement by the event K is available'. It corresponds to the stability properties of the system. In the case of the minimum fixed point, to represent the ultimate property for the system, and which is classified as Exisetence pattern. All attributes with the system is represented by logical aspects Mu as shown in Table 1.

$$\begin{aligned} \nu Z. [K]ff \wedge [-]Z &= true \\ &\wedge [K]ff \wedge [-]\mathit{true} \\ &\wedge [K]ff \wedge [-]([K]ff \wedge [-]\mathit{true}) \\ &\ldots \end{aligned}$$

**Table 1.** Property specification pattern for Modal Mu Logic.

| Pattern type | Num. |
|---|---|
| Absence | 6 |
| Universally | 16 |
| Existence | 20 |
| Bounded Existence | 1 |
| Response | 6 |
| Precedence | 0 |
| Response Chain | 0 |
| Precedence Chain | 0 |
| UNKNOWN | 0 |
| Total | 43 |

## 3 Pattern based specification system

When you check the properties of Fairness using the specification pattern is a required attribute 'when you approach the car go across the car'. And the 'crossing the car or train it should occur indefinitely'. If you specify the attributes Bounded Existence pattern can be expressed as follows:

$$\nu X.[car](\mu Y.\langle - \rangle tt \wedge [-ccross]Y) \wedge [-]X)$$

And, "If the car is gone (Q) should always be followed by (R) occurs passing train." When you apply the pattern Precedence it can be specified as follows:

$$\mu Z.([Q]ff \vee ([-R]ff \wedge [-]Z \wedge \langle - \rangle tt))$$

Therefore, the specification equity properties with infinite recursion property of causality can be specified as follows:

$$\mu X.\nu Y1.([Q]ff \vee ([-ccross](\nu Y2.(R \wedge X) \wedge [-ccross]Y2)) \wedge [-ccross]Y1$$

The more complex attribute specifications, pattern-based approach is more efficient.

For example on CWB-NC (The concurrency Work Bench of the New Century), Let us apply the substantive requirements of the property specifications 'railway ringer system' [5]. The meaning of the attributes 'occurs sometime (Eventually) silent.', If the structure is a complex system specifications are as follows:

prop eventually_silent = not ( min A1 = ( not ( ['det:0] ( not ( max D = ( min E1 = ( not ( not ( min F(not(['comm_out:2,'stat_out:2]ff)V<'recovered:0>F)))/\ ( <'tick:4>D V <-'tick:4>E1 ) ) ) ) ) ) V <->A1 ))

If the specification of the above complicated expression patterns using the specification can be simply represented as a set relationship of the pattern is as follows.

**Existence.** ~Universally(~Response(~BoundedExistence~(~Existence `recovered:0 or (comm_out:2 or ~statout:2)and tick:4-ff or tick:4)))))

This study was conducted to analyze the physical property requirements of the specification and the CWB-NC ECW (Edinburgh Concurrency Workbench) The results are shown below.

1) Safety = { Absence, Universality }
 (1) Absence = { AG(!Φ), EG(!Φ), AG(!K), EG(!K) }
  - AG(!Φ) = SA : νZ.¬Φ∧[-]Z
  - EG(!Φ) = SE : νZ.¬Φ∧<->Z
  - AG(!K) = AA : νZ.[K]ff∧[-]Z
  - EG(!K) = AE : νZ.[K]ff ∧<->Z
 (2) Universality = { AG(Φ), EG(Φ), AG(K), EG(K) }
  - AG(Φ) = SA : νZ.Φ∧[-]Z
  - EG(Φ) = SE : νZ.Φ∧<->Z
  - AG(K) = AA : νZ.[K]tt ∧[-]Z
  - EG(K) = AE : νZ.[K]tt ∧<->Z

2) Cyclic = { Universality(Response) }
 (1) Universality = { AG[Φ→AX(AF(ψ))], EG[Φ→EX(EF (ψ))], AG[K→AX(AF(L))], EG[K→EX(EF(L))] }
  - AG[Φ→AX(AF(ψ))] = SA : νZ.Φ([-](ψ ∨ <->tt ∧ [-]Z)) ∧ [-]Z
  - EG[Φ→EX(EF (ψ))] = SE : νZ.Φ(<->(ψ ∨ <->tt ∧ <->Z)) ∧ <->Z
  - AG[K→AX(AF(L))] = AA : νZ.[K]([-](<->tt ∧ [-L]Z)) ∧ [-]Z
  - EG[K→EX(EF(L))] = AE : νZ.[K](<->(<-> tt ∧ <-L> Z )) ∧ <->Z

3) Linveness = { Existence }
 (1) Existence = { A F(Φ), E F(Φ), A F(K), E F(K) }
  - A F(Φ) = SA : μZ.Φ∨(<->tt∧[-]Z)
  - E F(Φ) = SE : μZ.Φ∨<->Z
  - A F(K) AA : μZ.(<->tt∧[-K]Z)
  - E F(K) AE : μZ.(<->tt∧<-K>Z)

4) Counting = { Bounded Existence }
 (1) Bounded Existence = { AG F(Φ), EG F(Φ), AG F(K), EG F(K) }
  - AG F(Φ) = SA : νX.μY.<->Y∨(<->X∧Φ)
  - EG F(Φ) = SE : νX.μY.<->Y∨([-]X∨Φ)
  - AG F(K) = AA : νX.μY.[-]([<->tt∧[-K] Y]∨[-]X
  - EG F(K) = AE : νX.μY.<->([<->tt∧<-K>Y] ∨<->X

5) Until = { Until, Precedence, Chain Precedence }

(1) Until = { A[Φ U ψ], A[Φ W ψ], E[Φ U ψ], E[Φ W ψ], A[K U L], A[K W L], E[K U L], E[K W L] }
- A[Φ U ψ] = SA : μZ.ψ∨(Φ∧<->tt∧[-]Z)
- A[Φ W ψ] = SE : μZ.ψ∨(Φ∧<->tt∧<->Z)
- E[Φ U ψ] = SA : νZ.ψ∨(Φ∧[-]Z)
- E[Φ W ψ] = SE : νZ.ψ∨(Φ∧<->Z)
- A[K U L] =AA : μZ.[-(K∪L)]ff∧<->tt∧[-L]Z)
- A[K W L] = AE : νZ.[-(K∪L)]ff∧[-L]Z)
- E[K U L] = AA : μZ.[-(K∪L)]ff∧<->tt∧<-L> Z)
- E[K U L] = AE : νZ.[-(K∪L)]ff∧<-L>Z)

(2) Precedence = { A[!Φ W ψ], E[!Φ U ψ], A[!K W L], E[!K U L] }
- A[!Φ W ψ] = SA : μZ.ψ∨(!Φ∧ [-]Z)
- E[!Φ U ψ] = SE : νZ.ψ∨(!Φ∧<->tt∧[-]Z)
- A[!K W L] = AA : νZ.[-(-K∪L)]ff∧[-L]Z)
- E[!K U L] = AE : μZ.[-(-K∪L)]ff∧<->tt∧<-L > Z)
- Chain Precedence = { !E[!ΦU(Ψ&!Φ&EX(EF(γ)), !E[!Φ W(Ψ&!Φ&EX(E F(γ)), !E[!K U (L&!K & EX (EF (J)), !E[!K W (L&!K&EX(EF(Jγ)) }
- !E[!ΦU(Ψ&!Φ&EX(EF(γ)) = SA : μZ.(ψ∧!P∧<->(ϒ∨<-> Z)∨ (! Φ ∧<->tt∧[-]Z)
- !E[!Φ W(Ψ&!Φ&EX(E F(γ)) = SE : μZ.(ψ∧!P∧<->(ϒ∨<-> Z) ∨ ( ! Φ∧ [-]Z)
- !E[!K U (L&!K & EX (EF (J)) = AA : μZ.([L∧!K]tt∧<->(<->tt∨<-J>Z)∨(<->tt∧[-K]Z)
- !E[!K W (L&!K&EX(EF(Jγ)) = AE : μZ.([L∧!K]tt∧<->(<->tt∨<-J>Z)∨(<->tt∧<-K>Z)

6) Fairness = { Response, Chain Response }

(1) Response = { AG(Φ→AF(ψ)), EG(Φ→EF(ψ)), AG(K→ AF(L)), EG(K→EF(L)) }
- AG(Φ→AF(ψ)) = SA : νX.Φ(μY.Ψ∨(<->tt∧[-]Y)∧[-]X
- EG(Φ→EF(ψ)) = SE : νX.Φ(μY.Ψ∨<->Y)∧<->X
- AG(K→ AF(L)) = AA : νX.[K](μY. (<->tt∧[- ]Y)∧[-]X
- EG(K→EF(L)) AE : νX.[K](μY.<-L>Y)∧<->X

(2) Chain Response = { AG(Φ→AF(ψ&AX(AF(γ)), EG(Φ→EF(Ψ&EX(EF(γ)), AG(K→AF(L&AX(AF(J)), EG(K→EF(L&EX(EF(J)) }
- AG(Φ→AF(ψ&AX(AF(γ)) = SA : νX.Φ(μY.(Ψ∧[-]γ)∨<->tt∧[-]Y)∧[-]X
- EG(Φ→EF(Ψ&EX(EF(γ)) = SE : νX.Φ(μY.(Ψ∧<->γ)∨<->Y)∧<->X
- AG(K→AF(L&AX(AF(J)) = AA : νX.[K](μY.([-L]ff∧[-][-J]ff)∨<->tt∧[-]Y)∧[-]X
- EG(K→EF(L&EX(EF(J)) = AE : νX.<K>(μY.([-L]ff∧[-][-J]ff)∨<->Y)∧<->X

Lamport is largely classified as safety and finality of property [6]. However, this classification is too comprehensive and may need to be broken down more. Mana has offered a little more detailed classification structure for linear temporal logic [7]. However, this classification is mainly on the syntax of temporal logic formulas. The syntax is more preferable than the classification means as the size of the expression go large. So Dewyer is complementary to the two above classification presents a detailed and even semantic-based taxonomy. His classification CTL, LTL has been well applied aspects Mu inadequate in logic that we have to target. The Stirling was classified as greater safety and finality, fairness, circulation, recovery of property emergence of Modal Mu logic [8]. But we should classify the structure is covers all five properties. So it would help the creation and understanding aspects of Mu If you are using a logical classification structure presented.

## 4. Result and future works

This paper provides a single framework that encompasses all of the logical specification through the property specification pattern research for Modal MU logic. As a result, the Modal Mu made it possible to analyze and understand complex logical expressions and facilitate state and behavior were the center of the property and the analysis and representation of the more complex Modal Mu logic through the extension of the detailed specifications of the actual classification of the path to the center of. Also it enables the study of new and advanced specification language approach. Future research is as follows. First, we will proceed with the collection, classification, and statistical work of the actual specifications for ensuring the safety of the specification pattern. Second, prove the regularity between attributes, patterns, meaning, and expression patterns for the expansion of inventory and will apply it to a component based methodology. Third, to implement a pattern based specification of the property inventory system for the automatic Modal Mu logic.

**Reference**

[1] Clarke E M et al. 1999 *Model Checking* MIT Press.
[2] Emerson E A 1997 *Model Checking and the Mu-Calculus* Proceedings of the DIMACS Symposium on Descriptive Complexity and Finite Models, pp. 185—214.
[3] Corbett J C et al. 2000 *A Language Framework For Expressing Checkable Properties of Dynamic Software* Proceedings of the SPIN Software Model Checking Workshop, LNCS 1885.
[4] Dwyer M B et al. 1998 *Property Specification Patterns for Finite-State Verification* Proceedings of the Workshop on Formal Methods in Software Practice.
[5] Cleaveland R *The Concurrency Workbench of the New Century* http://www.cs.sunysb.edu/~cwb/
[6] Lamport L 1977 *Proving the correctness of multiprocess Programs* IEEE Transactions on Software Engineering, **SE-3(2)** 125-143.
[7] Manna Z et al. 1992 *The Temporal Logic of Reactive and Concurrent Systems* SpringerVerlag.
[8] Stirling C 1996 *Modal and Temporal Logics for Processes* Lecture Notes in Computer Science **1043** 149-237.

**Acknowledgment**

This work was supported by the Korea Institute of S&T Evaluation and Planning grant funded by the Korean Government (No. KISTEP-2016-BB16010)

IOP Conf. Series: Materials Science and Engineering **317** (2018) 012056 doi:10.1088/1757-899X/317/1/012056

# Path Planning for Robot based on Chaotic Artificial Potential Field Method

**Cheng Zhang**[1]

[1]City Institute, Dalian University of Technology, Dalian,116600, China

**Abstract.** Robot path planning in unknown environments is one of the hot research topics in the field of robot control. Aiming at the shortcomings of traditional artificial potential field methods, we propose a new path planning for Robot based on chaotic artificial potential field method. The path planning adopts the potential function as the objective function and introduces the robot direction of movement as the control variables, which combines the improved artificial potential field method with chaotic optimization algorithm. Simulations have been carried out and the results demonstrate that the superior practicality and high efficiency of the proposed method.

## 1. Introduction

The path planning for Robot under unknown environments is referred to search an optimal or sub-optimal path which is a safe collision-free path from the starting point to the objective point [1]. It generally falls into global planning and local planning [2]. The global planning includes: visibility graphic method, tangent graphic method, penalty function method and enumeration algorithm etc [3]. The local planning mainly includes: artificial potential field method, fuzzy logical method, genetic algorithm, simulated annealing algorithm and ant colony optimization etc [4]. The artificial potential field method is frequently used in local path planner. It has the advantages of simple structure, less computation and good capabilities of obstacle avoidance except for local optimum and shaking before obstacles etc. Aiming at the above, combining improved artificial potential method with chaotic optimization algorithm, the path planning for robot based on chaotic artificial potential field method is presented. It effectively solves the path planning for robot under unknown environments.

## 2. Improved artificial potential method

Artificial potential field method is frequently used in local path planner, the principle of which is that robot is regarded as the particle in Virtual Force Field and its movement controlled by resultant forces which is composed of repulsive force from obstacles and attractive force from the objective point. The path generated by the artificial potential field method is generally smooth and safe [5], but this method has local optimum etc.

### *2.1. Repulsive force potential function.*

The repulsive force from obstacles acts on robot in virtual force field. Potential function used in the algorithm is shown as (1):

$$U_{rep}(x)=\begin{cases}0.5\eta\left[\dfrac{1}{\rho(x,x_0)}-\dfrac{1}{\rho_0}\right]^2\rho^2(x,x_g) & \rho(x,x_0)\le\rho_0\\ 0 & \rho(x,x_0)>\rho_0\end{cases}\quad(1)$$

In (1),$\eta$ is repulsive force coefficient; $\rho(x,x_0)$ is the shortest distance between robot and obstacles; $\rho(x,x_g)$ is relative distance between robot and the objective point; $\rho_0$is the constant, within $\rho_0$ the repulsive force from obstacles acts on robot but no repulsive force beyond $\rho_0$. The repulsive force is negative gradient of its potential function accordingly as shown in (2):

$$F_{rep}=\begin{cases}F_{rep1}+F_{rep2} & \rho(x,x_0)\le\rho_0\\ 0 & \rho(x,x_0)>\rho_0\end{cases}\quad(2)$$

$$\left\|F_{rep1}\right\|=\eta\left[\frac{1}{\rho(x,x_0)}-\frac{1}{\rho_0}\right]\left[\frac{\rho^2(x,x_g)}{\rho^2(x,x_0)}\right]\quad(3)$$

$$\left\|F_{rep2}\right\|=\eta\left[\frac{1}{\rho(x,x_0)}-\frac{1}{\rho_0}\right]^2\rho(x,x_g)\quad(4)$$

In 2, $F_{rep1}$ and $F_{rep2}$ stand for two components forces of $F_{rep}$, The direction of $F\ rep1$ is pointing from the obstacles to robot and the direction of $F\ rep2$ is pointing from robot to the objective point.

### *2.2. Attractive force potential function*

The attractive force from the object point acts on robot in potential field. Potential function being used in the algorithm is shown (5):

$$U_{att}(x)=0.5\kappa\rho^2(x,x_g)\quad(5)$$

In (5), $\kappa$ is attractive force coefficient; $\rho(x,x_g)$ is relative distance between robot and the object point. Attractive force is negative gradient of its potential function, the direction of which is pointing at the object point with corresponding attractive force as shown in (6):

$$F_{att}=-\kappa\rho(x,x_g)\quad(6)$$

In conclusion, the resultant force is shown as (7):

$$F=F_{rep}+F_{att}\quad(7)$$

### *2.3 Variable repulsive force coefficient*

Because traditional artificial potential field method is that all obstacles has the same repulsive effect, $\eta$ of all obstacles is the same. According to the rule of the movement for robot, obstacles in the direction of movement have greater repulsive force than ones in the other direction. The idea is introduced to artificial potential field method and $\eta$ is redefined as shown in (8):

$$\eta_V=\eta|\sin\beta\sin\alpha+\cos\beta\cos\alpha|\quad(8)$$

In (8), $\alpha$ is the direction angle (that is the angle between the direction of movement for robot and Y axis) and $\beta$ is the direction angle of obstacles ( that is the angle between position of obstacles and Y axis) and $\eta$ is the constant repulsive force coefficient. The variable repulsive force coefficient is introduced and the (9) is shown.

$$F_{rep}=\eta_V F_{rep}/\eta\quad(9)$$

Computing the repulsive force by using the above method can weaken repulsive forces of obstacles for robot except for directions of movement and reduce shaking in the process of movement.

## 3. Chaotic artificial potential field method

Chaos is a universal phenomenon in nonlinear system. Chaos phenomenon has stochastic property, ergodicity and regularity. In the optimization area, the ergodic property can be used as an optimization mechanism to escape from local optimums[6]. Chaos has been a kind of novel global optimization technique. People pay much attention to the research of the optimization method based on the chaotic search [7]. By improving the method or combining it with other methods, such as simulated annedling, genetic algorithm, artificial potential field method. The optimization performance has been improved greatly.

When chaotic optimization algorithm is chosen, dynamical systems that generate chaotic variables should be first selected, the following Logistic system is frequently used in all kinds of established system [8].

$$X_{n+1} = \mu X_n(1 - X_n) \tag{10}$$

In (10), $\mu$ is control parameter ($0 \leq \mu \leq 4$, $X \in [0,1]$). When $\mu$ goes to 4, the system is in chaos state. The ergodic property can be used to escape from local optimums.

Based on a research of artificial potential field method and chaotic optimization algorithm, chaotic optimization algorithm is applied to improved artificial potential field method and the path planning for Robot based on chaotic artificial potential field method is proposed. Potential function is the objective function and control parameter is the direction of movement for robot in the algorithm. Based on the path planning of chaotic artificial potential field method, chaotic optimization algorithm is applied to compute the direction of next step for robot and decides next sub-target point and so on. The path planning for local optimums is implemented in real time.

## 4. Simulation experiment

Simulation environments is the grid map generated by simulation soft with the measurements of 30 by 20. In the figure S point is the starting point and G point is the objective point. Black grids are obstacles and these are the regions where robot can't pass through. Robot can pass in white grid regions. Simulation experiment parameter is shown as follows: $\rho_0$ is 15, K is 2 and $\eta$ is 10. The step size is 1 for robot. The simulation results are shown from figure 1 to figure 3.

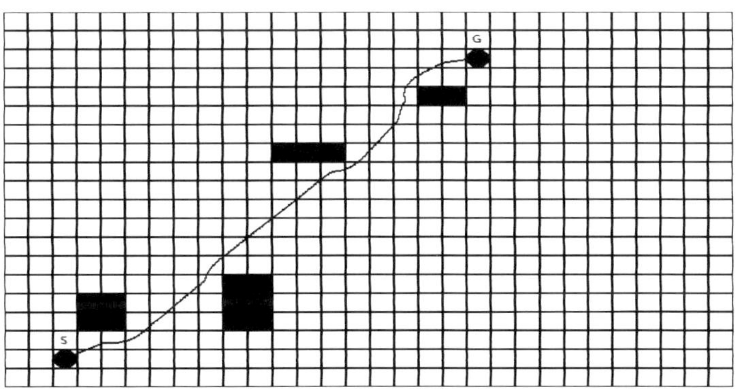


**Figure 1.** Conditions with approaching to the starting point and objective point for obstacles.

When obstacle is adjacent to the starting and objective point, the algorithm is applied on the path planning for robot to avoid the obstacles in figure 1. The robot can well avoid the obstacle and no shaking when the obstacle is close to the starting point. There are no conditions that t obstacles push away the robot that can't get to the objective point when it is close to the objective one. The algorithm has good ability of obstacles avoidance for other obstacles.

IOP Conf. Series: Materials Science and Engineering **317** (2018) 012056 doi:10.1088/1757-899X/317/1/012056

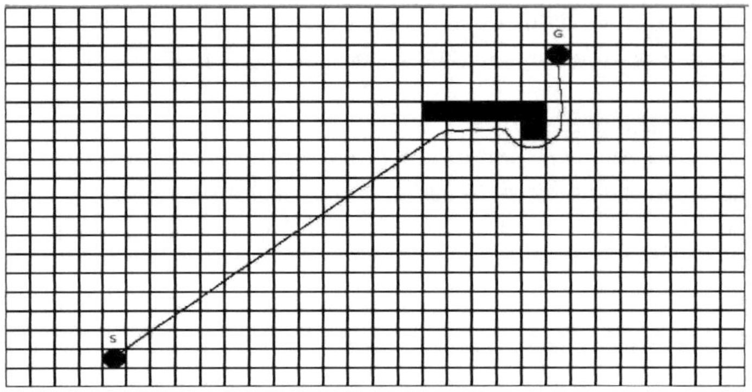


**Figure 2.** Condition with local optimums

The algorithm is applied on the path planning for robot to avoid the obstacles when there are easily local minimums in condition with the concave-shaped obstacles in figure 2. The robot can avoid well the obstacles, no shaking and no unattainable targets. Local optimums in traditional artificial potential field method are solved in the algorithm.

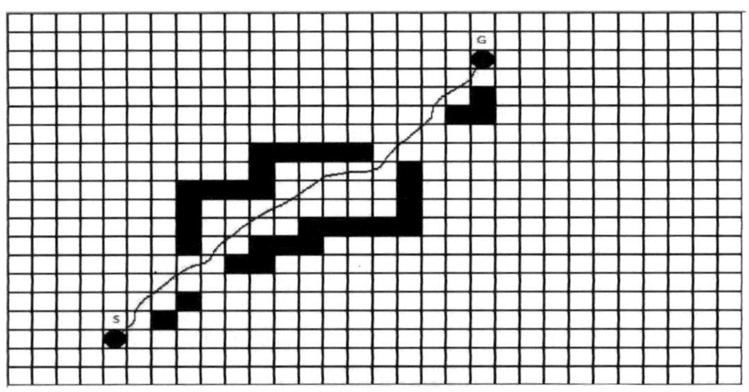


**Figure 3.** Conditions with one or more closer obstacles.

In figure 3 the algorithm is applied on the path planning for robot to avoid the obstacles in conditions with one or more closer obstacles. The robot doesn't bypass the obstacles but to pass between them. The problem that robot can't find the path in condition with near the obstacles in traditional artificial potential field method are solved in the algorithm.

## 5. Conclusions

Combining the improved artificial potential field method with chaotic optimization algorithm, the path planning for Robot based on chaotic artificial potential field method is proposed in this paper. The best path planned by the algorithm is more real and more effective obstacles avoidance under unknown environment. According to the simulations results, local optimums in traditional artificial potential field method are solved in the algorithm. Robot can find the path to avoid the obstacles between the closer obstacles and the path is more smooth and practical.

## References

[1] Shi C and Sun H 2005 *Robot* **27** 152.
[2] Latombe JC 1991 *Robot Motion Planning* (Boston: Kluwer Academic).
[3] Ouyang Y X and Yang S G 2014 *Control Engineering of China* **21** 134.
[4] Ge S S and Cui Y J 2000 *IEEE. Trans. Rob. Autom* **5** 615.
[5] Hu Y and Zhang Q 2012 *Adv. Mater. Res.* **562** 937.
[6] Strogatz S H 2001 *Nolin. Dyn. Chaos* (Boulder: Westview).
[7] Li B and Jiang W 1997 *Contr. Theor. Appl.* **14** 613.
[8] Jovanovic V 2000 *Int. J. Numer. Mesh. Engng* **48** 137.

IOP Conf. Series: Materials Science and Engineering **317** (2018) 012057 doi:10.1088/1757-899X/317/1/012057

# Zipf's Law and the Frequency of Kazak Phonemes in Word Formation

**Ruiqing Xin[1], Yonghong Li[1] and Hongzhi Yu[1]**

[1]Key Lab of China's National Linguistic Information Technology, Northwest Minzu University, Lanzhou, 730030, China

**Abstract.** Zipf's Law is the basis of the principle of Least Effort, and is widely applicable in all natural fields. The occurring frequency of each phoneme in all Kazak words has been counted to testify the application of Zipf's law in Kazak. Due to the limitation of the sample size, deviation is unavoidable, but overall results indicate that the occurring frequency and the reciprocal rank of each phoneme in Kazak words formation are in line with Zipf's distribution.

## 1. Introduction

Modern Kazak is a common language for all Kazak people in the world. At present there are mainly two writing systems of Kazak. One is in Cyrillic for Kazakhstan, the other one is in Arabic for China. Although the writing systems are different, the pronunciations of the phonemes are almost the same, and all Kazak people can communicate well. In China, there are altogether 33 phonemes in Kazak, among which 9 are vowels and 24 are consonants [1]. The phonetic system of Kazak in China is shown in table 1.

**Table 1.** Kazak phoneme system

| | Kazak | IPA | | Kazak | IPA |
|---|---|---|---|---|---|
| **Front vowels** | ە | e | **Voiced consonants** | ب | b |
| | ٵ | æ | | د | d |
| | ٶ | ø | | گ | g |
| | ٷ | y | | ق | q |
| | ٸ | i | | ج | ʤ |
| **Back vowels** | ا | ɑ | | ۆ | v |
| | و | o | | ز | z |
| | ۇ | u | | ع | ʁ |
| | ى | ə | | م | m |
| **Voiceless consonant** | پ | p | | ن | n |
| | ت | t | | ل | l |
| | ك | k | | ڭ | ŋ |
| | چ | ʧ | | ر | r |
| | ف | f | **Approximant** | ي | j |
| | س | s | | ۋ | w |
| | ش | ʃ | | | |
| | ح | x | | | |
| | ھ | h | | | |

Great progress has been made in the research in Kazak phonetics. *Modern Kazak* (Kazak version) written by Geng Shimin & Li Zengxiang and *Modern Kazak* (Kazak version) composed by the Language Committee of Xinjiang Province both make a complete description of Kazak phonetics including the biological and physical properties of the phonemes, the principles of phonetic harmony, the phenomena of phonetic changes in speech, so on and so forth; while *Kazak Grammar* compiled by the Institute of Linguistics of the Education and Science Department of the Republic of Kazakhstan offers the most detailed depiction of Kazak phonetics [2]. But in the formation of words, what functions does each phoneme play? What is the occurring frequency of each phoneme? These are basic but interesting topics. Zipf's Law will be taken to analyze the occurring frequency of each phoneme in all Kazak words.

## 2. Zipf's Law

The principle, which means to get the greatest achievements with the least effort, has been known as the principle of Least Effort, or the Economy Principle, and it has been taken as one primary rule to monitor human's behavior. The principle was firstly raised by George Kinsley Zipf, based on what is called Zip's Law.

Zipf found that in a long discourse, the rank of a word (simplified as "r") which indicates the ranking of the occurring frequency of a word in the discourse, and the occurring frequency of the word (simplified as "f") have a certain relationship. He explained the relationship with the equation "r×f=c", in which "c" is "constant". That is to say, the rank of a word and its occurring frequency in the discourse are in inverse ratio. The higher the occurring frequency is, the smaller the rank is. The word which occurs most frequently in the discourse is ranked in the first place.

The most attractive is that the result of "r×f" is a constant. Zipf analysed the words in *Ulysses*, and got the amazing finding—the product of the rank and occurring frequency of each word is almost the same, that is 26, 000 approximate, and he draw the doubly logarithmic chart according to the research results, which is shown in figure 1[3].

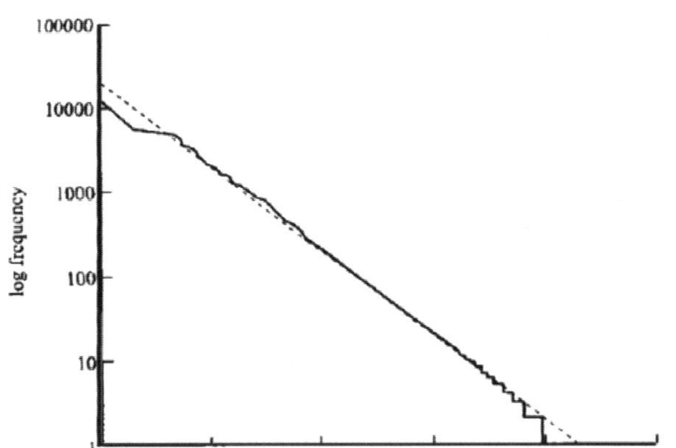


**Figure 1.** The Doubly Logarithmic Chart of Words in *Ulysses*.

In the chart, the line associating the frequency and the rank of the words is almost a straight line of 45 degrees. Of course, the result is just an average assessment. The relation between "r" and "f" is known as Zipf's Law.

Starting from Zipf's Law, and with further researches, Zipf raised the principle of least effort. The understanding could be that the writer is always trying to express the most with the fewest words, and thus, some words which are more favored by the writer occur more frequently in the discourse.

## 3. Research design

Since Zipf's Law has been reported to be applicable in many fields, such as information science, computer science, economy, so on and so forth. Whether it is also applicable to phonemes in word formation of Kazak is the focus of this paper.

This research is possible thanks to the Internet Kazak Corpus—Kazak-Chinese Dictionary uploaded by Li Jing [4] and *Kazak-Chinese Dictionary* complied by Nurbek Abken [5]. To guarantee the validity of the research, each entry of word in the corpus was examined according to *Kazak-Chinese Dictionary*. A new corpus—Kazak Phoneme Corpus was set up with all the Kazak words elicited from the Internet Kazak Corpus—Kazak-Chinese Dictionary. Because in Kazak words with front vowels, there is a front vowel mark "ء" before the words, while the symbols of the vowels in the words are the same with the back vowels counterparts, e.g. "ءمشە", it is complicated to count the frequency of all the vowels. Thus, all phonemes were turned into IPA before the division, and every word was divided into phonemes with the function of "LEFT( )" in Excel 2007. Entries with different definitions but the same spellings were screened and omission was done to excessive words with the same spelling, in order to guarantee the precision of the phoneme frequency in word formation. The sample of the new corpus is shown in figure 2. The number of all the words in the corpus is 49622.

| WN | IPA | KAZAK | Clas | Chinese Definition | FV | P1 | P2 | P3 | P4 | P5 | P6 |
|---|---|---|---|---|---|---|---|---|---|---|---|
| N48801 | ilw I | I ءىلۋ | noun. | 〈名〉（亲家之间的）礼物，礼品，赠品 ءىلۋ الۋ 接受礼物 ｜ ءىلۋ بەرۋ 赠送礼品。 | ء | i | l | w | | I | |
| N48803 | əlwde | ىلۋدە | adj. | ىلۋدە ءبىر 少有的，稀罕的，罕见的 ىلۋدە ءبىر كەزدەسەتىن نارسە 罕见之物。 | | ə | l | w | d | e | |
| N48804 | ilwli | ءىلۋلى | adj. | 〈形〉挂着的，悬挂着的；吊着的 ءىلۋلى كيىم 挂着的衣服。 | ء | i | l | w | l | i | |
| N48805 | ələgw | ىلىگۋ | verb1. | ىلىك II 的不定式。 | | ə | l | ə | g | w | |
| N48806 | ələgəs | ىلىگىس | verb. | 〈动〉争吵，闹别扭 ول مەكەۋى سوزگە ىلىگىسىپ قالدى 他俩争吵起来了。 | | ə | l | ə | g | ə | s |
| N48807 | əlgəsw | ىلگىسۋ | verb1. | ىلگىس 的不定式。 | | ə | l | g | ə | s | w |

※ WN=word number, FV= front vowel mark, P=phoneme place

**Figure 2.** Sample of Kazak Corpus.

## 4. Results analysis

### *4.1. Distribution of Kazak phonemes in word formation*

With the function of "COUNTIF ( )", the occurring frequency of each phoneme was counted and then the phonemes were ranked according to the frequency. The result is shown in table 2.

**Table 2.** The frequency and rank of Kazak phonemes.

| Kazak | Phoneme | Frequency | Rank | Kazak | Phoneme | Frequency | Rank |
|---|---|---|---|---|---|---|---|
| ا | a | 49585 | 1 | ۇ | u | 6812 | 18 |
| ى | ə | 38698 | 2 | ز | z | 6601 | 19 |
| ت | t | 25058 | 3 | ج | ʤ | 5951 | 20 |
| ل | l | 21784 | 4 | پ | p | 5465 | 21 |
| ە | e | 20972 | 5 | ڭ | ŋ | 4028 | 22 |
| ر | r | 19520 | 6 | ع | ʁ | 3616 | 23 |
| ق | q | 17969 | 7 | گ | g | 2579 | 24 |
| س | s | 16129 | 8 | ٸ | i | 2210 | 25 |
| ن | n | 13049 | 9 | ٵ | æ | 1289 | 26 |
| د | d | 12962 | 10 | ٷ | y | 617 | 27 |
| ۋ | w | 11948 | 11 | ح | x | 561 | 28 |
| ي | j | 11622 | 12 | ف | f | 445 | 29 |

| | | | | | | | |
|---|---|---|---|---|---|---|---|
| ك | k | 10951 | 13 | ٶ | ø | 397 | 30 |
| ش | ʃ | 9819 | 14 | ۆ | v | 241 | 31 |
| و | o | 8955 | 15 | ھ | h | 106 | 32 |
| م | m | 8971 | 16 | چ | ʧ | 37 | 33 |
| ب | b | 7802 | 17 | | | | |

Based on the data in table 2, a curve of Kazak phonemes' frequency was drawn in figure 3. Although the curve in figure 3 is not the same as that in figure 1, the general tendency of the distribution is in a triangle with an angle of about 45 degrees, and with the help of extension lines the most steady part of the curve can be a line with an angle of about 45 degree.

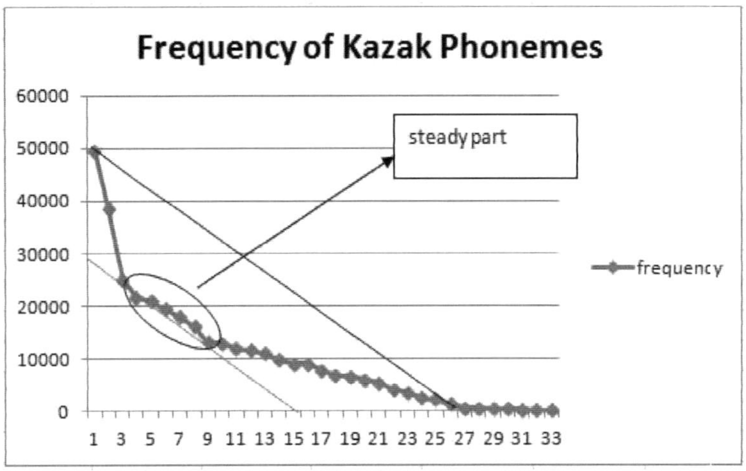


**Figure 3.** Curve of Kazak phonemes' frequency.

*4.2. Application of Zipf's Law in Kazak phonemes' distribution*

In figure 4 is the product curve of frequency and rank. According to the equation of Zipf's law "r×f=c", the average product of frequency and rank is a constant. Therefore, the product curve of Kazak phonemes' frequency and rank should almost be a line, which has been testified in figure 5, the logarithmic curve of the product of frequency and rank.

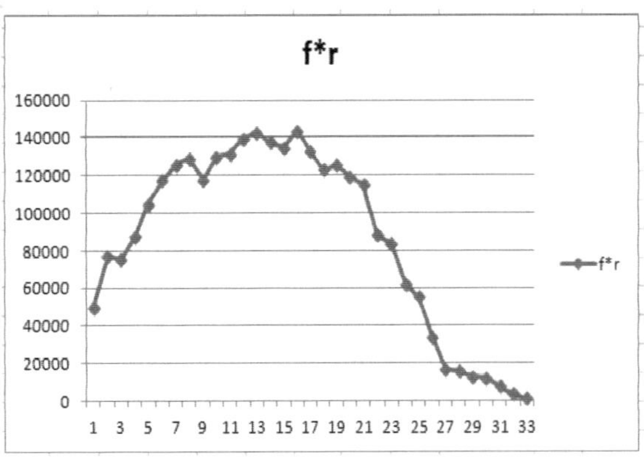


**Figure 4.** Product curve of frequency and rank.

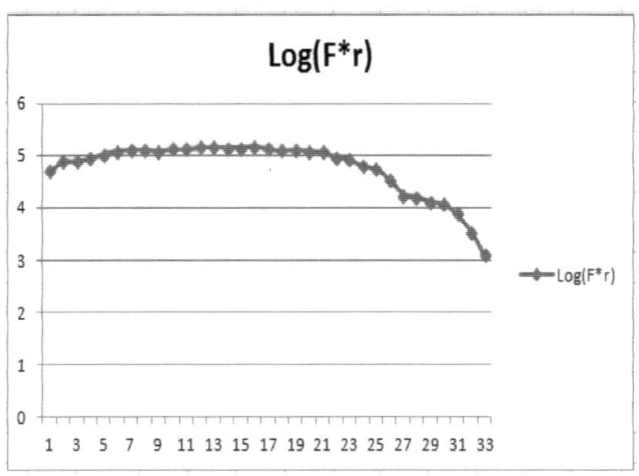


**Figure 5.** Logarithmic curve of the product of frequency and rank.

Data in table 2 and the curve in figure 3 have almost proved the application of Zipf's law in Kazak phonemes' distribution. The result is not exactly the same as Zipf's study of *Ulysess*, one of the possible reasons is that the sample of *Ulysess* is large, and the power of balance is great. Besides, the frequency and the rank are in inverse ratio, so as a matter of fact, in any sample, both the variables ranked at the first places and those ranked at the last places should not be taken as examples for the law, because according to the equation "r×f=c", there will be two extremes in a sample—the sample is filled with the variables occurring most frequently and without any variables occurring least frequently. It is believed this is why not every variable was taken into consideration in the research of *Ulysess*. This explains the reason why only the most steady part of the curve, which is driven from the average data, is reliable in this research.

## 5. Conclusion

Zipf distribution is the result of balance and average, the two extremes of any distribution frequency in rank will not meet the law, which is why only the steady parts of the above curves are taken for study. This research has proved Zipf distribution is working in Kazak phonemes' occurring frequency in word formation, while more researches are expected.

## References

[1] Geng S M & Li Z X 1985 *An Introduction to Kazak*. Beijing: The Ethnic Publishing House.
[2] Zhang D J 2009 *J. Yili Normal University (Soc. Sci. Ed.)*, **3** 25-39.
[3] Jiang W Q 2005 *J. Tongji University (Soc. Sci. Ed.)* **16(1)** 87-95.
[4] Li J 2015 Internet Kazak Corpus — Kazak-Chinese dictionary. http://vdisk.weibo.com/s/taoCn8PUHOqk4.
[5] Nurbek A 2014 *Kazak-Chinese Dictionary*. Beijing: The Ethnic Publishing House.

## Acknowledgements

This paper is based on part of the research findings of Northwest Minzu University Graduate Innovation Project *The Acoustic Properties of Kazak Consonants* (Project No. YXM2016002).

IOP Conf. Series: Materials Science and Engineering **317** (2018) 012058 doi:10.1088/1757-899X/317/1/012058

# Removal of Copper(II) Ions in Aqueous Solutions Using Tannin-Rich Plants as Natural Bio-Adsorbents

**J. Paksamut[1,*] and P. Boonsong[1]**

[1]Department of Social and Applied Science, College of Industrial Technology, King Mongkut's University of Technology North Bangkok, Bangkok, 10800, Thailand

**Abstract.** In this study, the purpose of our interest is to investigatethe adsorption behavior of copper (II) ions in aqueous solution using some tannin-rich plants as natural bio-adsorbents such as mangosteen peels (*Garciniamangostana* L.), cassava leaves (*Manihotesculenta* Crantz) and Thai copper pod leaves (*Sennasiamea* (Lam.)) as powder form in different dosage of adsorbent plant materials.The adsorption capacities at different pH of solution and contact time were performed.All the experiments in this studywere chosen at room temperature by batch technique. From the experimental results showed that cassava leaves gave better adsorbent properties than mangosteen peels and Thai copper pod leaves. The increasing dosage of all adsorbents and contact time have been found to increase adsorption capacities. In this respect, the adsorption capacities depend crucially on the adsorbents and contact time. The optimum pH of copper (II) ions adsorption was pH4. According to this work, it was observed that bio-adsorbent materials from tannin-rich plants could be used to remove copper (II) ions from aqueous solutions.

## 1. Introduction

One of the most important environmental problem issues is the presence of heavy metals contaminated in water, because heavy metals are not biodegradable [1]. The increasing heavy metals usage in industrial activities has caused the existence of them in water [2]. The heavy metals often found from industrial effluent are copper. Copper is a transition metal which belongs to period 4 and group IB of the periodic table [3]. Some toxic of copper in the body has a potentially damaging protein, lipid and DNA [4]. The excess copper compound in the body may affects on aging, mental illness and Alzheimer's diseases [5]. Removal of metal ions from wastewater in an effective manner has become an important issue.Adsorption technology has recently become a real alternative to traditional wastewater treatment due to its relatively simplicity and efficiency.Nowadays, agricultural materials are receiving as adsorbents for the removal of pollution from water. Different bio-adsorbents are developed from agro-wastes and used for heavy metals removal.One class of such materials is that of the tannins, a term which covers many families of chemical compounds. Tannins are large and complex molecules, weak acid and astringent taste, which they can be found in various parts of some plants. Tannins are divided in two types. One is the condensed tannin that is found in the bark and cores of the plants. The other is the hydrolysable tannin that can be found in leaves and sheath [6]. Tannins have shown to be potential alternative for the removal of heavy metal ions from aqueous solution, especially lead, iron, chromium and copper. Tannins exhibit extremely high metal ion chelating ability because metal ions can generate highly reactive phenolic hydroxyls of tannins molecules [7]. Mangosteen peels, cassava leaves and Thai copper pod leaves are agricultural waste materials, low cost and high amount of tannins, which they can be used as bio-adsorbents [8, 9].

The purpose of research was to studies the bio-adsorption of copper (II) ions from aqueous solution. We have selected three tannin-rich plants sources such as mangosteen peels, cassava leaves, Thai copper pod leaves as bio-adsorbents. The influences of adsorbent doses, contact time, and pH were studied in batch experiments at room temperature for selected the best bio-adsorbent materials and the most removable metals of optimum adsorption condition.

## 2. Materials and Methods

### 2.1. Preparation of adsorbent materials

Three types of agricultural waste materials have been employed to use as bio-adsorbent materials in order to remove copper (II) ions from aqueous solution. Firstly, the bio-adsorbents, such as mangosteen peels, cassava leaves and Thai copper pod leaves, were washed with distilled water in order to remove the impurities and leave to dry at room temperature. Then, they were cut into small pieces and dried in hot airoven at 70°C for 6 hours. After that, the dried bio-adsorbents were ground into powder by electrical blender in order to obtain small particle size.

### 2.2. Preparation of copper (II) ions in aqueous solutions

In general, copper (II) ions was used as the adsorbate. For this work, wecompared different five concentrations in order to detect the adsorption capacities for analytical standard calibration curve. Firstly, clear stock solutions of 1 mol/L copper (II) ionsin solution was prepared by dissolving 249.28 g of copper (II) sulphate pentahydrate ($CuSO_4 \cdot 5H_2O$) in 1L of distilled water. Then, clear stock solutions of 1 mol/L copper (II) ions solution was diluted with distilled water in concentration of copper (II) ions in aqueous solutions as 0.02, 0.04, 0.06, 0.08 and 0.10 mol/L, respectively.

### 2.3. Methods

*2.3.1. Preparation of standard calibration curve.* This experiment was used different five concentrations of copper (II) ions in aqueous solutions to make a series of calibration standard solution such as 0.02, 0.04, 0.06, 0.08 and 0.10 mol/L. The absorbances of these standards were measured to create a calibration curve by spectrophotometry method. The maximum wavelength of copper (II) ions solutions have been detected at 822 nm.

*2.3.2. Investigationofadsorbents dosage.* The various dosage each of different bio-adsorbents, such as 0.5, 1.0, 1.5 and 2.0 g, were added 10 ml of 0.10 mol/L copper (II) ions in aqueous solution at room temperature.

*2.3.3. Investigation of pH.* The various of pH were chosen to use around 2-5. These experiments were combined with 2 g of each bio-adsorbents and 10 ml of 0.10 mol/L copper (II) ions in aqueous solution. The pH of the solution was adjusted by using 0.10 mol/L hydrochloric acid and 0.1 mol/L sodium hydroxide.

*2.3.4. Investigation of contact time.* Different contact time on adsorption behavior of copper (II) ions in aqueous solution were used at 30, 60, 90, 120 minutes and 24 hours of 2 g of each bio-adsorbents and added 10 ml of 0.10 mol/L copper (II) ions in aqueous solution at room temperature.

## 3. Results and discussion

Adsorption method was employed to remove copper (II) ions from aqueous solution, which corresponds to bio-adsorbents dosage, pH and contact time. The results in this work will be discussed in the following section.

### *3.1. Standard calibration curve of copper (II) ions in aqueous solutions*

The calibration curve produced a linear relationship between the absorbance values and concentrations of copper (II) ions in aqueous solution. The standard calibration curve and the equation of copper (II) ions in aqueous solutions are shown in figure 1.

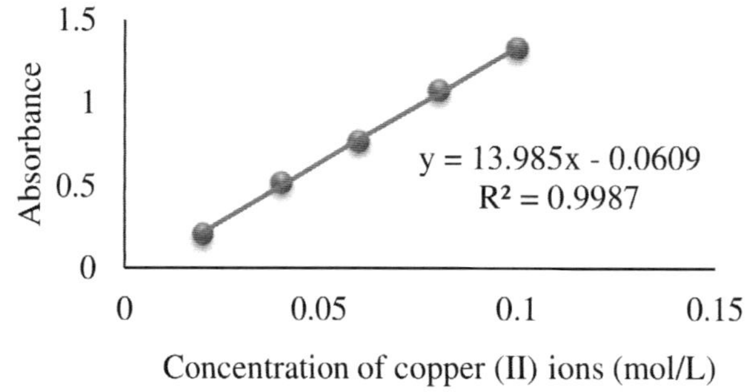


**Figure1.** Standard calibration curve of absorbance values and concentration of copper (II) ions.

*3.2. Effect of adsorbents dosage*

The effect of adsorbentsdosagewasemployed to investigatedifferent dosage of bio-adsorbents from 0.5-2.0 g in 10 ml of 0.10 mol/L copper (II) ions in aqueous solution at normal pH and room temperature. The results have been clearly demonstrated that the percentage removal of copper (II) ions from aqueous solution in creased when increasing the adsorbents dosage (figure 2). The percentage removal of copper (II) ions from aqueous solution for calculating is written as:

$$\%\text{removal of copper (II) ions from aqueous solution} = (C_i - C_e)/C_i \times 100$$

With regard to calculating percentage removal equation, $C_i$ and $C_e$ (mol/L) were represent the initial and equilibrium concentrations of copper (II) ions in aqueous solution, respectively [10]. In term of the experimental data, it has been shown that the removal efficiency and specific uptake of copper (II) ions in aqueous solution depend on quantity and sort of the bio-adsorbents. The percentage removal of copper (II) ions in aqueous solution increased when increasing bio-adsorbents dosage. In this respect, the percentage removal of copper (II) ions in wastewater corresponds to surface area of bio-adsorbents [11]. Moreover, tannin rich-plants had high metal binding capacity [7]. The optimum condition of bio-adsorbents dose with contain 0.10 mol/L copper (II) ions in aqueous solution of ratio as 2:10 were chosen for further adsorption experiments.

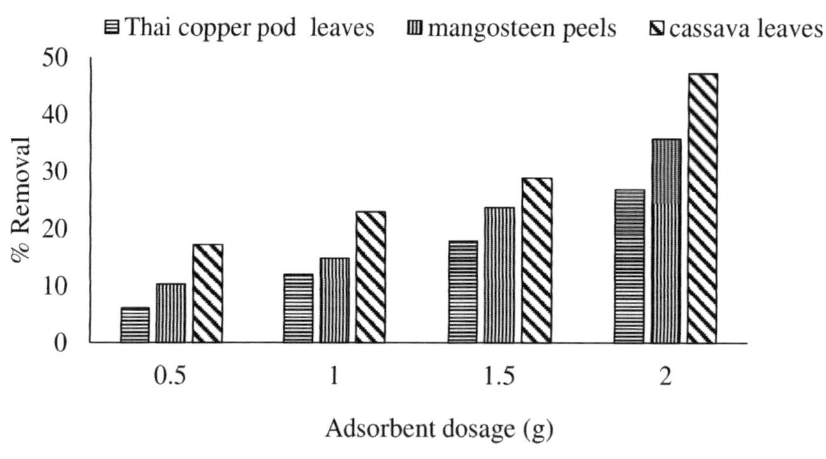


**Figure 2.** Effect of adsorbents dosage on percentage removal of copper (II) ions from aqueous solution.

*3.3. Effect of contact time*

Effect of contact time was investigated for varying in the range of 30 to 120 minutes and overnight as shown in figure 3. With regard to experimental data, when increasing contact time of bio-adsorbents with 0.1 mol/L of copper (II) ions in aqueous solution, high percentage removal of copper (II) ions from aqueous solution was found. For earlier studies, it was concluded that, the percentage removal of copper (II) ions from aqueous solution clearly confirm that it depends on bio-adsorbents dosage, pH and contact time. The rate of reaction of copper removal was very rapid in 30-120 minutes, thereafter the rate of copper removal constant. The adsorption rate after 120 minutes and equilibrium state because the binding site was shortly become limited and the remaining vacant surface sites were difficult to be occupied by copper (II) ions because of the driving force of mass transfer between the adsorbate and active site surface of bio-adsorbent in aqueous adsorption system decreases with time pass [5].

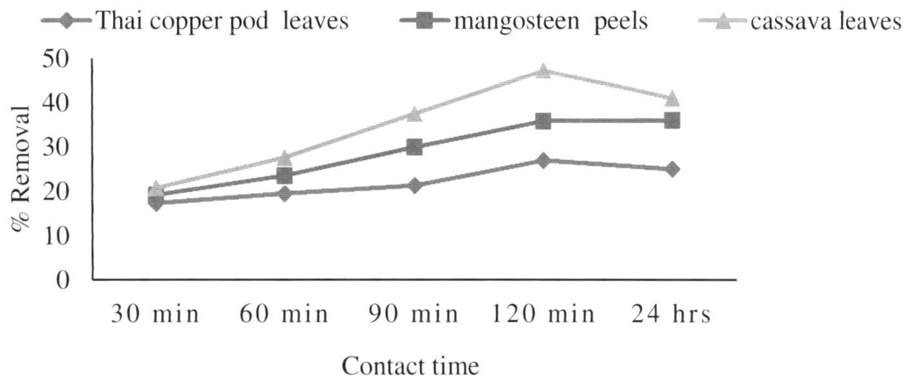


**Figure 3.** Effect of contact time on percentage removal of copper (II) ions from aqueous solution.

*3.4. Effect of pH*

Effect of different pH was investigated in the range of pH 2 to pH 5 in order to study the effect of percentage removal of copper (II) ions from aqueous solution as shown in figure 4. For the experimental data, it has been demonstrated that the percentage removal of copper (II) ions from aqueous solution in creased when using pH about 2 to 4. Moreover, the adsorption capacities decreased in pH 5. This was exported by considering and adopting some limitations of copper introduces a limiting at pH 5. At higher pH, its precipitation begins in the form of insoluble hydroxide as $Cu(OH)_2$. Therefore the removal was not complete by adsorption [5, 12]. The maximum adsorption efficiency in pH 4 may be due to the interaction of copper (II) ions with surface functional groups present in the bio-adsorbents.

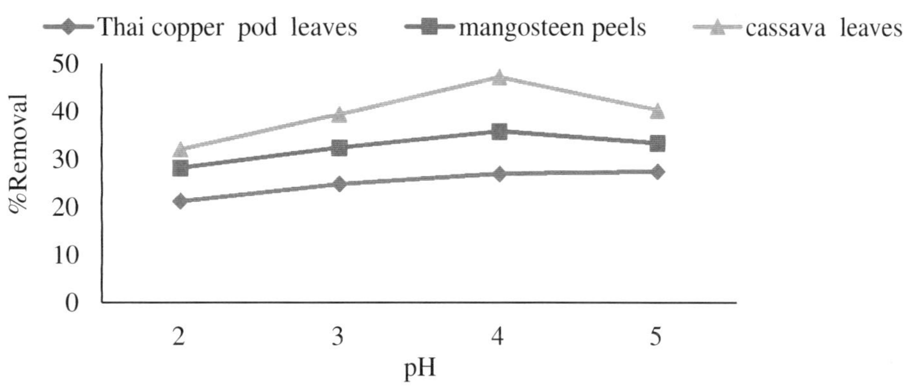


**Figure 4.** Effect of pH on percentage removal of copper (II) ions from aqueous solution.

IOP Conf. Series: Materials Science and Engineering **317** (2018) 012058 doi:10.1088/1757-899X/317/1/012058

## 4. Conclusions

As mentioned earlier, in the case of the batch experimental results was reported that the adsorption behaviour of heavy metal ions was a potentially attractive technique for investigation of removal of heavy metal ions from wastewater. According to this study, it has been demonstrated that tannin-rich plants, such as mangosteen peels, cassava leaves and Thai copper pod leaves, were natural bio-adsorbent materials. The bio-adsorbents were low-cost, easily prepared and high-affinity for heavy metal ions. Based on the experimental data, the optimum conditions with contain 2 g of bio-adsorbents dosage in 10 ml of 0.1 mol/L of copper (II) ions in aqueous solution, the pH of 4, the contact time of 120 minutes were chosen, which correspond to high percentage for removal of copper (II) ions from aqueous solution. With regard to bio-adsorbents, cassava leaves were shown the highest percentage removal of copper (II) ions from aqueous solution. In this respect, cassava leaves could be applied as an alternative low-cost bio-adsorbent for removal of heavy metal ions in wastewater.

## References

[1] Farhan S N 2015 *DJES* **8** 835-45.
[2] Innocent OBOH 2009 *LJS* **14** 58-65.
[3] Ahamed J A et al. 2012 *Arch. Appl. Sci. Res.* **4(3)** 1532-9.
[4] Moroncola B A et al. 2016 *Int. J. Multidiscip. Res. Dev.* **3** 22-8.
[5] Hossian M A et al. 2012 *Int. J. Geomate* **2** 227-34.
[6] Thakur S S et al. 2014 *Int. J. Chemtech. Res.* **6** 3628-34.
[7] Oseni M O et al. 2016 *Int. Res. J. Pure. Appl. Chem.* **12(1)** 1-17.
[8] Kasanlavit W 2015 *Rev. Integr. Bus. Econ. Res.* **4** 37-41.
[9] Burkill H M 1994 *The useful plants of west Tropical Africa* (Royal Botanic Gardens Kew).
[10] Rana K et al. 2014 *IJAERS* **1** 55-9.
[11] Prabha R T et al. 2014 *IJSR* **3** 705-9.
[12] Ahsan H 2007 *Pak. J. Anal. Environ. Chem.* **8** 21-5.

IOP Conf. Series: Materials Science and Engineering **317** (2018) 012059 doi:10.1088/1757-899X/317/1/012059

# Method for Evaluating the Corrosion Resistance of Aluminum Metallization of Integrated Circuits under Multifactorial Influence

**Kolomiets V. I.**[1]

[1]National Research Moscow State University of Civil Engineering, 129337, 26, Yaroslavskoye Shosse, Moscow, Russia

**Abstract.** The influence of complex influence of climatic factors (temperature, humidity) and electric mode (supply voltage) on the corrosion resistance of metallization of integrated circuits has been considered. The regression dependence of the average time of trouble-free operation $t$ on the mentioned factors has been established in the form of a modified Arrhenius equation that is adequate in a wide range of factor values and is suitable for selecting accelerated test modes. A technique for evaluating the corrosion resistance of aluminum metallization of depressurized CMOS integrated circuits has been proposed.

## 1. Introduction

Integrated circuits (IC), manufactured on the basis of complementary transistors with the structure of metal-oxide-semiconductor (CMOS IC), found an exceptionally wide application in products of industrial electronics and computer engineering. The main advantages that have determined their wide application are extremely low power consumption and a wide range of supply voltages (up to 15 V). However, these advantages, on the other hand, also determined the low corrosion resistance of CMOS IC metallization, that required the development of special technological measures to increase it, as well as methods of extreme tests for evaluation of the effectiveness of these measures.

For assessment of the corrosion resistance of IC metallization, determined directly by the design and technological features of the IC manufacturing process, the focus should be on corrosion processes in the IC metallization having a free internal chamber. For this purpose it is reasonable to apply extreme tests methods in conditions of the influence of high temperature and humidity, which, combined with the supply voltage [1], make it possible to reproduce the electrolytic corrosion mechanism typical for the majority of corrosion failures of CMOS IC during operation.

## 2. Methods

As an indicator of the IC resistance to electrolytic corrosion of aluminum metallization, as a rule, the average time to failures $t$ is used [2]. For the optimal selection of test conditions, it is of great interest to detect the quantitative relationships between the time $t$ and the influencing factors, i. e. the formulation and analysis of the mathematical model, for example, in the form of a regression equation [3].

$$t = f(x_1, \ldots, x_m, \mathrm{A}), \tag{1}$$

where $x_1, \ldots, x_m$ – relevant influencing factors ($m$ – their total quantity);
$\mathrm{A} = [a_1, \ldots, a_m]$ – $m$-dimensional vector of regression coefficients.

Since the given matter is solvable, as a rule, by an experimental method, to solve the matter, it is necessary to:

1) create a plan of active tests [4];
2) select the type of the mathematical model (1);
3) conduct multifactorial tests of the IC in accordance with the plan, creating $N$ combinations of $m$ factors each time and determining the corresponding value of the average time between failures;
4) find the vector **A** from the condition of minimum dispersion [5]

$$D = \sum_{i=1}^{N} (t_i^{э} - t_i)^2 / (N-M) \rightarrow min, \tag{2}$$

where $t_i^{э}$, $t_i$ – the values of the time between failures with the $i$- th combination of influencing factors, determined, respectively, experimentally and according to the mathematical model (1);

5) to assess the adequacy of the model (1) using, for example, the F-criterion [6]; if the result is positive, to use the obtained mathematical model (1) in solving applied matters posed in the study, and if it is negative, to go to a new kind of mathematical model, to conduct additional tests (to increase the number $N$), or (and) to introduce new influencing factors (to increase the value $m$), then to repeat all actions again.

It should be noted that for certain combinations of influencing factors, when the time $t_i^{э}$ is too long and there is no possibility to increase the duration of the tests, it is possible to use for some experiments the calculated equivalent of the time between failures [7].

## 3. Discussion of the Results

In accordance with the above mentioned, the matter of the model formulation (1) for the dependence of the time to failures $t$ of the depressurized integrated microcircuits of the CMOS IC of the 564 series (the Russian analogue of CD4000A) under the influence of temperature $x_1$, humidity $x_2$ and supply voltage $x_3$ $(m=3)$, was solved, for which multifactorial tests were carried out. Their results for different supply voltages are presented in the table.

During the tests, the average values of the time to failures $t_i^{э}$ were determined, and in a number of cases when the failure did not occur within 1,000 h, the value of $t_i^{э}$ was estimated analytically [7].

**Table 1.** The time (h) of 564 series integrated circuits between failures at voltages of 5; 10; 15 V.

| $x_2$, % | $x_1$, °C | | | | | | | | | | | |
|---|---|---|---|---|---|---|---|---|---|---|---|---|
| | 55 | | | 85 | | | 100 | | | 120 | | |
| | 5 V | 10 V | 15 V | 5 V | 10 V | 15 V | 5 V | 10 V | 15 V | 5 V | 10 V | 15 V |
| 25 | 20600 | 12200 | 7210 | 4850 | 3010 | 1850 | 2560 | 1610 | 1020 | 1190 | 762 | 675 |
| 85 | 3910 | 2320 | 1370 | 953 | 588 | 364 | 506 | 317 | 201 | 235 | 150 | 133 |
| 90 | 2680 | 1580 | 933 | 629 | 388 | 239 | 332 | 208 | 131 | 156 | 99 | 88 |
| 98 | 532 | 314 | 186 | 126 | 78 | 48 | 67 | 42 | 26 | 31 | 20 | 18 |

Assuming that the proceeding processes are largely subject to the Arrhenius law [8], the results were mathematically processed and a mathematical model was obtained:

$$\lg t = -5.8+0.0725x_3+1.05 \lg(100-x_2) + (2{,}800-54x_3)/x_1 \tag{3}$$

The Fisher test [6] confirmed the adequacy of this model with a confidence coefficient of at least 95%. Its accuracy is illustrated at figure 1.

Calculations by the model showed that the time to failure (which is proportional to the rate of corrosion processes) is mainly determined by the relative humidity and temperature of the medium. To a lesser extent, the rate of corrosion depends on the value of the supply voltage.

Methods for assessing the corrosion resistance of aluminum metallization of the CMOS IC, developed on the basis of the results of multifactorial tests, provides for accelerated testing of depressurized IC in unsaturated water vapor at elevated temperature with simultaneous electrical voltage supply.

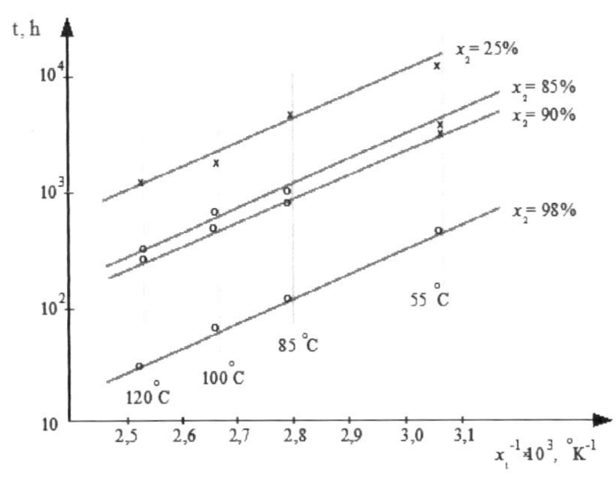


(a)

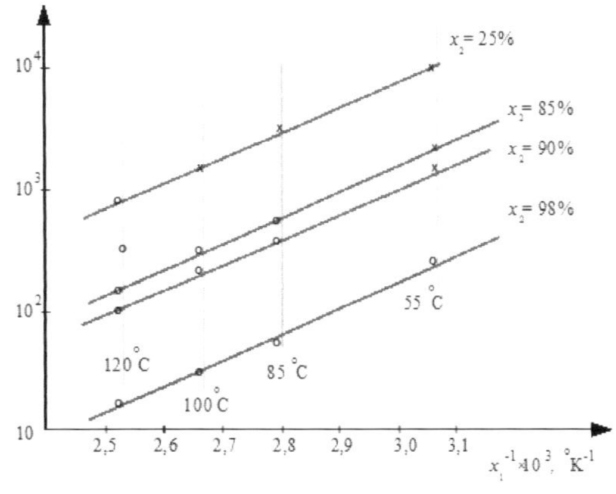


(b)

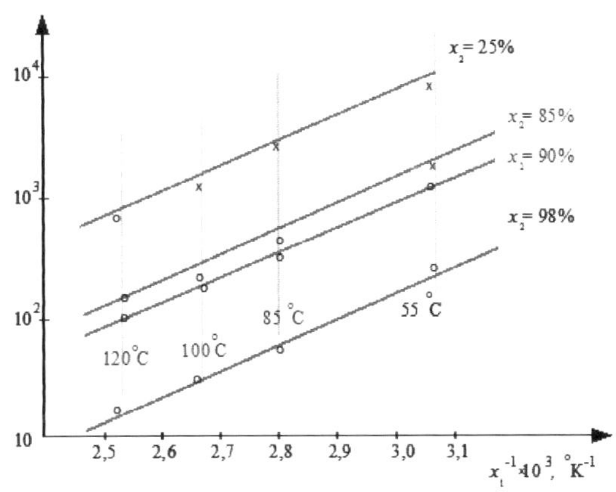


(c)

**Figure 1.** Dependences of the time between failures of 564-series integrated circuits on temperature and humidity at supply voltages 5(a), 10(b), 15V(c):

**o** – experiment, **x** – projected equivalent, — – mathematical model (3).

The basic modes are:

-85% relative humidity, 85 °C temperature (it is realized using the chamber of heat and humidity);

-85% relative humidity, 120 °C temperature (it is realized using the autoclave).

The duration of the tests is determined by their purpose, the capabilities of the laboratory base, the design and technological features of the IC, the selected modes.

## 4. Conclusions

The application of the theory of test planning and regression analysis has made it possible to obtain the adequate mathematical model connecting the indicator of corrosion resistance with the influencing factors and taking into account the physics of the proceeding processes. On the basis of the model obtained, the methods for evaluating the corrosion resistance of aluminum metallization of CMOS IC have been proposed, which are applicable when adjusting the technological processes for the IC manufacturing. The methods are easy to implement, have sufficient efficiency and statistical stability.

## Reference

[1] Luis A, Escobar and William Q. Meeker. 2006 *Statist. Sci.* **21**(4) 552–77.
[2] John W O 1996 *Semiconduct. Sci. Tech.*, **11(2)** 155–62.

[3] Radchenko S G 2005 *Stable Methods for Estimating Statistical Models: Monograph* Kiev: PP Sansparel, pp. 504.
[4] Voznesensky V A 1981 *Statistical Methods of Experiment Planning in Technical and Economic Studies,* Moscow: Finansy i statistika, pp. 263.
[5] Rebrova I A 2010 *Planning of the Experiment: The Educational Guidance* Omsk: SibADI, pp. 105.
[6] Bolshev L N, Smirnov N V 1983 *Tables of Mathematical Statistics* Moscow: Nauka, pp. 416.
[7] Kartashov G D 1976 *The Electronic Equipment* Series 8. **6(48) pp.** 7–17.
[8] Epifanov G I, Moma Yu A 1979 *Physical Basics of Technology and Design of Radio-Electronic and Computer Equipment* Moscow: Sovetskoe radio, pp. 352.

IOP Conf. Series: Materials Science and Engineering **317** (2018) 012060 doi:10.1088/1757-899X/317/1/012060

# Improvement of the Assignment Methodology of the Approach Embankment Design to Highway Structures in Difficult Conditions

**Y Chistyy[1], E Kuzakhmetova[2], Z Fazilova[3] and O Tsukanova[4]**

[1]Head of the Department of Buildings and Constructions on Transport at Russian University of Transport (MIIT) Moscow, Russia, Candidate of Technical sciences
[2]Professor of the Department of Buildings and Constructions on Transport at Russian University of Transport (MIIT) Moscow, Russia, PhD in Technical sciences
[3]Associate professor of the Department of Transport construction at Russian University of Transport (MIIT) Moscow, Russia, PhD in Technical sciences.
[4]Chief technologist, Invest Capital Consulting ltd, Engineer

**Abstract.** Design issues of junction of bridges and overhead road with approach embankment are studied. The reasons for the formation of deformations in the road structure are indicated. Activities to ensure sustainability and acceleration of the shrinkage of a weak subgrade approach embankment are listed. The necessity of taking into account the man-made impact of the approach embankment on the subgrade behavior is proved. Modern stabilizing agents to improve the properties of used soils in the embankment and the subgrade are suggested. Clarified methodology for determining an active zone of compression in the subgrade under load from the weight of the embankment is described. As an additional condition to the existing methodology for establishing the lower bound of the active zone of compression it is offered to accept the accuracy of evaluation of soil compressibility and determine shrinkage.

## 1. Introduction

In all developed countries, including in the Russian Federation, transport is one of the basic branches of the economy, as well as an important part of the industrial and social infrastructure. Opportuneness, safety and environmental friendliness of transport services in freight and passenger traffic have a great economic importance.

Traffic conditions on the roads are improved due to complex of actions, which includes: improvement of methods for calculating the design of road elements; planning activities; increase of transport-operational qualities of road surfaces; equipment of roads by means of the organization of movement, arrangement of roads.

It is not difficult to see that only if the first event is properly executed all the others are effective. The authors focus on the development of the improvement of the existing assignment methodology of the approach embankment design to highway structures. A special document describing the procedure for the construction design of approach embankment, has not yet been developed. Moreover, the design procedure should include: compilation of a design model, a sustainability prognosis and weak subgrade shrinkage, calculation methods of actions to ensure stability and accelerate subgrade shrinkage. There are only a few methodological documents that solve particular issues of individual design of the approach embankment to the highway structures. The main paper for the standard design is the document [1]. In addition there are a number of works, among them [2]. The main on individual design of approach embankments is ODM [3].

To increase the stability of the approach embankment on a weak subgrade, vertical reinforcing piles are used to ensure a smooth reduction in the embankment shrinkage towards the highway structure. To accelerate the settlement (consolidation) of the approach embankment on a weak subgrade, additional technological measures (temporary loading of the embankment) or constructive (vertical drains or drainage slots, substitution of foundation soil, etc.) are applied.

Despite this, the smoothness of the transition from the deformed embankment to the more rigid structure of the span structure is not ensured, especially in the case of the use of weak subgrade.

## 2. Methods

It is obvious that, in general, it is possible to provide the required reliability of the above-mentioned actions to ensure stability, accelerate the settlement and exclude inadmissible deformations of the approach embankment only with a systematic approach to solving this and related problems. The system approach includes:

1. improvement of the legal framework;
2. improvement of the regulatory and technical base;
3. scientific and technical support of all stages of the construction of the facility and its reconstruction.

We will consider in more detail the improvement of the regulatory and technical base. At the same time, it is necessary to take into account new scientific research, innovative developments in new building materials, equipment and technologies. Furthermore, it should be considered with the expansion of the geography of the construction of transport facilities in difficult conditions, including specific soils.

To understand the essence of improving the regulatory and technical basis, taking into account all of the above, it is advisable to consider the stages of development of the theory of designing structures. Conceptually, it can be distinguished the following stages in the development of this theory: the first is a separate calculation of the structure and grounds (typical design, 40-50 years of the previous century); the second is a joint calculation of the structure and the criteria (individual design, 80-90 years of the previous century); the third is the joint calculation of the structure and the grounds, taking into account the man-made impact of the subgrade structure (individual design, the beginning of the 2000s). Meanwhile, as a calculation system was used "structure-ground". As a result, a pleiad of scientists gave suggestions on calculations for the design of objects features of physical and mechanical properties of various types of soils [4, 5].

## 3. Results

In recent years, the man-made impact of construction projects on soils at their ground has significantly increased. This is due to increased load from the structure and transport, with increasing intensity and speed of movement of vehicles. In this regard, by Kuzakhmetova E. K. it is proposed to switch from the system "structure-ground" to the system "structure - man-made ground". The latter provides for an assessment of the change in the natural working conditions of soil to man-made using new criteria [6, 7]. According to this system, to implement the design of the structure as a whole system, it is necessary to compile a geotechnical model. Such model of the system “structure-man-made ground” includes:

1. structure with designed parameters and a soil profile of the subgrade soil;
2. assessment of the zone of influence of the structure on the subgrade, i.e. active zone of compression;
3. indicators of the composition, condition and properties of soils of different layers of the subgrade, the groundwater level in the natural state, and after the construction of structures (in the active zone of compression);
4. drainage conditions, both in the natural state and after compaction of grounds under load from the weight of the structure (in the active zone of compression).

In the process of individual design of the linear transport facility geotechnical model should be prepared for both the linear sections, and for concentrated structures of a facility.

Consider the features of designing and compiling a geotechnical model in the case of a combination of a non-rigid and rigid system on the example of the approach embankment to the bridge.

Let's remind, that in the traditional approach for smooth entrance of the car on the bridge while junction construction of it with embankment it is proposed to follow the requirements:

1. to ensure the proper density of soils of the subgrade throughout its height;
2. to provide reliable drainage of surface water from the top and the mass of the embankment using drainage backfill behind the bridge bearing and in abutment, drainage course under the surface with side trays and impervious protection coating and roadside within junsction;
3. to weather the subgrade until the pavement surfacing for at least a year during which the main subgrade and ground settlement occurs;
4. to lay approach slab of a length sufficient to cover zones of formation of local subsidence and to ensure a smooth junction between the roadway of the bridge and the road surface. Such a construction is shown in figure 1.

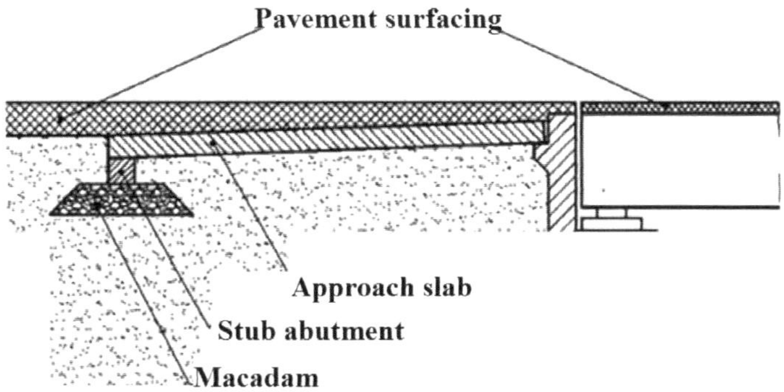


**Figure 1.** Approach embankment design.

In the opinion of the authors, it is necessary to add to the above listed requirements: the application of new innovative suggestions for ensuring the stability of the structure. This additional requirement includes, at least, the implementation of:

1. introduction of new building materials, including organic and inorganic additives for improving the properties of soils;
2. clarifying zone in the subgrade, which should be taken into account when calculating the design of the approach embankment.

In accordance with the studies of O.A. Tsukanova technology of ground stabilization with chemical or polymer-mineral stabilizers is useful layer-by-layer embankment filling behind bridge backwall and pavement surfacing on the approach embankment.

Replacement the traditional approach slab design based on macadam to the ground mixed with the chemical stabilizer will increase the strength properties of the layer and shorten the time of work execution. A high level of water resistance of stabilized grounds makes it possible to increase the stability of the structure due to the restriction of water ingress into the mass of the embankment.

Nowadays, there are a lot of agents for ground stabilization, each of which has advantages and disadvantages. Application area of stabilizing agents is predetermined by the type of soil, the road-climatic zone and the specific-conditions-of-use of the structure. It should be noted that all stabilizing agents are introduced into the soil together with organic and inorganic binders or are complex materials based on them. So, experience suggests that more effective is chemical («Kompozit STM» and SoilGrip, manufacturer – Lam-Torg ltd, Moscow) and polymermineral stabilizing agents («Nicoflok», manufacturer – Nikel ltd, Saint Petersburg) work with clay soils, and with sandy – mineral («Mobet Dortsem», manufacturer Biiskkhimstroymaterialy ltd, Biysk). Polymer stabilizing agents («TechniSoil», manufacturer - Dorogi Budushchego ltd, Moscow) it is advisable to stabilize dry soils [8]. For ground maturing, stabilized with mineral and chemical agents, it is necessary to achieve optimum soil dampness with layer-by-layer compaction. Thus, the availability of modern road construction equipment and the right approach to the selection of the new generation of stabilizing agents are currently making a method of improving the properties of grounds (or their stabilizing) with various additives that are most expedient, economic and technological.

As for the definition of active zone of compression, it is a very important stage of individual design. To assign correctly the design of the approach embankment is possible only with an accurate assessment of the impact zone of the zone of its influence on the ground base. The existing approach of defining active zone of compression is that its lower bound corresponds to the level on which $\sigma_z = 0,2\ \sigma_{gr}$, where $\sigma_z$- stress from the weight of the embankment, $\sigma_{gr}$ – stress of own weigh if the ground. These criteria were proposed in the 60-80 years of last century, and focused mainly on tough and plastic clay consistency. They are also transferred to the updated documents [9, 10].

## 4. Discussion

Therefore, the competency of this approach for weak organic and organomineral soils was assessed by Kuzakhmetova E. K. Studies have shown that it is necessary to impose additional conditions for the assessment of the lower bound of the active zone of compression.

As a criterion for the additional condition, it is proposed to take the accuracy of estimating the soil compressibility and determining shrinkage. For this purpose, the permissible error in the settlement is assigned, for example 5% of its final number, that is, $S_t=0{,}05S_{end}$. Rate of strain $\lambda$ (weak mass) is calculated, corresponding $S_t$. Further on the consolidation-test curve of the form $\lambda=f(P)$ there is such a section on which, with a slight load increment $\Delta$P rate of strain increment $\Delta\lambda$ does not exceed $\lambda t$. On the consolidation-test curve this area has an inclination angle equal to $tg\alpha = \frac{\Delta\lambda}{\Delta P}$. Load, at which the rate of strain is obtained $\lambda_t$, will correspond to the load on the desired level. Knowing its result, it is possible to set a refined the lower bound of the active zone of compression, after which the settlement is less than the accepted accuracy of its determination.

The above methodology will be considered using the example of an automobile road III technical grade. The height of the approach embankment is 4 meters, width on top is 12 meters, slope embedment is 1:2. At the subgrade embankment there is a muddy ground with medium dampness, its density is 1,7 t/m$^3$ and thickness is H=6 meters. Consolidation-test curve of the soil is shown on the figure 2.

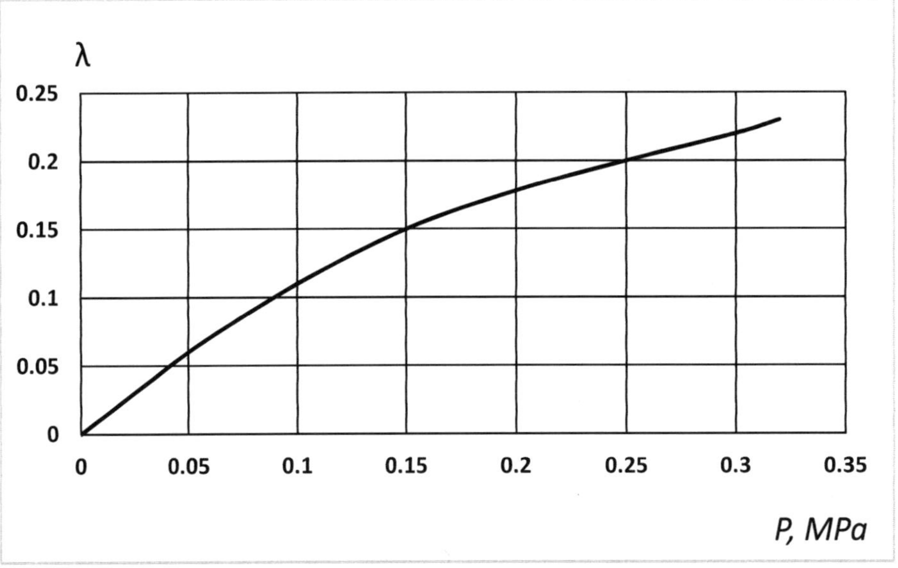


**Figure 2.** Consolidation-test curve of mud ground with density 1,7t/m$^3$.

Settlement of the weak subgrade under the load of the weight of the embankment and its own weight of soil is calculated by the formula $S_{end}=\lambda_i \times H=2,16$ m. The accuracy of the settlement estimate is 0,1 m. Hence $\lambda_t = 0,016$.

As an example, we determined the active zone of compression using the existing technique for the above conditions and for two different densities of mud ground (1,7 t/m$^3$ и 1,0 t/m$^3$). The results are shown in figure 3.

The graph shows that with decreasing soil density active zone of compression increases. For density of soil 1,7 t/m$^3$ the active zone of compression number is 12,7 m, for density of soil 1,0 t/m$^3$ – 20 m. We specify the lower bound of the active zone of compression (12,7m), on which the load acts 0,21 MPa. For this purpose, we find the section on the consolidation-test curve with increment $\Delta\lambda$ equal to 0,016. This site is close to the load 0,21 MPa. As a result, it was found that $\Delta_t$ corresponds to a load $P$=0,15 MPa. Knowing this load, it is calculated the refined lower bound of the active zone of compression, which turned out to be 9 m.

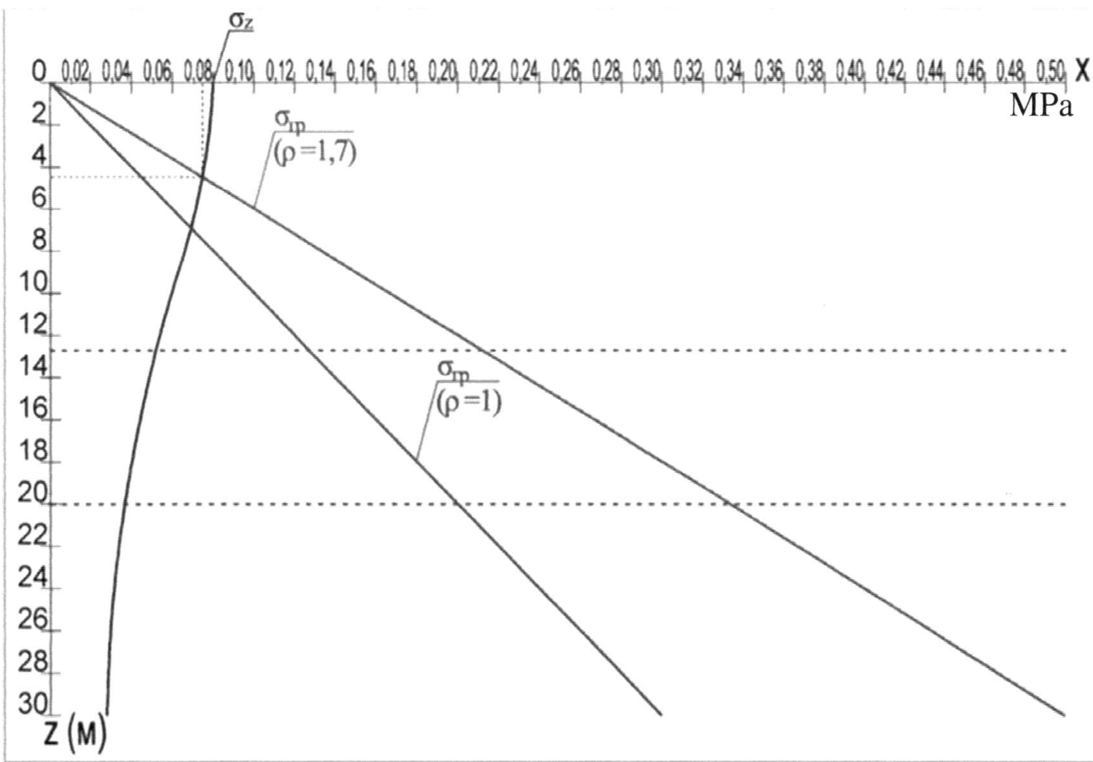


**Figure 3.** Definition active zone of compression according to the existing methodology.

Such methodology for determining lower bound of the active zone of compression in the subgrade under the load from the weight of the approach embankment, based on the including of the additional condition (accuracy of settlement estimation) increases the accuracy of finding the active zone of compression with all the ensuing consequences.

Thereby, the proposed approach to refine the active zone of compression, that is man-made impact zone of the embankment on its subgrade, is very important for increasing the reliability of assigning additional measures to accelerate the settlement of the embankment, as well as to ensure its stability. Consequently, this approach makes it possible to ensure operation capacity of pavement surfacing, and therefore traffic safety.

IOP Conf. Series: Materials Science and Engineering **317** (2018) 012060 doi:10.1088/1757-899X/317/1/012060

## References

[1] Mintransstroy SSSR AB-82 1991 *Series 3.503.1-96 Junction of highway bridges with approach embankment*, Moscow: Mintransstroy SSSR AB-82, pp.124.

[2] Starishkin V, Fazilova Z 2017 *Areas of transitional stiffness in the zone of abutment to bridges with a ballast and ballastless track structure. The introduction of modern design and advanced technology in the track facilities.* Paper presented at the 11th Scientific-practical Conference with International Participation, Moscow, MIIT, pp. 16-21.

[3] ODM 218.2.069-2016 2016 *Recommendations on the design of subgrade approaches on a weak subgrade to highway structures*, pp. 61.

[4] Dobrov E 2013 *Soil Mechanics* Academy, pp. 256.

[5] Galperin A, Cheskidov V, Borodina Y, Demidov A 2016 *The study of non-stationary hydrogeomechanical processes in clay soils*, Mountain information and analytical bulletin, **5** 195–205.

[6] Kuzakhmetova E 2011 *Improvement of outlook methodology for the shrinkage of the subgrade system "structure-ground* Bases, Foundations and Mechanics of Soils **5**, pp. 16-21.

[7] Kuzakhmetova E 2017 *Development of technogenic direction in engineering geotechnics for the purpose of accounting the impact of the structure to the grounds during designing.* Proceedings of the Ways to ensure joint work of the engineering structure and specific soils: the 11th Scientific-practical Conference, Moscow, MIIT, pp. 13-7.

[8] Kuzakhmetova E, Tsukanova O, Podgornyy A 2014 *STO 40065513-001-2016 Execution of polymeric coatings and restoration of asphalt-concrete coatings of highways with the use of polymer binder «Technisoilljj-5»*. Moscow: MIIT.

[9] Ministry of Regional Development RF (2012) SP 34.13330-2012 *Highways. Updated version of SNiP 2.05.02-85 * (with Amendment No. 1)*. Moscow: Ministry of Regional Development RF

[10] Ministry of Regional Development RF (2011) *SP 35.13330-2011 Bridges and pipes. Updated version of SNiP 2.05.03-84**. Moscow: Ministry of Regional Development RF.

IOP Conf. Series: Materials Science and Engineering **317** (2018) 012061 doi:10.1088/1757-899X/317/1/012061

# Investigation of the Influence of Acoustic Oscillation Parameters on the Mechanism of Waste Rubber Products Combustion

**R F Shakurov[1], O R Sitnikov[1], A I Galimova[1] and A F Sabitova[1]**

[1]Kazan National Research Technical University named after A.N. Tupolev – KAI, Kazan, Russia

**Abstract.** The article presents an analysis of the used methods of recycling of waste rubber products. The worn out tires are exposed to natural decomposition only after 50 - 100 years, and toxic organic compounds used in the manufacture constitute a danger to the environment. It contemplates a method of recycling waste rubber products in devices where pulsating combustion is realized. The dependence of the influence of acoustic pulsation parameters on the combustion mechanism of waste rubber products and on the composition of combustion products was experimentally investigated and established. For this purpose, the setup scheme based on the Rijke effect is optimized. The resonance pipe is coaxially embedded in the shaft. The known mathematical model of finding the combustion zones in the Rijke pipe, corresponding to the gas flow oscillations with the maximum amplitude, is applied to the chosen scheme. Investigations were carried out for three positions of the grate relative to the lower section of the experimental pipe, in which 1st, 2nd, 3rd modes of oscillation are formed. There are favorable conditions arise for the secondary combustion of mechanical particles entrained in the gas flow in the tube. The favorable conditions for afterburning also include the fact that through the upper section of the resonant pipe, the ambient air, caused by the features of the standing wave, is mixed into the gas stream. A comparative analysis of the change of gas concentration composition along the length of the resonance tube is carried out. It is established that the basic mode of oscillations contributes to the reduction of nitrogen oxides, in comparison with the oscillations occurring simultaneously at several harmonics, considering the main one. The results of research for the three positions of the grate in relation to the lower section of the installation are presented in tabular form, in which 1, 2, 3 modes of oscillation are formed. The analysis of experimental results confirms that the content of harmful compounds in the gas emissions below the maximum allowable norms.

## 1. Introduction

Nowadays the problem of disposing the large quantities accumulated industrial and household waste is acquired special urgency. Therefore, the question about using of secondary resources with maximum efficiency arises more and more acutely. Using out-of exploitation used tires and waste rubber as an alternative fuel is very perspective. Burning tires is approaching commercial fuels like black oil by quantity of generated heat [1].

Out-of exploitation used tires are stockpiled publicly or thrown in a landfill or buried in the ground. So they are exposed to natural decomposition only after hundreds of years, and organic compounds with toxic properties which used by manufacture are of great environmental hazard. Therefore, used recycling technology, including recovery of rubber products, should be ecologically clean and economically advantageous.

Thus, the reduction or even exclusion of harmful emissions into the atmosphere is an important issue in the field of environmental protection and ecology in general.

Various methods of recycling waste rubber products are used currently. In particular, the crumb rubber used in road construction in order to reduce the noise of automobile vehicle wheels. It is possible to obtain powders, short fibers, different dispersion degree crumbs of them, and use them as additives (or bases) in the manufacture of new products [2, 3].

However, current physical methods of tire recycling into rubber crumb are characterized by complicated processing technology and high energy costs. For example, the process of mechanical grinding of tires, especially with metal cord, require high energy expenditure, and pressure-destruction tire recycling technology, despite its apparent simplicity, is very complicated.

The drawbacks of low temperature processing used tires are the complexity of delivery, storage and the high cost of liquid nitrogen, complexity and high energy production in general.

Chemical methods for recycling used automobile tires, in particular at high temperatures, consist in thermal decomposition (destruction) of polymers in a given environment and receive processing products for various purposes. Liquid and gaseous pyrolysis products can be used not only as a fuel. They can be used as softeners for rubbers regeneration. Pitch pyrolysis resin is a good softener which can be used itself or in admixture with other components. The heavy fraction pyrolysate as an additive to bitumen is used in road building and can enhance its elasticity, resistance to cold and moisture. However, the recycling of used tires at high temperatures is very complex and energy-intensive process as mentioned above [4, 5].

## 2. Materials and Methods

Using unstable, namely pulsating (vibrating) combustion mode - one of the most promising ways to solve the problem of recycling waste rubber products. The acoustic oscillations arising in continuous media can intensify a series of processes in thermal systems. It is known that heat and mass transfer processes (mixtures formation, evaporation, heat transfer from the gas to the walls, heat exchange stream by reacting with heated bodies and others) are significantly accelerated and particle agglomeration and coagulation of the aerosol improves in the vibrating flow. Moreover, the calorific intensity of furnace volume is increased and completeness of fuel combustion compared to evenly combustion mode is improved by pulsating combustion. Vibrations cause additional air entering to the combustion chamber in the installations with mechanical or aerodynamic valve. These advantages can be used in power installations with moderate forcing combustion process in which the amplitude of acoustic oscillations will not be so large to cause any serious consequences. The prospects for the application of pulsating combustion for the combustion of solid, liquid and gaseous fuels (including alternative) are obvious.

A number of authors found [2, 5, 6] that the pulsating combustion is achieved by high combustion efficiency. It is achieved at rather low excess of air. As is known, the rate of combustion of condensed fuel, solid or liquid, is limited by the rate of oxygen mass transfer processes to the burning surface and the outflow of combustion products from it. This object is successfully solved in a pulsate system where fuels heavier particles cannot follow of fluctuations of the medium, whereby the area around the particles released from the combustion products and oxygen falls therein periodically.

In modern power installations intensification of the combustion process is carried out by using special blower-draft devices, by selection of optimum constructive scheme, by the influence of different frequencies fluctuations to the combustion zone and others. However, these methods complicate the construction of furnaces. And desired effect is not always achieved. Therefore, the most important direction in solving this problem is the development and creation of heat power installations, in which simplicity and reliability of the construction would be combined with a high calorific intensity of the combustion chamber, combustion efficiency, low concentration of carcinogenic substances in the waste gases [7, 8].

The most promising are the heat power installations (flue device), which is being implemented pulsating combustion [9].

Objective: to experimentally investigate and establish the dependence of the effect of the acoustic pulsations parameters on combustion mechanism of waste rubber products and the composition of the combustion products.

1. For this purpose, installation scheme based on Rijke effect have been optimized. The resonance tube is mounted coaxially to the shaft. The ratio of cross-section areas of the resonance tube and shaft corresponded to 1:1, 36.

2. The known mathematical model for finding the combustion zone in a Rijke tube corresponding to oscillations of the gas stream with the maximum amplitude was applied to the chosen scheme.

The formula for determining the location of the grate in the tube with maximum amplitude of oscillations will take the form [10]:

$$\frac{x}{L} = \frac{1+4m}{4n}, \tag{1}$$

where
m = 0,1,2,3 .......
n - harmonic number
L - tube length
x – grates position relative to the lower slice installation.

3. The composition of the waste gas was determined by gas analyzer DAG-16.

Below are shown in tabular form the results of studies for three grate positions relative to the lower slice of installation in which are formed 1, 2, 3 oscillation modes. When X = L / 4 acoustic waves generated modes 1, with X = L / 8 – 1 and 2 modes simultaneously, with X = L / 12 - 1, 2, 3 modes simultaneously.

**Table 1.** Average concentrations of carcinogenic substances in the waste gases at X = L / 4.

| measuring points | α | $CO_2$ % | CO ppm | NO ppm | $NO_2$ ppm | $NO_x$ ppm |
|---|---|---|---|---|---|---|
| 1 | 1.2 | 2.6 | 1359 | 169 | 7 | 176 |
| 2 | 1.4 | 2.8 | 1133 | 146 | 22 | 168 |
| 3 | 1.5 | 2.4 | 965 | 112 | 3 | 115 |

**Table 2.** Average concentrations of carcinogenic substances in the waste gases at X = L / 8.

| measuring points | α | $CO_2$ % | CO ppm | NO ppm | $NO_2$ ppm | $NO_x$ ppm |
|---|---|---|---|---|---|---|
| 1 | 1.2 | 2.8 | 1251 | 197 | 22 | 176 |
| 2 | 1.4 | 2.5 | 1109 | 163 | 21 | 179 |
| 3 | 1.5 | 2.2 | 829 | 121 | 6 | 141 |

**Table 3.** Average concentrations of carcinogenic substances in the waste gases at X = L / 12.

| measuring points | α | $CO_2$ % | CO ppm | NO ppm | $NO_2$ ppm | $NO_x$ ppm |
|---|---|---|---|---|---|---|
| 1 | 1.2 | 3.8 | 1212 | 257 | 22 | 179 |
| 2 | 1.3 | 2.6 | 1082 | 181 | 21 | 202 |
| 3 | 1.6 | 2.0 | 741 | 151 | 6 | 151 |

## 3. Results

It should be noted, the zone of pressure pulsation with maximum amplitude is located in the middle part of the oscillation circuit than the location of the grate at a distance ¼ of the contour length from of the lower slice. This pressure epure is typical for the pressure epure of a standing wave in the tube opened on both sides when oscillations in it excited with the only difference being that the pressure antinode in the present case a few displaced from the center of the oscillating circuit. It is explained by the speed of convective flow in the resonant circuit. It should be noted that the position of the grate according to the ¼ length of the oscillating circuit from its lower end, promotes the formation of the acoustic oscillations with maximum amplitude in the contour.

As shown in the tables 1,2,3 concentration of carbon dioxide $CO_2$ in the gas stream compared with concentrations of carbon monoxide CO and nitrogen oxide NO as it approaches the outlet tube slice is slightly reduced. This indicates that there are favorable conditions in the tube for afterburning of mechanical particles entrained in the gas stream. The fact, that throw the upper slice of the resonance tube ambient air caused by the features of standing wave mixes into the gas flow, is a favorable condition of afterburn to. Undoubtedly, the above factor enriched waste gas with oxygen, contributing to a more active afterburn of mechanical particles. The air ratio in the flow increases, and the concentration of carbon monoxide CO and nitrogen oxide NO reduces with increasing distance from the zone of combustion. At the same time it should be noted, only a slight change in the concentration of carbon dioxide $CO_2$, when would be expected a significant reduction in its concentration along the length of the resonance tube by analogy with other gases.

## 4. Contribution/Originality

From comparison analysis of the gas composition concentration changes along the length of the resonance tube also follows that the main oscillation mode contributes to lower nitrogen oxides (compared with vibrations occurring simultaneously at several harmonics, considering basic). Thus, the flow oscillations with a frequency $f = 84$ Hz nitrogen oxide concentration is 169 ppm, for flow fluctuations which occur simultaneously with the main higher harmonics, the concentration of nitrogen oxides is 257 ppm [11, 12, 14].

## 5. Conclusion

Analysis of the results of experimental studies confirmed that in applications where pulsating combustion is implemented, the content of harmful compounds in gas emissions greatly reduced. The prospects for the application of pulsating combustion for the combustion of solid, liquid and gaseous fuels (including alternative) are obvious.

## References

[1] Babkin Yu L 1965 *Chambers of pulsating combustion as a furnace device* Thermal Engineering (Moscow: MAIK Publishing House Nauka / Interperiodika) **9** 23-8.

[2] Babkin Yu L, Zhirnov B C 1968 *Pulsational intensification of combustion of a pulverized-coal torch* Pulsational combustion: Mater, to scientific and technical. Conf. (Chelyabinsk: WoF Publishing House of VTI).

[3] Buchman S V, Krylova N P 1968 *Investigation of the effect of velocity pulsation on the combustion of fine coal particles* Pulsatory combustion: Mater, to scientific and technical. Conf. (Chelyabinsk: WoF Publishing House of VTI).

[4] Podymov V N, Severyanin B C, Shchelokov N M 1879 *Applied research of vibration combustion* (Kazan: KazGU) p 132.

[5] Chiu H H, Hwang J S 1994 *Transition duality and histeresis of a combusting droplet* Proc. of the Zel'dovich memorial (Moscow: Russian Section of the Combustion Institute) **2** 164–7.

[6] Demirbas A, Al-Sasi B O, Nizami A S 2016 *Energ. Sour. Part A: Recov. Util. Envir. Effect. (Taylor and Francis Inc.)* **38(17)** 2487-93.

[7] Dubrovsky O V 1959 *Experimental study of pulsational combustion of liquid fuel in combustion chambers of stationary gas-turbine units* Thermal Engineering (Moscow: MAIK Publishing House Nauka / Interperiodika) **6** 56-61

[8] Egorov M Yu, Yegorov Ya V 2005 *Bull. Perm Nat. Res. Polytech. Univ. Mech. (Perm)* **13** 101-9.

[9] Katsnelson B D, Severyanin B C, Lyskov V Ya 1968 *Investigation of pulsating combustion of solid fuel* Pulsational combustion: Mater, to scientific and technical. Conf. (Chelyabinsk: Publishing House of the All-Union State Technical University) pp. 43-49.

[10] Pavlov G I, Shakurov R F 2003 *Bull. Kazan State Tech. Univ. A N Tupolev. (Kazan)* **2** 13-6.

[11] Pavlov G I, Kochergin AV, Sitnikov O R, Galimova A I, Shakurov R F, Kochergina K A, Garmonov S Yu 2011 *Bull. Kazan Tech. Univ.* **19** 174-9.

[12] Sabitova A F, Shakurov R F 2015 *Environmental and economic aspects of the use of waste rubber products as an alternative fuel* YOUNG ELPIT 2015: Ecology and Life Safety of Industrial Transport Complexes Scientific (Samara: ed. Vasiliev AV) pp. 281-4.

[13] Severyanin B C 1975 *Burning of a Particle of Fuel in a Pulsating Flow,* Thermal Engineering (Moscow: MAIK Publishing House Nauka / Interperiodika) **6** 144-7.

[14] Shakurov R F 2001 *Investigation of the combustion mechanism of solid combustible substances in the Rijke pipe* Abstract of thesis. dis. for the degree of Candidate of Technical Sciences (Kazan) pp. 21.

# Research and Practice on New Technology for Architectural Green Environment in Cities

**Zhang Wu[1], Jeung hvung Cho[1]**

[1]Pukyong National University, South Korea

**Abstract.** The importance of urban development has become a topic that has been discussed in all industries for a long time. How to make rational use of existing limited resources for redevelopment has become the primary issue in the future construction of a city. Designers have introduced green three-dimensional environmental design for a city into modern urban design. At present, Japan and South Korea focus on development of green three-dimensional environmental projects for cities, in which application of green three-dimensional building design is particularly prominent. This article learns from successful cases on urban three-dimensional environment design in Japan and Korea and makes profound discussion about how new city-model agriculture develops in China for the purpose of solving the problem of urban construction in China in the aspects of theory and Practice.

## 1. Background

When dealing with environmental problems, domestic enterprises usually take the way of post processing, and then develop "clean production technology" through engineering or internal improvement activities. Recently, environmental protection in domestic engineering and environmental protection problems on products are spreading rapidly. The whole world is actively tackling the problem of environmental pollution and sustainable development. Under the circumstances of actively carrying out activities of environmental pollution and sustainable development, solving environmental problems in industrial construction is a shortcut to fundamentally solve the global environmental problems.

However, most domestic business owners, product designers, environmental leaders and others cannot fully understand the concept and necessity of environmental protection design, or they doubt about the effectiveness of such design. In other words, some enterprises and related people believe that environment is the product of business operation. It is absolutely wrong to develop social economy and then to improve ecological environment for greater economic benefits. It is absolutely wrong not to pay attention to problems about environmental protection. However, in order to protect products and enterprises in their countries, the advanced countries such as Europe and America use environmental protection design as a trade barrier to protect products and enterprises in their countries. In Asia, Japan and South Korea are examples. In order to create this goal, Korean and Japanese governments and enterprises also regard environmental protection design as an axis of business operation. The governments and enterprises in Japan and South Korea don't only develop economy, but also increase efforts for protection of ecological environment construction. Since the middle of 1990s, green agricultural products have been more and more popular. Due to expansion of eco-friendly agriculture-practicing farms and expansion of the market scale, the social interests are increasing. Since 2000, green agricultural products in South Korea increase steadily at the rate about 40% each year. In 2013, the market of environmental-friendly agricultural products in South Korea reached 31 trillion won, accounting for about 7 to 8% of the total agricultural output. Due to worry about safety of agricultural products, the consumption behaviors on eco-friendly agricultural products have been changing recently.

## 2. Example analysis on modern three-dimensional green design

### *2.1. Korean hanging garden*

The Seoul consortium in South Korea designed and built a hanging garden in 2017. The garden was made up of a 936-meters long overpass, which started to be built in May 2015. The hanging garden is actually an overpass, which includes economic business circles and public spaces. This hanging garden was invested and created jointly by Ben Kuipers (Danish landscape architect) and KECC (local HA landscape partner in Korea). A part of the hanging garden has plants placed as high as 16 meters from the ground. The garden also includes parks and botanical gardens. With the experience to make creation, the designers also increase fun brought by economic circles and public space facilities for public after users' experience. Designers allow the audience to participate in the aspects of urban cultural and commercial levels in a smart manner and make radiation effect to the surrounding. It has become the landmark of Seoul City. The hanging garden is a good project to promote economic development (See table 1).

**Table 1.** Seoul Hanging Garden.

| | | |
|---|---|---|
| Structure map: A 936-meters long structure transformed from an overpass, with 16 small pavilions added as coffee shops, shops, exhibition halls, flower shops, trampolines, foot bath rooms, stages and children's theatres and visiting guidance center. |  | 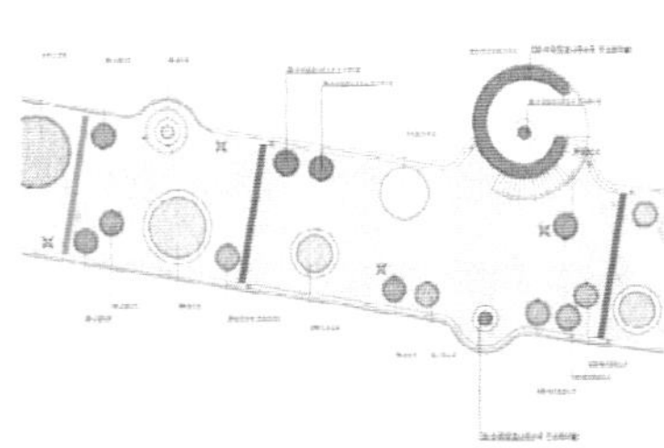 |
| Overlook: 52 plants including trees, shrubs and flowers are grown in 645 basins. About 160 species and subspecies are collected. The park includes 24000 plants (trees, shrubs and flowers). | 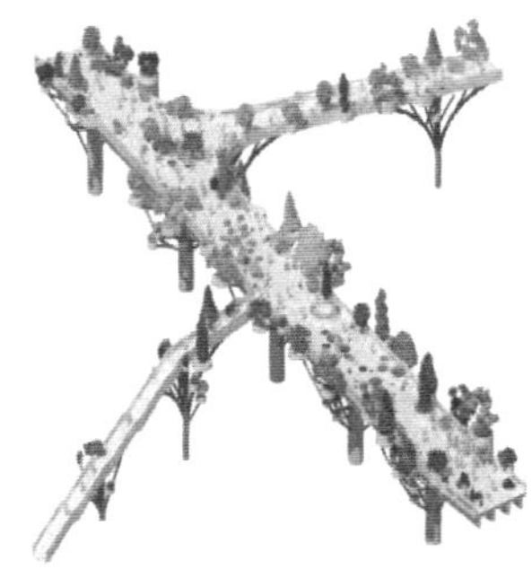 | 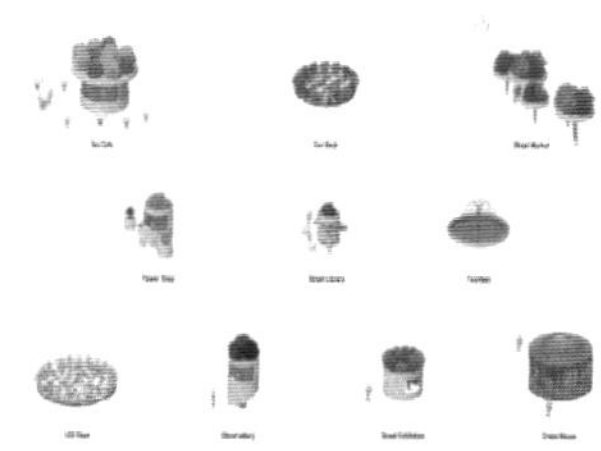 |
| Real picture: In day time, pure white building structures and plants with flowers changing colors in four seasons show full of vitality. The design of blue lighting at night is a great luster for the night. Lighting art is also displayed. |  |  |

This project of Hanging Garden will serve as new reference for the future development of buildings in green three-dimensional environment in South Korea. The hanging garden aims to expand its scope to the peripheral of the overpass, and to transform it into economic space and public space with rising environmental protection space and harmonious environmental development strategies. Interior design of the hanging garden increases the quantity of pedestrian streets and green squares so as to provide more space for building small gardens, green access, green roofs, green parking fields and terrace projects. These are part at the peripheral. More and more improved infrastructure will be convenient for residents, driving economy, saving land and vertically developing the new trend of modern urban architecture design by using existing resources.

**Table 2.** ACROS Fukuoka [5].

| | | |
|---|---|---|
| Structure Diagram<br>The Fukuoka Symphony Hall at floor 1, the party hall at underground floor 2, the international conference hall at floor 4, the rounded hall at floor 1.<br>The exchange gallery at floor 2, seminar room at floor 1/2, large conference room at floor 7, exercise room at floors 1-5and underground floor 2, Cultural Tourism and Information Plaza at floors 1-2, Craftsman Gallery at floor 2, exchange area at floor 1, Cultural Exhibition Corridor at floor 2, Akuros Fukuoka Ticket Center at floor 2, parking at underground floor 3. | 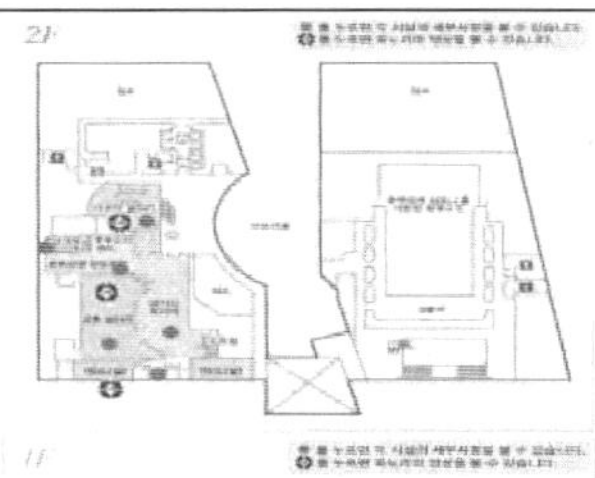<br>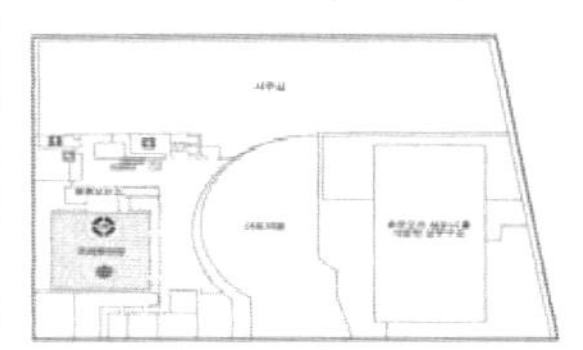<br><br>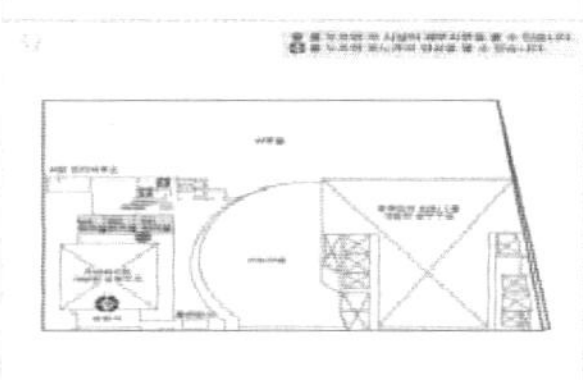 | 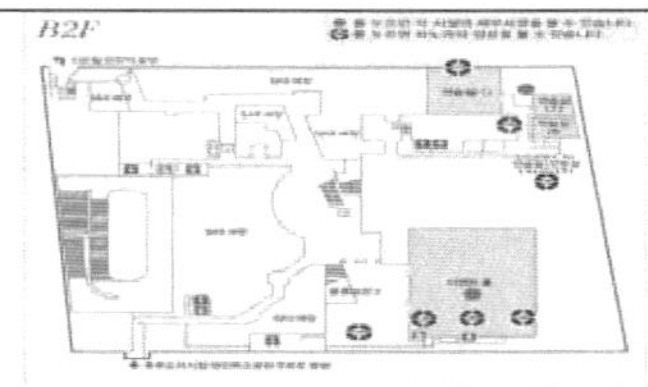<br>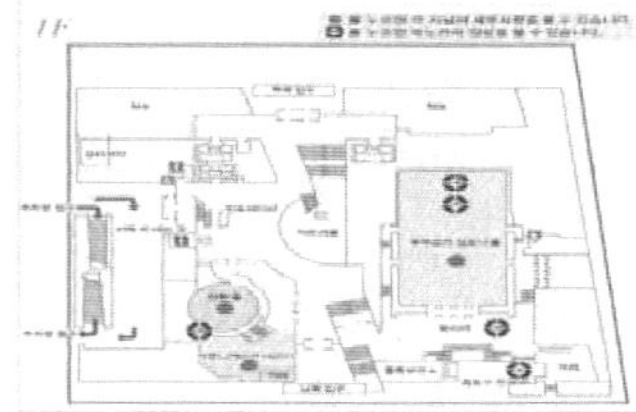<br>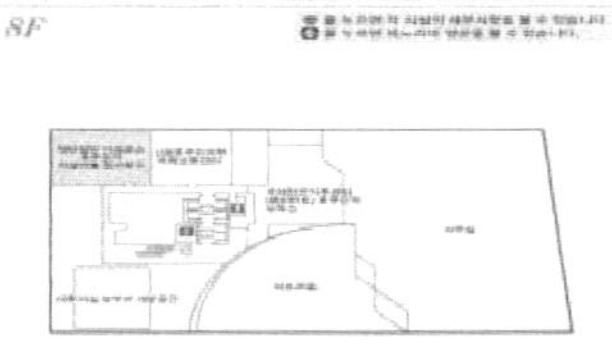<br>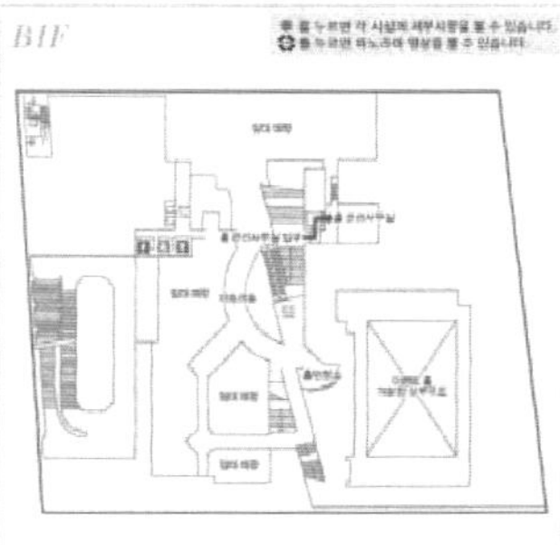 |
| Structure Diagram:<br>This uses unique staircase models of the b uilding, including various facilities such as the halls, stores, and office rooms. | 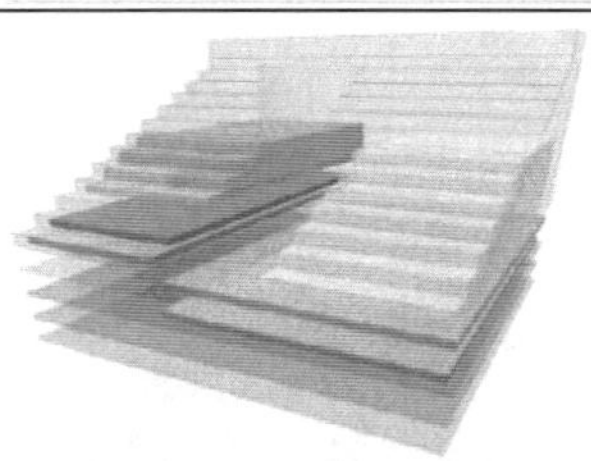 |  |
| Relativity image:<br>The garden roof is covered with 5000 plant s, a total of 76 varieties, 60 meters above the ground. |  |  |

### *2.2. Famous green roof in Fukuoka, Japan*

ACROS Fukuoka located in Fukuoka is one of the shocking representative works of green buildings in Fukuoka, Japan. It attracts people because of its unique design and more green space. One side of ACROS Fukuoka faces the most bustling commercial street in Fukuoka City and it is a glass wall, which makes it like a traditional office building. The other side is a huge green roof with layers downward. The layers are covered with 35000 plants. The landscape terrace extends to form a garden.

The building is 60 meters high above the ground. Its green roof is the side facing the sun and being covered with green vegetation. The covering plants include many species and they can change their colors with seasons. Just with these green plants, the whole building maintains a constant temperature level, which obviously reduces its energy consumption and forms a good ecological cycle. The designing thought and design plan for the building were jointly completed by the Argentina architects Emilio Ambasz and Associates. Such a large structure with a large leisure space brings people a unique visual feast and physical enjoyment

Acros Fukuoka was established with a whole design. It is not that a roof was built before the garden. It's a roof garden that is conceived in advance and is combined with the building when it is designed. It is a real green building. The architectural form is designed with a three-dimensional design concept.

### *2.3. Application and display of architectural agriculture in modern green design*

Because of the continuous development of modern agriculture, environmental problems such as environmental damage, water pollution, air pollution and soil pollution gradually appeared. Living environment has gradually become a bad cycle of environment, resulting in the phenomenon of heat islands with air temperatures rising. In the present developing society, what an efficient and fast production mode can better adapt to the development of the present era? In a limited space and increasingly harsh environment, it is necessary to establish a good commercial space for good environmental circulation in internal parts of buildings. In the urban farm system, a large number of production crops are produced, and development of architectural agriculture is necessary for development of urban agriculture. Architectural agriculture refers to the use of the relationship between organisms to make full use of space for combination of different species, obtainment of benefits and avoidance of harm, multi-species coexistence and multi-level configuration, then forming a three-dimensional agricultural management mode including three-dimensional plantation, three-dimensional breezing and three-dimensional plantation-breeding with energy recycling of multi-level materials. Now the buildings with a three-dimensional agricultural scale in a city are bound to be the direction of future agriculture.

### *2.4. Advantages of three-dimensional agriculture*

In a unit area of land (water) or in certain areas, three-dimensional planting, three-dimensional breeding or three-dimensional compound planting with breeding are made. The process skillfully uses artificial input to improve an energy cycling efficiency, material conversion rate and the second-property material output for an agriculture mode with multi-species coexistence, multi-levels configuration, multi-time interleaving and multi-levels quality and energy transforming. Stereoscopic agriculture utilizes virtuous circulation relationship between organisms to make full use of space and to combine different biological groups for promoting profits, avoiding damages, co-existence with many species, configuration at multiple levels and recycling of multi-level material energy, so as to achieve efficient utilization of unit area. 1. Stable supply should be ensured to reach sustainable production in a whole year, not affected by climate change, natural disasters and other external environment variables; 2. Continuous improvement of quality should be made to ensure the production efficiency. 3. Improvement of food safety awareness (excluding pesticides). 4. Potential is great for future growth. 5. Through scientific management, high added value in agriculture should be ensured. 6. The practical research found that local food is effective to reduce carbon dioxide. 7. People can enjoy the charm of the city and the leisure life of beauty. 8. Construction of eco-friendly buildings for experience, education and tourism. 9. Through construction in the aspect of environmental protection, agricultural plantation technology should be ensured to enhance national competitiveness. 10. Factory employment rate should be improved to create a good efficiency for factories.

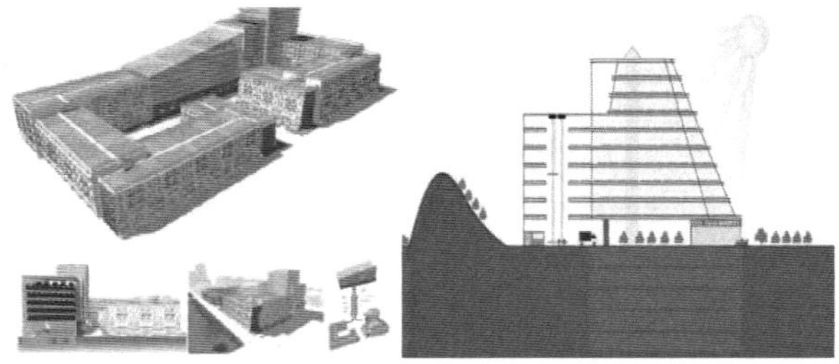

**Figure 1.** Architectural Agricultural Simulation Chart.

## 3. Conclusion

Stereoscopic agriculture is a system with ecological functions. The existing resources are reused. Such repeated development forms a benign circular ecological environment. The technology of stereoscopic agriculture has been used for environmental agricultural facilities, but it hasn't yet produced food in stereoscopic agriculture in the tall buildings in cities. The development of Stereoscopic agriculture is not only suitable for tall buildings. Upper parts of comprehensive buildings such as schools, restaurants and hospitals can become a variety of different fields for crop growing. These fields can provide fresh vegetables and fruits for people in the city to prevent modern health problems due to agriculture. Architectural agriculture can produce food without agricultural chemicals throughout the year. Fish and poultry can also be cultured indoors. Because the agriculture doesn't use large agricultural machinery and trucks to transport grain from the countryside to cities, it reduces use of large amounts of fossil fuels and discharge of greenhouse gases.

## Reference

[1] Takamasamoto, 2008 *Electroniceng.* 8 64-8.
[2] Seoul Hanging Garden -- *Source of Inspiration for Development of South Korea in the Future* (Landscape in China).

IOP Conf. Series: Materials Science and Engineering **317** (2018) 012063 doi:10.1088/1757-899X/317/1/012063

# Axial Compression Behavior of a New Type of Prefabricated Concrete Sandwich Wall Panel

**Xie Qun[1], Wang Shuai[1], Liu Chun[2]**

[1]School of Civil Engineering and Architecture University of Jinan, Jinan 250022
[2]Kunshan Eco-building Technology Co, Ltd, Kunshan, Jiangsu 215300

**Abstract.** A novel type of prefabricated concrete sandwich wall panel which could be used as a load-bearing structural element in buildings has been presented in this paper. Compared with the traditional sandwich panels, there are several typical characteristics for this wall system, including core columns confined by spiral stirrup along the cross-section of panel with 600mm spacing, precast foamed concrete block between two structural layers as internal insulation part, and a three-dimensional (3D) steel wire skeleton in each layer which is composed of two vertical steel wire meshes connected by horizontally short steel bar. All steel segments in the panel are automatically prefabricated in factory and then are assembled to form steel system in site. In order to investigate the structural behavior of this wall panel, two full-scale panels have been experimentally studied under axial compressive load. The test results show that the wall panel presents good load-bearing capacity and integral stiffness without out-of-plane flexural failure. Compared to the panel with planar steel wire mesh in concrete layer, the panel with 3D steel wire skeleton presents higher strength and better rigidity even in the condition of same steel ratio in panels which verifies that the 3D steel skeleton could greatly enhance the structural behavior of sandwich panel.

## 1. Introduction

Concrete sandwich panel is a widely used prefabricated wall element in residential and office buildings which is usually composed of two concrete layers separated by an insulation layer. Within each concrete layer there is a steel wire mesh and a serial of continuous rebar trusses as transverse connectors running along the full height of panel are adopted to attach concrete layers together which could provide resistance to horizontal shear. Sandwich panel has several advantages such as light self-weight, heat insulation and fast erection [1]. The sandwich wall panel system was initially developed and manufactured as non-load-bearing cladding or compartment elements in office building in U.S in the 1990s. Since the sandwich panels exhibit fully comparable performance to those of precast concrete solid panels, an interest in sandwich panels using as load-bearing structural elements in low-story buildings has been observed in the past decade years. A. Benayoune [2] has investigated the behavior of precast concrete panels subjected to eccentric compressive load and the results show that the load-bearing capacity of panels will nonlinearly raise with the increase of height-to-thickness ratio. Finally a semi-empirical formula has been proposed for strength prediction. Junsuk Kang [3] has adopted FEM method to study the structural performance of concrete sandwich panel with foam insulation under axial compression. Fabrizio Gara [4] has illustrated experimentally and numerically both the flexural and compressive behavior of fabricated sandwich panels without shear connectors. The aim is to determine the behavior of panel with different slender ratios under the axial and eccentric load. The result indicate that the strength of panels will decrease with the increase of slender ratios. With the development of housing industrialization and prefabricated building construction in

China recently, some studies have been carried out on the structural behavior of concrete sandwich panel currently. Mohamad N [5, 6] has used foamed concrete as an alternative material to produce a new type of precast sandwich panel and a double symmetrical truss-shaped connectors are employed in this panel. The experimental and theoretical results of axial and eccentric compression show that the slender ratio H/t has significant effect on the axial behavior of panel. Compared with the panel with single connectors the panel with double connectors presents better compressive strength and ductility. Olsen M D [7] has investigated the structural performance of concrete sandwich panels with carbon fiber connectors and the results show that the carbon fiber connector can effectively minimize the influence of thermal bridge due to its excellent property of heat isolation. Woltman G D [8] has adopted glass fiber reinforced polymer (GFRP) connectors as an alternative to metallic connectors and the load-slip behavior of sandwich panel has been established through a serial of shear tests. Hopkins P [9] has developed an innovative fiber reinforced polymer (FRP) shear plate connector and studied its effects on the flexural behavior of insulated concrete sandwich panels in terms of stiffness, strength, and applicability for constructions. Liew J Y R [10] has proposed a new concept for designing composite structures comprising a lightweight concrete core sandwiched between two steel plates which are interconnected by J-hook connectors. Tests confirms that the shear transfer capability of J-hook connector is superior to the conventional headed stud connector.

Based on the existing research results, as a structural system traditional sandwich panels are only employed in buildings with three stories below due to the limitation of load-bearing capacity, stiffness and connection type. In order to the possible application for multi-story buildings even high-rise buildings, a new prefabricated sandwich panel has been presented in this work, compared with the common sandwich panels, there are several typical characteristics for this new panel which includes core columns confined by spiral stirrup along the panel cross-section with 600mm spacing, foamed concrete block between external structural layers as internal insulation layer, and self-compact fine aggregate concrete applied in external layers with high flow performance. To improve the strength and stiffness of panels, a 3D steel wire skeleton in each layer which is composed of two vertical steel mesh connected with horizontal steel bar has been employed. A detailed configuration of panel is shown in figure 1. All steel elements of this panel are mechanically and automatically manufactured in factory, then transported to construction site for the overall steel skeleton assembly. Concrete in layers and core columns is needed to be casted in site. The thickness of concrete layer is adjustable according to the strength requirement of structural system and the existence of 3D steel wire skeleton and core columns is expected to contribute to greatly enhance the structural performance of panel. A test of full-scale specimens subjected to axial compressive load has been carried out in order to understand the actual behavior of this new sandwich panel.

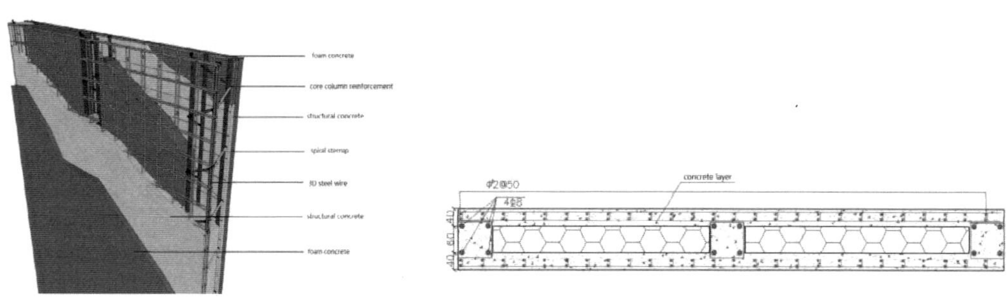


**Figure 1.** Section of the new fabricated sandwich panel.

## 2. Test program

With the purpose of performance comparison, two full-scale panels are designed with same height and width but different panel thickness, wire diameter as well as steel wire type, for specimen ZY1 planar steel wire mesh is used in concrete layers while a 3D steel wire skeleton is employed for specimen ZY2. The compressive strength of concrete is 36.8 MPa and the yield strength of rebar is 417 MPa. The detailed information of specimens is shown in table 1.

**Table 1.** Specimens information (unit:mm).

| No. | Height | Width | Thickness Concrete layer | Insulation thickness | Total | Steel wire mesh | Diameter of steel wire | Slenderness ratio | Reinforcement ratio |
|---|---|---|---|---|---|---|---|---|---|
| ZY1 | 2400 | 1300 | 25 | 60 | 110 | planar | 3 | 21.8 | 1.17% |
| ZY2 | 2400 | 1300 | 40 | 60 | 140 | three-dimension | 2 | 17.1 | 0.97% |

All the specimens are vertically installed with an unconstrained end on the top and a fixing end on the bottom. Two 2500kN hydraulic jacks are setup at the top beam of specimen whose function was to transmit the centralized load to a uniformly distributed load along the full width of specimen. The axial load has been applied monotonically with a constant increment of 50kN. A lateral support system consisting of a serial of 50mm-diameter solid steel rollers and 20mm-thick steel angles is installed close to the side face of top beam to restrain lateral movement. The test setup is presented in Figure 2.

**Figure 2.** Test setup.

## 3. Experimental results

In the early stage of loading, the strains for both steel reinforcement and concrete have a linear increase with the growth of load, and the vertical and horizontal displacements slowly develop within a very small value. For ZY1 the first visual crack occurs in the mid-height of panel under the axial load of 1060 kN. Under load of 1280 kN another inclined crack appears at the corner of panel bottom which develops obliquely upwards following the load increment. With the increase of axial loads, several new cracks vertically occur in the middle of panels and the existing diagonal cracks extend and widen quickly. At the ultimate load of 1420kN a sudden failure occurs with a main vertical crack almost thorough the panel height. The specimen ZY-2 had similar structural performance with ZY-1 in the process of loading. The peak strength was 4280 kN with severely local concrete crush in the top of concrete layer. For two specimens there were no massive cracks found in the surface of panel even in the ultimate state. The main test results were listed in table 2.

**Table 2.** Experimental results.

| Specimen No. | ZY-1 | ZY-2 |
|---|---|---|
| Crack load | 1060 kN | 4000kN |
| Ultimate load | 1420 kN | 4284kN |
| Maximum lateral displacement | 3.2mm | 8.9mm |
| Failure mode | Vertical through crack | Local concrete crush |

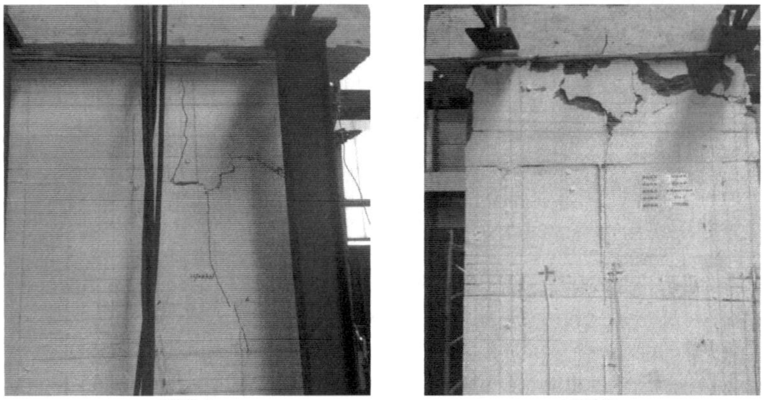

ZY-1 ZY-2

**Figure 3.** Failure modes.

## 4. Results analysis

### *4.1. Load-deformation analysis*

The lateral deflections obtained from the data of five LVDTs along the specimen height under axial load have been presented in figure 4. For ZY-1 the lateral deflections approximately kept a linear growth in the initial loading stage. While the load exceeds 800kN lateral deformation began to develop towards inverted direction since the excessive displacement of panel top triggered the restriction function of lateral bracing system. After then with the load increased, the lateral displacement at HD5 which is near the bottom end of panel gradually decreased back to zero in the final failure, while the displacement at HD1 which is near the loaded end of panel developed much faster than other measured points. The farther distance away from the panel top, the less lateral displacement was expected which could be attribute to the second-order effect. Similar to the deformation characteristic of ZY-1, the load-lateral deflection curves of ZY-2 developed almost elastically in the early stage of loading. Compared with ZY-1, the curves of ZY-2 measured at different positions show a monotonic development towards one direction mainly due to the smaller slenderness ratio.

### *4.2. Load-strain analysis*

The load-concrete strain curves of ZY1 were presented in figure 5 (a). There are two measured points in the mid-height of specimen and one at the bottom corner. All the data recorded by concrete strain gauges indicated that the full cross-section of specimen was in compression. However due to the lateral flexure caused by high slenderness ratio even under the axial compression, there actually were a concrete layer of specimen in the combined action of tension and compression and another layer in pure compression. Based on this analysis, the strains of CC3 and CC7 in tensile layer had smaller values than that of CC4 in compressive layer under the same axial load. All concrete strains developed within relatively small values during the loading program and the peak strains is about 1500με which was much lower than the ultimate compressive strain of concrete. In figure 5 (b) there were four load-concrete strain curves for different measure points of ZY2. Compared with ZY1, the growth of concrete strains in ZY2 was almost proportional to the increment of axial load. The strain linearly developed up to 4000με and much exceeded the ultimate concrete strain in uniaxial compression which verified that the excellent confinement of 3D steel skeleton system.

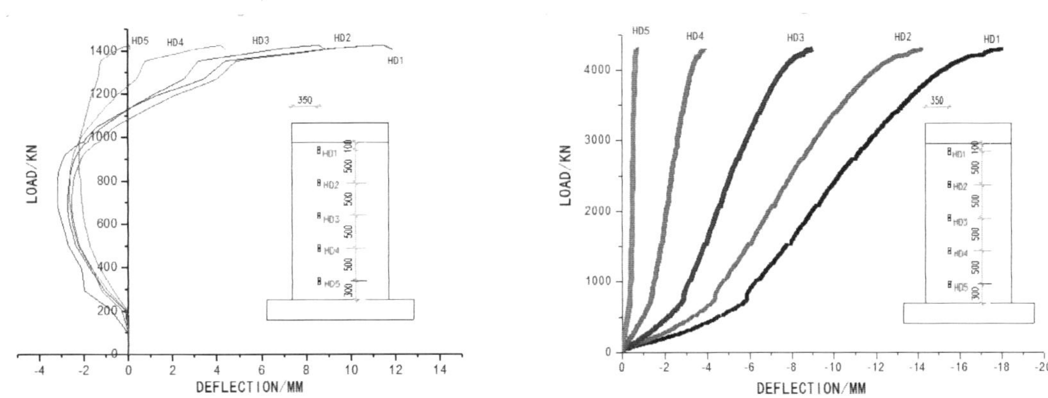


(a) ZY-1 (b) ZY-2

**Figure 4.** Load-lateral displacement curves.

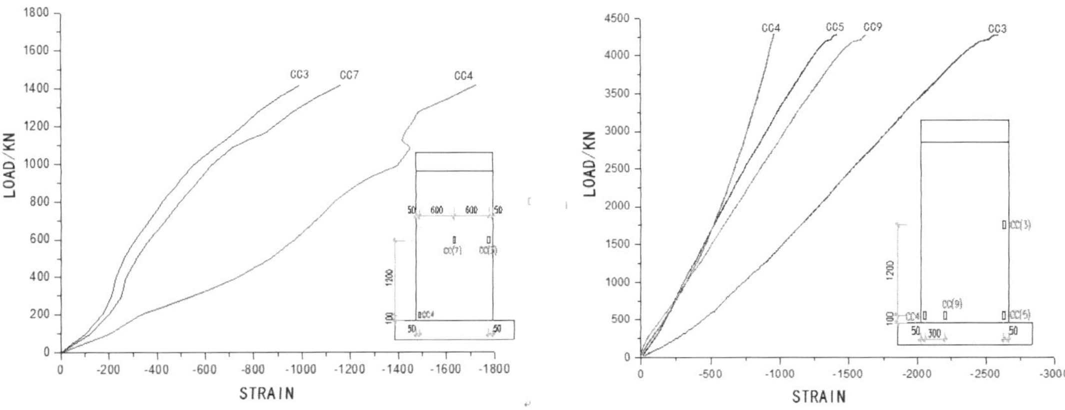


(a)ZY-1 (b)ZY-2

**Figure 5.** Load-concrete strains curves.

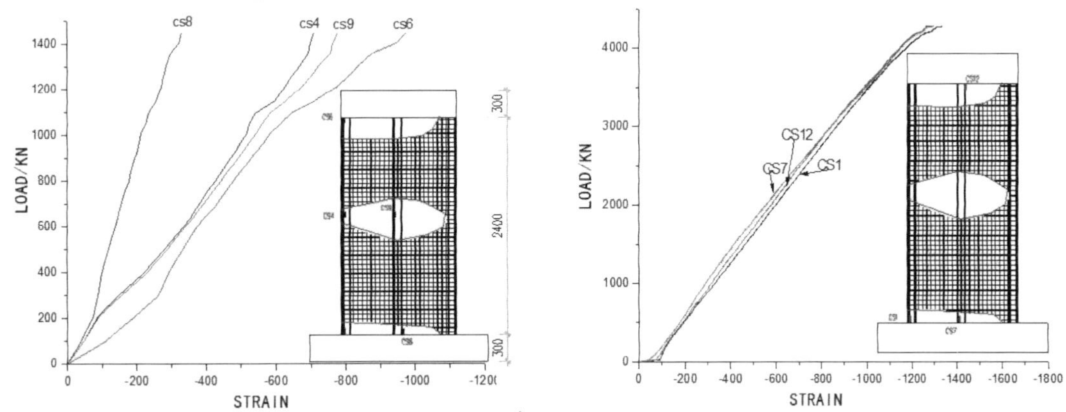


(a)ZY-1 (b)ZY-2

**Figure 6.** Load-reinforcement strains curves.

## 5. Conclusions

The structural behavior of a novel prefabricated sandwich panel under axial load has been experimentally investigated in this work. Some conclusions could be drawn as following:

1) A brittle failure occurred with a main vertical crack almost thorough the full height for specimen ZY-1 and severe local concrete crush in the panel top appeared in the final state for specimen ZY-2. No massive cracks were found in the surface of panels during the whole process of loading.

2) Compared with ZY-1 with planar steel wire mesh, the specimen ZY-2 with 3D steel wire skeleton presented better structural performance with greatly higher strength and robustness due to the smaller value of slenderness ratio even with less reinforcement ratio.

3) The lateral deflections almost kept a linear growth in the initial loading stage and due to the existence of core columns there was no obvious out-of-plane deflection generally seen in traditional sandwich panel subjected to axial compression. The strain curves both for concrete and reinforcement approximately presented a proportional growth with the development of axial load.

## References

[1] Palermo M, Ricci I, Silvestri S, Gasparini G, Trombetti T, Foti D, Ivorra S 2014 *Eng. Struct.* **76** 75-89.

[2] Benayoune A, Samad A A A, Ali A A A, Trikha D N 2007 *Constr. Build. Mater.* **21** 677-85.

[3] Kang J 2015 *Composites Part B* **68** 153-61.

[4] Gara F, Ragni L, Roia D, Dezi L 2012 *Eng. Struct.* **37** 193-204.

[5] Mohamad N, Wan I G, Abdullah R 2017 *Constr. Build. Mater.* **134** 185-98.

[6] Mohamad N, Muhammad H M 2011 *Adv. Mater. Res.* **335-336** 1107-16.

[7] Olsen M D 2006 *Concr. Int.* **28** 33-45.

[8] Woltman G D, Tomlinson D G, Fam A 2011 *A Comparative Study of Various FRP Shear Connectors for Sandwich Concrete Walls* Advances in FRP composites in civil engineering. Springer, Berlin, Heidelberg pp. 237-240.

[9] Hopkins P, Norris T, Peña N 2014 *Development of FRP plate shear connector for insulated concrete sandwich panels* The International Conference on Frp Composites in Civil Engineering pp. 133-45.

[10] Liew J Y R, Sohel K M A 2009 *Eng. Struct.* **31(5)** 1166-78.

**Acknowledgement**

This work was supported by the Shandong Provincial Key Research and Development Program of China (Grant no.2017GSF22106) and the Natural Science Foundation in Shandong Province of China (Grant no. ZR2014EL037).

IOP Conf. Series: Materials Science and Engineering **317** (2018) 012064 doi:10.1088/1757-899X/317/1/012064

# Biofuel Production from Jatropha Bio-Oil Derived Fast Pyrolysis: *Effect and Mechanism of CoMoS Supported on $Al_2O_3$*

**T Rodseanglung[1, 3*], T Ratana[2, 3], M Phongaksorn[2, 3] and S Tungkamani[2, 3]**

[1]Department of Social and Applied Science, College of Industrial Technology, King Mongkut's University of Technology North Bangkok, Thailand
[2]Department of Industrial Chemistry, Faculty of Applied Science, King Mongkut's University of Technology North Bangkok, Thailand
[3]Research and Development Center for Chemical Engineering Unit Operation and Catalyst Design (RCC), King Mongkut's University of Technology North Bangkok, Thailand

**Abstract.** The aims of this research was to understand the $CoMo/Al_2O_3$ sulfide catalyst effect to remove oxygen-containing and nitrogen-containing molecules from Jatropha bio-oil derived fast pyrolysis converted to biofuels via hydrotreating process. The activity and selectivity of CoMo/ $\gamma$-$Al_2O_3$ sulfided catalysts in hydrodeoxygenation (HDO) of Jatropha bio-oil derived fast pyrolysis was evaluated in a Parr batch reactor under 50 bar of $H_2$ atmosphere for 2 h at 300 320 and 340 $^oC$. It appeared that the $CoMo/Al_2O_3$ sulfide catalyst have high performance in activity for promoting the fatty acid, fatty ester, fatty amide and fatty nitrile compounds were converted to paraffin/olefin (Diesel range), this could be the CUS site on supported $Al_2O_3$ catalyst. The difference in selectivity products allowed us to propose a reaction scheme.

## 1. Introduction

The condensed pyrolysis liquid of bio-oil derived fast pyrolysis is a promising sign of renewable energy source, which can be substituted for petroleum fuels owing to its composition that looks like petroleum fuels, having an advantage of its short span and the zero net carbon emissions because it can be used to produce completely a biomass. This could solve the reduction of a greenhouse gas. However, these liquids is a complex mixture that mainly includes acid, ketone, phenols and ester etc, their high oxygen content cause to the poor thermal, high viscosity and corrosion [1]. Moreover, the bio-oil derived fast pyrolysis from Jatropha residues can present the nitrogen-containing compounds such as pyridine and amide groups in bio-oil, leads to concern with $NO_x$ when the combustion of engine and catalyst deactivation occurred. Consequently, the conversion of condensed pyrolysis liquids to biofuels is needed to reduce the oxygen and nitrogen content via hydrodeoxygenation (HDO) and hydrodenitrogenation (HDN) reaction.

The hydrodeoxygenation (HDO) is one of the methods for bio-oil upgrading, of which commonly hydrodeoxygenation catalysts is widely studied such as supported noble reduced metal sulfided CoMo and sulfided NiMo supported on $\gamma$-$Al_2O_3$ catalysts [2]. However, one of the defects of supported noble catalysts is very costly and more percentage of metal leading to be active. Then the found refineries oil using the sulfide CoMo and NiMo supported on $\gamma$-$Al_2O_3$ catalysts are widely utilized for hydrotreatment process [3]. It was reported that the $ZrO_2$ and $TiO_2$ supports can promote the HDS activity due to the good redox properties, however, both are as poor as mechanical properties and low surface area. In addition, hydrodenitrogenation (HDN) reactions are more difficult to remove and poisons the catalysts [4]. Then we needs to the acidity sites in the catalysts for the improvement of promoting the mild hydrocracking [5].

IOP Conf. Series: Materials Science and Engineering **317** (2018) 012064 doi:10.1088/1757-899X/317/1/012064

## 2. Experimental

### *2.1. Catalyst preparation and characterization*

The supported catalyst Mo/$\gamma$-$Al_2O_3$ was prepared by the sol-gel method, after that the all catalysts were dried at 50 °C for 48 h and calcined at 550 °C for 4 h. The calcined catalysts were prepared by using incipient wetness impregnation of 3% solution of cobalt nitrate, drying and calcination as same the supported catalysts condition and denoted as 310CMA. Before each test, the catalysts were sulfided ex-situ at 400 °C for 4 h under $H_2S/H_2$ 5% (v/v). The specific surface area, pore volume and average pore diameter of synthesized catalysts was determined in oxide-form by nitrogen adsorption using a BELSORP: MINI II instrument and the acidity was determined by temperature programmed desorption of ammonia ($NH_3$-TPD) using a BELCAT-B instrument. The Powder X-ray diffraction (XRD) patterns of the catalyst was performed on Rigaku TTRAX III an instrument for confirmed the phase identification of catalyst. Finally, the morphology of Mo sulfide catalysts was evaluated by transmission electron microscopy (TEM) on a JEOL JEM-2100 transmission electron microscope operated at 200 kV.

### *2.2. The jatropha bio-oil derived fast pyrolysis*

The Jatropha bio-oil derived fast pyrolysis-ESP fraction (JBP) samples used in the current study were provided by Thailand Institute of Scientific and Technological Research (TISTR), Thailand. The water content of JBP was determined for using a Karl Fischer titration method on an automatic titrator and the total acid number was measured by using an automatic potentiometric titrator (Metrohm 888 Titrando equipped). The elemental analysis of oxygen nitrogen carbon and sulfur content in bio-oil are measured in percent of weight by CHN628 Series Determinators. The chemical composition of JBP was diluted in tetraline and analyzed before hydrotreatment reaction with GC-MS and GC-FID.

### *2.3. Catalytic activity test*

The high temperature and pressure batch reactor (Parr 4848) was used for hydrotreatment reaction of JBP. These samples extracted with 1:5 ratio of bio-oil: tetralin solvent were weighed 20 g with 0.4 g of catalysts, persulfide carried out at 300-340 °C under hydrogen pressure initial 50 bar for 2 h. The liquid products were analyzed by Agilent 7890 gas chromatograph (GC) equipped with the flame ionization (FID) detector interfaced with an Agilent model 5975 mass spectrometer (MS).

## 3. Results and discussion

### *3.1. Catalyst characterization*

The results in table 1, found that the specific surface area of 310CMA was 231.2 $m^2/g$, and pore diameter was 3.97 nm, including the total amount acid sites was 1.05 mmol/g.cat, due to the $\gamma$-$Al_2O_3$ support its high acid sites, stability and surface area [6].

**Table 1.** The synthesized catalysts properties.

| Catalysts | Total surface area ($m^2$/g cat) | Pore diameter (nm) | Pore volume ($cm^3$/g) | Acidic (mmol $NH_3$/g.cat) | NO absorbed (μmol/g) |
|---|---|---|---|---|---|
| Commercial | 246.7 | 6.90 | 0.43 | 0.86 | - |
| 10CMA | 248.8 | 3.75 | 0.23 | 1.12 | 49.12 |
| 310CMA | 231.2 | 3.97 | 0.25 | 1.05 | 58.05 |

Figure 1a displays the XRD patterns of the calcined catalysts, then this research was focused on the diffraction patterns of catalyst support 310CMA exhibiting the characteristic peaks at the 2θ angle of 37.5°, 46° and 66.7° that indicated the $\gamma$-$Al_2O_3$ crystalline, while TEM images (figure 1b) display the $MoS_2$ slabs and $MoS_2$ stack oriented on support as a dark line, that was presented short slab length (2-3 nm) and low stacking of 2-3 stacks, as a result, its high dispersion, and low stacking of $MoS_2$ increased the number of rim site as the HYD active site for olefins [6, 7].

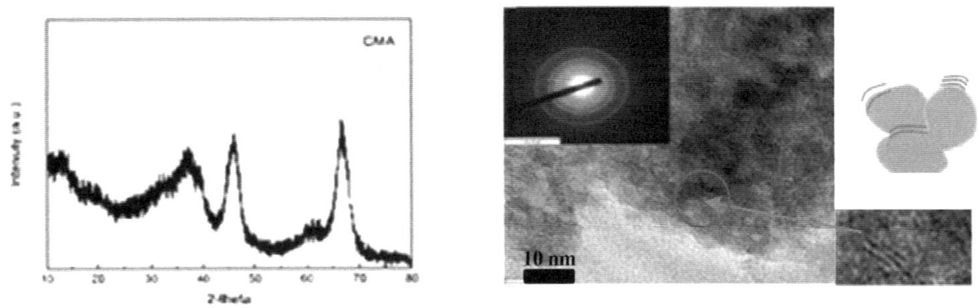


(a) (b)

**Figure 1.** (a) The XRD pattern and (b) TEM image of 310CMA.

### *3.2. Physical properties of Jatropha bio-oil derived fast pyrolysis*

The feed derived fast pyrolysis Jatropha bio-oil has total acid number of 84.37 mg KOH/g due to their contain large amount of fatty acid compounds and, the water contents of these samples were presented as 3.19 %. The elemental compositions were present as C H O and N elements were 69.27 9.86 14.60 and 6.27% respective, they're decomposed from lignocellulose hemicellulose and N element decomposed from protein are displayed in table 2. As the result, the oxygenate and nitrogen content in this bio-oil more than diesel fuel, their high oxygen content cause to the poor thermal, high viscosity and corrosion [7]. It needs to upgrading by removed the oxygen via HDO reaction.

**Table 2.** Properties of Jatropha bio-oil derived fast pyrolysis.

| Properties | This bio-oil | Diesel [8] |
|---|---|---|
| HHV (MJ/kg) | 33.92 | < 0.02 |
| Water content (%) | 3.19 | |
| Total acid number (mg KOH/g) | 84.37 | - |
| Elemental composition (received basis, wt.%) | | |
| % C | 69.27 | 86.0 |
| % H | 9.86 | 13.6 |
| % O | 14.60 | 0 |
| % N | 6.27 | < 0.05 |

### *3.3. Chemical Properties of Jatropha bio-oil derived fast pyrolysis*

The products in JBP presented a complex mixture, mainly including organic acid ketone phenolic aromatic naphthene N-heteroatom alcohol ester fatty-acid fatty-amide paraffin or olefin and fatty-nitrile as shown in figure 2, then we are classified as two groups are gasoline and diesel range products, followed by a number of carbon atom in the products [9].

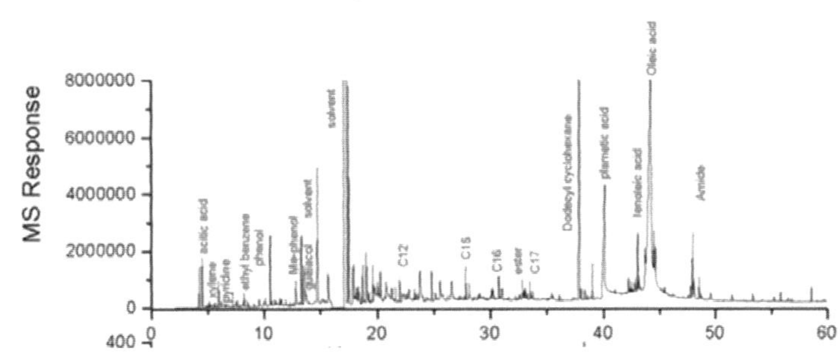


**Figure 2**. GC-MS chromatogram of Jatropha bio-oil derived fast pyrolysis.

### *3.4. Catalytic activity test on Jatropha bio-oil derived fast pyrolysis*

Figure 3a shows the selectivity of catalytic reaction of bio-oil as gasoline range that includes organic acid alcohol/ketone phenolic aromatic naphthene and N-heteroatom compounds, then the increase of the reaction temperature from 300 to 340 °C found that its has a slight difference products in all catalysts because the large molecules of fatty acid hindered catalytic hydrotreating on active sites [13]. While the yield products of diesel range (fatty acid, fatty ester, fatty amide, fatty nitrile and paraffin/olefin), it has a large products exchange as seen in figure 3b. The reaction pathway involves esterification ammonolysis decarboxylation decarbonylation and dehydration, then each pathway through the hydrogenation (HYD) of unsaturated fatty acid to saturated fatty acid, which the reaction can be written as in scheme 1.

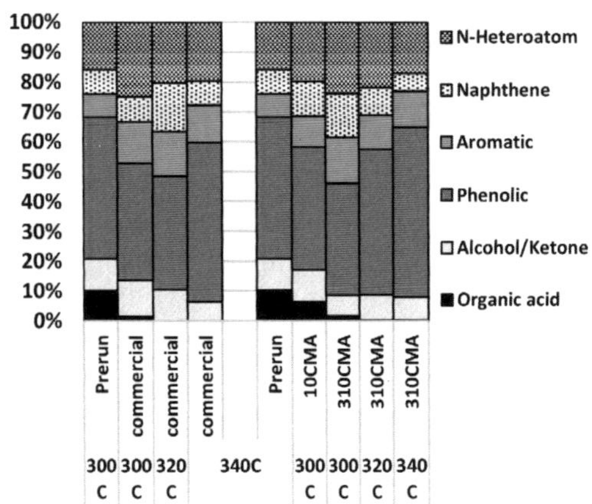


(a) Gasoline range products

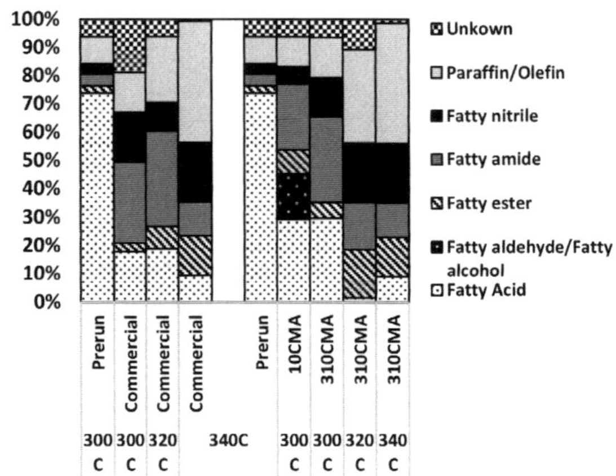


(b) Diesel range products

**Figure 3.** Product selectivity obtained with the $CoMoS/Al_2O_3$ of Jatropha bio-oil derived fast pyrolysis at 300 320 and 340 °C.

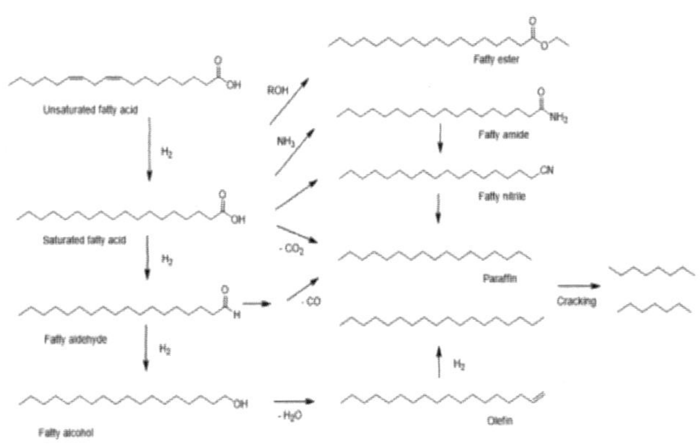


**Scheme 1**. Reaction network of deoxygenation of JBP over sulfide catalysts.

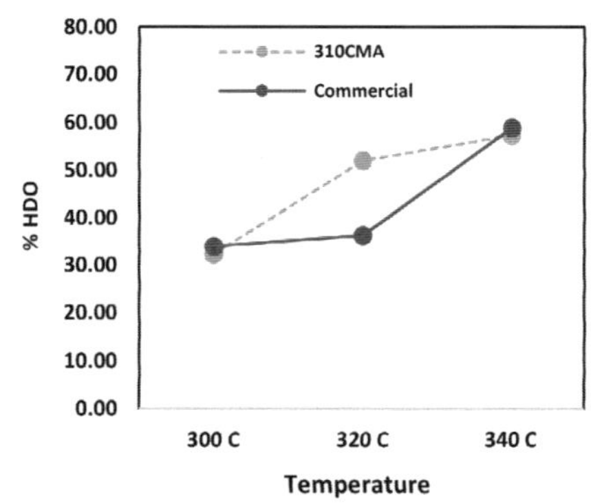


**Figure 4.** HDO conversion of commercial and 310CMA catalysts.

As in figure 3b, the 10CMA show fatty aldehyde/fatty alcohol are intermediately and give low the remove oxygen-containing at 300 °C compared with commercial catalysts, whereas the fast pyrolysis with 310CMA catalyst products such as fatty acid compounds are decreasing compared with crude Jatropha bio-oil derived fast pyrolysis, whereas fatty amide fatty ester can be covered to via hydrogenation, and its can reacted with nitrogen compounds are formation to fatty amide, while some reaction of fatty acid can be produced the paraffin /olefin via deoxygenation pathways and not found them immediately of fatty aldehyde/fatty alcohol, then increasing temperature at 320 °C appears the higher activity for converted fatty acid to paraffin/olefin and the increasing temperature is 340 °C, it's dramatically decreased of fatty amide compounds., there are summarized in figure 3b. This results the 310CMA promoted deoxygenation of fatty acid, fatty ester and fatty amide to paraffin/olefin, due to the important site for HDO reaction is the HYD active site, which the $Al_2O_3$ has a large amount of rim site shown in of sulfide $MoS_2$ catalysts [10, 11]. In addition, the HDO percentage of the 310CMA catalyst compared with the commercial catalysts, it's found that the 310CMA was presented the high HDO activity than the commercial catalysts at 320-340$^o$C as shown in figure 4.

The HDN of this research was found that there was nothing removed, because of the amide compounds transformed to nitrile through hydrogenation, However, all catalysts cannot apparently remove nitrogen due to nitrogen tolerance limitation of the catalysts [6].

## 4. Conclusion

From all results, we could assume that the main pathways for the removed oxygen content in JBP of the CoMoS base on $Al_2O_3$ catalysts reacted undergo acid site and HYD active sites leading to the cracking reaction isomerization hydrogenation direct-oxygenation decarboxylation and decarbonylation reaction as fatty acid → ester → paraffin or olefin.

## Reference

[1] Peng J, Chen P, Lou H and Zheng X M 2009 *Bioresour. Tech.* **100** 3415.
[2] Yang Y, Gilbert A and Xu C 2009 *Appl. Catal.* A: General **360** 242.
[3] Ho T C 2004 *Catal. Today* **98** 3.
[4] Sintarako P, Praserthdam P, Thammongkol V, Pokacharoenwatjana B, Yuanglamyai W and Inthiwong C 2015 *Catal. Commun.* **62** 89.
[5] Ardiyanti A R, Gutierrez A, Honkela M L, Krause A O I and Heeres H J 2011 *Appl. Catal.* A: *General* **407** 56.
[6] Rodseanglung T, Ratana T, Phongaksorna M, Tungkamaniaa S 2015 *Energ. Proc.* **79** 378.
[7] Ramírez J, Macías G, Cedeño L, Gutiérrez-Alejandre A, Cuevas R and Castillo P 2004 *Catal. Today* **98** 19.
[8] Fan Y, Shi G, Liu H and Bao X 2009 *Appl. Catal. B: Environ.* **91** 73.
[9] Fisk C A, Margan T, Ji Y, Crocker M, Crofcheck C and Lewis S A 2009 *Appl. Catal. A: General.* **358** 150.
[10] Zhao C et al. 2009 *Angew. Chem. Int. Ed.* **48** 3987.
[11] Soo-Young N 2014 *Renew. Sustain. Energ. Rev.* **40** 1108
[12] Zhang D, Duan A, Zhao Z, Wan G, Gao Z, Jiang G, Chi K and Chuang K H 2010 *Catal. Today* 149 62
[13] Goodwin V, Yoosuk B, Ratana T and Tungkamaniaa S 2015 *Energy Procedia* **79** 486

## Acknowledgment

The authors would like to thank Japan International Cooperation Agency (JICA) and the National Institute of Advanced Industrial Science and Technology (AIST) for the financial support and training course. The authors also would like to acknowledge Research and Development Center for Chemical Engineering Unit Operation and Catalyst Design (RCC), College of Industrial Technology, KMUTNB for the support of research facilities. Thailand Institute of Scientific and Technological Research (TISTR), Thailand were provided Jatropha bio-oil derived fast pyrolysis.

IOP Conf. Series: Materials Science and Engineering **317** (2018) 012065 doi:10.1088/1757-899X/317/1/012065

# Research on Reasons for Repeated Falling of Tiles in Internal Walls of Construction

**LiBin Xu[1], Shangwei Chen[1], Xinzhou He[1] and Guoliang Zhu[1]**

[1]Guizhou Construction Science Research and Design Institute of CSCEC, Guiyang, China

**Abstract.** In view of the quality problem of repeated falling of facing tiles in some construction, the essay had a comparative trial in laboratory on cement mortar which is often used to paste tiles, special tile mortar and dry-hang glue, and measured durability of tile adhesive mortar through freezing and thawing tests. The test results indicated that ordinary cement mortar cannot meet standards due to reasons like big shrinkage and low adhesive. In addition, the ten times of freezing and thawing tests indicated that ordinary cement mortar would directly shell and do not have an adhesive force, and moreover, adhesive force of special tile mortar would reduce. Thus, for tiles of large size which are used for walls, dry-hang techniques are recommended to be used.

## 1. Introduction

As a decorative material, tiles are widely used in decoration engineering, but there are quality accidents due to hollowing and falling. Although there are many reports on relevant issues in China, they just talked in generalities and were not strongly targeted [1][2]. Entrusted by a construction company, the writer analyzed reasons for repeated falling of tiles for internal decoration of a building constructed by the company. The construction company suspected the tilts had problems with the reason that although it had been repaired several times, the phenomenon would appear again in the next year. Through filed investigation, the following phenomena were discovered: ① falling tilts were in staircases and elevators which had a bigger temperature difference; ② the tiles had a big size- 600mm*600mm*10mm, thus they were heavy; ③ tiles fell down in blocks and the back side of the tiles hardly had mortar; ④ the bonding mortar was cement mortar which was modulated by workers according to their experience and rebound strength of entity mortar was bigger than 30MPa. According to the investigation, the writer planned to adopt ordinary cement mortar, special tile mortar and dry-hang glue to have put-off tests for comparison, and moreover, he measured durability of bonding effects of tiles through freezing and thawing so as to analyze reasons for falling of tiles.

## 2. Test raw materials and methods

### *2.1. Test raw materials*

Cement: PO.32.5 cement, 28d compressive strength: 40.8MPa

Sand: river sand, fineness module: 2.8

Special tile mortar: homemade and constituted by cement, superfine compound powder, quartz sand (below 1mm), water reducing agent, retarder, super absorbent polymers (re-dispersible latex powder) and admixture according to certain ratio

IOP Conf. Series: Materials Science and Engineering **317** (2018) 012065 doi:10.1088/1757-899X/317/1/012065

Dry-hang glue: two-component AB dry-hang glue, produced by Shenzhen Jiajiehe Building Materials Co. LTD.

### *2.2. Test methods*

Considering that tiles of the project fell down in blocks, workers in the construction progress might wet up back side of the tiles and then smeared mortar, or directly smear mortar. Thus, technically, the following three working conditions should be considered: ① directly smear mortar in the back side of tiles; ② directly smear mortar in the back side after sock tiles in water for 24h; ③ drain tiles after socking them in water for 24h and then smear mortar in the back side, and the detailed testing program can be seen the table 1. Have mixing ratio experiment according to materials provided by the construction party, and the proportion of cement, sand and water is 3:1:0.9. After molding, smear them in processed small tiles (size of test piece: 95mm*45mm*10mm) and have pull-out test 28d later.

**Table 1.** Testing program.

| Number | Mixing proportion | | | Wetting of tile surface |
|---|---|---|---|---|
| | Cement (g) | Sand (g) | Water (g) | |
| J1 (common mortar) | 1500 | 500 | 450 | Dry smear |
| J2 (common mortar) | 1500 | 500 | 450 | Directly smear mortar after socking in water |
| J3 (common mortar) | 1500 | 500 | 450 | Drain surface after socking in water, and then smear mortar |
| Number | Special tile mortar | | Water | Wetting of tile surface |
| T1 (special mortar) | 1000 | | 450 | Dry smear |
| T2 (special mortar) | 1000 | | 450 | Drain surface after socking in water, and then smear mortar |
| Number (dry-hang glue) | Proportion (A:B) | | | Wetting of tile surface |
| S1 | 1:1 | | | Dry smear |

In addition, the field investigation indicated that main positions where tiles fell down were staircase and elevator with significant temperature changes. Therefore, the essay selected whole pieces of tiles as experiment objects and smear mortar in the back side of tiles. After standard curing of 28d, it had 10 times of freezing and thawing cycle tests: low temperature: -5 degree; freezing: 8h; thawing: 16h.

## 3. Test results and discussion

### *3.1. Adhesion comparison of different materials to tiles*

According to the requirements of Adhesive for ceramic tiles and investigation JC/T 547-2005, anti-pulling strength of cementitious adhesive is bigger than 0.5MPa. According to the testing program in table 1, the test results are as shown in table 2.

**Table 2.** Test results of small tiles.

| Number | Pulling damage load (KN) | | | Average value (KN) | Anti-pulling strength |
|---|---|---|---|---|---|
| J1 | 1.248 | 1.872 | 1.664 | 1.59 | 0.37 |
| J2 | 1.160 | 1.352 | 1.072 | 1.19 | 0.28 |
| J3 | 2.464 | 1.920 | 2.400 | 2.26 | 0.53>0.5 |
| T1 | 3.432 | 4.824 | 2.752 | 3.67 | 0.86>0.5 |
| T2 | 2.752 | 3.408 | 3.600 | 3.25 | 0.76>0.5 |

IOP Conf. Series: Materials Science and Engineering **317** (2018) 012065 doi:10.1088/1757-899X/317/1/012065

| | | | | | |
|---|---|---|---|---|---|
| S1 | 10.082 | 6.360 | 7.812 | 8.08 | 1.89 |

Analysis by analyzing results of pulling tests:

① For anti-pulling strength of ordinary mortar, through dry smear and directly smear after socking in water, adhesive force of mortar cannot meet standard requirements and the reasons include: water retention of ordinary mortar is weak. Tiles will absorb a small quantity of water after dry smear, causing decline of adhesion between mortar and tiles; after socking in water, if smearing mortar when there are water in surface, there will be a "weak layer" between tiles and mortar. After socking in water, if smearing mortar after draining tile surface, adhesive force will be improved, which is also proved by the test. To sum up, construction technology has a big influence on cohesive force of tiles and ordinary mortar.

② For anti-pulling force of special mortar, both dry smear or (smear after draining surface of socked tiles) can meet standard requirements, and moreover, anti-pulling strength of dry smear is higher. Cause analysis: special mortar often has admixtures like Dispersible adhesive powder which can enhance adhesive force and water retention capacity of special mortar without wetting tile surface. If tile surface is wet, there will be a "weak layer", thus reducing adhesive force of special mortar.

③Comparison of whole pieces of tiles in adhesive force: binding force of dry-hang glue > binding force of special tile mortar > binding force of ordinary mortar.

④ According to the pulling damage test (see in table 3), we can know that common mortar hardly adheres to tiles and it directly crack in bonding parts of tiles; a little special mortar adheres to and cracks in the mortar layer; while adhesive force of dry-hang glue is relatively strong and parts of damages are structural adhesive which is used in later pulling damage test.

**Table 3.** Adhesion of different materials to tiles.

| Number | Picture | Description |
|---|---|---|
| 1 |  | Cracks of ordinary mortar and tiles are in binding parts of mortar and tiles. The back side of tiles hardly has mortar. |
| 2 |  | Cracks of special mortar are in mortar layers. The back side of tiles has mortar. |

| | | |
|---|---|---|
| 3 |  | Cracks of dry-hang glue are in non-binding parts of tiles and dry-hang glue. |

⑤ Although strength of ordinary mortar which is made by field workers is higher (rebound strength of mortar reaches to 30MPa), its shrinkage and transformation are big. After molding, there will be cracks in surface of molded small test pieces (see in figure1), causing a weak binding force between mortar and tiles. However, since special tile mortar is added with water and additives such as thickening agent, the adhesive stress will be strengthened and there will be no cracks (see in figure1).

Common mortar special mortar

**Figure 1.** Adhesion comparison of ordinary mortar and special mortar to tiles.

*3.2. Influence of freezing and thawing test on adhesive force*

Hollowing and falling of tiles often happen in places which have big temperature differences such as corridor and elevator, and they are not thoroughly solved through repeated repair, thus the writer designed a freezing resistance test for adhesive force. The test results are as shown in table 4.

**Table 4.** Test results of small tiles.

| Number | | Pulling damage load (KN) | | | Average value (KN) | Anti-pulling strength (MPa) |
|---|---|---|---|---|---|---|
| 1After 10 times of freezing and thawing | DJ1 | 1.400 | 1.352 | / | 1.38 | 0.32 |
| | DJ2 | / | / | 0.452 | 0.45 | 0.11 |
| | DJ3 | 0.888 | 2.456 | / | 1.67 | 0.39 |
| | DT1 | 2.792 | 1.720 | 2.720 | 2.41 | 0.56>0.5 |
| | DT2 | 2.856 | 3.112 | 2.824 | 2.93 | 0.69>0.5 |
| | DS1 | 4.312 | 4.544 | 3.851 | 4.24 | 0.99 |
| Note: / stands for that mortar directly falls down from tile surface, which cannot be measured. | | | | | | |

According to table 4, we can see that after freezing and thawing cycles, adhesive force of common mortar has a sharp drop, and some even fall on its own, which is in line with actual problems in engineering. Although adhesive force of special mortar also declines, it can still meet standard requirements. In addition, adhesive force of dry-hang glue also declines, but its freezing resistance will not be assessed considering that it belongs to organic products.

In addition, since mortar and density and drying shrinkage of tiles are different, the writer selected whole pieces of tiles after adhered with mortar for freezing and thawing tests, and the test results are as shown in table 5 and figure 2.

**Table 5.** Test results of big tiles.

| Number | Pulling damage load (KN) | | | Average value (KN) | Anti-pulling strength (MPa) |
|---|---|---|---|---|---|
| J1 | / | / | / | Hollowing, unable to be measured | / |
| J2 | / | / | / | Hollowing, unable to be measured | / |
| T1 | / | 0.288 | / | 0.288 | / |
| T2 | 1.040 | 0.656 | 0.576 | 0.757 | 0.18 |

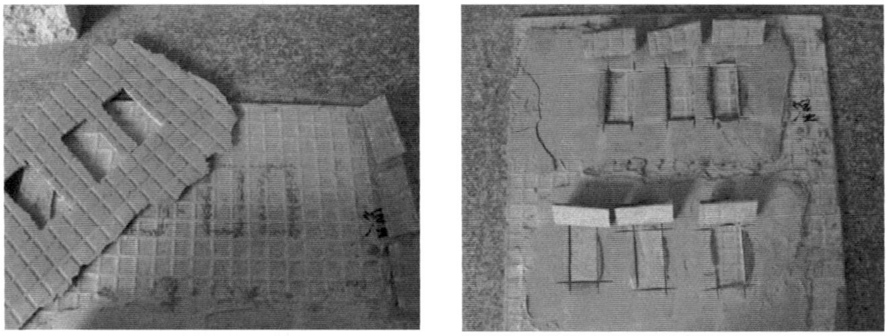

Common mortar special mortar

**Figure 2.** Adhesion of mortar and tiles (whole pieces of tiles).

Compared with small test pieces, when using whole pieces of tiles, adhesive force of mortar will have a big drop – common mortar will directly shell and the back side of tiles hardly has mortar; and special tile mortar also shells. Cause analysis: there are big differences in transformation of mortar and tiles. When temperature has fierce changes, transformation differences will cause rapid decline of adhesive force of mortar.

## 4. Conclusion

Conclusions through comparative tests:

① Common mortar can meet demands of tiles on adhesive force in earlier stages by adopting proper construction technology (smear mortar when tile surface gets wet and the water is not drained), but its long-term performance cannot be guaranteed.

② Since special tile mortar is added with water retention and tackifying components, it can realize ideal adhesion effects by dry smear in construction.

③ For tiles with a big size, it is suggested to use dry-hang technology to guarantee durability of adhesive effects of tiles.

④ Since falling of tiles is a long process and that the writer adopted the method of freezing and thawing to measure adhesive effects of mortar and tiles will have a certain regularity in earlier stages, a large number of tests are needed to verify.

## References

[1] Zhang Q H 2017 *Inner Mongolia Sci. Tech. Econ.* **3** 146.
[2] Liu F L 2004 *Inner Mongolia Coal Econ.* **1** 46-7.

## Acknowledgements

The project is funded by Qian Ke He GZ Zi [2015] 3013, CSCEC4B-2016-KT-17 and CSCEC-2016-Z-38

IOP Conf. Series: Materials Science and Engineering **317** (2018) 012066 doi:10.1088/1757-899X/317/1/012066

# Carrier Transport of Silver Nanowire Contact to *p*-GaN and its Influence on Leakage Current of LEDs

**Munsik Oh[1], Jae-Wook Kang[2], and Hyunsoo Kim[1*]**

[1]School of Semiconductor and Chemical Engineering, Semiconductor Physics Research Center, Chonbuk National University, Jeonju, Chonbuk 561-756, Korea
[2]School of Flexible and Printable Electronics, Polymer Materials Fusion Research Center, Chonbuk National University, Jeonju, Chonbuk 561-756, Korea

**Abstract.** The authors investigated the silver nanowires (AgNWs) contact formed on *p*-GaN. Transmission line model applied to the AgNWs contact to *p*-GaN produced near ohmic contact with a specific contact resistance ($\rho_{sc}$) of $10^{-1}$~$10^{-4}$ $\Omega\cdot cm^2$. Noticeably, the contact resistance had a strong bias-voltage (or current-density) dependence associated with a local joule heating effect. Current-voltage-temperature (*I-V-T*) measurement revealed a strong temperature dependence with respect to $\rho_{sc}$, indicating that the temperature played a key role of an enhanced carrier transport. The local joule heating at AgNW/GaN interface, however, resulted in a generation of leakage current of light-emitting diodes (LEDs) caused by degradation of AgNW contact.

## 1. Introduction

Recently, silver nanowires (AgNWs) have been extensively investigated in the field of organic semiconductor optoelectronic devices because of their intriguing properties including high optical transmittance, low sheet resistance, flexibility, and low cost process [1-7]. Very recently, our group [8] first demonstrated group III-nitride LEDs fabricated with AgNW transparent conductive electrodes (TCEs). This was feasible because AgNWs could form a reasonable ohmic contact to *p*-GaN. Consequently, the LEDs fabricated with AgNWs produced 56% brighter light output power than the reference LEDs. This suggests that the AgNWs are quite promising for use as next-generation transparent conductive electrodes [9, 10]. However, for obtaining better performance, it is still necessary to address the contact properties of AgNWs on *p*-GaN, whereas in-depth studies on this subject are very rare.

In this connection, we investigated contact properties of AgNWs contact formed on *p*-GaN. For this purpose, a transmission line model (TLM) method was used [11, 12]. Since the direct electrical probing of AgNWs contact is impossible owing to its thin and weak network structures, the specifically designed TLM structure was used as shown in Fig. 1. The current-voltage (*I-V*) curves of contact were analyzed using universal TLM method, giving bias-voltage-dependent or current density-dependent $\rho_{sc}$ values. The carrier transport at contact was further analyzed by using current-voltage-temperature (*I-V-T*) measurement. To investigate the reliability issues of LEDs fabricated with AgNW contacts, *I-V* curves were also investigated.

## 2. Experimental procedure

The AgNW films were formed using conventional spin-coating process (for 40 s at 800 rpm). The AgNW films were observed by scanning electron microscopy (SEM), revealing that their overall thickness was approximately 70 nm, as shown in figure 1. For the contact study, a TLM patterns having 150 μm × 200 μm contact pads and gap spacing of 10, 20, 40, 60 μm were used. Specifically, 20 nm-thick $SiO_2$ films were deposited on the top *p*-GaN layer of a commercially available LED wafer by an *e*-beam evaporator, and then selectively wet etched (using buffered oxide etchants) to exposed *p*-GaN layer surface for AgNWs contact. Then, a 10 nm-thick Pt pad was selectively deposited on the $SiO_2$ layer, followed by selective coating of the AgNWs onto the both exposed *p*-GaN layer and Pt pads. A conventional photolithographic lift-off technique was used for selective deposited Pt pads and AgNWs contact. The electrical properties of contacts were measured using a parameter analyzer (HP4156A), in which the chuck temperature varied from 250 to 350 K. The optical transmittance and the sheet resistance of the AgNW was measured using a UV/VIS spectrometer (V-670EX) and a four-point probe system (CMT-SR1000N).

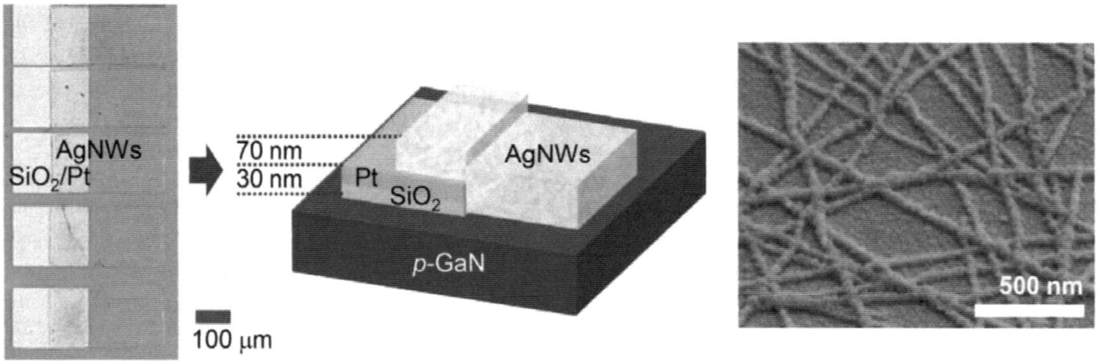


**Figure 1.** Optical microscopic and schematic cross-section views of AgNW TLM pad and SEM image of AgNW films.

The LEDs were also fabricated using AgNWs and conventional method to evaluate the leakage currents. For example, the rectangular mesa (200 μm × 500 μm) was defined by dry etching to a thickness of ~1.0 μm to expose the *n*-layer using an inductively-coupled plasma reactive ion etching system, on which a Ti/Al/Ni/Au (30/70/30/70 nm) layer was deposited as an *n*-electrode by an *e*-beam evaporator. Rapid thermal annealing was performed at 550°C for 1 min in ambient $N_2$ to form an *n*-type ohmic contact. To form AgNW TCEs on the *p*-layer, a Ti/Au (20 nm/10 nm) probing pad was formed on the mesa, followed by selective AgNW coating on the exposed *p*-layer and Ti/Au probing pads by means of a lift-off technique. The spin-coating process was performed in the last process step to minimize the possible contamination of AgNWs by additional photolithographic processing. To implement our study, commercially available LEDs wafers were used; these were grown on *c*-plane sapphire substrates by a metalorganic chemical vapour deposition. The structure of the LEDs comprised 2.0 μm of undoped GaN, 3.5 μm of *n*-GaN, 5-period GaN/InGaN multiple quantum well (MQW) active regions with 450 nm-emission, a 0.024 μm *p*-AlGaN electron blocking layer, and a 0.14 μm *p*-GaN layer. The fabricated LEDs were evaluated using a parameter analyser and optical spectrometer (Ocean Optics-USB2000+). The detailed fabrication and characterization procedures can be found elsewhere [8].

## 3. Results and discussion

In the SEM images of AgNW films coated on a sapphire substrate, a typical squashed arrangement with an average diameter of ~30 nm and average length of a few ten micrometers was observed as shown in figure 1. Therefore, the AgNWs are likely to have very small contact area. Consistently, the AgNW contact to *p*-GaN showed a relatively poor *I-V* curves, i.e., nonlinear and not steep, as shown in figure 2. To obtain $\rho_{sc}$ values, a universe TLM method proposed by Piotrzkowski*et al.* [13] was used since the classical TLM method could not be used in case of having nonlinear *I-V* curve.

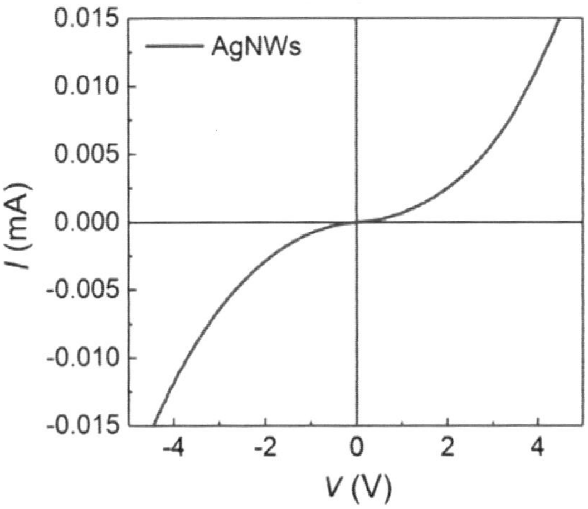


**Figure 2.** *I-V* curve of AgNW TLM pad, as measured from adjacent contact pad with the spacing of 10 μm.

Figure 3(a) and 3(b) shows the bias voltage and current density (*J*) dependent $\rho_{sc}$ values. It is quite evident that the $\rho_{sc}$ values drop significantly with increasing *V* or *J*. For example, the $\rho_{sc}$ values were $2.0 \times 10^{-3}$ $\Omega \cdot cm^2$ at $V = 3.5$ V and $1.5 \times 10^{-4}$ at $V = 5.0$ V. This indicates that the higher bias induces more current flow by reducing $\rho_{sc}$ values. Under higher bias condition, this might be attributed to the local joule heating effect, leading to a $\rho_{sc}$ reduction. Furthermore, this is self-reinforcing process since the reduced $\rho_{sc}$ value cause more current to flow and generate heat.

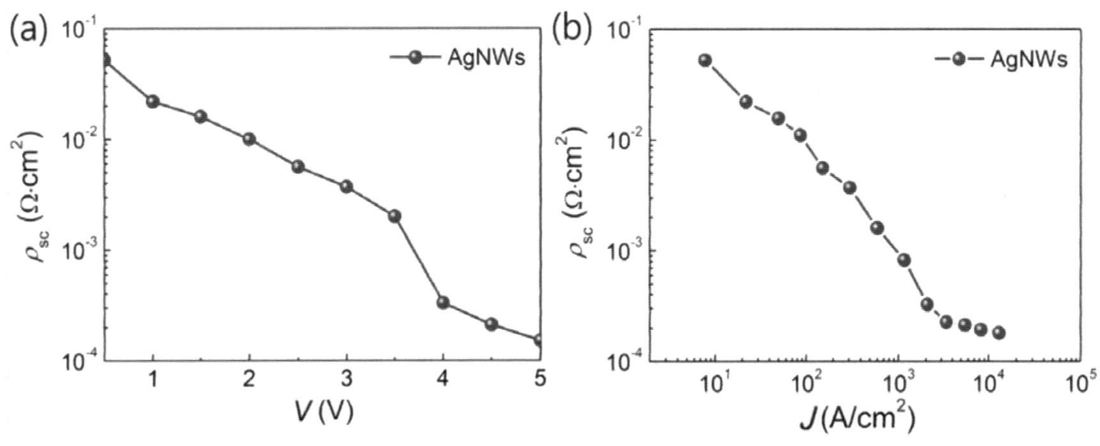


**Figure 3.** $\rho_{sc}$ of AgNW electrodes versus (a) bias voltage and (b) current density.

To investigate the carrier transport mechanism, the temperature dependent $\rho_{sc}$ values were plotted as shown in figure 4. It is shown that all bias dependent $\rho_{sc}$ values decrease monotonously with increasing temperature. This indicates that predominant factor determining $\rho_{sc}$ values is the temperature.

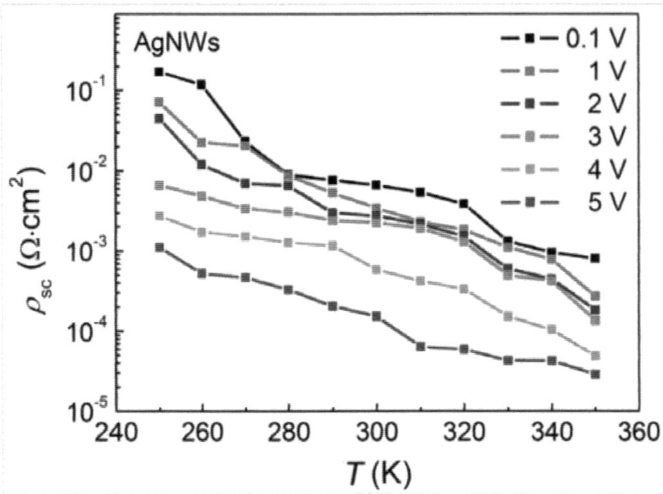


**Figure 4.** $\rho_{sc}$-$T$ curves of AgNW electrodes of various bias voltages.

Figure 5 shows the *I-V* curves of the LEDs fabricated with AgNW and Ni/Au electrodes. The insets show the schematic of AgNW (left) and reference (right) LEDs, respectively. It should be noted that the LED fabricated with AgNW had much higher forward and reverse leakage currents than the reference LED. This is attributed to the degradation of *p-n* junction, presumably due to a significant current flow underneath small-area AgNW contact.

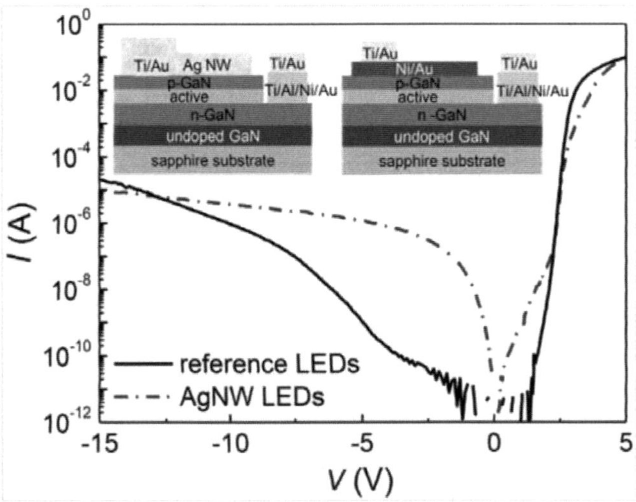


**Figure 5.** *I-V* curves of AgNW LEDs and reference LEDs. The inset of Fig. 5 shows the schematic diagram of AgNW and reference LEDs.

Despite these observed forward and reverse leakage currents, the LEDs fabricated with AgNW showed 1.6 times higher EL intensity than the reference LEDs as shown in figure 6. This is due to the combined effects of the enhanced light extraction associated with increased optical transmittance, i.e., the optical transmittance at 450 nm was 96.5 and 81.1 % for the AgNW and Ni/Au, respectively, and the improved current spreading associated with decreased sheet resistance, i.e., the sheet resistance was 11.7 and 11.3 Ω/sq for the AgNW and Ni/Au, respectively [8]. Indeed, this suggests that, once a novel TCE structure is designed to suppress the evolution of leakage current or to guarantee the reliability of LEDs, the AgNW will practically replace current TCE such as Ni/Au or ITO. This is presently under investigation.

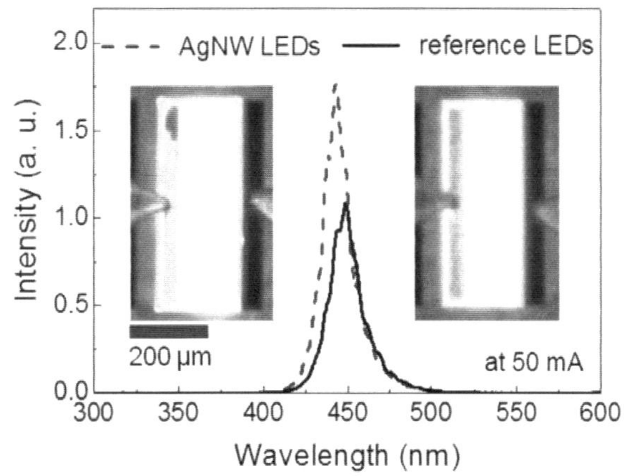


**Figure 6.** The room-temperature EL spectra of AgNW LEDs and reference LEDs. The inset shows the EL images taken at 3 mA.

## 4. Conclusions

The electrical characteristics of AgNWs contact formed on *p*-GaN was analyzed. The contact resistance dropped significantly with increasing bias-voltage or injected current. This could be due to a local joule heating effect associated with a significantly reduced contact cross-section. *I-V-T* measurement further revealed that the key parameter of ohmic contact was temperature. However, a local joule heating resulted in junction degradation and hence an evolution of leakage currents, suggesting that novel AgNW structure should be developed for practical application.

## References

[1] Sun Y, Gates B, Mayers B and Xia Y 2002 *Nano Lett.* **2** 165-8.
[2] Lee J Y, Connor S T, Cui Y and Peumans P 2008 *Nano Lett.* **8** 689-92.
[3] Hu L, Kim H S, Lee J Y, Peumans P and Cui Y 2010 *ACS Nano* **4** 2955-63.
[4] Garnett E C, Cai W, Cha J J, Mahmood F, Connor S T, Christoforo M G, Cui Y, McGehee M D and Brongersma M L 2012 *Nat. Mater.* **11** 241-9.
[5] Song M, You D S, Lim K, Park S, Jung S, Kim C S, Kim D H, Kim D G, Kim J K, Park J, Kang Y C, Heo J, Jin S H, Park J H and Kang J W 2013 *Adv. Funct. Mater.* **23** 4177-84.
[6] Kang Y J, Kim D G, Kim J K, Jin W Y and Kang J W 2014 *Org. Electronics* **15** 2173-7.
[7] Nam S, Song M, Kim D H, Cho B, Lee H M, Kwon J D, Park S G, Nam K S, Jeong Y, Kwon S H, Park Y C, Jin S H, Kang J W, Jo S and Kim C S 2014 *Sci. Rep.* **4** 4788-1-4788-7.
[8] Oh M, Jin W Y, Jeong H J, Jeong M S, Kang J W and Kim H 2015 *Sci. Rep.* **5** 13483-1-13483-7.
[9] Oh M, Kim M S and Kim H 2016 *Adv. Electron. Mater.* **2** 1600154-1-1600154-7.
[10] Baek M, Oh M, Parida B, Kim M S, Ryou J H, IEEE Senior Member and Kim H 2017 *IEEE Trans. Electron. Dev.* **64** 3690-5.
[11] Park Y and Kim H 2011 *Appl. Phys. Express* **4** 085701-1-085701-3.
[12] Choi Y, Jung E, Kim Y, Oh S, Yoo M C and Kim H 2014 *ACS Appl. Mater. Interf.* **6** 18673-82.
[13] Piotrzkowski R, Litwin-Staszewska E and Grzanka Sz 2011 *App. Phys. Lett.* **99** 052101-1-052101-3.

## Acknowledgments

This research was supported in part by the Basic Science Research Program through the National Research Foundation of Korea (NRF), funded by the Ministry of Education (2017R1A2-B4007182) and in part by the Development of R&D Professionals on LED Convergence Lighting for Shipbuilding/Marine Environments (Project No: N0001363) funded by the Ministry of Trade, Industry & Energy (MOTIE, Korea).

# Building Energy Consumption Pattern Analysis of Detached Housing for the Policy Decision Simulator

**Jiyoun Lim[1] and Seung-Eon Lee[1]**

[1]Building and Urban Research Institute, Korea Institute of Civil Engineering and Building Technology, 283 Goyangdae-Ro, Goyang-Si, Gyeonggi-Do, 10223, Korea

**Abstract.** The Korean government announced its plan to raise the previous reduction goal of greenhouse gas emission from buildings by 26.9% until 2020 on July 2015. Therefore, policies regarding efficiency in the building energy are implemented fast, but the level of building owners and market understanding is low in general, and the government service system which supports decision making for implementing low-energy buildings has not been provided yet. The purpose of this study is to present the design direction for establishing user customized building energy database to perform a role to provide autonomous ecosystem of low-energy buildings. In order to reduce energy consumption in buildings, it is necessary to carry out the energy performance analysis based on the characteristics of target building. By analysing about 20-thousand cases of the amount of housing energy consumption in Korea, this study suggested the real energy consumption pattern by building types. Also, the energy performance of a building could be determined by energy consumption, but previous building energy consumption analysis programs required expert knowledge and experience in program usage, so it was difficult for normal building users to use such programs. Therefore, a measure to provide proper default using the level of data which general users with no expert knowledge regarding building energy could enter easily was suggested in this study.

## 1. Introduction

After raising the target of greenhouse gas emission reduction in buildings to 26.9%, the Korean Government is implementing the stricter regulation and introducing incentives to increase the building energy efficiency. To ensure that the national policy is successfully established timely, focusing on energy retrofit of existing building since they take up most of buildings is more needed than focusing on new buildings [1]. To improve the energy performance of existing building, it is important to easily assess the energy performances of the buildings and to accurately and quantitatively predict the performances of these energy reducing elements during the energy retrofit decision making stage [2].

EU and advanced countries such as U.S. have developed and provided a reliable national system to support decision making of relevant personnel including public building energy policy makers and building managers on efficient building energy usage and maintenance. In case of Korea, the speed of policies for building energy efficiency is fast but the understanding of market is generally low, and the government service system to support decision making for effective public building energy management is still not provided yet [3].

As such, this study is intended to segment the energy usage based on the actual use of energy source by 507 detached houses using the existing nationwide energy database and national statistics and to analyse the energy consumption patterns for development of policy decision simulator.

IOP Conf. Series: Materials Science and Engineering **317** (2018) 012067 doi:10.1088/1757-899X/317/1/012067

## 2. Energy-related database

Building energy-related database is provided by various subjects such as building permit information and energy usage information. However, it is too complicated to detach the pattern of energy consumption from the sum of energy usage because building energy consumption is influenced by various factors and the pattern of energy usage also changes due to the interaction between those factors. Also, the level of influence from such factors also varies according to the purpose of building and it is very difficult to predict such level due to too many types of variables and wide variations. Currently, there are various government-affiliated organizations but they all collect and manage building-related information for different purposes to each other as below.

**Table 1.** Energy-related Database Status in Korea.

| Energy-related Database Status | | | |
|---|---|---|---|
| Ministry | Agency | Retained information | Purpose of information collection |
| Ministry of Land, Infrastructure and Transport | Korea Appraisal Boar[4] | Building energy consumption | Calculate energy consumption unit for each building purpose and establish greenhouse gas emission quantity inventory for home and business |
| | Seumteo (e-AIS)[5] | Building ledger and energy efficiency level certification | Basic statistical data for national building status analysis |
| | Spatial Information Industry Promotion Institute (V-world) | 3D shape of buildings | Provide the national spatial information open platform |
| | Korea Infrastructure Safety and Technology Corporation | Quarterly energy consumption | Review energy consumption and manage energy performance |
| Ministry of Strategy and Finance | Statistical geographic information service | 2D plane of buildings | Provide the open platform which converges spatial statistical information and various external information to support spatial decision making for small areas |
| Ministry of Trade, Industry and Energy | Korea Energy Agency | Monthly greenhouse gas emission and energy consumption | Establish the rational energy usage promotion plan for public agencies, check the result and manage the greenhouse gas emission and energy consumption targets |
| Ministry of Environment | Korea meteorological administration | Weather information for each area | Provide national weather service |

This study segmented the actual building energy consumption data by region, size and characteristics by analysing the factors included in the building ledger and analysed the actual usage by purpose using the energy pattern of each energy source extracted from the sampling data of the Energy Consumption Survey.

## 3. Analysis of energy consumption according to purpose by detached houses in Seoul

The data used in this study are the energy consumption according to electricity and city gas energy source from January 2012 to December 2014 of 507 case samples among the detached houses in Seoul and the building ledger data. Each case is the data case that matches the data in the building ledger provided by Seumteo and the monthly electricity energy consumption based on the billing. 1% data were randomly extracted from the detached housing-building ledger of Seoul for the analysis.

### *3.1. Electricity energy consumption analysis*

Figure 1 shows the base unit of monthly energy consumption by detached houses in Seoul. The average consumption base unit was 5.7kwh/m2 in 2012, 5.7kwh/m2 in 2013 and 5.4kwh/m2in 2014. During the analysed period, the monthly base unit tended to peak in August.

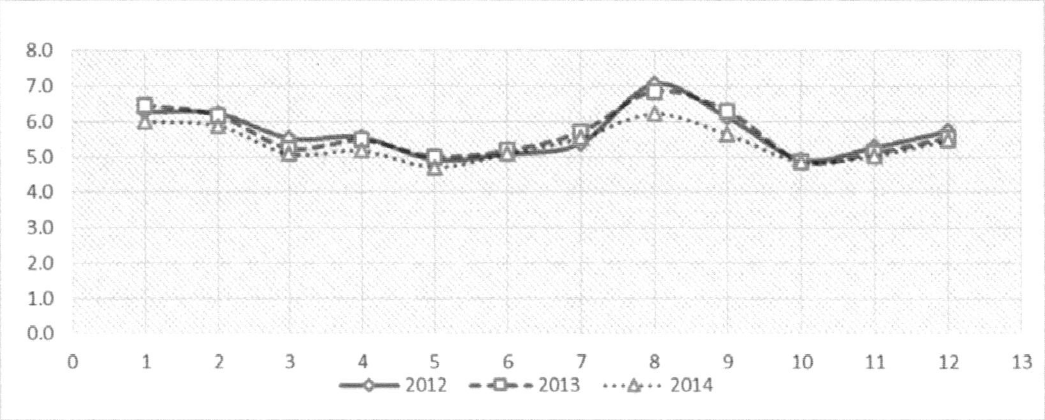


**Figure 1.** Electricity Energy Consumption-2nd Energy- of Detached House in Seoul.

The Energy Consumption Survey (2014) reported the annual electricity energy consumption by each detached house in urban regions in Korea was 318.67kwh [6] while the actual annual electricity energy consumption by each detached house case obtained for this study was around 1112kwh. The discrepancy is attributed to the fact that, although the average number of households per detached house in the origin data was two, the cases in the building ledger included the faulty registration of 0 household or more than 10 households in a detached house. Correcting the faulty registration, the annual energy consumption was 350~450kwh/household which was similar to the electricity energy consumption per household analysed by the Energy Consumption Survey.

### *3.2. Energy consumption disaggregation*

The energy consumption disaggregation is usually estimated with simulation or with the method of disaggregating by outside air temperature. Although there are cases of using the real-time aggregation system, it is difficult to consider them representative since some cases have fewer than 10 samples. As such, this study sampled the energy consumption for energy consumption disaggregation of detached houses by applying the ratio according to purpose of energy use extracted from the Energy Consumption Survey (2014).

The actual energy consumption patterns of 507 cases via the ratio of the detached housing energy consumption from the energy consumption survey are as below. Hot water & Cooking, Lighting, Building general power and etc. tend to be similar during the years however, Cooling and Heating have a trend to decrease the consumption by every year. The reason of the decreasing of cooling and heating energy consumption trend is the annual weather effect.

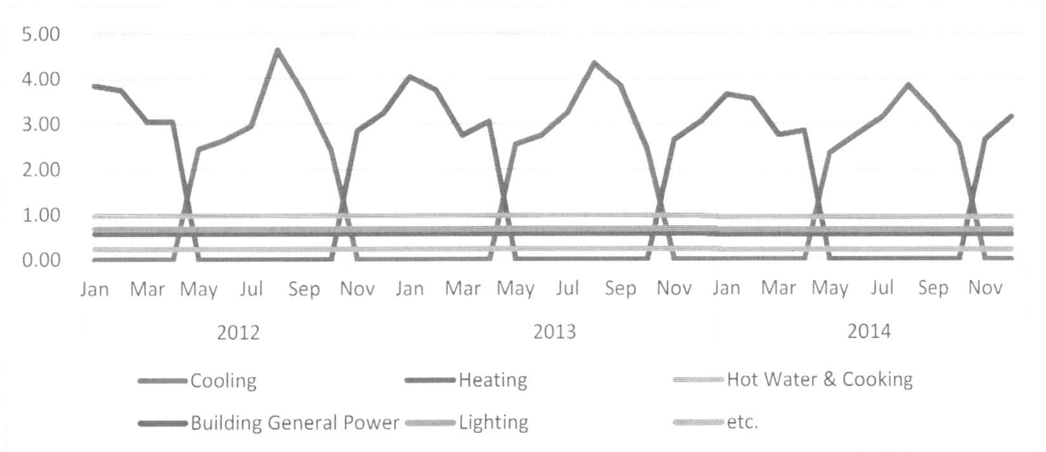


**Figure 2.** Disaggregation of electricity energy consumption of detached housing.

## 4. Conclusion and Future Work

In the research, the research team has been conducted on the disaggregation of energy consumption of detached housing in Seoul utilizing the energy use pattern ratio from the energy consumption survey in Korea (2014).

Energy consumption disaggregation can be more accurate with 1) non-weather dependent components, 2) weather dependent components, and 3) time and occupancy dependent components [7]. The Whole set of energy related data gathered and managed by the Korean government can be offered more specific energy usage patterns by conducting. The current work shows that the possibility of the level of detail of energy consumption breakdown. The accuracy of the disaggregation needs to consider about complexity of the building and comprehensive of the performance indicators of buildings. To improve the accuracy of energy disaggregation, the research team has to investigate the detailed performance indicators from the existing buildings for utilizing the baseline of the policy decision simulator.

## References

[1] Hensen J L M and Lamberts R 2011 *Building Performance Simulation for Design Operation* Taylor and Francis.
[2] Yoon S H, Kim Y J and Park C S 2014. *J. Archit. Inst. Korea* **30(10)** 189-97.
[3] Lim J Y, Kim D W and Lee S E 2016. *Energy Simulation Input Data Gathering of Public Buildings Utilizing by the National Database* ASIM2016.
[4] Building Energy Information System http://open.greentogether.go.kr/
[5] Seumteo 2015. http://www.eais.go.kr/
[6] MOTIE 2015. Energy Consumption Survey 2014. Korea Energy Economics Institute and Ministry of Trade, Industry, and Energy
[7] IEA EBC. 1987. IEA ECBC Annex11 Energy Auditing. 1987

## Acknowledgments

This research was supported by grant (17AUDP-B079104-04) from Architecture & Urban Development Research Program funded by Ministry of Land, Infrastructure and Transport of Korean government.

IOP Conf. Series: Materials Science and Engineering **317** (2018) 012068 doi:10.1088/1757-899X/317/1/012068

# Establishment of Low Energy Building materials and Equipment Database Based on Property Information

**Yumin Kim[1], Hyery Shin[1], Seung eon Lee[1]**

[1]Korea Institute of Civil Engineering and Building Technology, 283, Goyangdae-Ro, Ilsanseo-Gu, goyang-Si, Gyeonggi-Do, 10223, Korea

**Abstract.** The purpose of this study is to provide reliable service of materials information portal through the establishment of public big data by collecting and integrating scattered low energy building materials and equipment data. There were few cases of low energy building materials database in Korea have provided material properties as factors influencing material pricing. The framework of the database was defined referred with Korea On-line E-procurement system. More than 45,000 data were gathered by the specification of entities and with the gathered data, price prediction models for chillers were suggested. To improve the usability of the prediction model, detailed properties should be analysed for each item.

## 1. Introduction

The commercial websites for purchasing goods or comparing the prices of similar goods in Korea mainly focused on household items such as clothing and home appliances, however, the items which are handled in commercial websites are getting diverse so that the information of real estate are provided through the websites recently [1]. The services provided by those websites are usually the properties of the items, price comparison, consumers' review information and so on, allowing users of the websites to obtain information of the items. As the users of the websites increase, the manufacturers have to provide more information to the websites, and compete the price and quality of the product with similar manufacturers in the same field. When consumers purchase a product, they can obtain information regarding the product promptly through the websites for comparing the prices and purchase the same product at a low price. Due to such advantage, 51% of consumers purchased a product after price comparison through the websites for comparing the prices, and 45.4% among them considered product reviews when deciding whether or not to purchase the product [2].

The low energy building materials and equipment information portal in this study is established at national level with quasi-governmental agencies including Korea Energy Agency and Korea Appraisal Board. The portal is designed to integrate scattered database and provide reliable contents related to government systems and policies in order to allow portal service users to select a desired product rationally. Also, this portal leads healthy competitions between manufacturers and sellers through the product price and properties comparison so that reasonable prices can be established in the market. This paper suggests the methodology for gathering the data and the processed information from the gathered data

## 2. Low energy building materials and equipment database

### *2.1. Code system of low energy building materials and equipment database*

The data framework is necessary to gather enormous data for building materials and equipment database.

IOP Conf. Series: Materials Science and Engineering **317** (2018) 012068 doi:10.1088/1757-899X/317/1/012068

Korea On-line E-procurement system (www.g2b.go.kr) was established in 2002 to secure the transparency of public contract between the government, public institutes and private companies. The On-line E-procurement system has officially certified code system for gather the information of products which is supplied to the government or public institutes [3-5]. For operation and management of building materials and equipment database the code system of the Online E-procurement system was used in the building materials and equipment database after modified to expand the number of items for building materials and equipment database. The classification code of the database was same with Online E-procurement system up to sub-class, and the code for sub sub-class was recomposed according to the range of data to be established in this study. The established database for low energy building materials and equipment was classified into 5 levels for 165 codes as partly shown in table 1.

**Table 1.** Code system for building materials and equipment database.

| Major Class | | Middle Class | | Minor class | | Sub-class | | Sub-class | | Code name |
|---|---|---|---|---|---|---|---|---|---|---|
| CODE | ID | CODE | ID | CODE | ID | CODE | ID | CODE | ID | CODE |
| **30** | **Building materials** | 14 | Insulation | 15 | Insulation | 03 | Bid type insulation | 01 | Bid type insulation | 3014150301 |
| | | | | | | 08 | Fiber insulation | 01 | Fiber insulation | 3014150801 |
| | | | | | | 14 | Foam polystyrene insulation | 01 | Foam polystyrene insulation | 3014151401 |
| | | | | | | 15 | Extruded foam polystyrene insulation | 01 | Extruded foam polystyrene insulation | 3014151501 |
| **40** | **Pipes and Fluid control systems** | 10 | Heating, cooling and ventilation system | 17 | Cooling system | 01 | Cooler | 01 | Cooler | 4010170101 |
| | | | | | | 02 | Evaporator | 01 | Evaporator | 4010170201 |
| | | | | | | 03 | Evaporative cooler | 01 | Evaporative cooler | 4010170301 |
| | | | | | | 04 | Condensing uint | 01 | Condensing uint | 4010170401 |
| | | | | | | 08 | FCU | 01 | FCU | 4010170801 |
| | | | | | | 09 | AHU | 01 | AHU | 4010170901 |
| | | | | | | 10 | Reciprocating chiller | 01 | Reciprocating chiller | 4010171001 |
| | | | | | | 11 | Turbo chiller | 01 | Turbo chiller | 4010171101 |
| | | | | | | 12 | Screw chiller | 01 | Screw chiller | 4010171201 |
| | | | | | | 13 | Absorption chiller | 01 | Absorption chiller | 4010171301 |

### *2.2. Establishment of low energy building materials and equipment database*

Since data modelling is necessary for database establishment, the list and specifications of entities are suggested and database is established according to the list and specification of entities. The list of entities provides the information from the source of data such as period of update and the number of materials and equipment for each data item to be gathered.

Attribute information of materials and equipment database can be classified into general information and detailed information for each item. The general information field provides basic which should be included such as product model name, image, company name, price and specifications. Detailed information field contains additional information or specified performance or services which the manufacturers want to inform to the consumers. Figure 1 shows the list and specification of entities in materials and equipment database. With the specification of entities, more than 45,000 data of building materials and equipment were gathered. About 34,000 data were gathered from Korea Price Information (KPI) system which is the largest building material and equipment database in Korea, however KPI focuses on the price, performance data for materials and equipment were not contained for some items. About 14,000 data were gathered from the certified material and equipment database of Korea Energy Agency (KEA), on the other hand, KEA focuses on the performance, they do not provides price data.

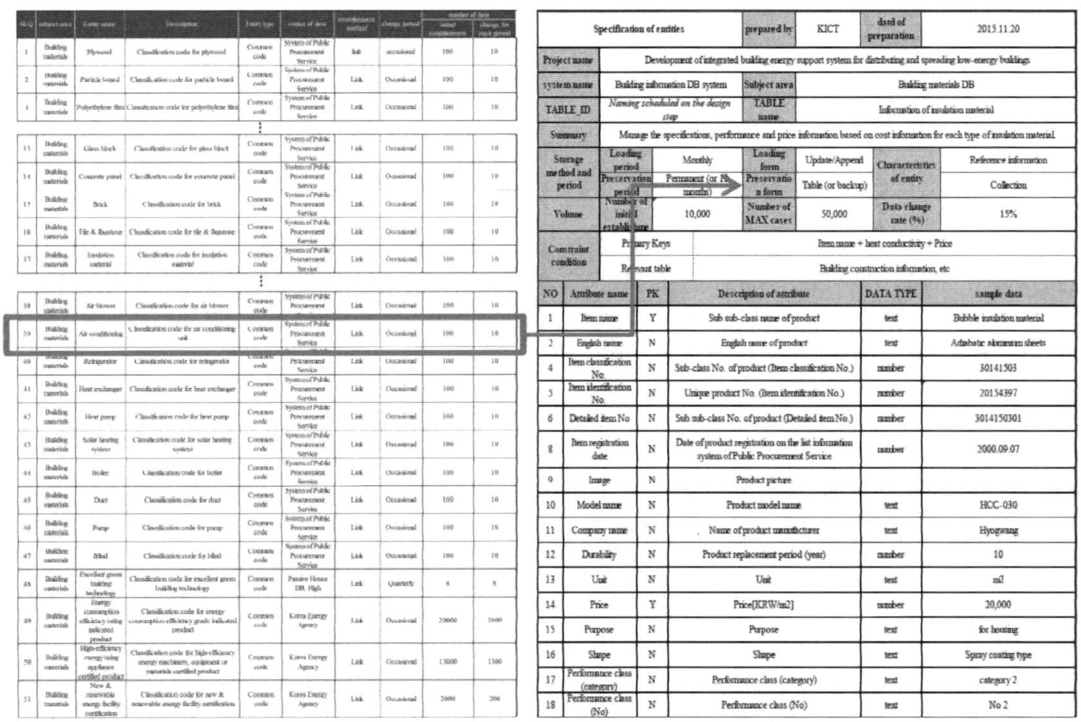

| | | | | | | |
|---|---|---|---|---|---|---|
| Specification of entities | | prepared by | KICT | date of preparation | 2015.11.20 | |
| Project name | Development of integrated building energy support system for distributing and spreading low-energy buildings | | | | | |
| System name | Building information DB system | Subject area | Building materials DB | | | |
| TABLE ID | Naming scheduled on the design step | TABLE name | Information of insulation material | | | |
| Summary | Manage the specifications, performance and price information based on cost information for each type of insulation material | | | | | |
| Storage method and period | Loading period | Monthly | Loading form | Update/Append | Characteristics of entity | Reference information |
| | Preservation period | Permanent (or Pk months) | Preservation form | Table (or backup) | | Collection |
| Volume | Number of initial establishment | 10,000 | Number of MAX cases | 50,000 | Data change rate (%) | 15% |
| Constraint condition | Primary Keys | Item name + heat conductivity + Price | | | | |
| | Relevant table | Building construction information, etc | | | | |

| NO | Attribute name | PK | Description of attribute | DATA TYPE | sample data |
|---|---|---|---|---|---|
| 1 | Item name | Y | Sub sub-class name of product | text | Bubble insulation material |
| 2 | English name | N | English name of product | text | Adiabatic aluminum sheets |
| 4 | Item classification No. | N | Sub-class No. of product (Item classification No.) | number | 30141503 |
| 5 | Item identification No. | N | Unique product No. (Item identification No.) | number | 20154397 |
| 6 | Detailed item No | N | Sub sub-class No. of product (Detailed item No.) | number | 3014150301 |
| 8 | Item registration date | N | Date of product registration on the list information system of Public Procurement Service | number | 2000.09.07 |
| 9 | Image | N | Product picture | | |
| 10 | Model name | N | Product model name | text | HCC-030 |
| 11 | Company name | N | Name of product manufacturer | text | Hyogwang |
| 12 | Durability | N | Product replacement period (year) | number | 10 |
| 13 | Unit | N | Unit | text | m2 |
| 14 | Price | Y | Price[KRW/m2] | number | 20,000 |
| 15 | Purpose | N | Purpose | text | for housing |
| 16 | Shape | N | Shape | text | Spray coating type |
| 17 | Performance class (category) | N | Performance class (category) | text | category 2 |
| 18 | Performance class (No) | N | Performance class (No) | text | No 2 |

**Figure 1.** List and specification of entities.

## 3. Price prediction models

The users of the database usually focus on the performance and the price because they want maximize the performance of the materials or equipment within affordable budget range. Therefore the database needs to provide price prediction model by the performance for the materials or equipment. Data for chillers such as absorption chiller, screw chiller and so forth are accumulated with specification of entities for a chiller. Figure 1 shows the cost variation of chillers according to refrigerating capacity. As the R2 value of the prediction model is less than 0.7, chillers were separated and analysed by the types. Figure 2 and 3 show the variation for absorption chillers and screw chillers respectively. Absorption chillers could be predicted by the model, however, screw chiller model should be separated by detail properties as shown in figure 3. Table 2 shows the equation of prediction models.

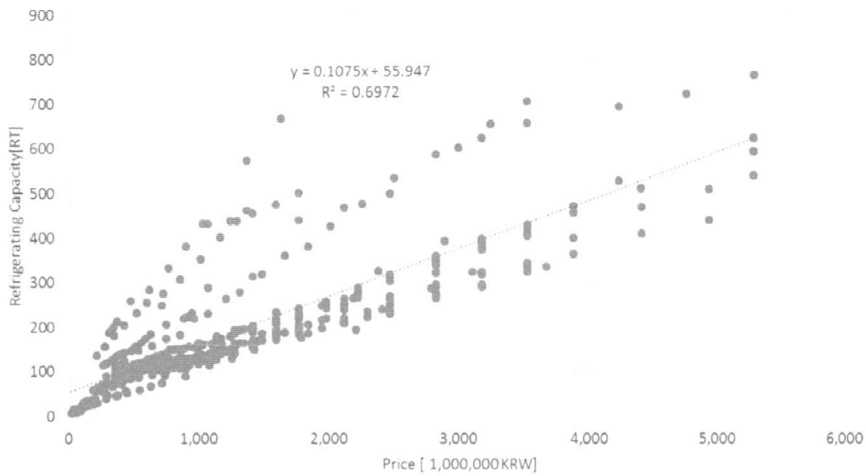


**Figure 2.** Relationship between price and performance of chillers.

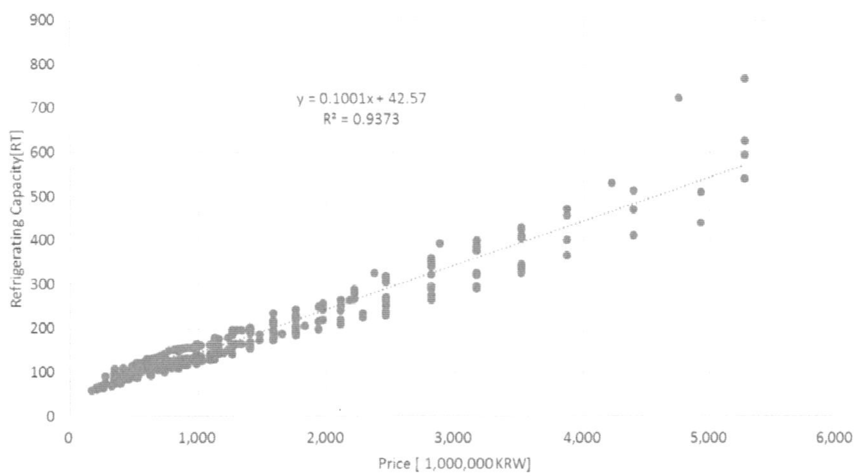


**Figure 3.** Relationship between price and performance of absorption chillers.

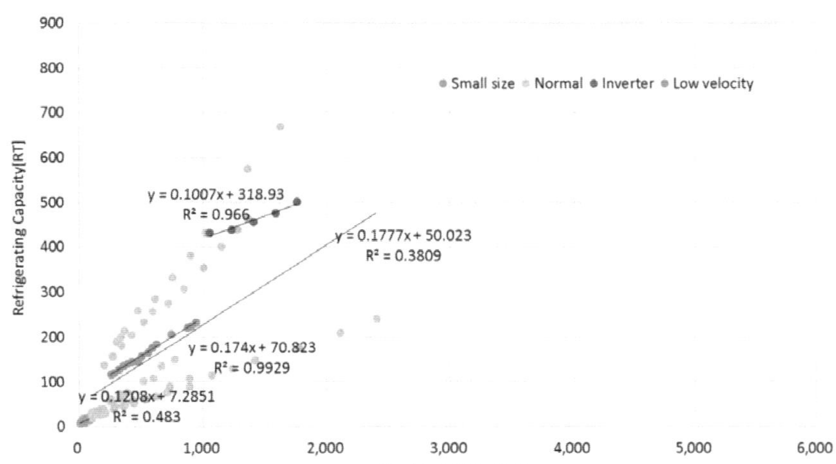


**Figure 4.** Relationship between price and performance of screw chillers.

IOP Conf. Series: Materials Science and Engineering **317** (2018) 012068 doi:10.1088/1757-899X/317/1/012068

**Table 2.** Prediction model for chillers.

| Item | Equation | R2 |
|---|---|---|
| Chiller | Y=0.1075x +55.947 | 0.6972 |
| Absorption chiller | Y=0.1001x+42.57 | 0.9373 |
| Small size screw chiller | Y=0.1208x+7.2851 | 0.4830 |
| Normal screw chiller | Y=0.1777x+50.023 | 0.3809 |
| Inverter screw chiller | Y=0.1007x+318.93 | 0.9660 |
| Low velocity screw chiller | Y=0.1740x+70.823 | 0.9929 |

## 4. Conclusions

The purpose of this study is to suggest the methodology for establishment of reliable low energy building materials and equipment and expansion of usability of the database by providing processed information from the gathered data.

The framework of the database was defined referred with Korea On-line E-procurement system for maintenance of the system and specification of entities was defined by the properties of the items. More than 45,000 data were gathered by the specification of entities and with the gathered data, price prediction models for chillers were suggested. To improve the usability of the prediction model, detailed properties should be analysed for each item.

## References


[1] Lee J 2016 *A Study on the Expansion of Database for Establishing Low Energy Building Materials Information Portal* CIB World Building Congress 2016.

[2] Ma M 2008 *Survey on the information provided by price comparison web sites* Policy Research Report of Korea Consumer Agency.

[3] Public Procurement Service (http://www.pps.go.kr/kor/index.do)

[4] Public Procurement Service 2015 *Press release of public procurement innovation plan for promoting economic revitalization* (available online http://www.pps.go.kr/bbs/selectBoard.do?boardId=PPS093&boardSeqNo=1638).

[5] Seo J 2009 *J. Korean Assoc. Reg. Inform. Soc.* **12** 105-26.


## Acknowledgments

This research was supported by grant (17AUDP-B079104-04) from Architecture & Urban Development Research Program funded by Ministry of Land, Infrastructure and Transport of Korean government.

IOP Conf. Series: Materials Science and Engineering **317** (2018) 012069 doi:10.1088/1757-899X/317/1/012069

# Numerical Model of High Strength Concrete

**R Z Wang[1], C Y Wang[2], Y L Lin[2]**

[1]National Center for Research on Earthquake Engineering, Taipei 10668, Taiwan.
[2]Department of Civil Engineering, National Central University, Taoyuan 32001, Taiwan.

**Abstract.** The purpose of this paper is to present a three-dimensional constitutive model based on the concept of equivalent uniaxial strain. closed Menetrey-Willam (CMW) failure surfaces which combined with Menetrey-Willam meridian and the cap model are introduced in this paper. Saenz stress-strain model is applied and adjusted by the ultimate strength parameters from CMW failure surface to reflect the latest stress or strain condition. The high strength concrete (HSC) under tri-axial non-proportional loading is considered and the model in this paper performed a good prediction.

## 1. Introduction

Darwin and Pecknold [1] proposed a 2D constitutive model to simulate general biaxial loading in 1977, the model supposed to work with FEM, so the model takes an incremental form rather than total form, and assumed concrete as an incrementally linear elastic material, besides Darwin and Pecknold first developed the concept of equivalent uniaxial strain, the purpose of equivalent uniaxial strain is to eliminate the Poisson's effect and simplify the procedure to attain true parameters. Elwi and Murray [2] extended this model to the 3D domain in 1979.

## 2. 3D Incremental constitutive model

Please follow these instructions as carefully as possible so all articles within a conference have the same style to the title page. This paragraph follows a section title so it should not be indented.

It is commonly accepted that concrete is treated as an initially isotropic material with anisotropic induced by stress changed. If the shear stress in neglected and only the principle stress are considered, the incremental constitutive relationship can be defined as follows:

$$d\sigma = D_L \, d\varepsilon \tag{1}$$

$$\begin{Bmatrix} d\sigma_1 \\ d\sigma_2 \\ d\sigma_3 \end{Bmatrix} = \frac{1}{\Omega} \begin{bmatrix} E_1(1-\nu_{23}\nu_{32}) & E_1(\nu_{21}-\nu_{23}\nu_{31}) & E_1(\nu_{31}-\nu_{32}\nu_{21}) \\ E_2(\nu_{12}-\nu_{13}\nu_{32}) & E_2(1-\nu_{13}\nu_{31}) & E_2(\nu_{32}-\nu_{31}\nu_{12}) \\ E_3(\nu_{13}-\nu_{12}\nu_{23}) & E_3(\nu_{23}-\nu_{21}\nu_{13}) & E_3(1-\nu_{12}\nu_{21}) \end{bmatrix} \begin{Bmatrix} d\varepsilon_1 \\ d\varepsilon_2 \\ d\varepsilon_3 \end{Bmatrix} \tag{2}$$

where $E_i$ represents the elastic modulus; $\nu_{ij}$ represent the Poisson's effect in $i$ direction caused by the stress in $j$ direction and $\Omega = 1-\nu_{12}\nu_{21}-\nu_{13}\nu_{31}-\nu_{23}\nu_{32}-\nu_{12}\nu_{23}\nu_{31}-\nu_{21}\nu_{32}\nu_{13}$. In order to derive the relationship between strain and equivalent uniaxial strain, the uniaxial constitutive relationship can be described as follows:

$$d\sigma_i = E_i d\varepsilon_i \tag{3}$$

By comparing (1) and (3) we can derive the relationship between principle strain and equivalent uniaxial strain as following form:

$$d\sigma_i = E_i d\varepsilon_{iu} = D_L d\varepsilon_i \tag{4}$$

$$\begin{Bmatrix} d\varepsilon_{1u}^{t_{n+1}} \\ d\varepsilon_{2u}^{t_{n+1}} \\ d\varepsilon_{3u}^{t_{n+1}} \end{Bmatrix} = \frac{1}{\Omega}\begin{bmatrix} (1-v_{23}v_{32}) & (v_{21}-v_{23}v_{31}) & (v_{31}-v_{32}v_{21}) \\ (v_{12}-v_{13}v_{32}) & (1-v_{13}v_{31}) & (v_{32}-v_{31}v_{12}) \\ (v_{13}-v_{12}v_{23}) & (v_{23}-v_{21}v_{13}) & (1-v_{12}v_{21}) \end{bmatrix}\begin{Bmatrix} d\varepsilon_{1}^{t_{n+1}} \\ d\varepsilon_{2}^{t_{n+1}} \\ d\varepsilon_{3}^{t_{n+1}} \end{Bmatrix} \tag{5}$$

## 3. Uniaxial stress-strain relationship

In this model, Saenz [3] uniaxial stress-strain relationship model (see figure 1) which can describe both ascending and post-peak softening response has adopted. The formulation can be written as follows:

$$\sigma_i = \sigma_{ic}\frac{K_i\left(\frac{\varepsilon_{iu}}{\varepsilon_{ic}}\right)}{1+A_i\left(\frac{\varepsilon_{iu}}{\varepsilon_{ic}}\right)+B_i\left(\frac{\varepsilon_{iu}}{\varepsilon_{ic}}\right)^2+C_i\left(\frac{\varepsilon_{iu}}{\varepsilon_{ic}}\right)^3} \tag{6}$$

$$\sigma_{if} = 0.85\sigma_{ic},\ \varepsilon_{if} = 1.41\varepsilon_{ic} \tag{7}$$

where $\sigma_{ic}$ represents ultimate stress and $\varepsilon_{ic}$ is corresponding ultimate strain; $\varepsilon_{iu}$ represents equivalent uniaxial strain; $\sigma_{if}$ and $\varepsilon_{if}$ represents softening stress and softening strain respectively.

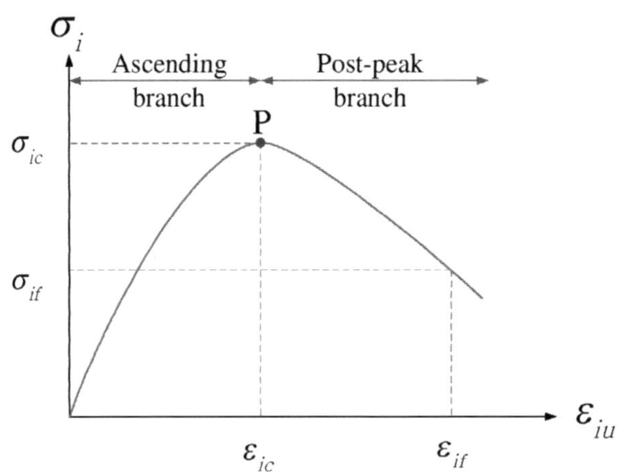


**Figure 1.** Saenz uniaxial stress-strain relationship model.

Equation (6) is to be used to derive the incremental elastic modulus by taking $E_i = \frac{d\sigma_i}{d\varepsilon_{iu}}$ :

$$E_i = \frac{E_0}{1+A_i\left(\frac{\varepsilon_{iu}}{\varepsilon_{ic}}\right)+B_i\left(\frac{\varepsilon_{iu}}{\varepsilon_{ic}}\right)^2+C_i\left(\frac{\varepsilon_{iu}}{\varepsilon_{ic}}\right)^3} \tag{8}$$

Because of the dearth of data on Poisson's ratio under triaxial stress conditions. Almost all of the empirical equation about the variation of Poisson's ratio are derived from the uniaxial compression tests. Kupfer's [4] equation is employed in this model and it represents in terms of strain ratio:

$$\nu = \nu_0 \left[ 1.0 + 1.3769\left(\frac{\varepsilon_{iu}}{\varepsilon_{ic}}\right) - 5.36\left(\frac{\varepsilon_{iu}}{\varepsilon_{ic}}\right)^2 + 8.586\left(\frac{\varepsilon_{iu}}{\varepsilon_{ic}}\right)^3 \right] \quad (9)$$

where $\nu_0$ is initial Poisson's ratio; $\varepsilon_{iu}$ represents equivalent uniaxial strain; $\varepsilon_{ic}$ represents the ultimate strain.

## 4. Analytical results

The basic material properties of high strength concrete in this experiment is defined below table 1, and both of tensile strength and tensile strain are an approximately tenth of compression condition.

**Table 1.** Material properties of HSC.

| Initial elastic modulus ($E_0$) | Initial Poisson's ratio ($\nu_0$) | Compression strength ($f_c'$) | Compression strain ($\varepsilon_c$) |
|---|---|---|---|
| 40000 (MPa) | 0.2 | 68 (MPa) | 0.002514 |

Two loading paths is considered in this paper, stress ratio is equal to 1/2 for path T-1 or 1/6 for path T-2 in the confining stage. The experimental results for different loading paths are listed in the following table (See figure 2 to figure 5):

**Table 2.** Triaxial strength of HSC under different loading path.

| Loading path | Confining pressure $\sigma_3$ (MPa) | Axial strength $\sigma_1$ (MPa) |
|---|---|---|
| Uniaxial | 0 | 67 |
| T-1 | 3.5 | 84.9 |
| | 7 | 99 |
| | 14 | 130.7 |
| T-2 | 14 | 132.7 |
| | 21 | 157.1 |
| | 28 | 179.9 |

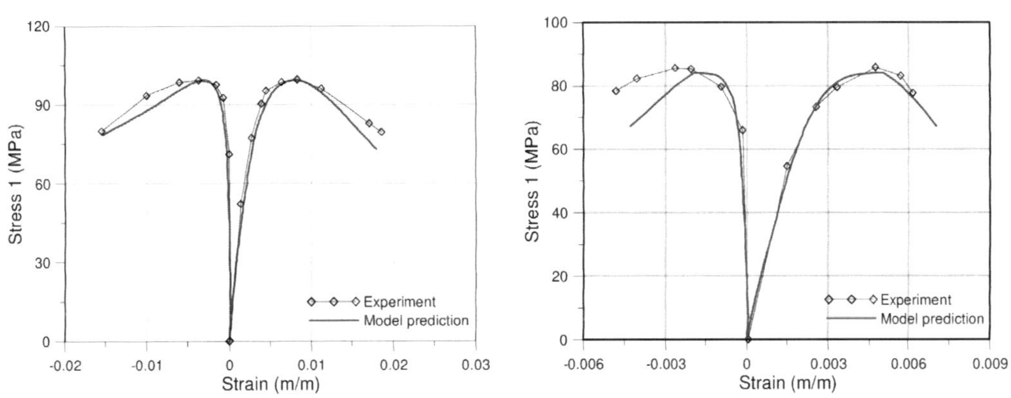


**Figure 2.** T-1 HSC Experiment Data vs. Analysis Confining = 0MPa (Left); 3.5MPa (Right).

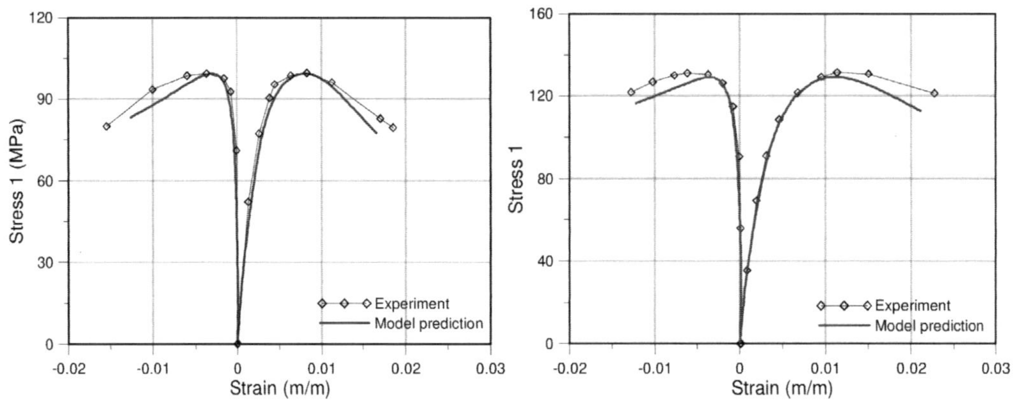


**Figure 3.** T-1 HSC Experiment Data vs. Analysis Confining = 7MPa (Left); 14MPa (Right).

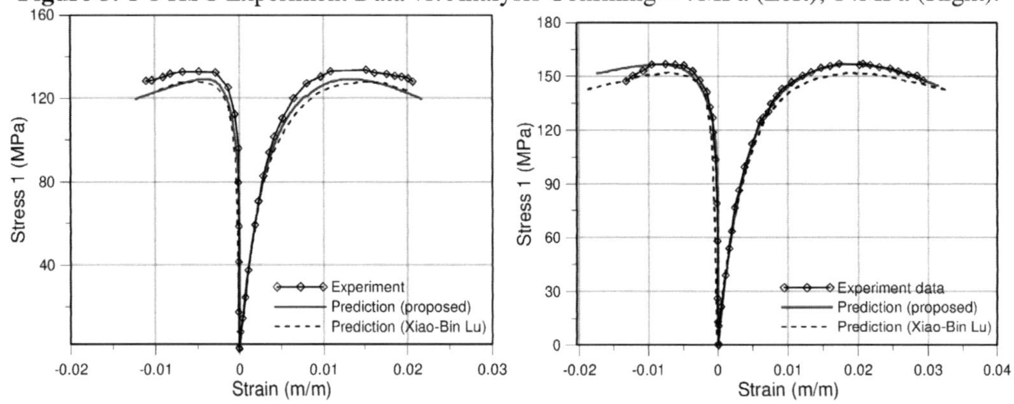


**Figure 4.** T-2 HSC Experiment Data vs. Analysis Confining = 14MPa (Left); 21MPa (Right).

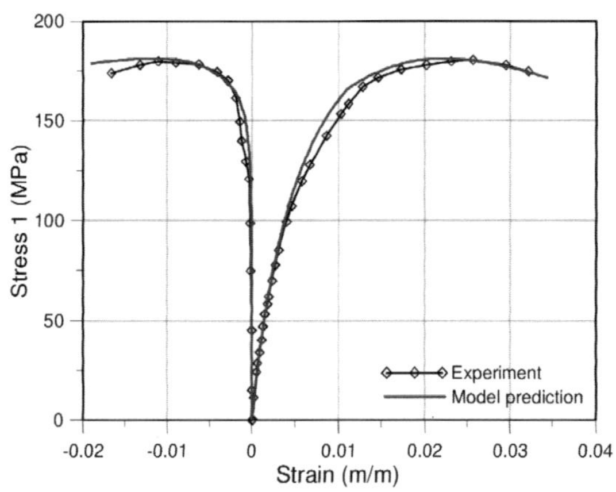


**Figure 5.** T-2 HSC Experiment Data vs. Analysis Confining=28MPa.

## 5. Conclusion

The three-dimensional equivalent uniaxial strain model of high strength concrete for non-linear finite element analysis of concrete structures under non-proportional load is presented in this paper. Two failure surfaces in principal stress space and equivalent uniaxial strain space are considered. It is postulated that the failure surface in equivalent uniaxial strain space has the same shape as principal stress space. The constitutive law is an orthotropic model. All the model modifications are based on loading path (Stress or Strain), not only the kinematic cap model but the softening failure surface. In another way, we believed the shape of material failure surfaces are influenced by the loading applied on the material.

## References

[1] Darwin D, Pecknold D A 1976 *J. Struct. Div.* **102**(2) 355-69.
[2] Elwi A A. Murray D W 1979 *J. Eng. Mech. Div.* **105**(4) 623-41.
[3] Saenz I P 1964 *ACI J.* **61(9)** 1229–35.
[4] Kupfer H B, Hilsdorf H K, Ru¨sch H 1969 *ACI J.* **66(8)** 656–66.

# Electrochemical Performance of Ni-MOFs for Supercapacitors

**Yujuan Li[1], Lili Song[1], Yinghui Han[1], Guangyou Wang[1]**

[1]School of Mathematics & Physics, North China Electric Power University; Baoding, China

**Abstract.** In this work, the Ni-MOFs of electrode material has been synthesized, characterized and studied for the electrochemical properties of electrode materials. The effects of the doping amount of Ni, calcination temperature and time were studied in detail. The results suggested that the electrochemical properties were obviously improved by the Ni-MOFs of electrode material and the best preparation conditions can also improve the electrochemical properties of electrode materials. These results open a way for the design of tailored MOFs as electrode materials for supercapacitors.

## 1. Introduction

Supercapacitors are considered as one efficient renewable energy storage device because of their high power density, rapid charging-discharging rate and long cycle life [1-3]. Thus, research efforts on supercapacitors are focused on achieving higher energy densities, and the development of new materials is of paramount importance to achieve this goal [4-6].

Currently, metal organic frameworks (MOFs) or porous coordination polymers, as a new type of electrode material, have attracted more and more attention. MOFs are new porous materials with high surface area, tunable pore size and incorporated redox metal centres, which are emerging as novel supercapacitor electrodes [7]. Díaz R [6] et al. pointed out that Co8-MOF-5 show electrical conductivities which are good enough to result in electrodes with nearly ideal supercapacitive behavior at rates up to 200 mV $s^{-1}$ in an electrolyte composed of 0.1 M tetrabutylammonium hexafluorophosphate ($TBAPF_6$) in acetonitrile. Yuan Xia [8] et al. reported the energy storage performances of the as-prepared double-layer nanospheres were further investigated in lithium ion battery and supercapacitor systems. Zhang et al. [9] obtained a hierarchical $Co_3O_4$ material by heating a Co-based MOF in air which exhibited an initial specific capacitance of 208 F $g^{-1}$ at 1 A $g^{-1}$. However, to the best of our knowledge, metal organic frameworks are rare in supercapacitors [10-12].

In the present work, it provided a method for the preparation of the Ni-MOFs of electrode material. The influence of the molar ratio of Ni-MOFs on the electrochemical properties was also investigated in this work. The preparation conditions had great influence on the electrochemical activity, such as the hydrothermal temperature, hydrothermal time and so on. In addition, it also studied the effect of reactive power on electrochemical activity.

## 2. Experimental section

### *2.1. Material*

All chemicals reagents were analytical grade and were used without further purification. Nickel(II) nitrate hexahydrate [$Ni(NO_3)_2 \cdot 6H_2O$] was provided by the Fuchen chemical reagent factory, in Tianjin, China. P-phthalic acid ($H_2BDC$) was supplied by the Guangfu institute of fine chemical engineering, in Tianjin, China. N, N-dimethylformamide (DMF), ethanol ($C_2H_5OH$) and other reagents were purchased from the Huaxin Reagents Co., in Tianjin, China. Ni-MOFs samples were prepared using hydrothermal method.

*2.2. Methods*

The material was prepared by the hydrothermal process. $Ni(NO_3)_2 \cdot 6H_2O$ (3.8g, 0.01mol) and p-phthalic acid (0.32g, 0.05mol) were dissolved in 60 ml solvent of $C_2H_5OH$, DMF and deionized water at room temperature with a given solvent volume ratio of $C_2H_5OH$ : DMF : deionized water = 1:1:1. The mixture was stirred for three hours on a magnetic stirrer. The mixture was transferred into a 100 mL Teflon-lined autoclave, which was sealed and maintained at a certain temperature and time.

The electrochemical measurements using a three-electrode electrochemical system, a 6 M/L KOH aqueous solutions as the electrolyte, Pt plate acted as the counter electrode and Ag/AgCl electrode served as the reference electrode. The working electrode consisted of active material, acetylene black and polytetrafluoroethylene (PTFE) with the mass ration of 8:1:1. Cyclic voltammetry and galvanostatic charge-discharge investigation were implemented using a CHI760E electrochemical workstation (ChenHua, Shanghai). The specific capacitance was calculated from the galvanostatic charge-discharge curves using the following equation [13-15]:

$$C = \frac{I * \Delta t}{m * \Delta V}$$

where I is charge-discharge current at a discharge time Δt (s), ΔV is dropout voltage, and m is the mass of active electrode materials.

## 3. Result and discussion

*3.1. The effect of the hydrothermal time on the electrochemical property*

The increase of hydrothermal time will increase the crystallization degree of the sample, and the grain will increase gradually. A series of experiments were carried out to reveal the relationship between the hydrothermal time and the electrochemical properties, the result was shown in figure 1. The representative cyclic voltammetry (CV) curves are shown in figure 1(a) with the hydrothermal time.

When the hydrothermal temperature was changeless, the electrochemical properties of the prepared samples increased first and then decreased with the increasing of the hydrothermal time. The specific capacitance values at different hydrothermal times are shown in figure 1(b). When the time was 48h, the electrochemical properties of the sample was the best. This is because the hydrothermal time shorter would lead the crystal particles surface of the sample covered by some residual organic matter and reduce participation reaction active sites; while the calcination time is too long, the porous structure and specific surface area of the sample would be destroyed. So in terms of the researched electrode materials, the optimal amount of hydrothermal time was 48h.

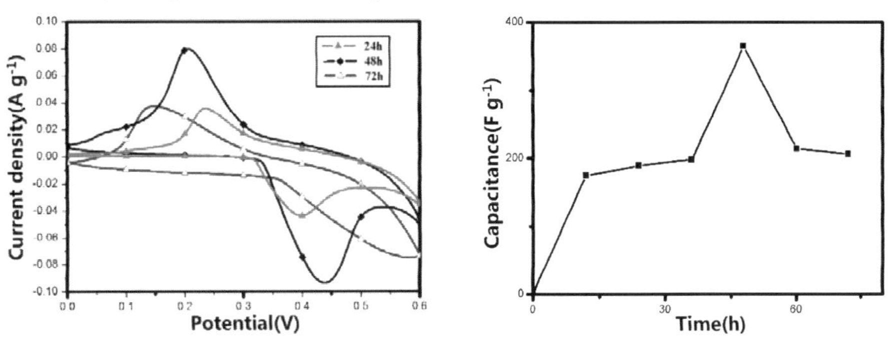


(a) (b)

**Figure 1.** (a) The CV curves at different hydrothermal times;
(b) The specific capacitance values at different hydrothermal times.

### *3.2. The effect of the hydrothermal temperature on the electrochemical property*

The hydrothermal temperature of material not only affects the final particle size, but also influences the crystal structure and crystal structure of the material. The effect of different hydrothermal temperature on the activity of the prepared catalyst was shown in figure 2. The materials were calcined for 48 hours. A series of experiments were carried out to reveal the relationship between the hydrothermal temperature and the electrochemical properties, which are shown in figure 2(a) and figure 2(b). The result was shown that the hydrothermal temperature would influence the degradation rate significantly, the electrochemical properties increased with the increase of hydrothermal temperature within the range of 50℃~200℃, however, when more than 100℃, the electrochemical properties will decline. This indicates that the electrochemical reaction rate of Ni-MOFs prepared at 100℃ is faster, the storage charge is more, and the specific electrical capacity is larger.

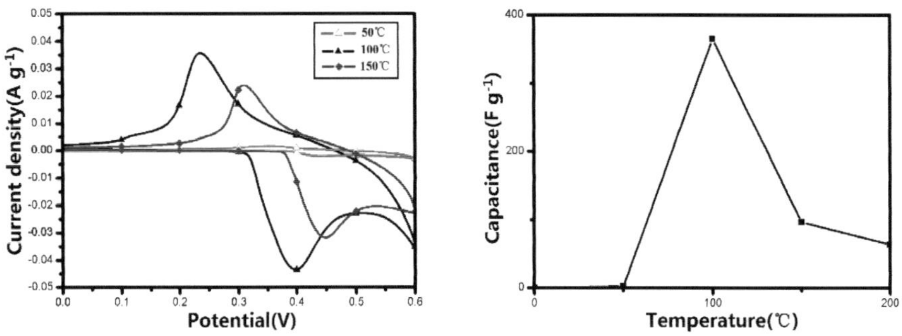


(a) (b)

**Figure 2.** (a) The CV curves at different hydrothermal temperatures;
(b) The specific capacitance values at different hydrothermal temperatures.

### *3.3. The effect of the scan rates and cycle life on the electrochemical property*

The plot of potential versus time at current densities is shown in figure 4. The electrochemical performance of the Ni-MOFs was evaluated in a three-electrode configuration with 6 M/L KOH electrolyte. The representative cyclic voltammetry (CV) curves are shown in Fig. 3(a) with the scan rates from 10 to 30 mV $s^{-1}$. So in terms of the researched electrode materials, the optimal amount of scan rates was 10mV $s^{-1}$. Fig. 3(b) is the variation of specific capacitance retention of Ni-MOFs electrodes versus cycle number in two electrode system. The capacitance retention of Ni-MOFs remains at 95.5% after 1000 charge-discharge cycles, showing excellent cycle stability [16].

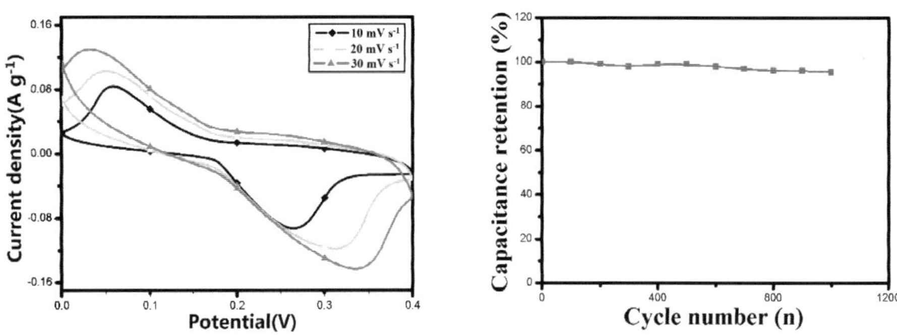


(a) (b)

**Figure 3.** (a)The cyclic voltammetry curves at different scan rates;
(b) Capacitance retention of Ni-MOFs electrode versus cycle number.

## 4. Conclusion

Ni-MOFs, a green and multifunctional material, promises to be a new technology for high-performance supercapacitors. Multiple factors affect the electrochemical properties of electrode materials. The best preparation conditions will lead to better electrochemical properties. Industrial feasibility analysis indicates that Ni-MOFs electrode material could be a promising alternative to conventional methods.

## References

[1] Wang Y, Xia Y, 2013 *Adv. Mater.* **25** 5536.
[2] Simon P, Gogotsi Y 2008 *Nat. Mater.* **7** 845.
[3] Yan J, Wang Q, Wei T, Fan Z 2014 *Adv. Energ. Mater.* **4** 157-64.
[4] Wang L, Chen D, Wang J, Liu G, Wu W, Liang G 2015 Rsc. Adv. **5** 99856-65.
[5] Venkateswara Rao Ch, Rambabu B 2010 Sol. State Ion. **181** 839-43.
[6] Diaz R, Orcajo M G, Botas J A, Calleja G, Palma J 2012 Mater. Lett. **68** 126-8.
[7] Xu J, Yang C, Xue Y, Wang C, Cao J, Chen Z 2016 *Electrochim. Acta.* **211** 595-602.
[8] Xia Y, Wang G, Zhang X, Wang B, Wang H 2016 *Electrochim. Acta.* **220** 643-53.
[9] Zhang F, Hao L, Zhang L, Zhang X 2011 *Int. J. Electrochem. Sci.* **6** 2943-54.
[10] Guo B, Yang Y, Hu Z, An Y, Zhang Q, Yang X, et al. 2017 *Electrochim. Acta.* **223** 74-84.
[11] Ke F S, Wu Y S, Deng H 2015 *J. Solid State Chem.* **223** 109-21.
[12] Ma L, Liu R, Liu L, Wang F, Niu H, Huang Y 2016 *J. Pow. Sour.* **335** 76-83.
[13] Yu L, Zhang G, Yuan C, Lou X W 2013 *Chem. Commun.* **49** 137-9.
[14] Zhang J, Jiang J, Li H, Zhao X S 2011 *Energ. Environ. Sci.* **4** 4009-15.
[15] Peng H, Ma G, Sun K, Mu J, Luo M, Lei Z 2014 *Electrochim. Acta.* **147** 54-61.
[16] Yu M, Zhang L, He X, Yu H, Han J, Wu M 2016 *Mater. Lett.* **172** 81-4.

## Acknowledgements

This research was supported by the Fundamental Research Funds for the Central Universities of China (NO.2015ZZD3), the National Natural Science Foundation of China (No. 51308212) and the Fundamental Research Funds for the Central Universities of China (No.13ZD18).

# Development and Application of On-line Monitor for the ZLW-1 Axis Cracks

**Yang Shi-jun[1], Yang Qian-hui[2], Jin Jian-guo[1]**

[1]School of Energy and Power Engineering, Northeast Electric Power University, Jilin 132012, China
[2]State grid Hangzhou Fuyang Power Supply Company, Hangzhou 311400, China

**Abstract.** This article mainly introduces a method that uses acoustic emission techniques to achieve on-line monitor for the shaft cracks and crack growth. According to this method, axis crack monitor is produced by acoustic emission techniques. This instrument can apply to all the pressure vessels, pipelines and rotor machines that can bear buckling load. It has the online real-time monitoring, automatic recording, printing, sound and light alarm, collecting crack information function. After a series of tests in both laboratory and field, it shows that this instrument is very versatile and possesses broad prospects of development and application.

## 1. Introduction

For the high-speed rotating machinery, especially some large prime mover like steam turbine, it will cause serious accident if the shaft appears cracks. According to incomplete statistics, So far large steam turbine has happened hundreds of accidents in the world which caused by rotor cracks, and some caused huge losses. However, the instruments we use now cannot react crack extension, during the period that shaft appears cracks and cracks extension, the vibration signal nearly has no change, that is to say there is no shock character appearing.

According to the field experience, when the depth of shaft crack is greater than one-third of shaft diameter, the stiffness of the shaft will reduce, then vibration signal starts to change, we find it is too late if we depend on the change of the vibration signal.

Nowadays acoustic detection methods have great development, and acoustic emission testing is an important method of non-destructive testing [1]. Because acoustic emission testing is a dynamic and non-destructive testing methods, and acoustic emission signals comes from the crack itself, we can know the process that the crack appears expansion and monitor the security online all the time [2, 3].

The instrument this article introduces is based on the acoustic emission techniques [4]. Its main function is to monitor the crack initiation and extension of shaft and shaft parts online.

## 2. The principle and method of acoustic emission monitoring

The principle of using acoustic emission monitoring shaft and rotor blades crack is that when shaft or rotor blade cracks or crack propagation extends, it would produce elastic waves, and this wave spreads through the shaft, then sensor receives acoustic emission signals of cracks. The signal is passed to the processor by electromagnetic induction. In order to detect crack shaft correctly and eliminate the interference signal. We install a filter behind the preamplifier and adjust the band to crack band.

## 3. The composition of acoustic emission diagnostic instrument

This instrument includes three parts: acoustic emission transducers, preamplifier and signal processors.

Sensors are the eyes of the whole instrumentation, the sensitivity of monitoring instrument is determined by it. We choose double-ended resonant piezoelectric sensor by comparison. This instrument not only has strong anti-interference ability and it is easy to make, but the configuration of secondary instrument is also very convenient.

We choose the voltage amplifier as preamplifier, this is exactly matched with the selected piezoelectric transducer. In order to improve the sensitivity, we choose an operational amplifier which has high input impedance, low drift, low noise, its magnification is 40dB.

Signal processor is the heart of the whole monitoring instrument. We choose SCM of MCS-51 type to analyze signals and complete a variety of control functions. When signal is communicated to the instrument, it will be enlarged 60dB. Compared with threshold voltage, we will get ringing pulse, then starts pulse counter by Counter named 8031. The ringing pulses which are noted per second need to be arranged, and we judge whether it has cracks or not. If it works properly, it will only show the current ringing rate values of acoustic emission, if not, it will cause sound and light alarm and print to record the time and acoustic emission count value. Operating personnels can call the data stored in the data storage, then analyze the situation the rotor runs and the trend of crack extension.

The CPU this instrument uses is SCM belonging to the series of 8051, it uses modular structure on the design, that is to say, This process which completes instrumentation functions will be divided into several modules, then these modules need to be united organically. Because this instrument is applied to monitor online, it cannot calculate too many data's, all the procedures are written in assembly language to improve the speed and occupy little space.

According to the function of this instrument, the monitoring program of this instrument can be divided into the following modules:

The main program module. It is used to Initialize 8031 and 8255, and ensure the working status of each port. Otherwise, it also includes system self-test, setting up various working conditions and calling subroutines;

The data processing module; It is used to deal with the collected acoustic emission rate values to ensure the working status of measured objects. If it works in danger, it will turn to sound and light alarm and print to record the time; If it works well, it will turn to show the datas. Otherwise, it also has a function to choose and store a representative value.

Interrupt service routine. It includes timer interrupt subroutine and out interrupt subroutine. Timer interrupt subroutine is used to read the datas from the counter and reload timer initial value T0. We need to judge whether it gets 1s or not, if it gets, the flag bit is 77H, if not, it will be interrupted and returned The out interrupt subroutine is to judge which button is pressed, then carry out the function of the module the button represents.

Digital tube display subroutine. Its function is to make the datas which have been treated display on the LED digital tube. In order to facilitate operating personnels to read meter, we take the method of zero elimination, that is beginning to show from the first number that is not zero. If all the numbers are zero, it will only display a zero in the end.

Print subroutine. It includes three parts: error print subroutine, curve print subroutine and data sheet printing subroutine. They can help operating personnels master the trend of crack growth.

The main program is combined with these modules, they control the instrument together. The main program also need to complete initialization of the 8031 and 8255 chip, the status of timers and counters need to be set, we also need to set the time constant of the timer, interrupt mode and so on. Its structure is shown in figure 1.

IOP Conf. Series: Materials Science and Engineering **317** (2018) 012071 doi:10.1088/1757-899X/317/1/012071

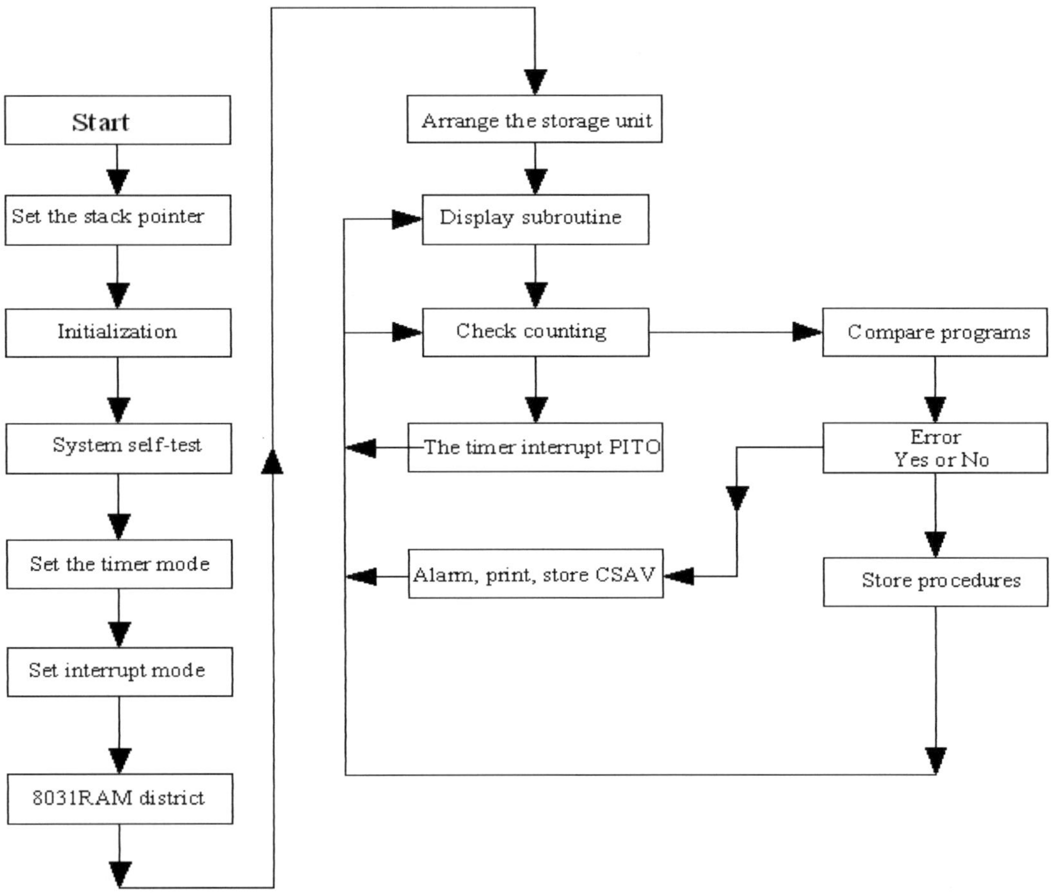


**Figure 1.** The main program block diagram.

## 4. The situation of practical application

### *4.1. The trial case in the laboratory*

The acoustic emission monitoring instrument of ZLW-1 shaft crack runs well in the laborary. In the trial operation, we do not find any trouble about the instrument. After that, we simulate the fracture experiments and test the sensitivity of the instrument.

In order to do the experiment, we take one specimen from the main steam pipe. Its material is 10CrMo910 steel. Firstly, we put it into a rectangular shape, in order to shorten the experimental time and get the experimental results as soon as possible, we make a crack in the middle. Secondly, we put the specimen on the hydraulic machine and load slowly. The sensor is coupled to the side of the specimen by magnet and turbine oil. At the beginning of the load, the specimen begins to change, but the instrument does not display obviously. When the crack of the specimen begans to expand (about 1mm), the instrument changes immediately, the digital tube shows 659(AE). We continue to load, the specimen cracks once again, it is about 2mm, the digital tube shows 643(AE), the third is 3mm and shows 1472(AE). Finally, the specimen fractures, and the digital tube shows 1881(AE).

### *4.2. The actual application in the scene*

The acoustic emission monitoring instrument of ZLW-1 shaft crack has been installed in 1# feed pump in Tong Liao power plant. The instrument runs well in the field. Though the pratical equipments do not have cracks, we can simulate axis cracks. When it has cracks, the instrument indicates a large value, then starts sound and light alarm and the printer prints the value at that moment. Through a series of analogs, it proves the instrument is applicable.

IOP Conf. Series: Materials Science and Engineering **317** (2018) 012071 doi:10.1088/1757-899X/317/1/012071

## 5. Conclusion

The shaft crack monitor instrument this article introduced can achieve to monitor rotor online all the time. In a series of laboratory tests, it detects the depth which is less than 1mm. However, the past diagnostic technique detects the cracks which has been extended 25%~40% depth of shaft diameter. Compared with the past, it has significant progress. These tests are basically close to the actual situation of the scene, so it can be directly transplanted to the turbine generator rotor and other non-rotating machinery.

## References

[1] Yang M W 2005 *Acoustic Emission Testing*. BeiJing. China Machine Press.
[2] Yang J F, Li L, Zhang B 2007 *NDT* **31(4)** 21-3.
[3] Zhang A P, Bao L 2003 *Chem. Eng. Mach.* **30(2)** 93-5.
[4] Yuan Z M, Ma Y K, He Z Y 1985 *Acoustic Emission Technique and Its Application*. China Machine Press.

## Acknowledgements

This work was supported by Department of Jilin Province Natural Science Foundation of China (20140101064JC).

# Study on the Tribological Properties of MC Nylon Composites Filled with Hydraulic Oil

**S Yuan[1,2], Y Li[1,2], J Wen[1,2], L Yin[1,2], Q Zhang[1,2]**

[1]Jiangsu Xuzhou Construction Machinery Research Institute,Xuzhou China
[2]State Key Laboratory of Intelligent Manufacturing of Advanced Construction Machinery, Xuzhou Construction Machinery Group, Xuzhou China

**Abstract.** Mechanical parts utilized in machinery, such as nylon slider and pulley, should have certain mechanical properties and good tribological properties, so that equipments' stability and smoothness can be assured. A kind of MC nylon (monomer cast nylon) composites filled with hydraulic oil was studied in this paper. The addition of hydraulic oil changed nylon's mechanical properties and tribological properties significantly, and improved the material's toughness and coefficient of friction. The composites have excellent strength, toughness and relatively low coefficient of friction when the content of the hydraulic oil is 4wt%.

## 1. Introduction

In the field of materials science, polymer materials, such as polyethylene (PE), polypropylene (PP) and nylon, have been used and researched commonly due to their good mechanical and processing characteristics [1-4]. Among the various polymers, nylon 6 (poly-amide) synthesized by anionic polymerization of caprolactam (CL) has been extensively investigated because of its high reaction rate, excellent strength, stiffness and wear-resistant properties.

At present, some parts on lots of machinery and equipments are made of nylon and other polymer materials. For example, the nylon slider, which is used to prevent the friction between telescopic booms, can improve the stability, and meanwhile support the pressure [5, 6]. Above all, nylon slider must have great mechanical properties, excellent friction and wear-resistant properties. On the commercial aspect, monomer casting nylons (MC nylon) can be directly produced from molten monomers (CL) in the mold, using sodium and diisocyanates as an activator and a chain initiator respectively [7]. Li Guolu et al. [3] studied the effect of the amounts of additives (catalysts, activators eg.) on mechanical properties, tribological properties of MC nylon. The article shows that MC nylon has excellent properties when the ratio of catalyst to activator is 0.002 to 0.003.

Pure MC nylon material has not met the requirements for machinery and equipment currently. Therefore MC nylon should be modified to improve its performance. Inorganic or organic fillers were added to improve the wear-resistant and self-lubrication properties of MC nylon, Zhang shihua et.al reported the trobological behavior of MC nylon composites filled with glass fiber and fly ash [8]. Nano calcium carbonate/graphite was used as reinforcing agents to modify MC nylon, which have improved the tribological properties of MC nylon significantly [9].

Hydraulic oil is an organic compound extracted from petroleum and contains alkanes, cyclic alkanes and other compounds. Nylon was a polymer of caprolactam. Both of them have good compatibility. In this article, a kind of hydraulic oil was added into MC nylon to improve the tribological properties of MC nylon. At the same time, the composite has great mechanical properties. This research can provide a reference for the study of the application of MC nylon.

## 2. Experiments

### *2.1. The reagents and the preparation process of MC nylon composites filled with hydraulic oil*

The reagents materials are shown in table 1. MC nylons filled with hydraulic oil were synthesized by anionic polymerization of molten CL with hydraulic oil in the mold.

**Table 1.** Reagents.

| Reagent | Name | Purity | Manufacturer |
|---|---|---|---|
| Monomer | Caprolactam C6H11NO (CL) | AR | Basf co., LTD |
| Catalyst | Sodium hydroxide NaOH | AR | Xuzhou kebao experiment instrument co., LTD |
| Activator | Toluene diisocyanate C9H6N2O2 (TDI) | AR | Chengdu aikeda chemical reagent co., LTD |
| Oil | Hydraulic oil | CP | Klueber Lubrication (Shanghai) Co., Ltd |

CL was dried for 8~12h in drying oven before experiment. It was heated in a synthesis device under nitrogen environment at 85℃. Then the hydraulic oil was added. With continuous heating, sodium metal, and tolylene diisocyanate (TDI) were added. The mixture was stirred at (135±5)℃ and then was casted into a mold immediately which where temperature was maintained at 170℃.

### *2.2. The scheme of hydraulic oil filled MCnylon*

The schemes of MC nylon filled with hydraulic oil is shown in table 2. The contents of hydraulic oil are from 2 wt%～8 wt%. The ratio of CL to NaOH to TDI is 1:0.003:0.002.

**Table 2.** The scheme of hydraulic oil filled MC nylon.

| Number | 1 | 2 | 3 | 4 |
|---|---|---|---|---|
| The contents of hydraulic oil | 0 | 2wt% | 4wt% | 8wt% |

### *2.3. Properties of the MC nylon*

- The morphology of MC nylon composites were tested by INSPECT S50 scanning electron microscopy (SEM) after the composites were covered with gold.
- The tensile strength, elongation, compression strength, bending strength and other mechanical properties were tested by ETM104C electronic universal testing machine. Five samples of each composite were tested.
- The friction tests were performed for 120min with a load of 200 N. The rotational speed was 200r/min. Three samples of each composite were tested.

## 3. Results and discussions

### *3.1. The morphology features of MC nylon filled with hydraulic oil*

After MC nylon composites were fractured brittlely in liquid nitrogen, the morphology of samples was observed through a scanning electron microscope (SEM). MC nylon composites filled with eight different contents of hydraulic oil are shown in Fig. 1. All of them have no obvious defects. The oil in composites looks like a regular globe, and distributed uniformly. With the increase of the content of oil from 2wt% to 8wt%, the sizes of oil were shown in table 3. The longest diameter of oil is 3.394μm. That's because a reunion phenomenon of the oil appeared in the matrix, which led to the growth of oil with the increase of the content of oil. Also, the oil distribute unevenly. As shown in figure 1, the minimum size of oil droplets is 3.071μm, which demonstrate a denser and more uniform distribution than others, more oil can't took bigger size of droplets for the MC nylon, and that would be difficult to disperse.

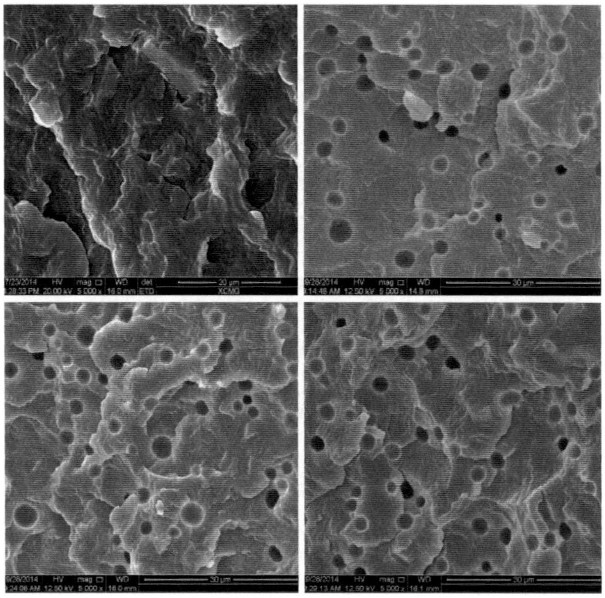

**Figure 1.** SEM morphology of MC nylon composites filled with different contents of hydraulic oil (a) 0; (b) 2wt%; (c) 4wt%; (d) 8wt%.

**Table 3.** The diameter of oil in MC nylon composites.

| **Number** | **1** | **2** | **3** | **4** |
|---|---|---|---|---|
| The contents of hydraulic oil | 0wt% | 2wt% | 4wt% | 8wt% |
| The diameter of oil | 0 | 3.226 | 3.394 | 3.071 |

### *3.2. Mechanical properties of MC nylon filled with hydraulic oil*

The mechanical properties of MC nylon filled with hydraulic oil are illustrated in figure 2. With the increase of hydraulic oil's content from 0wt% to 8 wt%, the composites' tensile strength, bending strength, and compression strength fluctuate downward drastically: specifically, from 84.0MPa to 49.5MPa, from 113MPa to 71MPa, from 132MPa to 39MPa, respectively. Similarly, the composites' elongation increase from 53.0% to 75.0%, but with a remarkable decline to 20.0% when the hydraulic oil's content reaches 4wt% at last.

All hydraulic oil particles deform simultaneously during the deformation of the plastic, which lead to an increase of the composites' deformation. Meanwhile, the formation of liquid holes cause internal stress concentration. Consequently, both the resistance of plastic deformation and the tensile strength of composites go down. The hydraulic oil dispersed in the matrix can transfer deformation energy well, which can trigger silver craze, terminate the crack and increase the toughness when materials are damaged. However, uneven dispersion appears along with the increase of the hydraulic oil, which leads to local stress concentration, enhancement of interface effect, and uneven stress deformation. As a result, the composites become unstable, and the toughness decrease, while the brittleness increase. As shown in figure 2, when the oil content is 4wt%, the comprehensive mechanical properties of MC nylon reduce significantly.

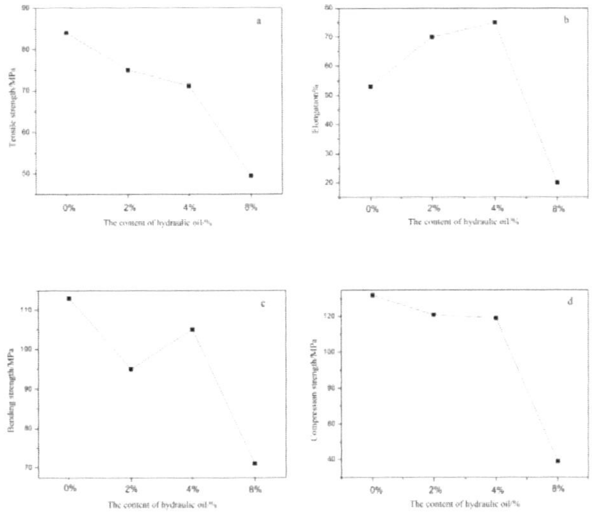


**Figure 2.** Mechanical properties of MC nylon composites filled with different contents of hydraulic oil (a) Tensile strength; (b) Elongation; (c) Bending strength; (d) Compression strength.

*3.3. Tribological properties of MC nylon filled with hydraulic oil*

The tribological properties of MC nylon filled with hydraulic oil are depicted in figure 3. The tribological properties of MC nylon improve significantly by the addition of hydraulic oil, i.e. the coefficient of friction decreases gradually with the increase of hydraulic oil content. The composites' coefficient of friction reduces to 0.16 when the hydraulic oil content is 8wt%, and the weight loss reduces to 2.4. The oil on the surface of composites overflows so that the matrix possess self-lubricating property while friction and wear are happening. As time goes, the amount of oil on the surface lessens. Then the oil inside overflows onto the surface gradually, which make it possible to self-lubricate with oil during the whole friction and wear period, and to reduce the coefficient of friction. Therefore, the lubrication enhances and the coefficients of friction of MC nylon composites filled with hydraulic oil are lower than that of pure MC nylon. At the beginning of friction, the weight losses of MC nylon composites are different due to the different contents of hydraulic oil there. After a period of time, nylon transfer film formed on the surface of the grinding ring. The weight loss of MC nylon was relatively low later. In general, the content of hydraulic oil has great influence on the wear weight loss.

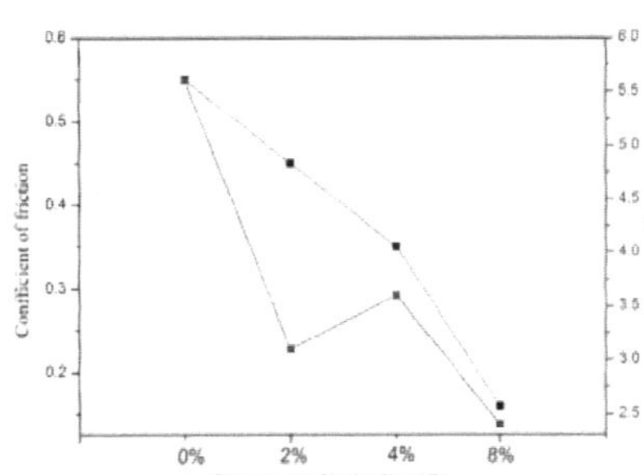


**Figure 3.** Tribological properties of MC nylon composites filled with different content of hydraulic oil.

## 5. Results

Within MC nylon filled with hydraulic oil, the oil distributed uniformly, and the size of the oil droplets increased with the raise of hydraulic oil's content.

The mechanical properties of MC nylon filled with hydraulic oil decreased obviously as a result of the addition of hydraulic oil. When the content of hydraulic oil was 4wt%, the mechanical properties of MC nylon reduced significantly.

The coefficient of friction of MC nylon filled with hydraulic oil reduced significantly attributed to the addition of hydraulic oil. The coefficient of friction of composites decreased gradually with the increase of the hydraulic oil.

MC nylon on heavy machinery must have excellent tribological properties as well as great mechanical properties (especially with good compression properties). To summarize, the MC nylon filled with hydraulic oil not only have great tensile strength and compression strength, but also has a lower coefficient of friction and wear weight loss when the content of oil is 4wt%

## References

[1] Al-Harthi M A 2014 *Polym. Compos.* **35** 2036-42.
[2] Azinfar B, Ahmad Ramazani S A, Jafariesfad N. 2014 *Polym. Compos.* **35** 37-44.
[3] Li G L, Wang K L, Cui Z P, Liu J J 1988 J. Tsinghua Univ. (Sci & Tech). **38(8)** 42-45.
[4] Mo Y M, Pan Y P, Wang X M, 2003 *J. Wuhan Univ. Tech. Mater. Sci. Ed.* **18(2)** 58-60.
[5] Qu J J, Luo Y X, Zhang Y Z, Cui L K, Gao W J, Yang C F 1994 *EPA*. **22(3)** 11-14.
[6] Liu Z W 1992 *EPA*. **20(4)** 13-15.
[7] Kang S C, Chung D W 2000 *Wear* **239** 244-250.
[8] Zhang S H H, Cui C H, Chen G 2011 *J. Wuhan Univ. Tech. Mater. Sci. Ed.* **27(2)** 290-5.
[9] Sun Y H, Tan S Z, Wang M G 2007 *EPA*. **35(9)** 56-9.

IOP Conf. Series: Materials Science and Engineering **317** (2018) 012073 doi:10.1088/1757-899X/317/1/012073

# Stress Corrosion Cracking Behavior of Hardening-Treated 13Cr Stainless Steel

Li-Bin Niu[1], Hisamitsu Ishitake[2], Sakae Izumi[3], Kunio Shiokawa[3], Mitsuo Yamashita[3], Yoshihiro Sakai[3]

[1]Shinshu University, Institute of Engineering, Academic Assembly, 4-17-1 Wakasato, Nagano 380-8553, Japan
[2]Shinshu University, Graduate student, Graduate School, Toshiba Corp., 1 Toshiba-Cho, Fuchu-Shi, Tokyo 183-0043, Japan
[3]Fuji Electric Co., Ltd. 1-1 Tanabeshinden, Kawasaki-Ku, Kawasaki 210-9530, Japan

**Abstract.** Stress corrosion cracking (SCC) behavior of the hardening-treated materials of 13Cr stainless steel was examined with SSRT tests and constant load tests. In the simulated geothermal water and even in the test water without addition of impurities, the hardening-treated materials showed a brittle intergranular fracture due to the sensitization, which was caused by the present hardening-treatments.

## 1. Introduction

13Cr martensitic stainless steel is widely used as a steam turbine blade material for thermal, nuclear and geothermal power plants due to its well mechanical properties and corrosion resistance. On the other hand, in the power plants, trace amounts of corrosive impurities such as chloride ions ($Cl^-$) and sulfate ions ($SO_4^{2-}$) present in the steam are easily carried over into the steam turbine with the steam, which tend to cause corrosion damages of the turbine materials [1-4]. Besides, erosion on the front edges of turbine blades in the wet region of the steam turbine due to the drain attack is also one of the problems [5, 6]. It is very important to improve corrosion resistance as well as erosion resistance of turbine materials.

As a method for suppressing erosion of steam turbine blades in power plants, hardening treatments such as quenching of the blade materials have been applied. However, the microstructure of the material changes due to the hardening treatments such as quenching, and the residual tensile stress occurs at the boundary portion of the heat treatment. For this reason, it is concerned that the stress corrosion cracking (SCC) occurs easily on the blade surfaces of the steam turbine during operation due to the corrosive impurities present in the steam, especially chloride ions ($Cl^-$). In addition to improving the erosion resistance of the blade material, reduction of the SCC susceptibility is also an important subject in prolonging the service life of the steam turbine. However, the influence of various hardening treatment conditions on the SCC susceptibility of blade materials has not yet been sufficiently clarified. In this work, stress corrosion cracking behavior of the materials of 13Cr stainless steel, which were hardening-treated under different conditions, was investigated in corrosive water environments.

IOP Conf. Series: Materials Science and Engineering **317** (2018) 012073 doi:10.1088/1757-899X/317/1/012073

## 2. Experimental

### *2.1. Materials and specimen*

13Cr martensitic stainless steel, which has been used as the steam turbine blade material of power plants, was used in this work. Its chemical composition is as follows (mass%): 0.21 C, 0.32 Si, 0.62 Mn, 13.36 Cr, 0.46 Ni, and the balance of Fe. The as-received steel plates were cut from the virgin steam turbine material. Two kinds of hardening treatment were performed in the present work, in which the steel plates were heated up to 1100°C followed immediately by water quenching (WQ) and air cooling (AC), respectively. After that, they were tempered at 250°C for 2h. The Vickers hardness were measured after the hardening treatments. The average hardness of the water quenched and the air cooled materials WQ and AC were HV522 (100gf) and HV535 (100gf), respectively, and both of them were higher than that of the as-received steel of HV307 (100gf).

The specimens, as shown in figure 1, were cut from the hardening-treated and also the as-received steel plates. The surfaces were abraded using 150-800 grit emery papers. Prior to testing, the specimens were degreased with acetone and rinsed with deionized pure water.

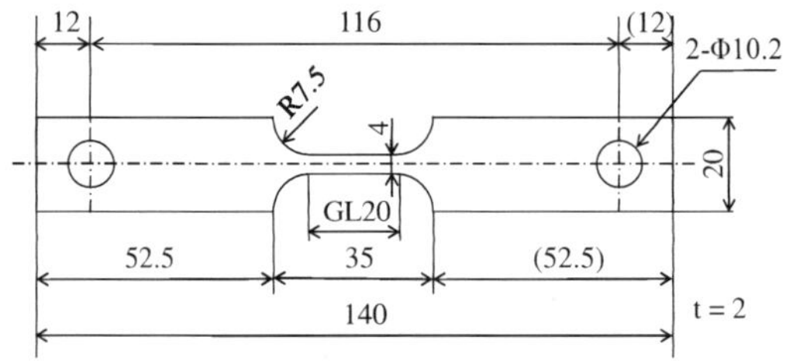


**Figure 1.** Specimen geometry (dimension in mm).

**Table 1.** Quality of the test waters.

| | $Cl^-$ (ppm) | $SO_4^{2-}$ (ppm) | $CO_2$ or $N_2$ | DO (ppm) | pH | Temp. (°C) |
|---|---|---|---|---|---|---|
| W1 | 10,000 | 50 | $CO_2$ bubbling | <1.0 | 3.9 | 90 |
| W2 | 0 | 0 | $N_2$ bubbling | <1.0 | 6.4 | 90 |

### *2.2. Test waters*

Considering the geothermal steam turbine in a severe corrosive environment, as a kind of test water the simulated geothermal water, named W1, was prepared from ion-exchanged pure water by adding the impurity ions of 10,000 ppm $Cl^-$ and 50 ppm $SO_4^{2-}$, and by continuously bubbling $CO_2$ gas into the water before and during test. For comparison another test water without the addition of the above impurities, named W2, was prepared from ion-exchanged pure water by continuously bubbling $N_2$ gas into the water before and during test to remove dissolved oxygen (DO). Table 1 shows the quality of the test waters. The temperature of the test waters was 90°C.

### *2.3. Experimental procedure*

Slow strain rate tensile (SSRT) tests were carried out on the specimens with a strain rate of $4.2 \times 10^{-7}$ $s^{-1}$ in the test waters. Constant load tests were also conducted up to 100h at a stress of 650MPa in the test waters. In SSRT tests and constant load tests, the changes in elongation of the specimen were recorded by a data logger.

After the above tests, the fracture surfaces of the ruptured specimens were observed with a scanning electron microscope (SEM). The gauge portion surfaces were also observed with SEM, after emery-polishing up to #1500 and etching with a solution consisted of 8.4 ml methanol, 1.5 ml hydrochloric acid and 0.1g picric acid.

## 3. Results and discussion

### *3.1. SCC behavior of the hardening-treated material*

The stress-strain curves of the hardening-treated and the as-received materials obtained from SSRT tests in the simulated geothermal water W1 are shown in figure 2. It can been seen that the three specimens exhibited different stress-strain behavior. For the as-received material, the specimen in the water W1 showed a total elongation of 6.45% with a 0.2% proof stress of 670MPa and a ultimate tensile strength of 798MPa. However, the two specimens of the hardening-treated materials showed a brittle deformation mode. For the hardening-treated material WQ with water quenching, even though the specimen showed the highest rupture stress of 946MPa, it ruptured with almost no plastic deformation. Furthermore, the specimen of the hardening-treated material AC with air cooling ruptured in the elastic deformation region with a lower rupture stress of 436MPa.

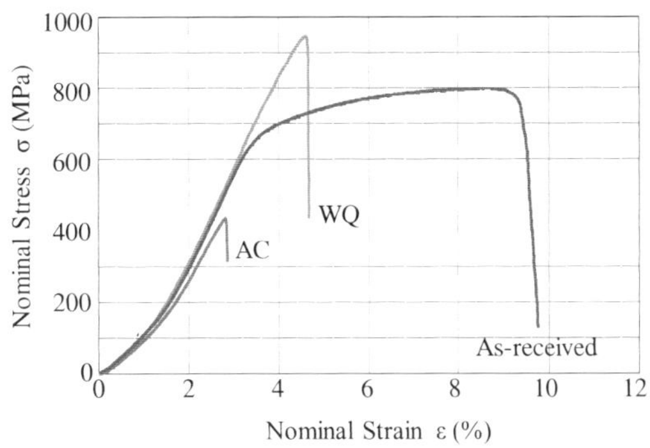


**Figure 2.** Stress-strain curves of the specimens obtained from SSRT tests in W1.

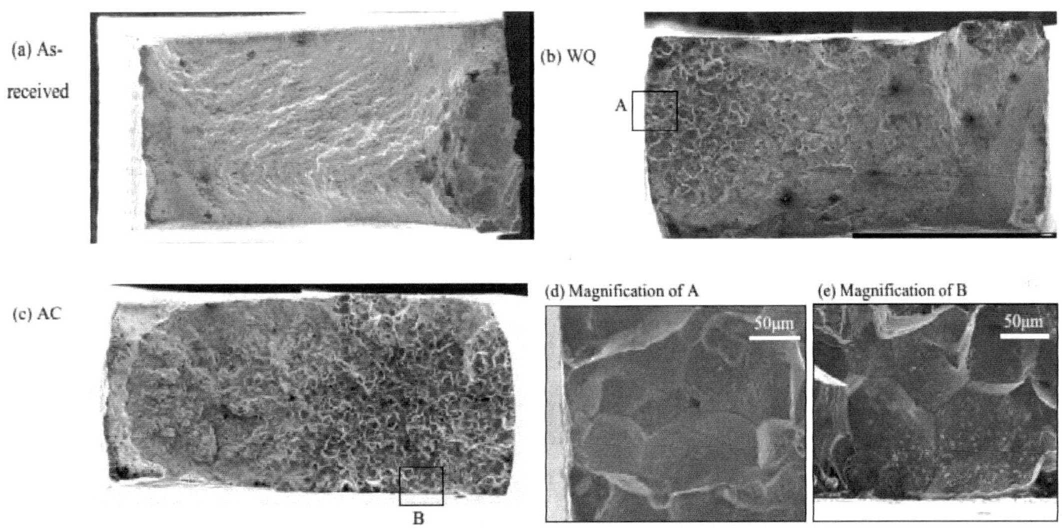


**Figure 3.** Fracture surfaces of the specimens ruptured in W1.

Figure 3 shows the overall fracture surfaces with partially magnified SEM images of the three specimens of the hardening-treated as well as the as-received materials ruptured in the simulated geothermal water W1. The fractography of the as-received material specimen presented a feature of ductile fracture mode. While the two specimens of the hardening-treated materials WQ and AC showed the mixed mode of ductile fracture with intergranular fracture regions. Particularly, it was observed that the intergranular fracture region on the AC specimen occupied more than half of the fracture surface. Therefore, it is considered that the intergranular crack propagation rate was faster in the AC specimen. Furthermore, as shown in the magnified SEM images of figure 3(d) and figure 3(e), clear grain facets were particularly observed near the specimen surfaces. By the SEM observation on the gauge portions, it was confirmed that many cracks formed on the gauge portion surfaces of the

WQ and AC specimens tested in the simulated geothermal water W1 (figure 4). From the above results, it is inferred that the intergranular cracks nucleation occurred at the surfaces of the WQ and AC specimens. It is considered that sensitization was caused by the present hardening-treatments, especially for the material AC with air cooling, so that the susceptibility of intergranular cracking increased.

The gauge portion surfaces after emery-polishing and etching, of the specimens ruptured in the simulated geothermal water W1, were examined in detail with SEM. No clear cracks were observed on the specimen surface of the as-received material. It is considered that the as-received material of 13Cr stainless steel has a superior corrosion resistance even though in the simulated geothermal water. However, for the hardening-treated materials WQ and AC, many cracks were observed on the gauge portion surfaces of the specimens tested in this water. Furthermore, it was found that most cracks formed in the specimen of the hardening-treated material AC with air cooling were in an intergranular mode along prior austenite grain boundaries (prior γ-GB). It is considered that the sensitization occurred in the hardening-treated material AC was forwarded by the long time air cooling, due to the formation of the Cr-depletion layers along the prior γ-GB in sufficient time. On the other hand, for the hardening-treated material WQ with water quenching it was found that the cracks formed not only in the intergranular mode but also in the transgranular one. Even so, it is considered that the fracture of the WQ specimen in the test water W1 was dominated by the intergranular cracking, which was the same as that of the AC specimen. As an example, figure 5 gives the SEM images near the crack tips on the gauge portion surfaces after emery-polishing and etching of the specimens ruptured in the simulated geothermal water W1. In addition, it is suggested that both the intergranular cracking (figure 5(a), figure 5(b)) and the transgranular cracking (figure 5(c)) occurred on the two hardening-treated materials were SCC (IGSCC and TGSCC), which were caused by the 10,000 ppm $Cl^-$ present in the water W1. It has been reported that both of IGSCC and TGSCC occur easily on the sensitized stainless steel in an aqueous solution of chloride [7].

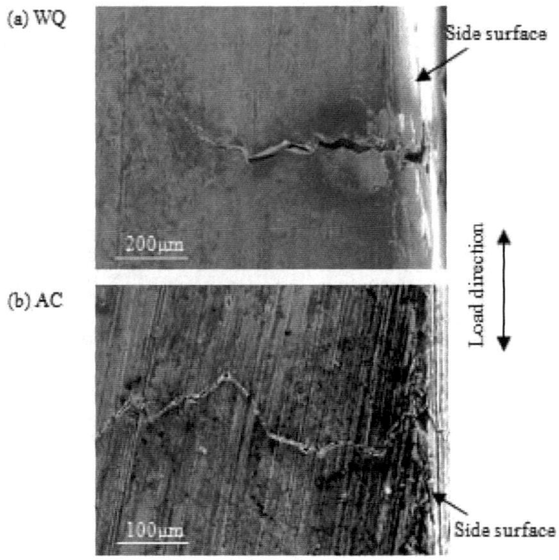


**Figure 4.** Cracks formed on the gauge portion surfaces of the specimens tested in the simulated geothermal water W1.

### 3.2. *Effect of impurities on cracking behavior*

To investigate the effect of corrosive impurities on the cracking behavior of the hardening-treated materials, SSRT tests were also conducted in the test water W2 without the addition of impurities. figure 6 shows the stress-strain curve of the hardening-treated material WQ with water quenching in the test water W2. For comparison, the stress-strain curve of the same hardening-treated material in the simulated geothermal water W1 was also plotted in this figure. As shown in figure 6, the WQ specimen in the test water W2 without the addition of impurities exhibited a brittle deformation mode. It showed a rupture stress of 741MPa, which is lower even than that of the WQ specimen in the simulated geothermal water W1.

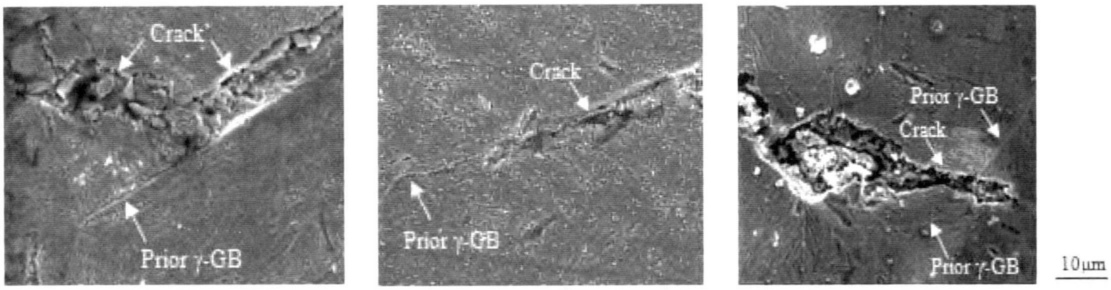


(a) AC (b) WQ (c) WQ

**Figure 5.** SEM images showing the crack tips on the gauge portion surfaces after emery- polishing and etching of the specimens ruptured in W1.

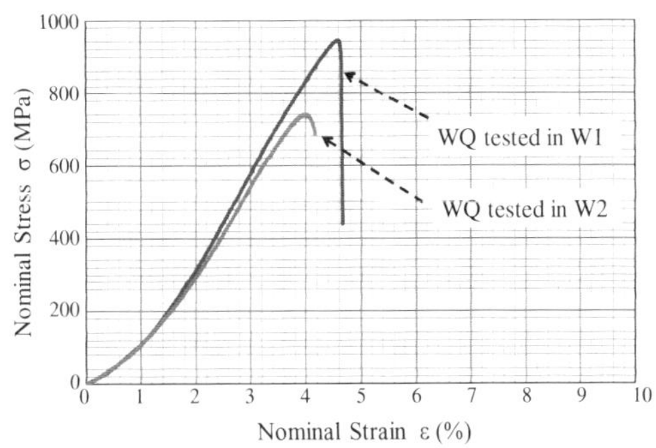


**Figure 6.** Stress-strain curves of the WQ specimens tested in the simulated geothermal water W1 and in the test water W2 without impurities.

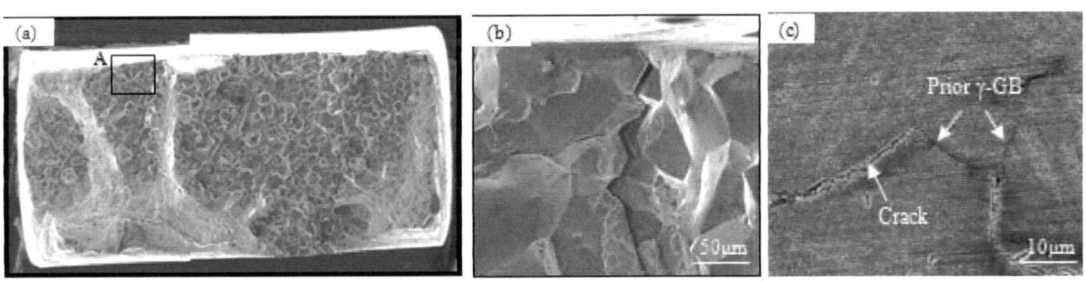


(a) Fracture surface; (b) Magnification of A; (c) Gauge portion surface

**Figure 7.** SEM images showing the fracture surface and the gauge portion surface after emery-polishing and etching of the WQ specimen ruptured in W2.

The overall fracture surface and a magnified SEM image of the fracture surface as well as the gauge portion surface after emery-polishing and etching of the specimen ruptured in W2 are shown in figure 7. Most of the fracture surface exhibited a feature of intergranular fracture mode. The cracks along prior γ-GB were observed on the gauge portion surface near the rupture surface. From the above results, it is assumed that the hardening-treated material WQ in the test water W2 without addition of impurities ruptured with other behavior than SCC.

Constant load tests were also conducted on the specimens of the hardening-treated material WQ in the simulated geothermal water W1 and in the test water W2 without impurities. Figure 8 shows the time variations of nominal strain of the two specimens loaded at 650MPa. The specimen in W1 ruptured in 8.3 h, while the specimen in W2 did not rupture up to 100h. The gauge portion surfaces of the two specimens after emery-polishing and etching were examined with SEM. It was found that the cracks along prior γ-GB formed on the specimen ruptured in W1, while no cracks were observed on the specimen tested in W2. It is considered that most of the cracks formed on the specimen in the constant load test in the simulated geothermal water W1 were intergranular SCC (IGSCC), which were the same as those formed in the above SSRT test and were coursed by the 10,000 ppm $Cl^-$ present in the water.

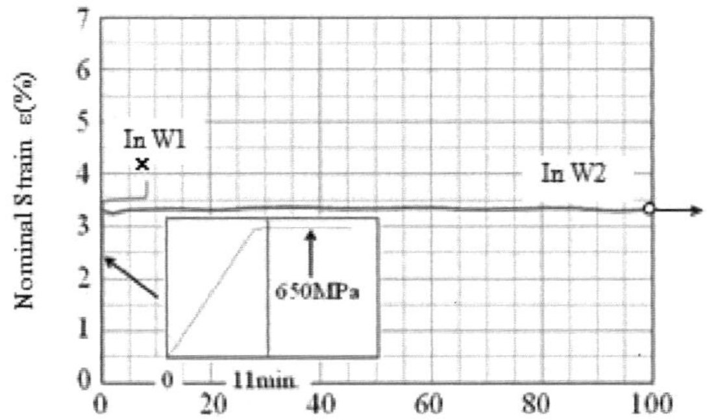


**Figure 8.** Time variations of nominal strain of the WQ specimens tested at a constant load of 650MPa in the simulated geothermal water W1 and in the test water W2 without impurities.

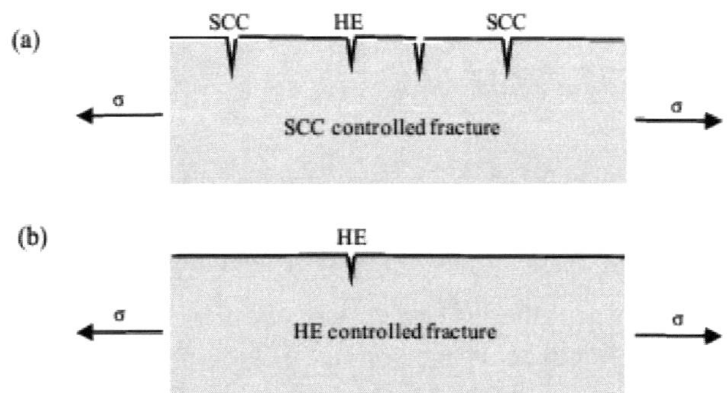


(a) In the simulated geothermal water W1
(b) In the test water W2 without addition of impurities
**Figure 9.** Schematic illustration of cracking behavior of the WQ specimens in SSRT tests.

As stated above, no cracks were observed on the WQ specimen in the constant load test in the test water W2 without addition of impurities. However, as shown in figure 7, the cracks along prior γ-GB were observed on the WQ specimen ruptured in the SSRT test in the W2. Therefore, it is wondered if the cracks formed in the SSRT test in the W2 without $Cl^-$ were IGSCC or not.

In both of the test water W1 and W2, the cathodic reactions should be mainly the reduction reaction of hydrogen ions ($H^+$), because the dissolved oxygen (DO) were removed by continuously bubbling of $CO_2$ gas (W1) or $N_2$ gas (W2). As atomic hydrogen (H) were generated and adsorbed on the specimen surfaces, it is considered that hydrogen embrittlement (HE) occurred on the specimens of the hardening-treated materials. It has been known that in high strength steels HE occurs easily even if a very small amount of hydrogen [8]. It has been also reported that in high strength steels HE occurs easily in an intergranular mode [9]. In this work, it is considered that the WQ specimen in the SSRT test in W2 ruptured unexpectedly due to the occurrence and rapid growth of HE cracking. On the other hand, for the present hardening-treated materials in the simulated geothermal water W1, it is considered that IGSCC controlled the cracking behavior due to the presence of the 10,000 ppm $Cl^-$, even though HE occurred also on the specimen surfaces. It is also assumed that for the specimen of the hardening-treated material WQ in SSRT test in the simulated geothermal water W1, due to many IGSCC occurred on the surface the plastic restraint was relaxed the specimen showed a higher stress to rupture than that in the test water W2 (figure 6). Figure 9 schematically illustrates the cracking behavior of the WQ specimen in the simulated geothermal water W1 and in the test water W2 without addition of impurities.

## 4. Conclusions

Stress corrosion cracking behavior of the hardening-treated materials of 13Cr stainless steel was examined with SSRT tests and constant load tests, in the simulated geothermal water and in the test water without addition of impurities. The results obtained are as follows.

(1) Both the two hardening-treated materials, WQ with water quenching and AC with air cooling, showed a brittle deformation mode in the test waters due to the sensitization caused by the hardening-treatments. Especially the hardening-treated material AC showed a more brittle fracture mode, as the sensitization was forwarded by the long time air cooling.

(2) In the simulated geothermal water, IGSCC with a part of TGSCC occurred on the hardening-treated material WQ with water quenching, while most of IGSCC occurred on the hardening-treated material AC with air cooling. In the test water without addition of impurities, the hardening-treated material WQ in the SSRT test exhibited a brittle fracture mode. It is considered that HE cracking occurred on the WQ specimen in the test water.

## References

[1] Bellows J C 1999 *PowerPlant Chem.* **1** 26-30.
[2] Jonas O 1985 *Mater. Perform.* **24** 9-18.
[3] Jonas O, Machemer L 2008 *Steam turbine corrosion and deposits problems and solutions* Proc 37th Turbomachinery Symposium, (Houston, USA, 2008), 211-28.
[4] Sakai Y, Yamashita M, Sakata M 2000 *Geotherm. Res. Counc. Trans.* **24** 521-6.
[5] Sakai Y 1999 *Therm. Nucl. Pow.* **50** 778-83.
[6] Kato H, Shiokawa K, Sakai Y 2004 Turbomach. **32** 599-606.
[7] Oki T, Okido M, and Kanamatsu H 1985 *Boshoku Gijutsu* **34** 546-51.
[8] Shiraga T 2011 *Zairyo-to-Kankyo* **60** 236-40.
[9] Nakasato F 1982 *Bull. Japan Inst. Met.* **21** 441-8.

IOP Conf. Series: Materials Science and Engineering **317** (2018) 012074 doi:10.1088/1757-899X/317/1/012074

# Effects of the TiC Nanoparticle on Microstructures and Tensile Properties of Selective Laser Melted IN718/TiC Nanocomposites

**Xiling Yao[1], Seung Ki Moon[1], Bing Yang Lee[2], Guijun Bi[2]**

[1]Singapore Centre for 3D Printing, School of Mechanical & Aerospace Engineering, Nanyang Technological University, Singapore, 639798
[2]Singapore Institute of Manufacturing Technology Singapore 638075

**Abstract.** The purpose of this paper is to investigate the effects of TiC nanoparticle content on microstructures and tensile properties of the IN718/TiC nanocomposites fabricated by selective laser melting (SLM). 0.5wt%, 1.0wt%, and 2.0wt% of TiC nanoparticles are added to the IN718 powders. The bulk-form IN718/TiC nanocomposites with different TiC contents are fabricated in-situ by SLM using the same process settings. The evolution of microstructures and tensile properties as the effect of changing the TiC content is studied using the optical microscopy, scanning electron microscopy, X-ray diffraction analysis, and tensile testing. The increase of TiC content refines the microstructure, promotes the formation of the cellular morphology, and reduces the size and continuity of Laves precipitates. Increasing the TiC content improves the yield strength and ultimate tensile strength but decreases the ductility. The grain refinement, dislocation bowing, dislocation punching, and the reduction in Laves precipitate contribute to the strengthening effect in the IN718/TiC nanocomposites.

## 1. Introduction

Selective laser melting (SLM) is one of the "Power Bed Fusion" additive manufacturing (AM) processes for fabricating metallic parts. In the SLM process, metal powders are rapidly melted by a high-energy laser and then re-solidify to form metallurgical bonding [1]. SLM has been used to fabricate bulk-form parts in many different metallic materials, including titanium [2], aluminum [3], iron [4], cobalt-chrome [5], and nickel alloys [6]. The directional solidification and the rapid cooling rate (106-7K/s) generate unique microstructures for selected laser melted metals, including highly-oriented dendrites and ultra-fine grains, which result in different (sometimes superior) mechanical properties compared with cast and wrought metals [7-9].

Among Ni-based superalloys, IN718 is a Ni-Fe-Cr austenitic ($\gamma$) superalloy with good weldability [10]. IN718 have found broad applications in gas turbines and petrochemical equipment, due to its superiority in both strength and temperature properties [11]. IN718 has been manufactured by the SLM process successfully in previous research [12-15]. In as-built IN718 via SLM, the growth of $\gamma'$ and $\gamma''$ precipitates is inhibited by the rapid solidification rate, while the $\gamma$ matrix and Laves precipitates are the major constituent phases [16]. IN718 fabricated by SLM displays dendritic microstructures, which undergo morphological changes at different laser energy density [14]. Compared with the cast material, the enhanced hardness, wear performance, and oxidation resistance was observed in the SLM-fabricated IN718 due to microstructure refinement [15].

Ceramic particle reinforced MMCs are important materials in engineering applications due to their superior properties such as high strength and wear resistance. Ni-based MMCs reinforced by different types of ceramic particles were fabricated by laser-powered AM processes, during which the unmolten ceramic particles were embedded in metal matrices upon melt pool solidification. Cooper et al. [17] mixed Al2O3, SiC, and TiC micro-sized particles with IN625 powders and fabricated the MMCs using additive layer manufacturing (ALM). Improved hardness was observed in SiC and TiC reinforced MMCs compared with the pure IN625. In a similar research [18], micro-sized TiC particles were added to IN718, and the MMC was fabricated by laser metal deposition (LMD). The microhardness and wear properties were improved by increasing the input laser energy density. Jia and Gu [19] applied the SLM process to fabricated IN718-based MMC reinforced by TiC nanoparticles. The effects of laser energy on the density, microhardness, and wear rate of the IN718/TiC nanocomposite were studied.

In this paper, fully-dense IN718/TiC nanocomposites were fabricated in-situ by the SLM process. Different weight percentages (i.e. 0.5%, 1.0% and 2.0%) of TiC nanoparticles were used, and the evolution of microstructures and tensile properties as the effect of changing the TiC nanoparticle content was investigated. Optical microscopy (OM), scanning electron microscopy (SEM), and X-ray diffraction (XRD) were applied to analyse the microstructural morphologies and crystal phases. Strengthening mechanisms in the SLM-fabricated IN718/TiC nanocomposites were also discussed.

## 2. Experimental Setup

### *2.1. Powder Preparation*

Gas atomized spherical IN718 powders were used as the matrix material in the selective laser melted nanocomposites. The powder diameters ranged from 21 to 55μm. The chemical compositions of the IN718 powders are listed in table 1. Spherical TiC nanoparticles had diameters ranging from 80 to 150 nm. Three levels of TiC nanoparticle content in weight percentage, i.e. 0.5 wt%, 1.0 wt%, and 2.0wt%, were added to the IN718 powders. The IN718/TiC mixtures in powder form were then blended in an Inversina Tumbler Mixer (Bioengineering, Inc.) at 70rpm for 8 hours before being loaded into the SLM machine as the build material.

**Table 1.** Chemical compositions of the IN718 powders.

| **Element** | **Ni** | **Cr** | **Nb** | **Mo** | **Co** | **Ti** |
|---|---|---|---|---|---|---|
| wt% | 50-55 | 17 -21 | 4.75-5.5 | 2.8-3.3 | 1.0 | 0.65-1.15 |
| **Element** | **Mn** | **Si** | **Al** | **Cu** | **Fe** | |
| wt% | 0.35 | 0.35 | 0.2-0.8 | 0.3 | Bal | |

### *2.2. Selective Laser Melting Process*

Sample fabrications were performed in an SLM 250HL machine manufactured by SLM Solutions GmbH. The SLM machine utilizes a ytterbium fiber laser with a wavelength of 1070nm and the maximum power of 300W. Argon gas was purged in the build chamber, and the oxygen level was reduced to 0.05% or lower during the process. Carbon steel plates are used as substrates, which were pre-heated to 100°C before the laser was activated in order to reduce the thermal stress at the substrate/part interface. During the SLM process, the laser scanned on the powders over the area where the part was being built. The hatch orientation rotated 79 degrees every layer to relieve the residual stress. As-fabricated parts were separated from the substrate via wire cutting at room temperature.

**Table 2.** SLM process parameters for specimen fabrication.

| **Laser power** | **Layer thickness** | **Scan speed** | **Hatch space** |
|---|---|---|---|
| 300 W | 0.05 mm | 800 mm/s | 0.13 mm |

Preliminary tests were conducted to identify the suitable SLM process parameters for producing fully dense parts in pure IN718 and IN718/TiC nanocomposites. The identified SLM process parameters are listed in table 2. The same set of parameters was used to fabricate all specimens for material characterizations described in the rest of this paper.

### *2.3. Material Characterizations*
Crystal phases were identified by X-ray diffraction (XRD) (D8 Discover, Bruker) with Cu Kα radiation (40 kV, 40 mA). X-ray scanning was from 30° to 95° with the step size of 0.05°. Specimens are polished down to 0.05μm finish. To reveal the microstructures, polished samples are electro-etched in 70% phosphoric acid at 10V for 15 seconds. Microstructures were observed using optical microscopy (OM, OLYMPUS MX51) and scanning electron microscopy (SEM; ZEISS EVO 50). Tensile coupons following the ASTM E8 standard were machined from SLM fabricated rectangular blocks by wire cutting, as shown in figure 1. The loading orientations of the tensile coupons were perpendicular to the SLM build direction (Z-axis) and parallel to the X-Y plane. Tensile tests were conducted using an Instron Universal Tester at room temperature. Yield strength (YS), ultimate tensile strength (UTS), and percentage elongation at fracture were obtained from the tensile testing results. Fracture surfaces of the broken tensile coupons were observed using the SEM.

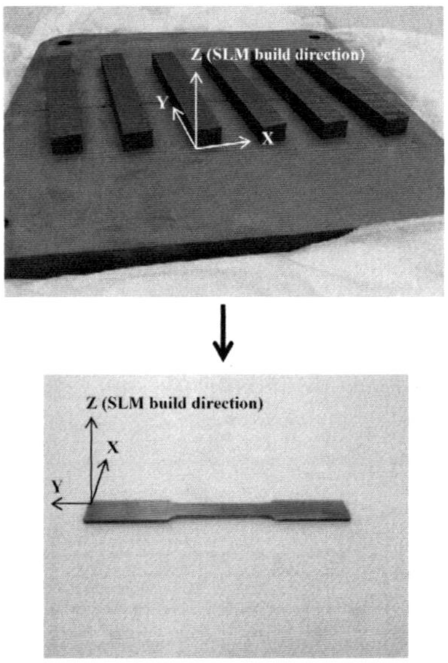


**Figure 1.** Tensile coupons machined from the SLM-fabricated rectangular blocks.

## 3. Results and Discussion
Bulk-form samples in pure IN718 and IN718/TiC nanocomposites with three different TiC nanoparticle contents (0.5, 1.0, and 2.0wt%) are fabricated from the powders by SLM. Densities of the specimens are measured to be higher than 99.9% of the theoretical values. The as-built samples display good surface finish and clear surface design features (i.e. the indented characters created in the CAD models). Cracks and distortions are absent.

### *3.1. Microstructures*
OM images taken from IN718/TiC nanocomposites with different TiC contents are shown in figure 2. Cumulatively-stacked scale-like melt pools are observed on the vertical plane parallel to the build direction (+Z). The concave shapes of the melt pools are formed by the Gaussian laser input in SLM. Small amounts of micro-sized pores caused by entrapped air bubbles appear at melt pool boundaries, which can hardly be avoided completely in SLM due to fast solidification.

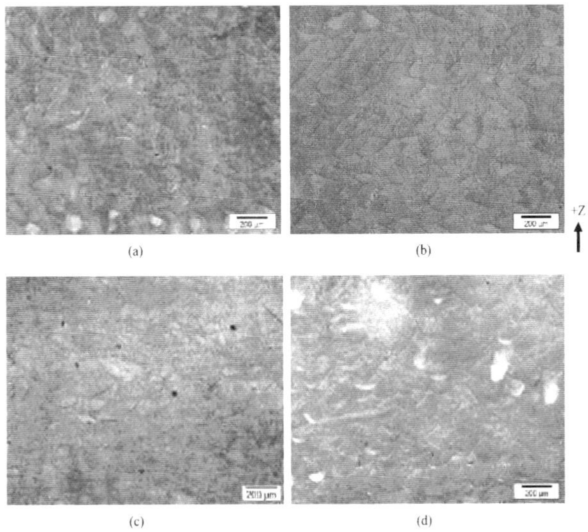

**Figure 2.** OM images of the (a) pure IN718, (b) IN718/0.5%TiC, (c) IN718/1.0%TiC, and (d) IN718/2.0%TiC nanocomposites.

Microstructure morphologies of pure IN718 and IN718/TiC nanocomposites with different TiC contents are shown in the SEM images (figure 3). Columnar microstructures oriented along the build direction are dominating in the SLM-fabricated pure IN718. The parallel dendritic grains with the spacing of about 2-3μm are formed by the highly directional cooling during the SLM process. The laser-induced heat flux in the melt pool orients vertically upwards, due to the vertical temperature gradient between the previously solidified layers and the current molten layer. With the addition of TiC nanoparticles, the SLM-fabricated IN718/TiC nanocomposites display a characteristic morphology dominated by equiaxed cellular microstructures with small cell aspect ratios. Long columnar grains still exist in IN718/TiC nanocomposites, but with decreased amount and size. In addition, finer microstructures are observed in the SLM-fabricated IN718/TiC nanocomposites compared with the pure IN718.

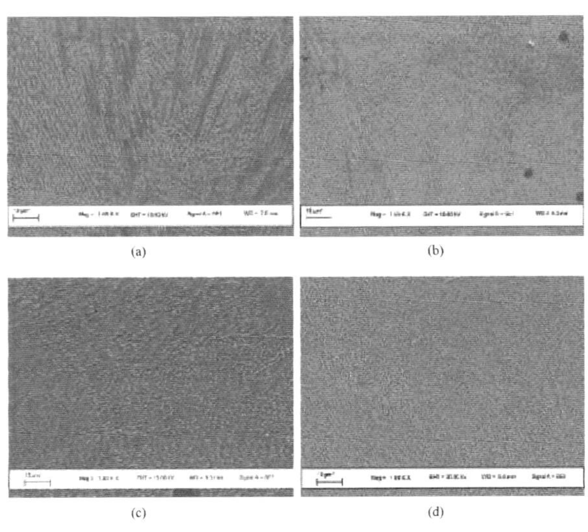

**Figure 3.** Low magnification SEM images of the (a) pure IN718, (b) IN718/0.5%TiC, (c) IN718/1.0%TiC, and (d) IN718/2.0%TiC nanocomposites

High magnification SEM images of the SLM-fabricated pure IN718 and IN718/TiC nanocomposites are shown in figure 4. The microstructure morphology of IN718/TiC nanocomposites varies with the TiC nanoparticle content. As the TiC content increase from 0 to 2.0%, the cell spacing decreases. As shown in figure 4, in the melt pool of the SLM-fabricated pure IN718, the solidification directions start from the melt pool boundary towards the center, being aligned with the thermal gradients caused by the Gaussian laser. In the IN718/TiC nanocomposites, as unmolten TiC nanoparticles are dispersed in the melt pool, they act as additional nucleation sites during crystallization and inhibit the grain growth. The higher solidification rate around the TiC nanoparticles contributes to the formation of finer microstructures in IN718/TiC nanocomposites [20]. The microstructure refinement effect of nanoparticles was also found in arc-welded materials [21].

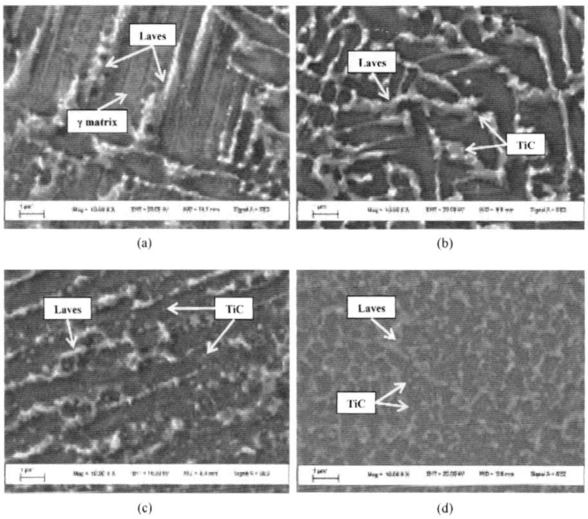


**Figure 4.** High magnification SEM images of the (a) pure IN718, (b) IN718/0.5%TiC, (c) IN718/1.0%TiC, and (d) IN718/2.0%TiC nanocomposites

The γ (fcc Ni-Fe-Cr) matrix and Laves (TCP, (Ni, Fe, Cr)2(Nb, Mo, Ti)) intermetallic compounds are the major constituents of the SLM-fabricated pure IN718 and IN718/TiC nanocomposites. Large and continuous Laves precipitates are observed in the pure IN718. As the TiC nanoparticle content increases in the IN718/TiC nanocomposites from 0.5 to 2.0%, the continuity of Laves gradually breaks down while the precipitates' size decreases. The formation of smaller and discontinuous precipitates in the IN718/TiC nanocomposites is caused by their higher solidification rates that cause insufficient Nb segregation [22]. Al, Ti, and Nb-rich precipitates such as the δ (orthorhombic Ni3Nb) phase and MC/M23C6/M6C carbides are not visible in the SEM images of the SLM-fabricated pure IN718 and IN718/TiC nanocomposites, due to inhibited atom (Al, Ti, Nb) segregation.

As shown in figures 4 (b), (c), and (d), spherical TiC nanoparticles (80-100nm diameter) are dispersed in the γ matrix of the IN718/TiC nanocomposites. Agglomeration of TiC nanoparticles has not occurred due to the high laser energy density (375J/m), calculated as the ratio of the laser power to scan speed. The high laser energy density results in intense Marangoni convection, low viscosity of the molten metal, and long circulating time for unmolten nanoparticles in the melt pool [23]. The avoidance of nanoparticle agglomerations reduces the detrimental micro-cracks formation in the matrix and hence prevents the deterioration of mechanical properties.

XRD analysis was carried out to identify crystal structures in the SLM-fabricated pure IN718 and IN718/TiC nanocomposites, as shown in the XRD spectra in figure 5. The γ crystal phase is found based on the ICDD database (PDF Number 00-035-1375, Ni-Cr-Fe). As the TiC nanoparticle content increases from 0 to 2.0%, no new peaks in the XRD spectra appear. Diffraction peaks corresponding to TiC cannot be seen, due to the small size of TiC nanoparticles in the matrix. Other crystal structures such as δ and MC carbides are not detected, indicating no or very little phase transition during the

SLM process. The strongest (200) peak in the pure IN718 indicates a preferable crystal orientation along the build direction as a result of laser-induced directional solidification (Amato et al. 2012). The prominence of the (200) peak remains in the IN718/0.5%TiC nanocomposite. However, as the TiC nanoparticle content increase to 1.0 and 2.0%, the (200) peak becomes weaker and it is no longer the most dominating peak. This change indicates a slightly more random distribution of crystal orientations in the IN718/TiC nanocomposites than in the pure IN718.

### *3.2. Tensile Properties*

As the result of the tensile test, the tensile properties of the SLM-fabricated pure IN718 and IN718/TiC nanocomposites with different TiC contents are listed in table 3. The characteristic tensile stress-strain curves of these materials are compared in figure 6. It can be observed that the SLM-fabricated IN718/TiC nanocomposites have higher yield strength (YS) and ultimate tensile strength (UTS) but lower elongate at fracture than the pure IN718. As the TiC nanoparticle content in the IN718/TiC nanocomposites increases from 0.5 to 2.0%, both YS and UTS improved while the elongation decreases.

The large and continuous Laves precipitates in the SLM-fabricated pure IN718 is detrimental to the material's strength. By increasing the fraction of TiC nanoparticles, the amount of Laves precipitates is decreased in the IN718/TiC nanocomposites. Therefore, the IN718/TiC nanocomposites suffer less from the weakening effect of Laves precipitates compared with the pure IN718.

As a trade-off of the strengthening effect of TiC nanoparticles in enhancing YS and UTS, the elongation at fracture decreases when the TiC content increases from 0 to 2.0%. After the tensile stress-strain exceeds the yield point, the matrix-TiC interfaces act as failure initiators where cracks may form and grow. Therefore, the addition of TiC nanoparticles increases the number of failure initiating sites after yielding and hence brings the final fracture earlier than the pure IN718.

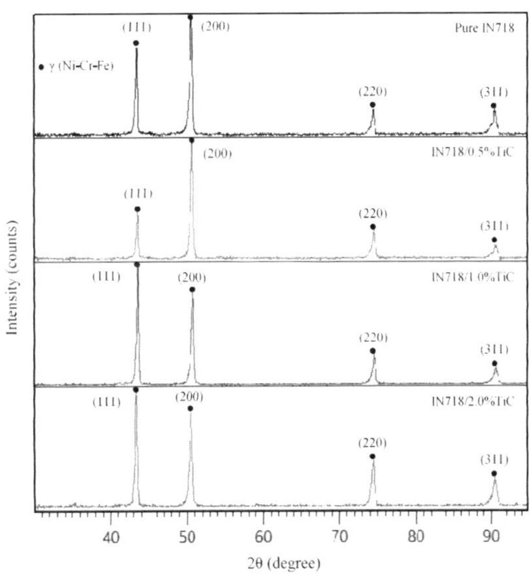


**Figure 5.** XRD spectra of the SLM-fabricated pure IN718 and IN718/TiC nanocomposites.

**Table 3.** Tensile properties of the SLM-fabricated pure IN718 and IN718/TiC nanocomposites.

| Materials | YS (MPa) | UTS (MPa) | Elongation (%) |
|---|---|---|---|
| Pure IN718 | 638.6 | 933.6 | 38.29 |
| IN718/0.5%TiC | 674.1 | 947.7 | 29.20 |
| IN718/1.0%TiC | 709.3 | 1023.0 | 17.27 |
| IN718/2.0%TiC | 800.7 | 1068.0 | 13.91 |

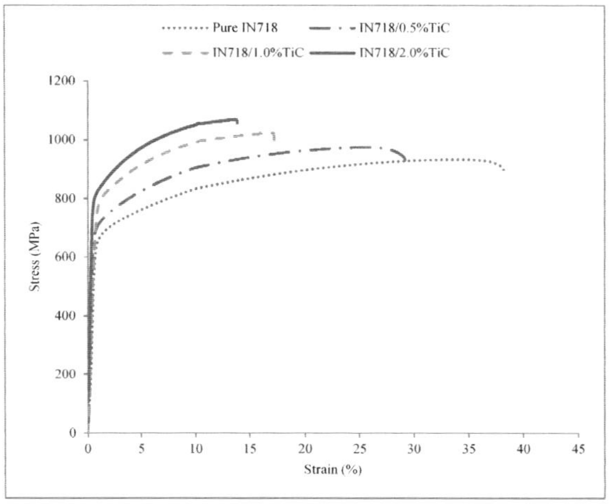


**Figure 6.** Tensile stress-strain curves of the SLM-fabricated pure IN718 and IN718/TiC nanocomposites

## 4. Conclusions

In this research, the SLM process was applied to fabricate IN718/TiC nanocomposites with different weight percentages (0, 0.5%, 1.0%, and 2.0%) of TiC nanoparticle. The effects of the TiC content on microstructures and the tensile properties of the SLM-fabricated IN718/TiC nanocomposites were investigated, and the findings are summarized as below:

1) Fully-dense specimens in the pure IN718 and IN718/TiC nanocomposite with different TiC contents can be fabricated using the same set of SLM process parameters. Surface defects or geometric distortions are absent.

2) Compared with the dominating columnar grains in the SLM-fabricated pure IN718, the microstructures of the IN718/TiC nanocomposites display finer cellular morphologies, and the cell spacing decreases with the increasing TiC nanoparticle content.

3) Compared with the large and continuous Laves precipitates in the SLM-fabricated pure IN718, the Laves precipitates decrease in size while the continuity breaks down as the TiC content increases in the IN718/TiC nanocomposites, due to higher solidification rate.

4) The increment of the TiC content from 0 to 2.0wt% increases both the yield strength and ultimate tensile strength but decreases the elongation at fracture.

In future research, the SLM process can be applied to fabricated IN718-based metal matrix composites with other reinforcements, such as oxides, borides, and carbon nanotubes (CNT). Apart from mechanical properties, thermal and chemical properties of SLM-fabricated IN718-based MMCs can also be studied in the future.

## References

[1] Kruth J P, Mercelis P, Vaerenbergh J V, Froyen L, and Rombouts M 2005 *Rapid Prototyp. J.* **11(1)** 26-36.
[2] Pawlak A, Szymczyk P, Ziolkowski G, Chlebus E and Dybala B 2015 *Rapid Prototyp. J.* **21(4)** 393-401.
[3] Kempen K, Thijs L, Van Humbeeck J and Kruth J P 2012 *Phys. Proc.* **39** 439-46.
[4] Song B, Dong S, Deng S, Liao H and Coddet C 2014 *Opt. Laser Tech.* **56** 451-60.
[5] Vandenbroucke B and Kruth J P 2007 *Rapid Prototyp. J.* **13(4)** 196-203.
[6] Carter L N, Essa K and Attallah M M 2015 *Rapid Prototyp. J.* **21(4)** 423-32.
[7] Brandl E, Heckenberger U, Holzinger V and Buchbinder D 2012 *Mater. Des.* **34** 159-69.
[8] Liu Z H, Zhang D Q, Sing S L, Chua C K and Loh L E 2014 *Mater. Charact.* **94** 116-25.
[9] Thijs L, Kempen K, Kruth J P and Van Humbeeck J 2013 *Acta. Mater.* **61(5)** 1809-19.

[10] Knorovsky G A, Cieslak M J, Headley T J, Romig A D and Hammetter W F 1989 *Metallurg. Trans. A* **20(10)** 2149-58.
[11] Zhang S and Zhao D 2012 *Aerospace Materials Handbook* CRC Press Boca Raton Florida US.
[12] Strößner J, Terock M and Glatzel U 2015 *Adv. Eng. Mater.* **17**(8) 1099-105.
[13] Amato K N, Gaytan S M, Murr L E, Martinez E, Shindo P W, Hernandez J, Collins S and Medina F 2012 *Acta. Mater.* **60(5)** 2229-39.
[14] Jia Q and Gu D 2014 *J. Alloy. Comp.* **585** 713-21.
[15] Jia Q and Gu D 2014 *Opt. Laser Tech.* **62** 161-71.
[16] Zhang D, Niu W, Cao X and Liu Z 2015 *Mater. Sci. Eng. A* **644** 32-40.
[17] Cooper D E, Blundell N, Maggs S and Gibbons G J 2013 *J. Mater. Proc. Tech.* **213(12)** 2191-200.
[18] Hong C, Gu D, Dai D, Gasser A, Weisheit A, Kelbassa I, Zhong M and Poprawe R *Opt. Laser Tech.* **54** 98-109.
[19] Jia Q and Gu D 2014 *J. Mater. Res.* **29(17)** 1960-9.
[20] Sing S L, Yeong W Y and Wiria F E 2016 *J. Alloy. Comp.* **660** 461-70.
[21] Fattahi M, Mohammady M, Sajjadi N, Honarmand M, Fattahi Y and Akhavan S 2015 *J. Mater. Proc. Tech.* **217** 21-9.
[22] Manikandan S G K, Sivakumar D, Rao K P and Kamaraj M 2014 J. Mater. Proc. Tech. **214(2)** 358-64.
[23] Yuan P, Gu D and Dai D 2015 Mater. Des. **82** 46-55.

**Acknowledgment**
This research was supported by SIMTech-NTU Joint Lab, National Research Foundation, Singapore, and Singapore Centre for 3D Printing.

IOP Conf. Series: Materials Science and Engineering **317** (2018) 012075 doi:10.1088/1757-899X/317/1/012075

# Using of Stone Flour from Some Mineral Rocks in Modern Concrete

**Moskvin Roman[1], Belyakova Elena[1*], Moroz Marina[1]**

[1]Russia, 440028, Penza, 28, German Titov St., Penza State University of Architecture and Construction

**Abstract.** There is shown the possibility of using mill ground rocks in SCC without deterioration of rheological properties of concrete mixtures. Obtained high-strength concrete of the new generation with high technical and economic indices and low unit costs per unit of cement strength.

## 1. Introduction

As history shows, the development of civilizations is directly related to the achievements in the field of construction. One of the reasons for the rapid development of our civilization is the invention in the mid-19th century of a building material such as cement concrete.

Modern concrete properties are significantly superior to products made from traditional concrete. Such concrete is called High Performance Concrete (HPC). The new generation concrete is usually referred to high-strength, ultra-high-strength, ultra-efficient, reactive-powdered concretes, self-compacting concrete, self-leveling and others. They consist of 7 or more optimally selected components. The properties of modern high-quality concrete significantly exceed traditional ones. In addition, they are diverse and can cover the entire scope of concrete use in construction: from ordinary and monolithic concrete with a compressive strength of not more than 50 MPa to high-quality and high-strength concrete 80-1800 MPa, as well as decorative concrete.

A feature of modern concrete is the presence in their composition of a significant amount of water-dispersed suspension (rheological matrix) consisting of cement, stone flour, fine sand and water with a syperplasticizer. A large amount of rheological matrix is necessary to ensure high mobility of such concrete mixtures. The large and fine aggregate floats as if "floating" in them, without encountering obstacles and resistance, whereas in old generation concrete, with a lack of matrix, the aggregate particles interfere with each other and for the laying of such concrete one it is necessary to apply vibrocompression or add water, increasing porosity and decreasing strength.

Increasing the volume of the rheological matrix can be achieved by increasing the proportion of finely divided components, such as cement or stone flour. Increasing the proportion of cement is not rational, because it will lead to its overspending and increase in the cost of concrete, as well as to an increase in shrinkage. Thus, it is more expedient to introduce stone flour in the optimum ratio with cement.

The most common in the Russian practice of making such concretes with stone flour is pulverized quartz, i.e. Ground quartz sand. Its advantages include high rheological activity, and in concrete, when it is used, there is an increase in density, a decrease in porosity. And the strength of concretes with this finely dispersed additive can be up to 150 MPa or more with an optimally selected recipe. With all its technological merits, the microquartz also has a significant drawback - an increased cost of production, associated with the washing of sand from clay particles, subsequent drying and grinding in mills to a high specific surface area. In this regard, pulverized quartz loses some of the advantages and is not in high demand in Russia.

## 2. Results and Discussions

Therefore, the aim of our research is to find an alternative to pulverized quartz in modern high-quality concretes. Its replacement will reduce the cost of such concrete, thereby making them more affordable in the market.

The novelty of the study is the proposed requirements for the rocks for the production of mineral filler used in the production of high-quality concrete.

1. The rocks must be dense in order to prevent the migration of water and dissolved syperplasticizer into the pores of the particles. This requirement is more applicable to rocks of sedimentary origin, which may have not only micro-, but also nanosporosity. The porosity of dolomites, limestones, dolomitized limestones varies depending on the deposit, which obliges the conduct of primary research on porosity. Thin-grained granites, basalts, diabases and the like, which are rocks of igneous origin, have only surface water absorption due to microcracks and surface wetting. Therefore, they do not require porosity control; When grinding internal cavities are destroyed.

2. The rocks must be dispersed and ground to a microscale level of 0 to 120 μm to form a flowable disperse system. Volcanic rocks in their majority are dense (cast or crystalline structure). Those volcanic rocks that are porous: ash, tuff, pumice, have a closed porosity with aluminosilicate pore walls and when grinding fine particles do not absorb water.

3. The rocks must have a rheological activity comparable to cement activity or close to it. An increase in the water requirement of a mixture of cement with a finely divided filler is allowed up to 10-15% in comparison with pure cement.

4. Fine particle filler particles must be positively charged, because all super- and syperplasticizers are anionic and negatively charged by functional groups, which contributes to the adsorption of syperplasticity polyions on the particle surface, thereby increasing the mobility of the mixture. On negative-charged particles, GP ions do not precipitate and their rheological effect is insignificant. Such rocks include acid rocks - quartz sand, diatomite, and flask.

Thus, we see as an alternative to the use of micro-quartz is the use of stone-crushing wastes of dense rocks and rocks that have better grinding than quartz sand.

One of the most important signs of the suitability of rocks for use as dispersed filler in powdered and powder-activated concrete of the new generation is the porosity index, determined by the water absorption (Wm) of the rock from which they are made. The water absorption of rocks depends on their porosity and pore nature: the number of open and communicating pores assessed by effective porosity, as well as by the pore size.

As the first suitability test, in accordance with the methodology developed by us, it is necessary to determine the water absorption of grains of crushed stone fraction 15-20 mm. Depending on the water absorption by mass after 48 hours, the crushed stone is divided into the following categories:

I category - water absorption no more than 2% - is the most suitable for obtaining concretes of all grades from M200 to M1200;

II category - water absorption from 2 to 4% - can be used for obtaining concrete grades from M200 to M1000;

III category - water absorption from 4 to 6% - can be used for concrete grades from M200 to M800.

An assessment of the suitability of rocks for water absorption showed that magmatic and dense carbonate rocks (dolomite, calcareous limestone) with a water absorption Wm = 0.09-0.8% are most suitable, but a flask and a diatomite are not suitable Wm = 60-120% .In the Penza region, a suitable rock is the Salovsky Sandstone (Wm = 1.4%).The Yessin limestone and the Nikolsky sandstone are conditionally suitable (Wm = 5-16%).Therefore, it was required to check their rheological activity in comparison with cement.

The final assessment of the quality of stone flour is to ensure the necessary dilution of the aqueous suspension from it in a mixture with cement and superplasticizer. The criterion for evaluation is the water-reducing effect (table 1). It is determined by the formula:

$$W_{eff} = \frac{(W/H)_{\text{unpl}}}{(W/H)_{\text{pl}}} \quad (1)$$

where $(W/H)_{unpl}$, $(W/H)_{pl}$ water-aggregate ratio of unplasticized and plastified slurries with the same dissolution from the Hegermann cone (280-320 mm).

**Table 1.** Rheological activity of flour from rocks by the water-reducing effect $W_{eff}$.

| Composition of a slurry | Type of stone flour | |
|---|---|---|
| | Limestone | Sandstone |
| Cement: Stone flour 100:0 | 2,10 | 2,10 |
| Cement: Stone flour 67:33 | 2,25 | 2,09 |
| Cement: Stone flour 50:50 | 2,37 | 2,05 |

As can be seen, limestone enhances the rheological properties of the suspension, and sandstone slightly worsens.

To assess the effectiveness of the use of stone flour in modern concrete, powder, powder-activated sand and gravelly concrete were made.

It is proposed to determine the quality of recipe and raw material selection by the specific consumption of cement per unit compressive strength (kg/MPa):

$$C^{sp}_{R_{\text{comp.str.}}} = \frac{C}{R_{comp.str.}} \quad (2)$$

This evaluation criterion is generalizing: technical, economic and environmental.

**Table 2.** Indicators of samples from powdered concrete.

| № | Cement, kg/m$^3$ | Type of stone flour | W/H | W/C | $\rho_{hum}$, [kg/m$^3$] | $R_{comp.str.}$, [MPa] | $C^{sp}_{R_{comp.str.}}$, [kg/MPa] | $W^3_m$, [%] |
|---|---|---|---|---|---|---|---|---|
| 1 | 681 | Dusty quartz | 0,124 | 0,386 | 2333 | 103,7 | 6,57 | 2,94 |
| 2 | 655 | Limestone (Issa) | 0,116 | 0,361 | 2302 | 110,5 | 5,93 | 2,11 |
| 3 | 680 | Dolomite | 0,106 | 0,328 | 2324 | 110,5 | 6,15 | 2,52 |
| 4 | 669 | Limestone (Togliatti) | 0,123 | 0,383 | 2318 | 113,9 | 5,87 | 2,97 |
| 5 | 677 | Sandstone (Nikolsk) | 0,154 | 0,479 | 2095 | 85,0 | 8,00 | 5,89 |
| 6 | 657 | Sandstone (Salovka) | 0,123 | 0,383 | 2234 | 103,7 | 6,34 | 2,69 |
| 7 | 660 | Diabase | 0,123 | 0,383 | 2305 | 107,1 | 6,16 | 2,81 |
| 8 | 690 | Granite | 0,115 | 0,347 | 2302 | 108,8 | 6,34 | 2,17 |

Note: Cement brand CEM I 52.5

Nomenclature:

W/H – water-aggregate ratio

W/C – water-cement ratio

$\rho_{hum}$ – density of humid concrete

$W^3_m$ – water absorption after 3 days

**Table 3.** Indicators of samples from powder-activated sand and gravel concrete

| № | Cement, kg/m$^3$ | Type of stone flour | W/H | W/C | $\rho_{hum}$, [kg/m$^3$] | $R_{comp.str.}$, [MPa] | $C^{sp}_{R_{comp.str.}}$, [kg/MPa] | $W^3_m$, [%] |
|---|---|---|---|---|---|---|---|---|
| Powder-activated sandy concrete | | | | | | | | |
| 1 | 623 | Sandstone (Nikolsk) | 0,128 | 0,379 | 2296 | 107,2 | 5,81 | 1,69 |
| 2 | 643 | Limestone (Issa) | 0,096 | 0,314 | 2303 | 105,4 | 6,10 | 1,76 |
| Powder-activated crushed stone concrete | | | | | | | | |

| | | | | | | | | |
|---|---|---|---|---|---|---|---|---|
| 3 | 310 | Limestone (Togliatti) | 0,064 | 0,496 | 2538 | 78,2 | 3,96 | 2,09 |
| 4 | 300 | Limestone (Issa) | 0,074 | 0,570 | 2445 | 68,0 | 4,41 | 2,47 |
| 5 | 247 | Limestone (Togliatti) | 0,069 | 0,650 | 2404 | 52,7 | 4,69 | 2,61 |
| 6 | 300 | Limestone (Togliatti) | 0,068 | 0,515 | 2451 | 90,1 | 3,33 | 2,37 |
| 7 | 294 | Sandstone (Nikolsk) | 0,069 | 0,650 | 2269 | 54,4 | 5,40 | 2,41 |
| 8 | 450 | Limestone (Togliatti) | 0,066 | 0,326 | 2370 | 97,8 | 4,60 | 2,89 |

As can be seen, the strength of powdered and powder-activated sand concretes was from 103 to 113 MPa on all rocks except for Nikolsky sandstone with a specific cement consumption of 5.9 to 6.6 kg / MPa. While this indicator for scraped concrete was 3.3-5.4 kg/MPa.

## 3. Conclusions

1. It has been established that ground rocks of various chemical and mineralogical composition and genetic origin are highly effective dispersed additives which, introduced in significant amounts from 40 to 100% of the mass of cement, allow producing concrete with compressive strength from 54 to 110 MPa depending on from the content of cement and the type of concrete.

2. It is established that the specific consumption of cement per unit compressive strength, which is a comprehensive evaluation criterion for all concretes produced, is in the range from 3.3 to 6.6 kg / MPa, which is much lower than in old and transitional generations with SP.

3. The water absorption of the new generation of concrete is determined, which ranges from 1.7 to 3.0%, depending on the type of concrete, which guarantees high water resistance, low water permeability and frost resistance, and determines their durability. Considering that self-packing concrete is a step towards the future, the transition to the technology of their production guarantees high technical and economic indicators of new-generation concrete.

## References

[1] Bazhenov Yu M 2007 *Technology of concrete. Textbook.* Moscow: Publishing House of the ASV, pp. 528.

[2] Kalashnikov V I 1996 *Basis of plasticization of mineral disperse systems for the production of building materials* Dissertation in the form of a scientific report for the degree of Doctor of Technical Sciences. Voronezh.

[3] Belyakova E A, Moskvin R N, Moroz M N, Belyakova V S 2014 *Mod. Sci. Res. Innov.* **12-1(44)** 204-7.

[4] Kalashnikov V I, Moskvin R N, Belyakova E A, Belyakova V S, Petukhov A V 2014 *Syst. Meth. Tech.* **2(22)** 113-8.

[5] Tarakanov O V, Kalashnikov V I, Belyakova E A, Moskvin R N 2014 Reg. Archit. Constr. **2** 47-53.

IOP Conf. Series: Materials Science and Engineering **317** (2018) 012076 doi:10.1088/1757-899X/317/1/012076

# Self-Assembly of ZnO Nanoplatelets into Hierarchical Mesocrystals and Their Photocatalytic Property

**Yongqiang Yang[1]*, Qinsheng Wang[1], Zheng Liu[1], Ling Jin[1], Bingxian Ou[1], Pengju Han[1], Qun Wang[1], Xiaobao Cheng[1], Wenjun Liu[1], Yu Wen[1], Yuan Liu[1], Weifang Zhao[1]**

[1]National Graphene Products Quality Supervision and Inspection Center (Jiangsu), Jiangsu Province Special Equipment Safety Supervision Inspection Institute, Branch of Wuxi, Wuxi, P. R. China

**Abstract.** In this work, a simple chemical procedure was developed for the preparation of mesocrystals consisiting of ZnO nanoplateletes. By simple mixing the aqueous solutions $Zn(NO_3)_2$, NaOH and ethanol at certain temperatures, the hierarchical mesocrystals with big at both ends but small in the middle were obtained. After being annealed in air at certain temperatures, the same structured ZnO mesocrystals were generated. The morphology, crystalline structure and chemical composition were characterized using SEM, XRD FT-IR and Raman. The photocatalytic properties of the ZnO mesocrystals were also investigated. It was illustrated that the ZnO mesocrystals show decent photocatalytic performance to the photodegradation of methyl blue.

## 1. Introduction

Mesocrystals, an ordered superstructure of crystals with mesoscopic size (1-1000 nm), have attracted great attention in the last few years due to their mesoscopic structure offering new opportunities for materials [1]. Within mesocrystals, the interaction between the building blocks, in most cases the small particle, are relatively weaker than the ion by ion or single molecule attachment which exists in conventional single crystals [2, 3]. The mesocrystals show typical hierarchical ordered structural features in which the individual small particles are packed in certain oriented aggregation manner, but preserve their inherent crystalline structure. Thus, the mesocrystals have chemical and physical properties differing from that of the single crystal counterparts. Additionally, owing to their unique size region, the mesocrystals should be easier to process or manipulate than the nanoparticles, which is advantageous for their preparation and application [4, 5].

ZnO, as a smart functional material, is a well-known wide band gap semiconductor. Both the bulk single crystal and nanocrystals have been widely studied and some of them have found practical applications s in solar cells, chemical sensors, optoelectronics, and photocatalysis [6-9]. Different mesocrystal structures of ZnO have been synthesized. For instance, Liu *et al.* obtained uniform-sized, core-shell ZnO mesocrystal microspheres using a facile one-pot hydrothermal method in the presence of water-soluble polymer poly (sodium 4-styrenesulfonate) [10]. Li *et al.* report on the fabrication of hollow ZnO mesocrystals that form via aggregation of nanometer-sized ZnO particles, and these particles was precipitated from the strongly hydrated tetrabutylammonium hydroxide[11]. Ye *et al.* prepared hierarchical ZnO mesocrystals from a precursor of synthetic layered basic zinc acetate nanobelts by a facile solution route in the presence of poly (vinyl pyrrolidone) [12]. Nevertheless, the high cost and the relatively complicated procedure of these methods somewhat limit their application for large scale preparation of ZnO mesocrystals. The properties of ZnO mesocrystals are also worth exploring further.

Herein, we reported a simple, rapid and cost-efficient way to synthesize one hierarchical ZnO precursor mesocrystals with big at both ends but small in the middle by simple mixing the aqueous solutions $Zn(NO_3)_2$, NaOH and ethanol at certain temperatures. After being annealed in air at certain temperatures, the same structured ZnO mesocrystals were generated. It was shown that the as-obtained ZnO mesocrystals have excellent photocatalytic activity and anti-optical bleaching properties to photocatalytic degradations of methyl blue.

## 2. Experimental Section

### *2.1. Preparation of ZnO mesocrystals*

In a typical experiment, 15 ml (0.5 mol/L) $Zn(NO_3)_2$ aqueous solution was mixed with 10 ml NaOH (5 mol/L) solution under stirring, after its complete dissolution, the clear precursor solution was obtained. Then, 10 ml ethanol was poured into the clear precursor solution under stirring, and the solution slowly turned milky white. The reaction temperature (10 °C) was adjusted using a cryogenic circulator in the process of reaction. After reaction for 30 min, the solid product was collected by centrifugation, and was washed using deionized water for three times, and then was dried at 80 °C. The as-prepared white solid product was finally annealed at 200 °C in air for 4 h.

### *2.2. Characterization*

The morphologies of as-prepared solid products were characterized using scanning electron microscope (SEM, Verios 460, FEI, America). The crystallinity was evaluated using X-ray powder diffraction (XRD) performed on a D8 Advance diffractometer (Bruker, German). The FTIR spectra were acquired on a Nicolet IS50 FTIR spectrometer (Thermo Fisher, America) with the samples milled in KBr wafer. Raman spectroscopy measurement was examined by laser confocal micro-Raman spectroscopy using LabRam HR Evolution (Horiba Jobin Yvon, Japan) with an excitation laser beam wavelength of 473 nm. The UV-vis absorption was measured using a UV-Vis spectroscope (Shimadzu UV2550, Japan).

### *2.3. Photocatalytic degradation experimental*

The photocatalytic degradations of methyl blue (MB) dye was carried in a photo-reactor equipped (Bilon, Shanghai). The reaction was carried out in quartz tubes under vigorous stirring, which was installed 6 cm away from the irradiation lamp. The lamps were equipped inside a cylindrical Pyrex tube cooled by circulated water. In a typical experiment, ZnO mesocrystals, 50 mg, were dispersed in 100 mL, methyl blue aqueous solution (15 mg/L), and was stirred for 20 min to reach an adsorption–desorption equilibrium of methyl blue onto the ZnO mesocrystals. The reaction was initiated by exposing the quartz tube to a mercury lamp (365 nm, 500 W) at room temperature (20 °C) under the same photoreaction conditions. The reaction process was followed using UV-Vis absorption spectroscopy measurement.

## 3. Results and discussion

As shown in figure 1a, the solid product of the reaction of $Zn(NO_3)_2$ and NaOH in the presence of ethanol at the 10 °C reaction temperature shows a hierarchical morphology with big at both ends but small in the middle. The as-obtained products assume sizes of ~5 μm in length, ~3 μm in big-end thickness and ~1μm in the middle thickness. From the high-magnification image (figure 1b and figure 1c), it can be found that each hierarchical structure is composed of many ordered nanoplatelets and exhibits typical mesocrystal character. The individual nanoplatelets within the mesocrystals assume sizes of ~800 nm in length and ~100 nm in thickness. The nanoplatelets are aggregated together in a side-by-side manner. Obviously, each mesocrystal could exist independently, and no obvious aggregation between the mesocrystals was observed. To elucidate the composition and the crystallinity of the mesocrystals, their XRD patterns were acquired and depicted in figure 1d. The main diffraction peaks of the hierarchical mesocrystals can be indexed to a rhombic phase of $Zn(OH)_2$ (JCPDS card No.

IOP Conf. Series: Materials Science and Engineering **317** (2018) 012076 doi:10.1088/1757-899X/317/1/012076

71-2215). Based on our previous work, a possible formation mechanism for the hierarchical mesocrystals was speculated [9]. Probably, $Zn(NO_3)_2$ reacted with NaOH in water at most 10 °C to slowly form first the nanocrystals (clusters) of $Zn(OH)_2$. Meanwhile, ethanol adsorbed onto certain crystalline special faces of the $Zn(OH)_2$ nanocrystals and might further block the growth of the $Zn(OH)_2$ nanocrystals along certain directions leading to the formation of nanoplatelets. On the other hand, the strong interactions between the ethanol molecules on different $Zn(OH)_2$ nanoplatelets drive them to assemble head-by-head and side-by-side way to finally form the hierarchical mesocrystals to reduce the surface energy.

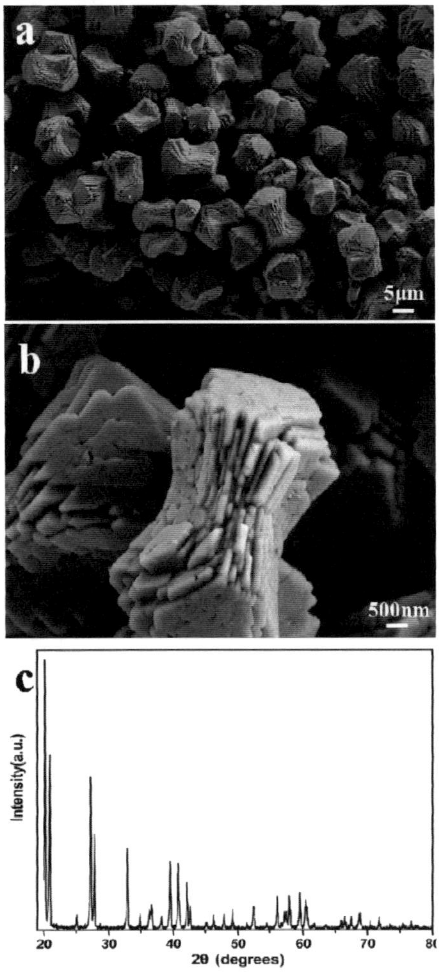


**Figure 1.** (a-b) SEM image of as-obtained hierarchical mesocrystals. (c) XRD patterns of the mesocrystals.

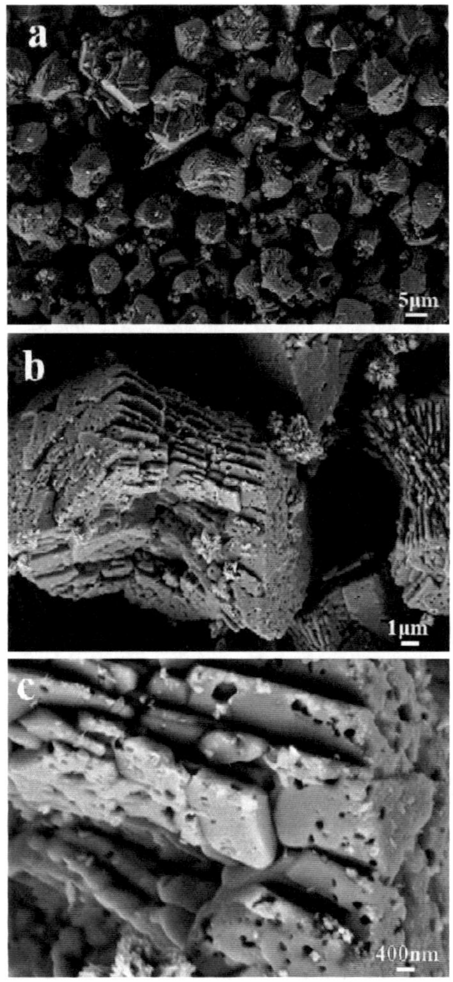


**Figure 2.** (a, b) Low-magnification and (c) high-magnification SEM image of annealed mesocrystals.

It is well known that $Zn(OH)_2$ could be easily decomposed at relatively higher temperature into $H_2O$ and ZnO, which may provide more attractive properties than the $Zn(OH)_2$. Thus, the powders of $Zn(OH)_2$ mesocrystals were annealed at 200 °C in air for 4 h. Though the hierarchical morphology with big at both ends but small in the middle and the mesocrystal character of the as-prepared products were both preserved properly after annealed treatment, the surface of the platelets appeared holes which suggested that the as-obtained products have experienced the recrystallization phenomenon (figure 2). The main reason is that the nanoplatelets possess lots of defects due to crystal could not achieve the full growth [13].

The as-obtained mesocrystals are composed of pure ZnO nanoplatelets. This was further confirmed by XRD. As shows in figure 3a, all diffraction peaks can be well indexed to a wurtzite structure (hexagonal) ZnO (JCPDS card No. 36-1451) [9]. The FTIR spectra of as-obtained products showed characteristic bands at 495 $cm^{-1}$, which is attributed to the absorption of Zn-O bond, also confirm the presence of ZnO [14]. Figure 3c represents the Raman spectrum of the samples at the range of 200 to 800 cm-1 at room temperature, which shows four peaks at 331, 381, 437 and 580 $cm^{-1}$. The peak at 438 $cm^{-1}$ corresponds to the optical $E_{2H}$ high frequency mode, which is the common Raman active peak for wurtzite hexagonal ZnO. These peaks at 331 and 381 $cm^{-1}$ are assigned to $E_{2H}$-$E_{2L}$ mode and $A_{1T}$ mode, respectively. The suppressed peak at 581 $cm^{-1}$ in the spectrum is attributed as $E_{1T}$ mode caused by impurities and defects. These results further confirm that wurtzite structure ZnO has very good crystal quality [15].

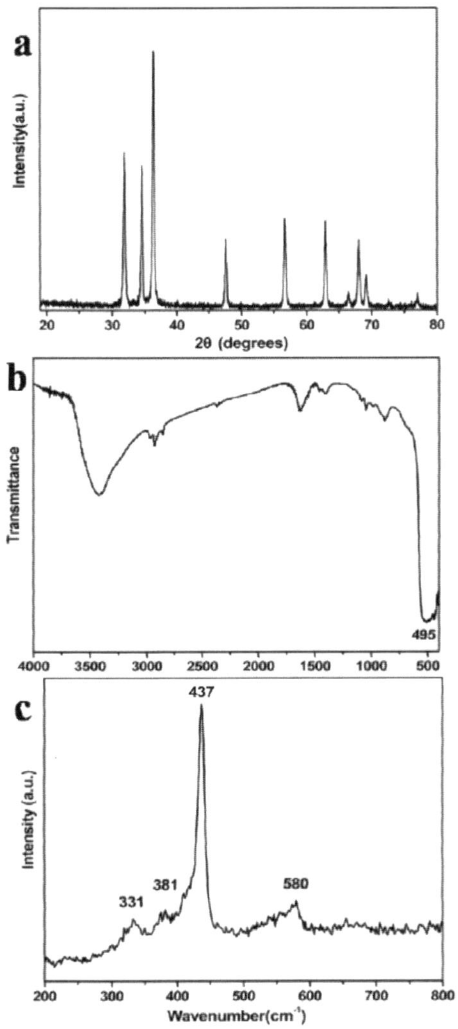


**Figure 3.** (a) XRD patterns, (b) FT-IR spectra and (c) Raman spectra of annealed mesocrystals.

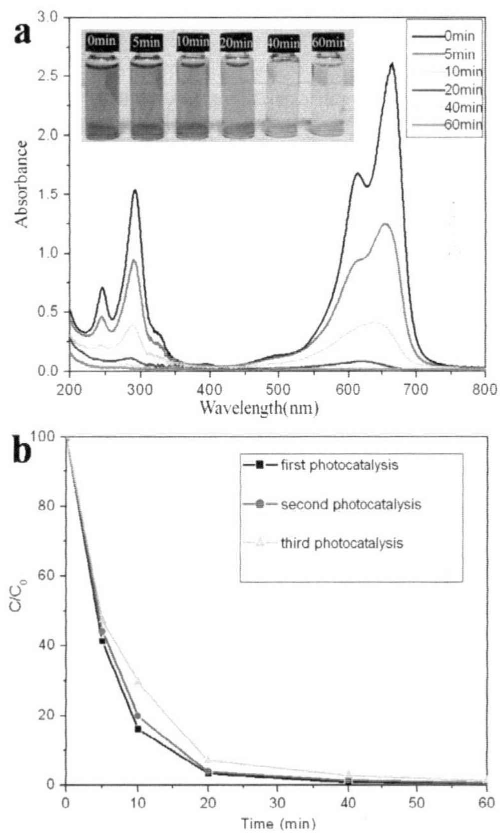


**Figure 4.** (a) UV-Vis absorption curves of the concentration of residual MB with different UV light irradiation time. (Inset) the color of the MB solution with different UV light irradiation time. (b) The photocatalytic efficiency curves of three photocatalytic cycles processing.

As a promising photocatalyst, ZnO has been studied for the photodegradation of some organic compounds. The MB, as a common cationic dye in textile industry, is one of the most notorious contaminants in aquatic environments owing to its huge amounts, slow biodegradation, and toxicity. In our study, the photocatalytic activity of the as-obtained ZnO hierarchical mesocrystals samples (50 mg) on the degradation of MB (initial concentration: 15 mg/L, 100 mL) under UV light irradiation was investigated. The absorption spectrum of MB solution was characterized by its characteristic absorption at 664 nm. Under dark conditions and with diffuse light 20 min, the adsorption–desorption equilibrium of methyl blue onto the ZnO mesocrystals were reach, then the light was turned on. The reaction time was scheduled to be 0, 5, 10, 20, 40 and 60 min, respectively. As shown in figure 4a, the ZnO mesocrystals samples were used for the photodecomposition of MB, and the absorption spectra decrease in intensity rapidly with the extension of exposure time. It revealed that the as-obtained ZnO mesocrystals samples exposure to UV light for 20 min resulted in over 95% degradation of MB, the concentration of MB decreased remarkably. The change in color of the MB solution was also suggested this result in the process of photocatalytic reaction (inset figure 4a). The stability and reusability of ZnO mesocrystals were studied in terms of performing the MB degradation reactions repeatedly three times. The photocatalytic efficiency of ZnO mesocrystals was preserved well (as shown in figure 4b) showing good optical stability. Additionally, the morphology and structure of the mesocrystals were well maintained during the reaction (figure 5).

Experiments also show that ZnO mesocrystals samples can be more readily separated from the slurry system by filtration or decantation after photocatalytic reaction due to their large weight and weak Brownian motion, and the solid catalyst can then be reused. After three repeated cycles of photodegradation of MB with the as-prepared ZnO, the catalyst retains its original photocatalytic activity (figure 4b), confirming ZnO mesocrystals are not photo-corroded or poisoned during the photocatalytic oxidation of the pollutant dye molecules. The most interesting point is that the morphology of the structure remains intact even after the catalysis, as shown in figure 5. The stability of a photocatalyst is important for its practical application. Therefore, the prepared hierarchical ZnO mesocrystals can be regarded as ideal photocatalysts for environmental purification on the industrial scale.

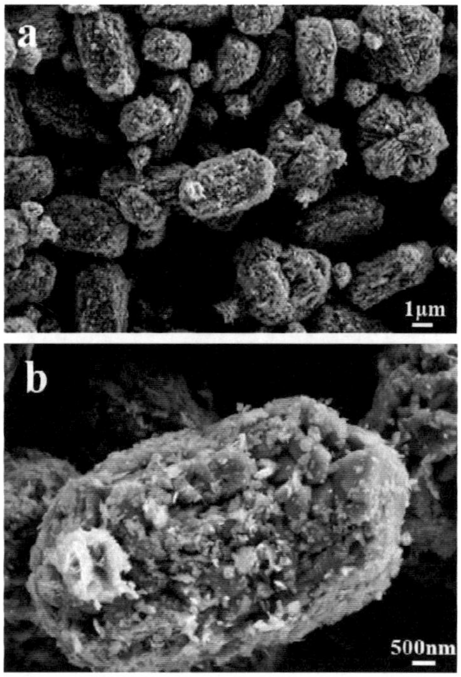


**Figure 5.** The SEM image of ZnO mesocrystal after three photocatalytic cycles.

## 4. Conclusion

The hierarchical ZnO mesocrystals with big at both ends but small in the middle and consisted of nanoplatelets have been prepared successfully, when the aqueous solutions of the $Zn(NO_3)_2$ and NaOH were mixed simply in the presence of ethanol at the 10 °C reaction temperature and followed by annealed in air. It was illustrated that the as-obtained ZnO mesocrystals have good stability and

IOP Conf. Series: Materials Science and Engineering **317** (2018) 012076 doi:10.1088/1757-899X/317/1/012076

photocatalytic performance in the photodegration of MB in water. The product exhibited the potential application of photocatalytic ability in pollutant organic compounds treatment. Given the simple and scalable production procedure, and the easy processing, we hope that the as-obtained ZnO mesocrystals may find practical application in photocatalysis and others.

## References

[1] Song R G, Colfen H 2010 *Adv. Mater.* **22** 1301-30.
[2] Colfen H, Antonietti M 2005 *Angew. Chem. Int. Ed.* **44** 5576-91.
[3] Colfen H, Mann S 2003 *Angew. Chem. Int. Ed.* **42** 2350-65.
[4] Song R Q, Colfen H 2010 *Adv. Mater.* **22** 1301-30.
[5] Nel A, Xia T, Madler L, Li N 2006 Sci. **311** 622.
[6] Hara K, Horiguchi T, Kinoshita T, Sayama K, Sugihara H, Arakawa H2000 *Sol Energ. Mater. Sol. Cell.* **64** 115-34.
[7] Rodriguez J A, Jirsak T, Dvorak J, Sambasivan S, Fischer D 2000 *J. Phys. Chem. B.* **104** 319-28.
[8] Huang M H, Mao S, Feick H, Yan H, Wu Y, Kind H, Weber E, Russo R and Yang P 2001 *Sci.* **292** 1897-9.
[9] Yang Y, Yang Y, Wu H, Guo S 2013 *CrystEngComm*, **15** 2608-15.
[10] Liu Z, Wen X D, Wu X L, Gao Y J, Chen H T, Zhu J, Chu P K 2009 *J. Am. Chem. Soc.* **131** 9405-12.
[11] Li Z, Gebner A, Richters J P, Kalden J, Voss T, Kubel C, Taubert A 2008 *Adv. Mater.* **20** 1279-85.
[12] Ye F, Peng Y, Chen G Y, Deng B, Xu A W 2009 *J. Phys. Chem. C.* **113** 10407-15.
[13] Huang C, Duan X 2014 *Phys. Stat. Solid. Appl. Mater. Sci.* **211** 1562-6.
[14] Bitenc M, Orel Z C 2009 *Mater. Res. Bull.* **44** 381-7.
[15] Phuruangrat A, Yhongtem T, Thongtem S 2009 *Mater. Lett.* **63** 1224-6.

## Acknowledgment

This work was financially supported by the Science Foundation of Jiangsu Province Special Equipment Safety Supervision Inspection Institute (KJ(Y) 2015012).

# Evaluation of the Possibility of Obtaining Tube-to-Tube Sheet Welded Joints of 15Cr5Mo Steel by Alternative Technological Process

**R.G. Rizvanov[1], D.Sh. Mulikov[1], D.V. Karetnikov[1], A.M. Fairushin[1], A.S. Tokarev[1]**

[1]The Chair "Technology of Oil Apparatus", FSBEI HE USPTU, Ufa, the Russian Federation

**Abstract.** This paper presents the results of the tests of joints of chrome-molybdenum steel, obtained by rotary friction welding. On their basis, conclusions were drawn about the weldability of this type of steel by friction welding, and also the applicability of this welding technology in the manufacture of heat exchange equipment.

## 1. Introduction

Heat exchanging equipment is a very significant part of the process equipment in the petrochemical and related industries. The share of heat exchange equipment in the petrochemical and petroleum industries is on average 50% [1].

The most widespread among others are shell-and-tube heat exchangers, as this type of heat exchanger has high efficiency, reliability and has a great variety of design.

An important component of the shell-and-tube heat exchangers is a tube bundle consisting of heat exchanging tubes, tube sheets and baffles. Useful life and reliability of heat exchangers of this type is largely determined by similar indicators of tube-to-tube sheet joints [2]. It is manufactured by rolling or rolling with welding according to GOST R 55601-2013 (state standard of Russian Federation) [3].

Rolled and welded joints are used for apparatuses that require increased strength and tightness of tube-to-tube sheet joints, particularly for heat exchangers operating at high pressures and temperatures, or when specific requirements for fire and explosion safety, toxicity or radioactivity of the working medium are imposed on heat exchange equipment [4].

In the manufacture of equipment used in high-temperature, corrosive environments and high mechanical stresses, heat resistant chromium and chrome-molybdenum steels of the martensitic class are widely used [5]. This steels are characterized by the ability for a long time to meet the performance required for the equipment and its components, at a relatively low cost.

However, the manufacture of welded equipment of these steels is difficult due to the susceptibility of this class of steels to harden. To prevent the formation of cold cracks during the welding process, preheating and heat treatment are necessary. Thus, for 15Cr5Mo steel (Russian analogue of X12CrMo5 steel), the preheating temperature is 300-350 °C and the heat treatment temperature is 740-760 °C [6].

The heat treatment operations are not only difficult to implement, but also energy-consuming, this significantly increases the costs of the manufacturers of heat-exchange equipment.

As a result, for today the use of steels of this class, including 15Cr5Mo, is complicated by the complexity of the assembly and welding process in the manufacture of oil refining and heat-exchange equipment using arc welding methods.

## 2. Experimental procedure and results

*2.1. Prerequisites for conducting research*

One of the alternative ways of reducing the laboriousness of manufacturing heat exchange equipment of hardenable steels is the use of welding methods with a high concentration of energy in the heating spot (laser welding [7], electron beam welding) or friction welding methods. The use of friction welding methods requires the reengineering of the design of the part elements of heat exchange equipment; although at the same time it significantly reduce the costs of manufacturing and the manufacturing time.

An example of the possible design of tube-to-tube sheet joints manufactured by rotary friction welding is shown in figure 1. After the tube 1 is rolled or fixed relative to the tube sheet 2, the edges are prepared at an angle 3. Then, with special tooling, the bushing 4 is rotated with the friction welding force and the forge force after the rotation is stopped. The internal diameter of the bushing 4 is equal to the flow diameter dt of the tube 1 [8]. The advantage of this method is that it can be used both in the manufacture and in the repair of heat exchangers, in particular of 15Cr5Mo steel. However, to date, there are no data on the properties of welded joints of 15Cr5Mo steel obtained by friction welding.

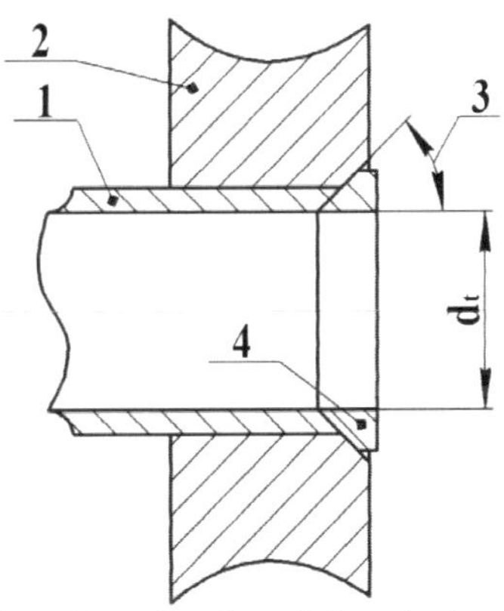


**Figure 1.** Scheme of the tube-tube sheet joint obtained by friction welding.

*2.2. Equipment and modes of welding*

To study the properties of these welded joints of 15Cr5Mo steel, tubes with an external diameter of 35 mm and a wall thickness of 6 mm were made. Rotary friction welding was carried out using a rotary friction welding machine "ПСТ-20Т" produced by company "КТИАМ" in Chelyabinsk. Welding modes are shown in table 1.

Then, the test samples were made by electrospark cutting of the welded joints of the obtained samples for static tensile, cyclic strength, impact, hardness tests, study of corrosion resistance and microstructure of various zones of welded joints.

**Table 1.** Modes of friction welding of samples

| Rotational speed, rpm | Force, kN | | Duration, sec. | |
|---|---|---|---|---|
| | *Friction welding* | *Forge* | *Friction welding* | *Forge* |
| 1250 | 50 | 80 | 7 | 2 |

*2.3. Static tensile test*

The evaluation of the static strength characteristics was carried out in accordance with the requirements of GOST 1497-84 and GOST 6996-66 at room temperature on cylindrical samples (type

IV in accordance with GOST 1497-84, type XXV in accordance with GOST 6996-66) on the tensile machine "Instron 5982." The values of the tensile strength $\sigma_В$, yield strength $\sigma_Т$, strain $\delta_5$ and reduction of area $\psi$ of samples of the base metal, the weld metal and the zone of thermomechanical influence, obtained as the average value of the results of the test of three samples, are shown in table 2.

It can be seen from table 2 that the characteristics of the static strength of the welded joint obtained are at a higher level than the characteristics of the base metal. The plastic properties of metal of the zone of thermomechanical influence on the basis of the values of the strain ($\delta_5$ = 24%) and the reduction of area ($\psi$ = 72%) can be considered high, especially taking into account that the welded parts were not subjected to thermal operations that are necessary to obtain the required level of the plastic properties when using fusion welding.

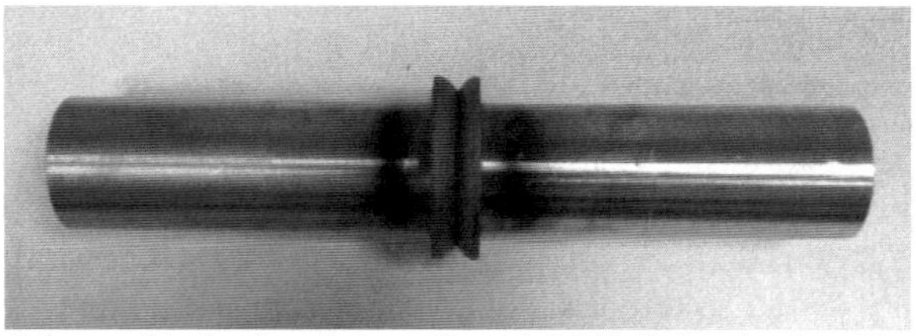

**Figure 2.** Photo of the welded tubes joint of steel 15Cr5Mo obtained by friction welding.

A significant increase in the tensile strength and yield stress of the zone of weld metal is observed, with a typical decrease in the values of the strain and reduction of area.

**Table 2.** Mechanical characteristics of samples.

| Mechanical characteristic | Base metal | Zone of thermomechanical influence | Weld metal |
|---|---|---|---|
| $\sigma_Т$, МПа | 268 | 393 | 1223 |
| $\sigma_В$, МПа | 522 | 558 | 1594 |
| $\delta_5$, % | 33 | 24 | 7,6 |
| $\psi$, % | 81 | 72 | 19 |

*2.4. Hardness test*

The allocation of the microhardness values along the cross section of the base metal, the zone of thermomechanical influence (TMI) and the weld metal is shown in figure 3.

It can be seen from the figure that the allocation of microhardness along the cross section of the welded joint corresponds to the Gaussian law with a maximum at the center of the weld, that is not typical for welded joints of 15Cr5Mo steel obtained by fusion welding [6].

The results of determination of the microhardness and photos of the macrostructure show that the width of the welded joint is approximately 2.8...3 mm for a 6 mm thick piece that is significantly smaller compared to the arc welding methods. Accordingly, this leads to a decrease in the volume of metal with a high level of stresses and the probability of formation of cold cracks.

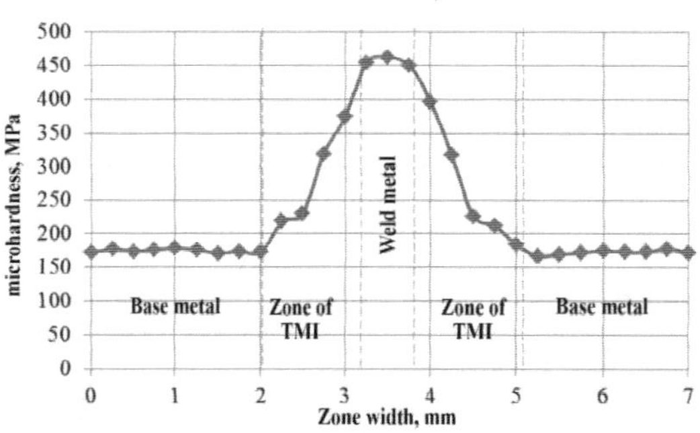


**Figure 3.** Allocation of microhardness along the cross section of the sample of steel 15Cr5Mo obtained by friction welding.

There are no microhardness rises in the width of the welded joint, which are characteristic for the heat-affected zone of welded joints obtained by arc welding. This is a factor that positively affects the welded joint resource due to the less heterogeneity of the mechanical properties along the cross section of the welded joint. In the authors' opinion, the values of microhardness increased in comparison with the base metal can be explained by the formation of an ultrafine-grained structure as a result of strain hardening during the thermodeformation cycle of friction welding (a similar opinion is expressed in [9]). Thus, it can be concluded that the increased values of microhardness in the zone of weld metal are observed not as a result of the formation of hardening structures, but as a result of a higher "density" of the metal structure of the welded joint. In [9] this fact is explained by an increase in the density of dislocations and grain-boundary hardening as a result of the formation of a highly deformed ultrafine-grained structure.

*2.5. Microstructure study*

Figure 4 shows micrographs of the microstructure of the characteristic zones of the welded joint obtained with a raster microscope with a 1000-fold magnification. A visual analysis of the micrographs shows that as a result of the thermodeformation welding cycle, the structure of the characteristic zones of the welded joint is more crushed than the base metal and increases the degree of its homogeneity. At the same time, the microstructure of the weld metal is more "disoriented" than the zone of thermomechanical influence and the base metal, which can be explained by the higher level of stresses and deformations arising from the contact of parts during welding and forging process. The above-mentioned character of the microstructure of the zone of weld metal partially explains the increase in the level of mechanical and plastic properties of the weld metal at increased values of microhardness.

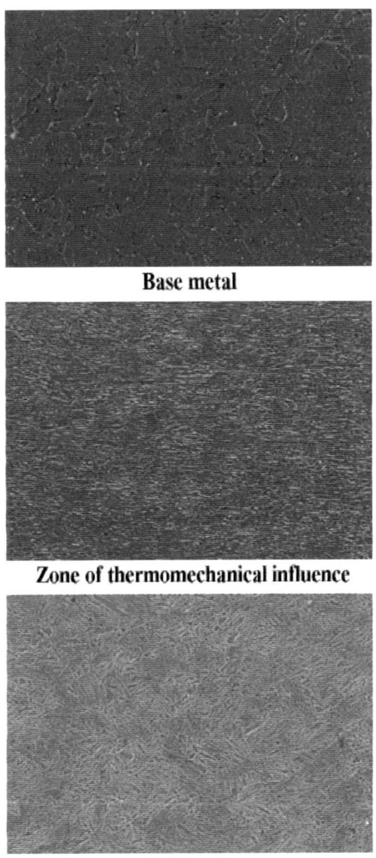


**Figure 4.** Microstructure of the characteristic zones of the welded joint (1000-fold increase).

IOP Conf. Series: Materials Science and Engineering **317** (2018) 012077 doi:10.1088/1757-899X/317/1/012077

### *2.6. Impact test*

Impact test was carried out with the copra with a vertically falling load "Instron CEAST 9350" on samples in accordance with GOST 6996 (three samples for each type of welding) with V-notch located at the center of the weld and a test temperature of 20 °C. Average values of impact toughness were as follows: for samples obtained by arc welding - 162 J/cm$^2$, for samples obtained by friction welding - 151 J/cm$^2$. Thus, the impact toughness of samples obtained by friction welding is close to the impact toughness of samples obtained by arc welding using the existing technology (using thermal operations).

### *2.7. Static bending test*

To assess the plastic properties of the welded joint, taking into account the susceptibility of 15Cr5Mo steel to form cold cracks, test for static bending of additional samples of rectangular section in accordance with GOST 1497-84 was carried out.

The tests for static bending were carried out until the surfaces were parallel. No visible cracks were detected after testing on samples of both types. This indicates that the used arc welding technology was implemented properly, and the friction welding technology has allowed obtaining welded joints with an acceptable level of plastic properties without the use of thermal operations.

Figure 5 shows photos of samples after static bending test.

On the sample obtained by friction welding, it can be seen that the weld metal has high strength properties, and as a result of bending the sample obtain a П-shape without any plastic deformation of the weld metal. The absence of cracks in this zone also indicates a favorable stress-strain state of the welded joint after welding.

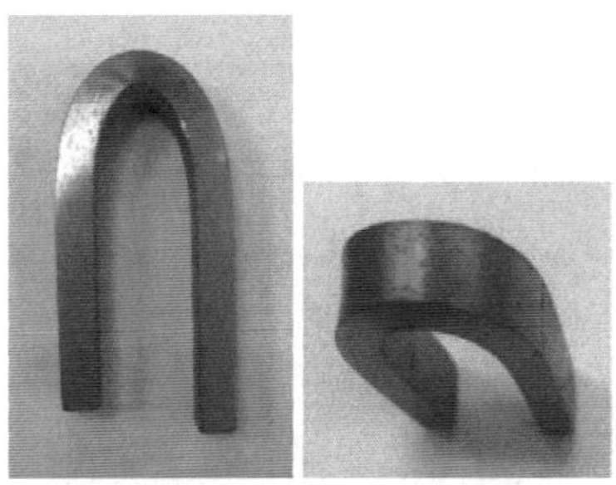

Sample obtained by arc welding

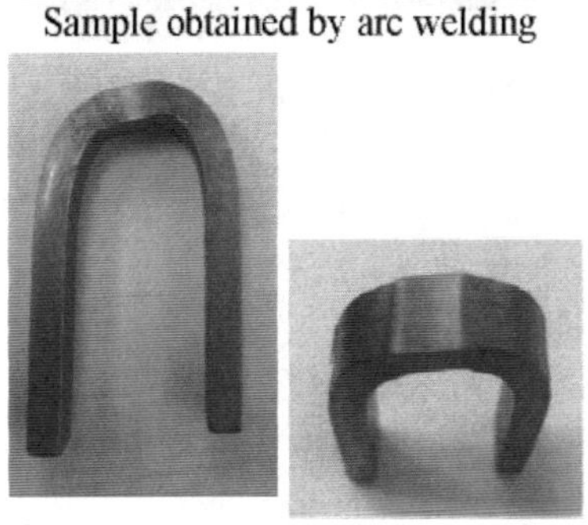

Sample obtained by friction welding

**Figure 5.** Samples after static bending test.

### *2.8. Corrosion resistance study*

To assess the corrosion rate of the welded joint obtained by friction welding, gravimetric tests were carried out by holding the samples in a 9% solution of sulfuric acid for 160 hours. Based on the results obtained, the corrosion rate was determined to be 36.8 g/m$^2$·h, which indicates that the welded joint has a high corrosion resistance in corrosive environments typical for operating conditions of 15Cr5Mo steel. In the opinion of the authors, the high corrosion resistance is due to the absence of thermal influence of the arc during the friction welding process, which has a positive effect on the microstructure of the weld metal and the residual stress level.

In addition, the electrode potential values were determined. Determining of the electrode potential values makes it possible to determine the corrosion resistance of welded joint zones and the most vulnerable areas.

The most favorable allocation of potentials is the case of equality of the electrode potentials of the weld and the base metal ($\varphi_{w.} = \varphi_{b.m.}$), when there is the process of general uniform corrosion. In real production conditions, it is difficult to achieve this equality, so it is preferable to displace the electrode potential of the weld to higher values. The anodic dissolution is transferred to the base metal while reducing the weld corrosion rate ($\varphi_{w.} > \varphi_{b.m.}$).

Due to the uneven corrosion of the welded joint, the mass index does not fully characterize its corrosion resistance. Therefore, the method of electrode potential was used to evaluate corrosion of welded samples.

Tests of the electrode potentials of the zones of the welded joints were carried out. The electrode potentials of the characteristic zones of the weld and the base metal were registered with a special probe. The value of the electrode potential was recorded with a voltmeter with a measurement error of ±5%. The results of the tests are shown in figure 6.

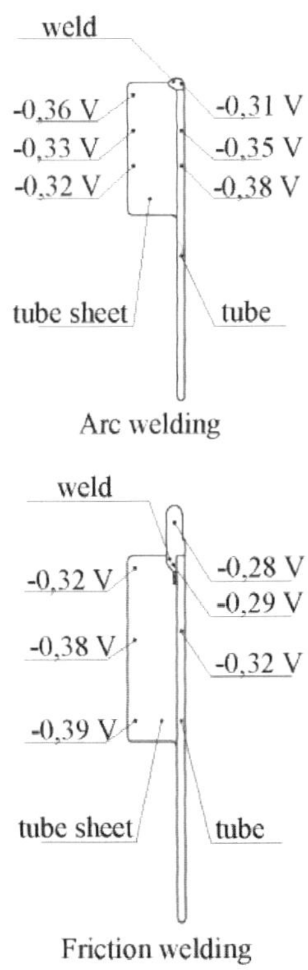


**Figure 6.** Results of measuring of the electrode potential values of samples.

The weld obtained by arc welding has a potential of -0.31 V (silver chloride electrode), and is more noble to the metal of both the tube and the tube sheet.

The potential of the weld of the sample obtained by friction welding has higher value than the potentials of both the tube and the tube sheet.

Thus, from the point of view of electrochemical corrosion, products obtained by friction welding are not inferior to corrosion resistance to products obtained by arc welding.

### *2.9. Cyclic strength test*

To assess the durability of the obtained welded joints, the samples of the base metal and the weld metal were subjected to a symmetrical cyclic tensile load with the beginning at zero, a frequency of 10 Hz and maximum amplitude equal to 0.6 of the yield strength of 15Cr5Mo steel. Tests, including the manufacture of samples with a diameter of 5 mm and length of 25 mm, were carried out in accordance with the requirements of GOST 25.502-79. As a result of the tests, the destruction of samples of both the base metal and welded metal was not observed at the number of load cycles of $10^5$. Thus, it can be concluded that the life of the welded joints of 15Cr5Mo steel with the number of load cycles of $10^5$ is not lower than samples of the base metal.

### *2.10. Evaluation of test results*

On the basis of the results, the weldability of 15Cr5Mo steel by rotary friction welding can be estimated as good; to obtain welded joints with a high level of strength and plastic properties, no additional thermal operations are required, and the life of the welded joints is at a good level. The introduction of friction welding into the technological process of manufacturing welded oil and gas equipment, in particular, tube bundles of heat exchangers of 15Cr5Mo steel, will require a significant upgrade of the friction welding equipment in connection with the design of tube bundles. However, it seems economically expedient because it allows avoiding energy-intensive thermal operations and reducing manufacturing time due to the higher degree of automation of the friction welding process.

## 3. Conclusion

1) 15Cr5Mo steel has good weldability by rotary friction welding, which allows obtaining welded joints with mechanical and plastic properties of the zone of thermomechanical influence at the level of the base metal without the use of thermal operations.

2) The use of friction welding for the production of tube bundles of heat exchangers of 15Cr5Mo steel is economically expedient, despite the need for a significant upgrade of the specialized welding equipment.

## References

[1] Berliner Yu I, Balashov Yu A 1976 *Technology of chemical and petroleum apparatus building* Moscow: Mechanical Engineering.

[2] Vikhman G L, Kruglov S A 1978 *Basic foundation of designing devices and machines of oil refineries* Moscow: Mechanical Engineering.

[3] GOST R 55601-2013 2014 *Heat exchangers and air coolers Attaching tubes to tube sheets General technical requirements* Moscow: Standartinform.

[4] Vladimirov A I, Peremyachkin V I 2001 *Repair of oil and gas refining and petrochemical equipment* Moscow: Oil and Gas.

[5] Zainullin R S 2000 *Resource-saving technologies in petrochemical apparatus building* Ufa: Transtech.

[6] Khalimov A A, Zharinova N V, Khalimov A G, Fairushin A M 2012 *Oil Gas Bus.* **3** 102-8.

[7] Tukaev R F, Ibragimov I G, Fairushin A M, Sisanbaev A V 2013 *Oil Gas Bus.* **5** 363-75.

[8] Karetnikov D V, Rizvanov R G, Fajrushin A M, Mulikov D Sh, Balandina A G 2014 *The method of joining the "tube-tube sheet" joint by friction welding*, patent RU 2012149757.

[9] Zakirova A A, Nagimov M I, Husnullin A M, Kashaev R M 2011 *Promis. Mater.* **12** 155–60.

IOP Conf. Series: Materials Science and Engineering **317** (2018) 012078 doi:10.1088/1757-899X/317/1/012078

# Transformation of localized necking of strain space into stress space for advanced high strength steel sheet

**Aeksuwat Nakwattanaset[1], Surasak Suranuntchai[1]***

[1]Department of Tool and Materials Engineering, Faculty of Engineering, King Mongkut's University of Technology Thonburi, 126 Pracha Uthit Road, 10140 Bangkok, Thailand

**Abstract.** Normally, Forming Limit Curves (FLCs) can't explain for shear fracture better than Damage Curve, this article aims to show the experimental of Forming Limit Curve (FLC) for Advanced High Strength Steel (AHSS) sheets grade JAC780Y with the Nakazima forming test and tensile tests of different sample geometries. From these results, the Forming Limit Curve (strain space) was transformed to damage curve (stress space) between plastic strain and stress triaxiality. Therefore, Stress space transformed using by Hill-48 and von-Mises yield function. This article shows that two of these yield criterions can use in the transformation.

## 1. Introduction

In recent years, the automotive industry focusses on energy savings. Advanced High Strength Steel (AHSS) is a crucial demand for the automotive industry and trend to increase caused by lightweight steels. The main defects of sheet metal forming in automotive industry is fracture caused by necking [1] forming limit curve (FLC) of sheet metals is popularly employed in automotive industries. Nevertheless, the behavior of advanced high strength steel (AHSS) grades with decreased ductility develop an issue concerning shear fracture. This issue could not be accurately predicted by the conventional forming limit curve (FLC). With this procession, the new forming criterion is cultivated, for example; the new representation of damage curve with relates to stress space between plastic strain and triaxility [2]. In this work, studied in strain space by using the test of Nakajima test and tensile tests of pure shear and combined loading specimens of Advanced High Strength Steel (AHSS) grade JAC780Y. Afterwards, analysis in transformation's calculated of forming limit curve (strain space) to be damage curve (stress space) by using anisotropic Hill-48 yield function and isotropic von-Mises yield function.

## 2. Measurement of strain based forming limit curve

### *2.1. Nakajima test*

The experimental forming limit data for JAC780Y advanced high strength steel (AHSS) sheets was obtained from the Nakajima test which accomplished by make use of the standard ISO 120004 - 2 [1, 2] and each simples repeated three times. All of the sheet samples' shape had 200 mm of length. However, the widths were varying between 55 to 200 mm. The thickness of the steel sheet was 1 mm. In this the experiments, the mean of an optical strain measurement system accumulated the mutilated samples' local strain histories. Generally, the evaluation of critical plastic strains for regular FLC occurred from estimated strain distributions at the stage of localized necking. This stage of localized necking could be characterized by local strain rates' development. In this case, the maximum rate of thickness reduction of principal in plane plastic strains were classified. Afterward, it's used as limiting strain values to fracture for the FLC (strain based).

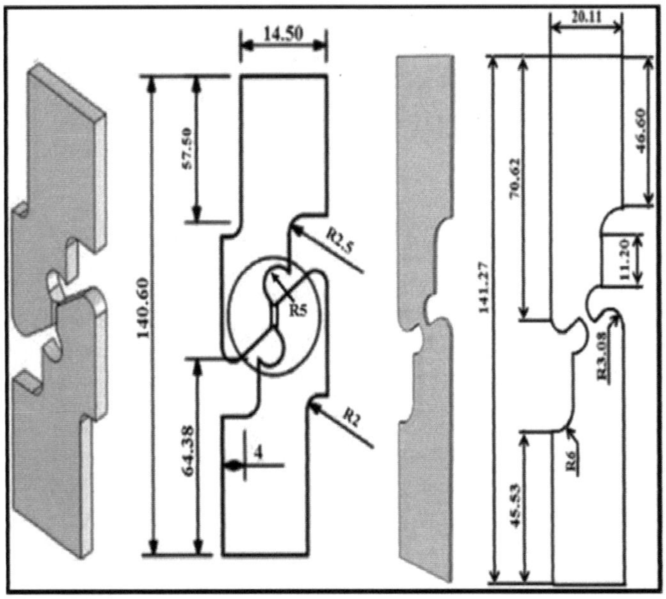


**Figure 1.** Pure shear/combined load sample [4].

### *2.2. Shear fracture limit test*

Pure shear's tensile tests and combines loading specimens were carried out for the receiving fracture beginning strains due to shear failure statue. Figure 1 show geometries of pure shear's using and combined loading sample [4]. In pure shear sample's case, the middle area of the sheet sample was controlled by closely pure shear stress. For the combined loading sample, the pure shear sample was lightly altered to produce a stress state revealing a combination of shear and tensile loading, as shown by Bao and Wierzbicki. [5]. The optical strain measurement system was also involved in experiment of calculation of local strain distributions developing in the critical field of both cases. Comparable with Nakazima test, the analytical maximum and minimum in level strains were obtained at the stage before crack appearance and used as shear fracture limit strains.

## 3. Measurement of stress based damage curve

### *3.1. Uniaxial tensile test*

The ASTM E8 standard specimen was used in uniaxial tensile test of the investigation of steel sheet activity. The test of sheet sample went through three loading directions; 0°, 45° and 90°. These directions became the rolling direction (RD). In the tension test, elongations within the gauge length of the test specimens regarding both sample length and width were determined. The calculation of *r*-values was undergoing by the ratio between valid width and thickness measured up to 15% of the total elongation.

### *3.2. Transformation into stress space damage curve*

Obtainable forming limit strains were transformed into stress triaxiality and plastic strain space under condition of anisotropic Hill'48 yield function and isotropic Von-Mises yield function [5]. The strain ratio ( $\alpha$ ) was defined by $\alpha = (\varepsilon_2 / \varepsilon_1)$ and the average normal anisotropy was expressed as $\bar{r} = (r_0 + 2r_{45} + r_{90})/4$ . The maximum and minimum principle strains could be transformed to stress triaxiality $\left(\eta_{Hill'48}\right)$ and effective plastic strain $\left(\bar{\varepsilon}_{Hill'48}\right)$ space combined with Von-Mises yield function by equation 1 and 2, with Hill'48 yield function by equation 3 and 4 [5].

$$\eta_{iso} = \frac{1+\alpha}{\sqrt{3}\sqrt{1+\alpha+\alpha^2}} \tag{1}$$

$$\bar{\varepsilon} = \frac{2\varepsilon_1}{\sqrt{3}}\sqrt{1+\alpha+\alpha^2} \tag{2}$$

$$\eta_{Hill'48} = \left(\frac{\sqrt{1+2\bar{r}}}{3}\right) \times \frac{1+\alpha}{\sqrt{1+\left[\frac{2\bar{r}}{1+\bar{r}}\alpha\right]+\alpha^2}} \tag{3}$$

$$\bar{\varepsilon}_{Hill'48} = \frac{1+\bar{r}}{\sqrt{(1+2\bar{r})}}\sqrt{\varepsilon_1^2+\varepsilon_2^2+\frac{2\bar{r}}{(1+\bar{r})}(\varepsilon_1\varepsilon_2)} \tag{4}$$

## 4. Results and discussion

### *4.1. Forming limit curve in strain space*

In the figure 2 show that in the Nakajima test and tensile test of shear and combined loading specimen, the determination was seen in the forming limit curve (strain based) of the examined steel grade JAC780Y. The strain based forming limit comprised three strain divisions. The first division correlated to the stress states between biaxial tension ($\varphi_1 = \varphi_2$) and plane strain ($\varphi_2 = 0$). The second domain was the scope from planes strain $(\varphi_2=0)$ to uniaxial tension ($\varphi_1 = -2\varphi_2$). The last division showed the stress state from uniaxial tension ($\varphi_1 = -2\varphi_2$) to pure shear ($\varphi_1 = -\varphi_2$). This third strain domain represented the shear facture of the steel's limit.

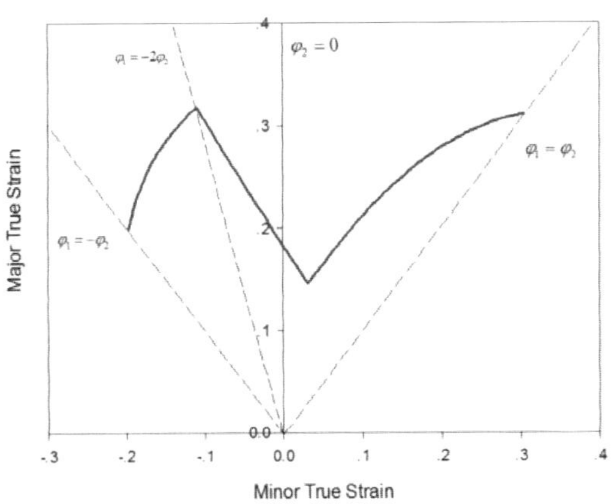


**Figure 2.** Forming limit curve (strain base) of with JAC780Y steel.

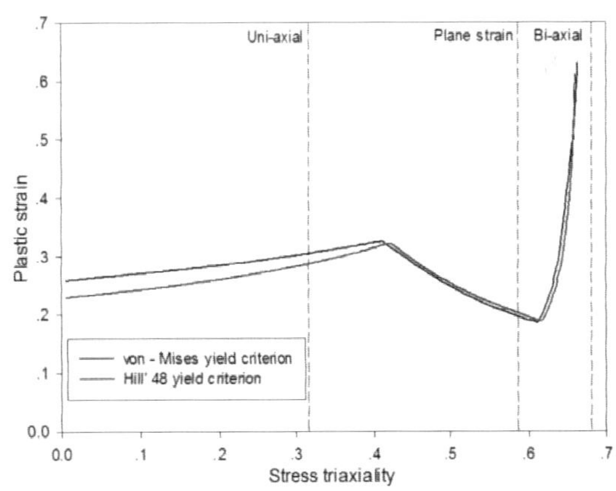


**Figure 3.** Damage curve of JAC780Y steel von-Mises and Hill'48 yield criterion.

**Figure 4.** Damage curve of JAC780Y steel with von-Mises yield criterion.

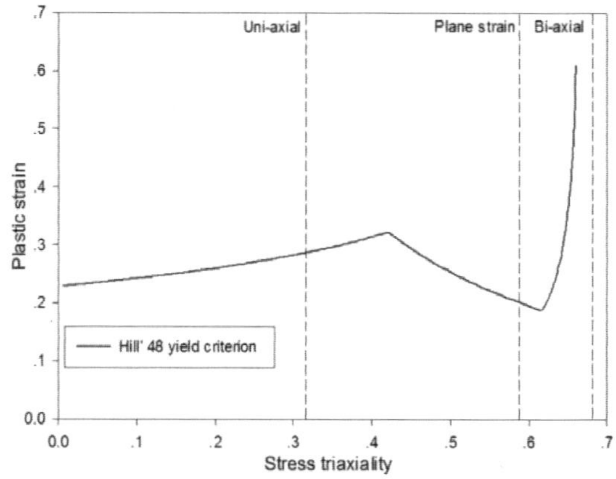


**Figure 5.** Damage curve of JAC780Y steel with Hill'48 yield criterion.

*4.2. Calculation of damage curve in the stress space*

The stress triaxiality and effective plastic strain space by numerical calculations came from the transformation of the strain based forming limit curve in figure 3. According to equation 1 and 2 for von-Mises yield criterion was displayed in figure 4 and equation 3 and 4 for Hill' 48 yield criterion in figure 5. At the same time, the stress based damage curve presented three separate divisions. The strain based forming limit curve show that the three areas correlated to the equal states of stress. Notice that the pure shear state, biaxial tension, plane strain and uniaxial tension were approximately showed by the triaxiality values of 0.667, 0.575, 0.33 and 0.In the section of pure shear - uni-axial tension hill'48 yield criterion was under the von - Mises yield criterion. As well as uni-axial tension - plane strain Hill'48 yield criterion was upper than von - Mises yield criterio along with plastic strain's axis. The beginning of section of plane strain - Bi-axial, von - Mises yield criterion had stress trixiality's value below than hill'48 yield criterion. The end of the Hill'48 yield criterion had stress trixiality's value was below von - Mises yield criterion along with the Stress trixiality's axis.

## 5. Conclusion

The current results show the strain and stress based damage curves from the AHS steel grade JAC780Y. The Nakajima test and tensile tests of pure shear and combined loading specimens were operated for receiving the forming limit curve. At that point, the strain based forming limit curve was converted to damage curve relationship of stress triaxiality and effective strain space under examined of the Hill'48 and von-Mises yield function. In future work, this damage curve relationship of stress triaxiality and plastic strain space will have examined on the geometries of automotive industry parts.

## References

[1] Kim S B, Huh H, Bok H H, Moon M B 2010 *J. Mater. Proc. Tech.* **211** 851-62.
[2] Gorji M, Berisha B, Hora P and Barlat F 2016 *Int. J. Mater. Form.* **9(5)** 573-84.
[3] Nakazima K, Kikuuma T and Hasuka K 1971 *Yawata Tech. Report* **284** 678-80.
[4] Lian J, Sharaf M, Archie F and Muenstermann S 2013 *Int. J. Damage. Mech.* **22(2)** 188-218.
[5] Bao Y and Wierzbicki T 2004 *J. Eng. Mater. Tech.* **126(3)** 314-24.
[6] Isik K, Silva M B, Tekkaya A E and Martin P A F 2014 *J. Mater. Proc. Tech.* **214** 1557-65.

## Acknowledgement

The authors would like to thank the Researchers and Research for Industry project of the Thailand Research Fund for financial support in this research. In addition, sincere thanks must be given to Thai Summit Automotive Co., Ltd for useful advice and providing the importance data about the experiment steel.

IOP Conf. Series: Materials Science and Engineering **317** (2018) 012079 doi:10.1088/1757-899X/317/1/012079

# Cold Forming of Ni-Ti Shape Memory Alloy Sheet

**Kaung-Jau FANN[1*] and Jhe-Yung SU[1]**

[1]Department of Mechanical Engineering, National Chung Hsing University, Taichung 40227, Taiwan

**Abstract.** Ni-Ti shape memory alloy has two specific properties, superelasiticity and shape memory effect, and thus is widely applied in diverse industries. To extend its further application, this study attempts to investigate the feasibility of cold forming its sheet blank especially under a bi-axial tensile stress state. Not only experiments but also a Finite Element Analysis (FEA) with DEFORM 2D was conducted in this study. The material data for FEA was accomplished by the tensile test. An Erichsen-like cupping test was performed as well to determine the process parameter for experiment setup. As a result of the study, the Ni-Ti shape memory alloy sheet has a low formability for cold forming and shows a relative large springback after releasing the forming load.

## 1. Introduction

Shape memory effect was found in Au-Cd alloy by Ölander [1]. The shape memory effect is actually generated by phase transformation between austenite at high temperature and martensite at low temperature [2]. The shape formed in austenite at high temperature will be recovered, if the part is deformed in martensite at low temperature and the deformation load is released, by raising the temperature, at which the austenite is stable. Shape memory alloys, which have the shape memory effect, possess a rubber-like behavior as superelasticity as well by phase transformation between austenite and martensite at high temperature where the austenite is table [3]. With these two distinguished properties, shape memory alloys can be implemented as dampers, devices, sensors, actuators, and so on [4]. However, not until Buelher and his colleague found the same effect in Ni-Ti alloy [5] would the application of this effect be expanded in the other industries [6], especially in medical devices [7].

Traditionally to secure or to memorize the shape of Ni-Ti shape memory alloy parts, they need further precipitation treatment at a relative high temperature. It cannot be avoided that tool sets are needed to clamp the parts to keep their shapes during the treatment at high temperature. Thus the thermomechanical process is commonly used to make Ni-Ti shape memory alloy parts by forming the alloys in hot state with a subsequent precipitation treatment, which is a cost intensive process and needs lots of investment for the tool sets [8]. Fann and Huang [9] therefore proposed a cold forming process, in that the shape memory alloys are first solid solutionized, then formed at room temperature, and thereafter aged for shape memory again at high temperature. Even this process would lose part of the shape formed at room temperature during the precipitation treatment, it is still worth to use it to form the shape memory alloy parts for saving the cost of the tool sets. This proposed process encourages researchers to explore the feasibility of forming Ni-Ti shape memory alloy sheets, which are usually formed at room temperature because of their geometrical characteristics to easily dissipate the heat. This article is therefore aimed to investigate the behavior of Ni-Ti shape memory alloy sheets during cold forming at room temperature, especially under a bi-axial tensile stress state.

## 2. Methods and setup

### *2.1. Material*

The material acquired for this study was a Ni-rich (50.0-50.8at% Ni) Ni-Ti shape memory alloy sheet having a thickness of 0.9 mm. The chemical compositions of the alloy sheet used in this study was determined by a glow discharge spectrometer and had 50.67at% Ni and 49.24at% Ti after converting weight percent (wt%) to atomic percent (at%). The content of Fe is as little as 0.10at% and can be regarded as a minimal contaminant during casting, so that the material can be considered as a pure binary Ni-Ti system.

*2.1.1. Tensile Test.* The Ni-Ti shape memory alloy sheet was placed in a heated furnace at 800°C for one hour and subsequently quenched in water, which served as a solid solution process affiliated to an annealing process [9]. Specimens are prepared according to the subsize defined by ASTM E8 [10] with a gauge length of 25 mm and width of 6 mm. Figure 1 shows the engineering stress strain curve obtained by tensile test after the unloading of an initial 12% tensile strain. This at 800°C solid solutionized Ni-Ti shape memory alloy sheet showed a low Young's modulus around 41 GPa at room temperature in austenite, started the martensitic transformation under the strain about 1.3% and a plateaus strength about 530 MPa. Once the strain reached about 11% the sheet metal started yielding and deformed plastically. After unloading at the stain of 12%, the sheet shows a further low Young's modulus about 21 GPa. This alloy could have a maximum elongation of 21% without necking, if its specimen was tested until failure. It means that this alloy has ductility to deform.

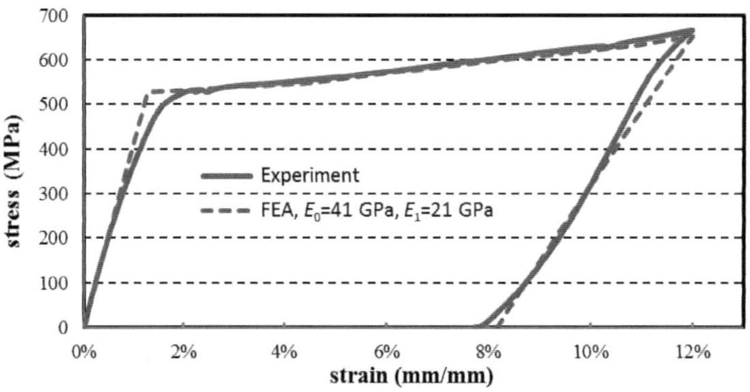


**Figure 1.** Stress strain curve of Ni-Ti shape memory alloy obtained by a tensile test and FEA.

*2.1.2. FEA of Tensile Test.* To prepare the material module for the Finite Element models later to analyze the deformation of the Ni-Ti shape memory alloy sheet during cold forming, an elastic-plastic Finite Element model was created to simulate the above mentioned tensile test of a 12% strain by DEFORM 3D. The commercial software DEFORM 3D can formulate the stress strain relation during the martensitic transformation because of its similarity to classic plasticity [11, 12]. The stress strain curve obtained from the FEA is shown as the dash line in figure 1. It matches very well the stress strain curve obtained by the experiment. Thus the idealized stress strain curve with the Young's modulus of 41 GPa for loading ($E_0$) and 21 GPa for unloading ($E_1$) as well as the software should be suitable for analyis of cold forming Ni-Ti shape memory alloy sheet.

*2.1.3. Experiment Setup of Cold Forming.* The cold forming of Ni-Ti shape memory alloy sheets in this study was a kind of stretch forming, in which the sheet was clamped around and its deformation came from the reduction of its thickness under a bi-axial tensile stress state. Figure 2 (a) shows the schematic experiment arrangement used in the study. A bottom die with a specified cavity depth was inserted into a shrink ring under an interference fit, so that the compression load of the punch could be borne by the bottom die with the support of the ring. The sheet blank was then clamped by a V shaped draw bead of the blankholder around the ring circumference at the diameter of 32 mm. Subsequently a hemispherical punch with a diameter of 30 mm descended with a traveling speed of 10 mm/min through the hole of the blankholder on the blank as well as into the cavity of the bottom die. Once the punch reached its bottom dead center, it returned upwards to its original top dead center and the blankholder as well as the formed sheet metal was then removed.

*2.1.4. Erichsen-Like Cupping Test.* The Erichsen-like cupping test was actually served as a pre-test for preparing the tools of experiments by seeing how deep the Ni-Ti shape memory alloy sheet could be formed under the experiment setup as above mentioned. Because the dimension for the cupping test deviated a little bit from the standard Erichsen cupping test [13], it was called in this study as Erichsen-like cupping test. Besides the punch diameter and the clamping method, the hole diameter in 28 mm of the shrink ring without the bottom die was different to the standard as well.

The Erichsen-like index from the five Erichsen-like cupping test results was located between 4.1 mm and 4.2 mm. The location of the fracture appeared always around the corner radius of the bottom die. It seems that the maximum strain occurred there.

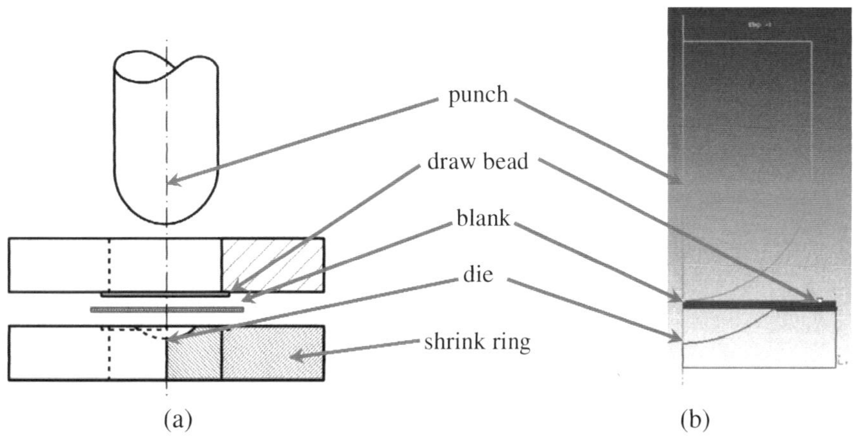


**Figure 2.** Cold forming Ni-Ti shape memory alloy sheet:
(a) schematic setup of experiment (b) rotation symmetrical model for FEA.

*2.1.5. Process Parameter of Cold Forming.* Based on the above obtained Erichsen-like index, the process parameter or the cavity depth of the bottom die for the cold stretch forming in this study was therefore set as 1 mm, 2 mm, 3 mm, and 4 mm, respectively. Assuming that the sheet was well clamped and the deformation of its portion contacting to the punch was homogenous and barely raised from the reduction of its thickness, the apparent strain e of the cold forming process is then defined as

$$e = \frac{a_1 - a_0}{a_0} \times 100\% \tag{1}$$

where $a_0$ and $a_1$ are the area of the thickness middle layer associated with deformation before and after cold forming. For the hemi-sperical punch used in this study, the apparent strain could be theoretically determined for the die cavity depth 1 mm, 2 mm, 3 mm, and 4 mm as 4.9%, 8.6%, 13%, and 17%, respectively. According to the stress strain curve in figure 1, the above listed apparent strains show that the deformations were at the stage of just matensitic transformation, midway of martensitic transformation, just plastic deformation, and midway of plastic deformation, respectively.

*2.1.6. Modeling for FEA of Cold Forming.* To detail the deformation of the Ni-Ti shape memory alloy sheet during cold forming, elastic-plastic rotation symmetrical Finite Element models were established by DEFORM 2D, in which the tress strain relation beyond the martensitic transformation was regarded as a plastic deformation based on [11]. Figure 2 (b) shows one of the models, which has the depth of the die cavity in 4 mm. The material for the FEA was set as that for figure 1. There were 4,559 elements and 4,965 nodes generated for the anlaysis model. The friction coefficient for the contact between the sheet and the punch as well as the die was set as 0.07, which is normally used in cold stamping process. The draw bead is fixed on the top of the blank.

## 3. Results and Discussion

### *3.1. Formed Profile*

The solid line shown in figure 3 is the profile of the cold formed sheet measured from a symmetry plane through the top of the dome with a coordinate measurement machine, while the dashdotted line is the die geometry having the forming depth of 4 mm regarded as target geometry to compare. It can be observed that the dome of the formed sheet is relative flat and low by springback, because the material has the superelastic property. Further compared with the dashed line obtained by FEA, in which the dome height is much higher than that from experiment. It can be attributed to that the boundary condition in Finite Element modeling might be deviated from the reality, especially the draw bead around the bottom die and the shrink ring.

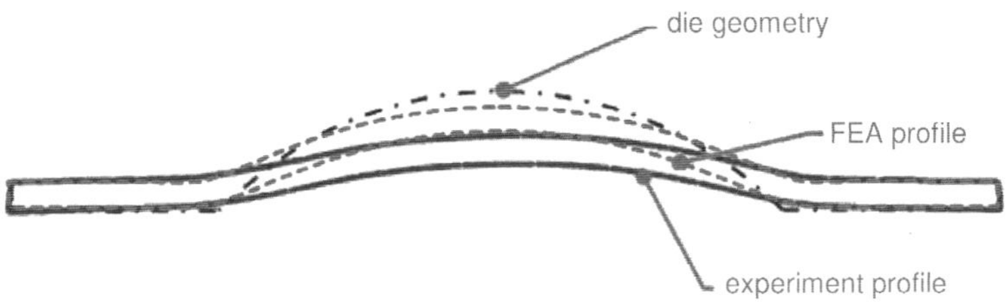


**Figure 3.** Profile of Ni-Ti shape memory alloy obtained by experiment and FEA in comparison to the die geometry.

*3.1.1. Forming Loads.* The solid lines in Figure 3 show the punch load over the stroke during cold forming the Ni-Ti shape memory alloy sheet to the bottom die with different depths. After reaching their maximum, the load drops quickly and then slows down as the punch returns upwards. At the depth of 0.16 mm, the punch load vanishes for cold forming with the bottom die having cavity depth of 1mm. That means that the deformation of the cold forming almost totally springs back. Forming with dies with deeper cavity had smaller springback. For the sheet formed with die cavity depth of 4 mm, the load vanished at 2.23 mm. Similar results could be found in the FEA results. The maximum load computed by FEA is about one and a half times that from experiment, because the modeling of the boundary conditions was moderately idealized in FEA as mentioned above.

*3.1.2. Springback Rate.* The springback rate $r$ after cold forming can be defined as

$$r = \frac{h_0 - h_1}{h_0} \times 100\% \tag{2}$$

where $h_0$ and $h_1$ are the die cavity depth and the dome height of the cold formed sheet after removing the forming load. According the data accessed from figure 4, it can be worked out from equation 2 that the springback rate for the depth of 1 mm, 2 mm, 3 mm, and 4 mm is 84%, 66%, 50%, and 44% from the experiments as well as 81%, 54%, 44%, and 34% from the FEA, respectively. It seems that the springback is still large for the maximum achievable dome height. Thus, cold forming might not be applicable for Ni-Ti shape memory alloy sheet, which has superelasticity and low formability.

## 4. Conclusions

This study has accomplished the tensile test of an at 800°C solid solutionized Ni-rich Ni-Ti shape memory alloy sheet at room temperature and formulated a material model for the FEA. Both of its

Young's moduli before and after martensitic transformation must be separately considered in the FEA of elastic-plastic forming process, because the modulus for loading is twice as high as that for unloading. This shape memory alloy sheet shows a relative low Erichsen-like index as 4.1 mm for the cupping test conducted in this study. Experiments for cold forming such alloy sheet were performed with the dies having depths of 1 mm, 2 mm, 3 mm, and 4 mm, respectively, which have low average strains of 4.9%, 8.6%, 12.6%, and 16.8%, respectively. As a result, both experiment and analysis show that cold forming might not be practical for Ni-Ti shape memory alloy sheet, because the springback is still too high for a maximum achievable deformation under bi-axial tensile stress state.

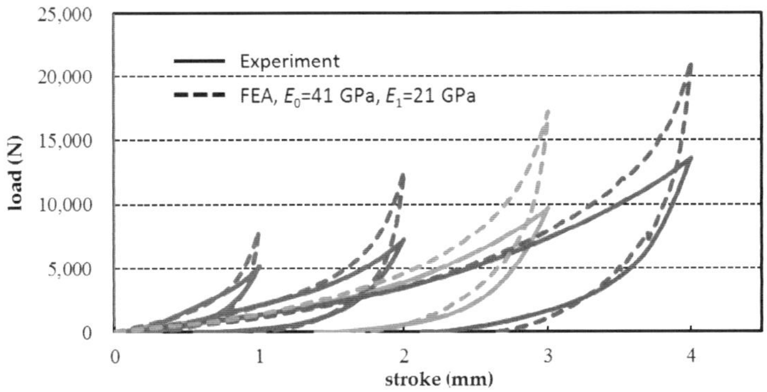


**Figure 4.** Punch load over stroke of Ni-Ti shape memory alloy obtained by experiment and FEA.

## 5. References

[1] Ölander A 1932 *Z. Krist.* **83A** 145-8.
[2] Chang L C and Read T A 1951 *T. Metall. Soc. AIME J. Metal.* **189** 47-52.
[3] Christian J W 1965 *Shape memory alloys*, in: J.W. Christian (Ed.), The Theory of Transformations in Metals and Alloys, Elsevier, Oxford, pp 1102–1106.
[4] Wayman C W and Duerig T W 1990 *An introduction to martensite and shape memory*, in: T.W. Duerig, K.N. Melton, D. Stöckel, C.M. Wayman, Engineering Aspects of Shape Memory Alloys, Butterworth-Heinemann, London, pp 3–20.
[5] Buehler W J, Gilfrich J V and Wiley R C 1963 *J. Appl. Phys.* **34** 1475-7.
[6] Hodgson D E, Wu M H and Biermann R J 1990 *Shape memory alloys*, in: ASM Handbook Committee, ASM Handbook, Volume 2: Properties and Selection: Nonferrous Alloys and Special-Purpose Materials, ASM International, Materials Park, pp. 897-902.
[7] Duerig R, Pelton A and Stöckel D 1999 *Mater. Sci. Eng. A* **273–275** 149–60.
[8] Fann K J and Hsu H C 2014 *Adv. Mater. Res.* **939** 430-6.
[9] Fann K J and Huang P M 2015 *Key Eng. Mater.* **661** 98-104.
[10] American Society for Testing and Materials 2011 Standard test method for tension testing of metallic materials, Designation: E8/E8M-11.
[11] Gao X, Huang W and Zhu J 2001 *Modeling and Numerical Simulation of Shape Memory Alloy Devices Using a Real Multi-Dimensional Model*, in: U. Gabbert, H.S. Tzou (Eds.), IUTAM Symposium on Smart Structures and Structronic Systems, Kluwer Academic Publishers, Dordrecht, The Netherlands, pp. 137-144.
[12] Fluhrer J, Deform 2D Version 8.1 User's Manual, Scientific Forming Technologies Corporation, Columbus, Ohio, USA.
[13] International Organization for Standardization, Metallic materials - Sheet and strip - Erichsen cupping test, ISO 20482:2013(E), 2013.

**Acknowledgement**
This research was financially supported by the Ministry of Science and Technology of the Republic of China with the grant coded MOST 104-2221-E-005-022.

**Institute of Physics**
Dirac House, Temple Back
Bristol BS1 6BE UK

ISSN: 1757-8981
ISBN 978-1-5108-5983-8

CURRAN ASSOCIATES INC.
proceedings.com